Nanotoxicology and Nanosafety 2.0

Nanotoxicology and Nanosafety 2.0

Editor

Ying-Jan Wang

MDPI • Basel • Beijing • Wuhan • Barcelona • Belgrade • Manchester • Tokyo • Cluj • Tianjin

Editor
Ying-Jan Wang
National Cheng Kung University
Taiwan

Editorial Office
MDPI
St. Alban-Anlage 66
4052 Basel, Switzerland

This is a reprint of articles from the Special Issue published online in the open access journal *International Journal of Molecular Sciences* (ISSN 1422-0067) (available at: https://www.mdpi.com/journal/ijms/special_issues/nanosafety).

For citation purposes, cite each article independently as indicated on the article page online and as indicated below:

LastName, A.A.; LastName, B.B.; LastName, C.C. Article Title. *Journal Name* **Year**, *Article Number*, Page Range.

ISBN 978-3-03936-748-1 (Hbk)
ISBN 978-3-03936-749-8 (PDF)

Contents

About the Editor . . . . **vii**

Yuan-Hua Wu, Sheng-Yow Ho, Bour-Jr Wang and Ying-Jan Wang
The Recent Progress in Nanotoxicology and Nanosafety from the Point of View of Both
Toxicology and Ecotoxicology
Reprinted from: *Int. J. Mol. Sci.* **2020**, *21*, 4209, doi:10.3390/ijms21124209 . . . . **1**

**Rong-Jane Chen, Yu-Ying Chen, Mei-Yi Liao, Yu-Hsuan Lee, Zi-Yu Chen, Shian-Jang Yan,
Ya-Ling Yeh, Li-Xing Yang, Yen-Ling Lee, Yuan-Hua Wu and Ying-Jan Wang**
The Current Understanding of Autophagy in Nanomaterial Toxicity and Its Implementation
in Safety Assessment-Related Alternative Testing Strategies
Reprinted from: *Int. J. Mol. Sci.* **2020**, *21*, 2387, doi:10.3390/ijms21072387 . . . . **5**

**Mingxi Jia, Wenjing Zhang, Taojin He, Meng Shu, Jing Deng, Jianhui Wang, Wen Li, Jie Bai,
Qinlu Lin, Feijun Luo, Wenhua Zhou and Xiaoxi Zeng**
Evaluation of the Genotoxic and Oxidative Damage Potential of Silver Nanoparticles in Human
NCM460 and HCT116 Cells
Reprinted from: *Int. J. Mol. Sci.* **2020**, *21*, 1618, doi:10.3390/ijms21051618 . . . . **29**

Zannatul Ferdous and Abderrahim Nemmar
Health Impact of Silver Nanoparticles: A Review of the Biodistribution and Toxicity Following
Various Routes of Exposure
Reprinted from: *Int. J. Mol. Sci.* **2020**, *21*, 2375, doi:10.3390/ijms21072375 . . . . **41**

**Jingcao Shen, Dan Yang, Xingfan Zhou, Yuqian Wang, Shichuan Tang, Hong Yin,
Jinglei Wang, Rui Chen and Jiaxiang Chen**
Role of Autophagy in Zinc Oxide Nanoparticles-Induced Apoptosis of Mouse LEYDIG Cells
Reprinted from: *Int. J. Mol. Sci.* **2019**, *20*, 4042, doi:10.3390/ijms20164042 . . . . **73**

Ye-Rin Jeon, Jin Yu and Soo-Jin Choi
Fate Determination of ZnO in Commercial Foods and Human Intestinal Cells
Reprinted from: *Int. J. Mol. Sci.* **2020**, *21*, 433, doi:10.3390/ijms21020433 . . . . **87**

**Melissa H. Cambre, Natalie J. Holl, Bolin Wang, Lucas Harper, Han-Jung Lee,
Charles C. Chusuei, Fang Y.S. Hou, Ethan T. Williams, Jerry D. Argo, Raja R. Pandey
and Yue-Wern Huang**
Cytotoxicity of NiO and Ni(OH)$_2$ Nanoparticles Is Mediated by Oxidative Stress-Induced Cell
Death and Suppression of Cell Proliferation
Reprinted from: *Int. J. Mol. Sci.* **2020**, *21*, 2355, doi:10.3390/ijms21072355 . . . . **105**

**Larry M. Tolliver, Natalie J. Holl, Fang Yao Stephen Hou, Han-Jung Lee, Melissa H. Cambre
and Yue-Wern Huang**
Differential Cytotoxicity Induced by Transition Metal Oxide Nanoparticles is a Function of Cell
Killing and Suppression of Cell Proliferation
Reprinted from: *Int. J. Mol. Sci.* **2020**, *21*, 1731, doi:10.3390/ijms21051731 . . . . **127**

**Fabian L. Kriegel, Benjamin-Christoph Krause, Philipp Reichardt, Ajay Vikram Singh,
Jutta Tentschert, Peter Laux, Harald Jungnickel and Andreas Luch**
The Vitamin A and D Exposure of Cells Affects the Intracellular Uptake of Aluminum
Nanomaterials and Its Agglomeration Behavior: A Chemo-Analytic Investigation
Reprinted from: *Int. J. Mol. Sci.* **2020**, *21*, 1278, doi:10.3390/ijms21041278 . . . . **145**

Marina P. Sutunkova, Svetlana N. Solovyeva, Ivan N. Chernyshov, Svetlana V. Klinova, Vladimir B. Gurvich, Vladimir Ya. Shur, Ekaterina V. Shishkina, Ilya V. Zubarev, Larisa I. Privalova and Boris A. Katsnelson
Manifestation of Systemic Toxicity in Rats after a Short-Time Inhalation of Lead Oxide Nanoparticles
Reprinted from: *Int. J. Mol. Sci.* **2020**, *21*, 690, doi:10.3390/ijms21030690 . . . . **159**

Zi-Yu Chen, Nian-Jhen Li, Fong-Yu Cheng, Jian-Feng Hsueh, Chiao-Ching Huang, Fu-I Lu, Tzu-Fun Fu, Shian-Jang Yan, Yu-Hsuan Lee and Ying-Jan Wang
The Effect of the Chorion on Size-Dependent Acute Toxicity and Underlying Mechanisms of Amine-Modified Silver Nanoparticles in Zebrafish Embryos
Reprinted from: *Int. J. Mol. Sci.* **2020**, *21*, 2864, doi:10.3390/ijms21082864 . . . . **175**

Sreeja Sarasamma, Gilbert Audira, Petrus Siregar, Nemi Malhotra, Yu-Heng Lai, Sung-Tzu Liang, Jung-Ren Chen, Kelvin H.-C. Chen and Chung-Der Hsiao
Nanoplastics Cause Neurobehavioral Impairments, Reproductive and Oxidative Damages, and Biomarker Responses in Zebrafish: Throwing up Alarms of Wide Spread Health Risk of Exposure
Reprinted from: *Int. J. Mol. Sci.* **2020**, *21*, 1410, doi:10.3390/ijms21041410 . . . . **195**

Sreeja Sarasamma, Gilbert Audira, Prabu Samikannu, Stevhen Juniardi, Petrus Siregar, Erwei Hao, Jung-Ren Chen and Chung-Der Hsiao
Behavioral Impairments and Oxidative Stress in the Brain, Muscle, and Gill Caused by Chronic Exposure of C_{70} Nanoparticles on Adult Zebrafish
Reprinted from: *Int. J. Mol. Sci.* **2019**, *20*, 5795, doi:10.3390/ijms20225795 . . . . **225**

Yung-Li Wang, Yu-Hsuan Lee, I-Jen Chiu, Yuh-Feng Lin and Hui-Wen Chiu
Potent Impact of Plastic Nanomaterials and Micromaterials on the Food Chain and Human Health
Reprinted from: *Int. J. Mol. Sci.* **2020**, *21*, 1727, doi:10.3390/ijms21051727 . . . . **249**

Uliana De Simone, Arsenio Spinillo, Francesca Caloni, Laura Gribaldo and Teresa Coccini
Neuron-Like Cells Generated from Human Umbilical Cord Lining-Derived Mesenchymal Stem Cells as a New In Vitro Model for Neuronal Toxicity Screening: Using Magnetite Nanoparticles as an Example
Reprinted from: *Int. J. Mol. Sci.* **2020**, *21*, 271, doi:10.3390/ijms21010271 . . . . **263**

Edward Suhendra, Chih-Hua Chang, Wen-Che Hou and Yi-Chin Hsieh
A Review on the Environmental Fate Models for Predicting the Distribution of Engineered Nanomaterials in Surface Waters
Reprinted from: *Int. J. Mol. Sci.* **2020**, *21*, 4554, doi:10.3390/ijms21124554 . . . . **291**

About the Editor

Ying-Jan Wang, Ph.D., graduated from the Department of Biochemistry, National Taiwan University, College of Medicine, Taipei, Taiwan. He currently serves as a distinguished professor at the Department of Environmental and Occupational Health, National Cheng Kung University, Medical College, Tainan, Taiwan. His laboratory's main research focus is understanding the molecular mechanisms responsible for environmental toxicants which are triggered by toxicity, as well as for hypersensitivity and carcinogenesis/cancer therapy. Ying-Jan Wang's research interests include the elucidation of the role of autophagy in regulating diverse biological processes, such as proliferation, programmed cell death and inflammation, which contribute to cytotoxicity, dermatitis and/or cancer therapy. In general, basal autophagy helps maintain homeostasis, while additional autophagy is induced in response to many different forms of stress. Thus, he hopes to explore whether autophagy acts as a pro-survival or pro-death player in the toxic response of environmental toxicants or cancer therapy. Along with his colleagues, Ying-Jan Wang discovered several regulatory modes and functions of autophagy. Firstly, they found that autophagy activation is a key player in the cellular response against nano-toxicity, in which endoplasmic reticulum (ER) stress caused by misfolded protein aggregation is involved in regulating the autophagic process. Secondly, they uncovered that using a combination of irradiation and anticancer drugs could be a potential therapeutic strategy for the treatment of malignant cancers, through the induction of both autophagy and apoptosis. Currently, Ying-Jan Wang is focused on exploring the roles of autophagy in the enhanced immune response of nanoparticles in related animal models. By revealing the regulation pathways of autophagy in environmental toxicants that are triggered by toxicity, our research may help in the development of novel and effective preventive strategies to combat several human diseases.

International Journal of
Molecular Sciences

Editorial

The Recent Progress in Nanotoxicology and Nanosafety from the Point of View of Both Toxicology and Ecotoxicology

Yuan-Hua Wu [1,†], Sheng-Yow Ho [2,3,†], Bour-Jr Wang [4,5,*] and Ying-Jan Wang [6,7,*]

1   Department of Radiation Oncology, National Cheng Kung University Hospital, College of Medicine, National Cheng Kung University, Tainan 704, Taiwan; wuyh@mail.ncku.edu.tw
2   Department of Radiation Oncology, Chi Mei Medical Center, Tainan 736, Taiwan; shengho@seed.net.tw
3   Graduate Institute of Medical Sciences, Chang Jung Christian University, Tainan 711, Taiwan
4   Department of Cosmetic Science and Institute of Cosmetic Science, Chia Nan University of Pharmacy and Science, Tainan 71710, Taiwan
5   Department of Occupational and Environmental Medicine, National Cheng Kung University Hospital, Tainan 704, Taiwan
6   Department of Environmental and Occupational Health, College of Medicine, National Cheng Kung University, Tainan 704, Taiwan
7   Department of Medical Research, China Medical University Hospital, China Medical University, Taichung 404, Taiwan
*   Correspondence: pochih.wang@msa.hinet.net (B.-J.W.); yjwang@mail.ncku.edu.tw (Y.-J.W.); Tel.: +886-6-235-3535 (ext. 5956) (B.-J.W.); +886-6-235-3535 (ext. 5804) (Y.-J.W.); Fax: +886-6-208-5793 (B.-J.W.); +886-6-275-2484 (Y.-J.W.)
†   These authors contributed equally to this work.

Received: 4 June 2020; Accepted: 10 June 2020; Published: 12 June 2020

Abstract: This editorial aims to summarize the 14 scientific papers contributed to the Special Issue "Nanotoxicology and nanosafety 2.0 from the point of view of both toxicology and ecotoxicology".

Recently, nanotechnology has been rapidly promoting the development of a new generation of industrial and commercial products and areas such as catalysts, sensors, environmental remediation, personal care products and cosmetics. In addition, nanomaterials also show great promise in the field of medicine in application such as imaging and drug delivery. However, the hazardous impact of nanoscale particles (NPs) on organisms have not been thoroughly clarified yet. In some cases, nanomaterials present unexpected risks to both humans and the environment. Assessments of the potential hazards associated with nanotechnology have been emerging, but substantial challenges remain because the huge amount of different nanomaterials cannot be effectively evaluated in a timely manner [1–3]. Although there exist numerous approaches to perform toxicity tests, common and reasonable biomarkers for toxicity evaluations are still lacking and need more investigation [4,5].

From the toxicological point of view, in terms of cytotoxicity, genotoxicity, hepatotoxicity, and embryo toxicity both in vitro or in vivo the toxic effects induced by metal or metal oxide NPs, such as AgNPs and ZnONPs are quite similar [4]. NPs-induced oxidative stress was considered as one of the initiators that can trigger disruption of mitochondrial membrane potential and induction of apoptosis and/or autophagy. It has been reported that AgNPs can induce ROS generation and apoptotic cell death in NCM460 and HCT116 human colon cells through regulation of p38, p53, Bax/Bcl-2 ratio and P21 [6]. Most in vitro studies have demonstrated the size- and surface coating-dependent cellular uptake of AgNPs. In vivo bio-distribution studies of AgNPs indicated that AgNPs could be accumulated in local as well as distant organs [7]. Intragastric exposure of mice to ZnONPs for 28 days disrupted the seminiferous epithelium of the testis and decreased the sperm density in the epididymis. Both apoptosis and autophagy can be induced in the testis tissues, meanwhile, up-regulation of the cleaved Caspase-8/Caspase-3, Bax, LC3-II, Atg 5, and Beclin 1 were found too. In vitro tests showed that ZnONPs could induce apoptosis and autophagy with the generation of oxidative stress in mouse Leydig cells [8]. ZnONPs were internalized into cells in both particle and ion form and was transported through intestinal barriers and absorbed in the small intestine primarily as Zn ions, whereas, a small amount of ZnO was absorbed as particles [9].

Toxic effects of nickel oxide (NiO) and nickel hydroxide (Ni(OH)$_2$) nanoparticles were analyzed in human lung (A549) and hepatocellular carcinoma (HepG2) cell lines. The results indicated nickel NPs were toxic to A549 cells but relatively nontoxic to HepG2 cells. Cytotoxicity was mediated by oxidative stress-induced apoptosis and suppression of cell proliferation [10]. In addition, transition metal oxide NPs including TiO$_2$, Cr$_2$O$_3$, Mn$_2$O$_3$, Fe$_2$O$_3$, NiO, CuO, and ZnO were analyzed in the same A549 cell culture model and the authors found that all NPs aside from Cr$_2$O$_3$ and Fe$_2$O$_3$ showed a time- and dose-dependent decrease in viability. The trend of cytotoxicity was in parallel with proliferative inhibition, revealing a strong correlation among viability, proliferation and apoptosis [11].

NPs may enter human body via various routes such as ingestion, inhalation and skin contact. After translocation with the help of the tblood and lymph systems, they are transported into other organs and lead there to systemic effects [12]. When considering the skin contact route, understanding the uptake mechanisms of ionic, elemental or metal NPs in combination with bioactive substances are important for the assessment of related health impact. High aluminum (Al) intake could lead to neurotoxicity and triggers concern about potential health risks such as Alzheimer's disease. Al NPs in cosmetic products raise the question about whether a possible interaction between Al and vitamin A/D metabolism might exist. Thus, the uptake and distribution of Al oxide (Al$_2$O$_3$) and metallic AlO NMs in the human keratinocyte cell line HaCaT were conducted. The results indicated that vitamin A and D exposure of cells affects the intracellular uptake of Al NPs and its agglomeration behavior [13]. When considering the inhalation route, systemic toxicity in rats after a short duration inhalation of lead oxide nanoparticles has been reported. Rats were exposed to lead oxide nanoparticle aerosols during 5 days for 4 h a day in a nose-only setup, demonstrating a lead oxide NPs retention in the lungs and the olfactory brain. Several lead-specific toxicological outcomes were found in the exposed rats, those included increase in reticulocytes proportion, δ-ALA urine excretion and the arterial hypertension's development [14].

From the ecotoxicological point of view, various NPs from industry might present possible risks to aquatic systems. Besides the metal or metal oxide NPs, carbon NPs such as fullerene (C$_{70}$) and nanoplastics has drawn great attention. Nevertheless, the comprehensive biological and toxicological effects of those NPs on the environment has still not been studied in detail, thus, it is important to analyze the effects of NPs on ecological systems as well as human health. The zebrafish test is an appealing in vivo model to assess the hazards of both conventional chemicals and NPs in ecotoxicology [15–17]. Identification of the targets and mechanisms for the deleterious effects of AgNPs in a zebrafish embryo model has been conducted. Exposure of AgNPs led to lysosomal activity changes and higher number of apoptotic cells distributed among the developmental organs of the zebrafish embryo. AgNPs of larger size exhibited different behavior from the smaller size, in which the embryo chorion play a pivotal role [15].

The toxicity of C$_{70}$ NPs in lung and skin cells have been demonstrated, whereas, the potential detrimental role in neurobehavior is largely unknown. Chronic effects of C$_{70}$ NPs exposure for two weeks on behavior and alterations in biochemical responses in adult zebrafish were performed. The results indicated a decreased locomotion, exploration, social interaction, mirror biting, as well as anxiety elevation and circadian rhythm locomotor activity impairment. Regarding the biochemical assays, the activity of superoxide dismutase (SOD), reactive oxygen species (ROS) and cortisol in the brain and/or muscle tissues increased significantly [17].

Currently, microplastics and nanoplastics are everywhere, contaminating our water, air, and food chain [18]. The toxicity of microplastic exposure on humans and aquatic organisms has been documented, but data on the toxicity and behavioral changes of nanoplastics in mammals are scarce.

Polystyrene nanoplastics (PS-NPs) was applied to investigate the neurobehavioral alterations, tissue distribution, accumulation and specific health risk of nanoplastics in adult zebrafish. The results demonstrated that PS-NPs accumulated in the gonads, intestine, liver, and brain and induced disturbance of lipid and energy metabolism as well as oxidative stress and tissue accumulation. Chronic exposure of high dose PS-NPs induced behavior alterations in their locomotion activity, aggressiveness, shoal formation, predator avoidance behavior and dysregulated circadian rhythm locomotion activity [16]. Thus, both embryo and adult zebrafish toxicity assays are suitable for investigating the metal and carbon-based NPs' ecotoxiciologial effects.

Alternative testing strategies are commonly used to assess the safety of chemicals, and many of these strategies were evaluated for the testing of NPs. In vitro testing was proposed as the principal approach with the support of in vivo assays to fill knowledge gaps, including tests conducted in non-mammalian species such as drosophila and zebrafish [3]. Human cell-based systems are recommended as relevant models to reduce uncertainty and to improve prediction of human toxicity. When compared with the transformed or immortalized cell lines, primary cells have similar characteristics as the original donor tissue, so more and more scientists are considering the use of primary cells in in vitro studies to establish more biologically representative models [19]. Recently, stem cell-derived models are under investigation as potential models for nanotoxicity assay. Stem cells can offer the opportunity to model the reproducibility of rare phenotypes in vitro [20]. For example, neuronal toxicity of Fe_3O_4 NPs has been demonstrated by using the in vitro differentiation of the human umbilical cord lining-derived-mesenchymal stem cells (hCL-MSCs) into neuron-like cells (hNLCs) as an ideal primary cell source of human origin [21]. The study demonstrated that hCL-MSCs can be easily differentiated into neuronal-like cells and the hNCLs are susceptible to Fe3O4NPs, thus, human primary cultures of neurons could be a new in vitro model for NPs evaluation [21].

A thorough understanding of the mechanisms by which NPs perturb biological systems is critical for a more comprehensive elucidation of their nanotoxicity and will also facilitate the development of prevention and intervention policies against adverse outcomes induced by NPs. The development of a good strategy for NM hazard assessment not only promotes a more widespread adoption of non-rodent or 3Rs principles but also makes nanotoxicology testing more ethical, relevant, and cost- and time-efficient. We hope that the abovementioned articles can provide updated knowledge regarding the nanotoxicology and nanosafety from the point of view of both toxicology and ecotoxicology.

Author Contributions: Writing, review and editing, Y.-J.W., B.-J.W., S.-Y.H. and Y.-H.W. All authors have read and agreed to the published version of the manuscript.

Conflicts of Interest: The authors declare no conflict of interest.

References

1. Lee, Y.H.; Cheng, F.Y.; Chiu, H.W.; Tsai, J.C.; Fang, C.Y.; Chen, C.W.; Wang, Y.J. Cytotoxicity, oxidative stress, apoptosis and the autophagic effects of silver nanoparticles in mouse embryonic fibroblasts. *Biomaterials* **2014**, *35*, 4706–4715. [CrossRef] [PubMed]
2. Chiu, H.W.; Xia, T.; Lee, Y.H.; Chen, C.W.; Tsai, J.C.; Wang, Y.J. Cationic polystyrene nanospheres induce autophagic cell death through the induction of endoplasmic reticulum stress. *Nanoscale* **2015**, *7*, 736–746. [CrossRef] [PubMed]
3. Chen, R.J.; Chen, Y.Y.; Liao, M.Y.; Lee, Y.H.; Chen, Z.Y.; Yan, S.J.; Yeh, Y.L.; Yang, L.X.; Lee, Y.L.; Wu, Y.H.; et al. The Current Understanding of Autophagy in Nanomaterial Toxicity and Its Implementation in Safety Assessment-Related Alternative Testing Strategies. *Int. J. Mol. Sci.* **2020**, *21*, 2387. [CrossRef]
4. Mao, B.H.; Tsai, J.C.; Chen, C.W.; Yan, S.J.; Wang, Y.J. Mechanisms of silver nanoparticle-induced toxicity and important role of autophagy. *Nanotoxicology* **2016**, *10*, 1021–1040. [CrossRef]
5. Jain, A.; Ranjan, S.; Dasgupta, N.; Ramalingam, C. Nanomaterials in food and agriculture: An overview on their safety concerns and regulatory issues. *Crit. Rev. Food Sci. Nutr.* **2018**, *58*, 297–317. [CrossRef] [PubMed]

6.   Jia, M.; Zhang, W.; He, T.; Shu, M.; Deng, J.; Wang, J.; Li, W.; Bai, J.; Lin, Q.; Luo, F.; et al. Evaluation of the Genotoxic and Oxidative Damage Potential of Silver Nanoparticles in Human NCM460 and HCT116 Cells. *Int. J. Mol. Sci.* **2020**, *21*, 1618. [CrossRef]

7.   Ferdous, Z.; Nemmar, A. Health Impact of Silver Nanoparticles: A Review of the Biodistribution and Toxicity Following Various Routes of Exposure. *Int. J. Mol. Sci.* **2020**, *21*, 2375. [CrossRef]

8.   Shen, J.; Yang, D.; Zhou, X.; Wang, Y.; Tang, S.; Yin, H.; Wang, J.; Chen, R.; Chen, J. Role of Autophagy in Zinc Oxide Nanoparticles-Induced Apoptosis of Mouse LEYDIG Cells. *Int. J. Mol. Sci.* **2019**, *20*, 4042. [CrossRef]

9.   Jeon, Y.R.; Yu, J.; Choi, S.J. Fate Determination of ZnO in Commercial Foods and Human Intestinal Cells. *Int. J. Mol. Sci.* **2020**, *21*, 433. [CrossRef]

10.   Cambre, M.H.; Holl, N.J.; Wang, B.; Harper, L.; Lee, H.J.; Chusuei, C.C.; Hou, F.Y.S.; Williams, E.T.; Argo, J.D.; Pandey, R.R.; et al. Cytotoxicity of NiO and Ni(OH)2 Nanoparticles Is Mediated by Oxidative Stress-Induced Cell Death and Suppression of Cell Proliferation. *Int. J. Mol. Sci.* **2020**, *21*, 2355. [CrossRef]

11.   Tolliver, L.M.; Holl, N.J.; Hou, F.Y.S.; Lee, H.J.; Cambre, M.H.; Huang, Y.W. Differential Cytotoxicity Induced by Transition Metal Oxide Nanoparticles is a Function of Cell Killing and Suppression of Cell Proliferation. *Int. J. Mol. Sci.* **2020**, *21*, 1731. [CrossRef] [PubMed]

12.   Gnach, A.; Lipinski, T.; Bednarkiewicz, A.; Rybka, J.; Capobianco, J.A. Upconverting nanoparticles: Assessing the toxicity. *Chem. Soc. Rev.* **2015**, *44*, 1561–1584. [CrossRef] [PubMed]

13.   Kriegel, F.L.; Krause, B.C.; Reichardt, P.; Singh, A.V.; Tentschert, J.; Laux, P.; Jungnickel, H.; Luch, A. The Vitamin A and D Exposure of Cells Affects the Intracellular Uptake of Aluminum Nanomaterials and Its Agglomeration Behavior: A Chemo-Analytic Investigation. *Int. J. Mol. Sci.* **2020**, *21*, 1278. [CrossRef] [PubMed]

14.   Sutunkova, M.P.; Solovyeva, S.N.; Chernyshov, I.N.; Klinova, S.V.; Gurvich, V.B.; Shur, V.Y.; Shishkina, E.V.; Zubarev, I.V.; Privalova, L.I.; Katsnelson, B.A. Manifestation of Systemic Toxicity in Rats after a Short-Time Inhalation of Lead Oxide Nanoparticles. *Int. J. Mol. Sci.* **2020**, *21*, 690. [CrossRef] [PubMed]

15.   Chen, Z.Y.; Li, N.J.; Cheng, F.Y.; Hsueh, J.F.; Huang, C.C.; Lu, F.I.; Fu, T.F.; Yan, S.J.; Lee, Y.H.; Wang, Y.J. The Effect of the Chorion on Size-Dependent Acute Toxicity and Underlying Mechanisms of Amine-Modified Silver Nanoparticles in Zebrafish Embryos. *Int. J. Mol. Sci.* **2020**, *21*, 2864. [CrossRef] [PubMed]

16.   Sarasamma, S.; Audira, G.; Siregar, P.; Malhotra, N.; Lai, Y.H.; Liang, S.T.; Chen, J.R.; Chen, K.H.; Hsiao, C.D. Nanoplastics Cause Neurobehavioral Impairments, Reproductive and Oxidative Damages, and Biomarker Responses in Zebrafish: Throwing up Alarms of Wide Spread Health Risk of Exposure. *Int. J. Mol. Sci.* **2020**, *21*, 1410. [CrossRef]

17.   Sarasamma, S.; Audira, G.; Samikannu, P.; Juniardi, S.; Siregar, P.; Hao, E.; Chen, J.R.; Hsiao, C.D. Behavioral Impairments and Oxidative Stress in the Brain, Muscle, and Gill Caused by Chronic Exposure of C70 Nanoparticles on Adult Zebrafish. *Int. J. Mol. Sci.* **2019**, *20*, 5795. [CrossRef]

18.   Wang, Y.L.; Lee, Y.H.; Chiu, I.J.; Lin, Y.F.; Chiu, H.W. Potent Impact of Plastic Nanomaterials and Micromaterials on the Food Chain and Human Health. *Int. J. Mol. Sci.* **2020**, *21*, 1727. [CrossRef]

19.   Joris, F.; Manshian, B.B.; Peynshaert, K.; De Smedt, S.C.; Braeckmans, K.; Soenen, S.J. Assessing nanoparticle toxicity in cell-based assays: Influence of cell culture parameters and optimized models for bridging the in vitro-in vivo gap. *Chem. Soc. Rev.* **2013**, *42*, 8339–8359. [CrossRef]

20.   Lynch, S.; Pridgeon, C.S.; Duckworth, C.A.; Sharma, P.; Park, B.K.; Goldring, C.E.P. Stem cell models as an in vitro model for predictive toxicology. *Biochem. J.* **2019**, *476*, 1149–1158. [CrossRef]

21.   De Simone, U.; Spinillo, A.; Caloni, F.; Gribaldo, L.; Coccini, T. Neuron-Like Cells Generated from Human Umbilical Cord Lining-Derived Mesenchymal Stem Cells as a New In Vitro Model for Neuronal Toxicity Screening: Using Magnetite Nanoparticles as an Example. *Int. J. Mol. Sci.* **2020**, *21*, 271. [CrossRef] [PubMed]

 International Journal of
Molecular Sciences

Review

The Current Understanding of Autophagy in Nanomaterial Toxicity and Its Implementation in Safety Assessment-Related Alternative Testing Strategies

Rong-Jane Chen [1,†], Yu-Ying Chen [2], Mei-Yi Liao [3], Yu-Hsuan Lee [4], Zi-Yu Chen [2],
Shian-Jang Yan [5], Ya-Ling Yeh [2], Li-Xing Yang [6], Yen-Ling Lee [7,†], Yuan-Hua Wu [8,*]
and Ying-Jan Wang [2,9,*]

1   Department of Food Safety/Hygiene and Risk Management, College of Medicine, National Cheng Kung University, Tainan 704, Taiwan; janekhc@gmail.com
2   Department of Environmental and Occupational Health, College of Medicine, National Cheng Kung University, Tainan 704, Taiwan; 101312123@gms.tcu.edu.tw (Y.-Y.C.); q781001@gmail.com (Z.-Y.C.); linn7627@hotmail.com (Y.-L.Y.)
3   Department of Applied Chemistry, National Pingtung University, Pingtung 900, Taiwan; myliao@mail.nptu.edu.tw
4   Department of Cosmeceutics, China Medical University, Taichung 651, Taiwan; yhlee@mail.cmu.edu.tw
5   Department of Physiology, College of Medicine, National Cheng Kung University, Tainan 701, Taiwan; johnyan@mail.ncku.edu.tw
6   Institute of Oral Medicine and Department of Stomatology, College of Medicine, National Cheng Kung University Hospital, National Cheng Kung University, Tainan 701, Taiwan; fingerzoo@gmail.com
7   Department of Hematology/Oncology, Tainan Hospital of Health and Welfare, Tainan 700, Taiwan; yenpig8291@gmail.com
8   Department of Radiation Oncology, National Cheng Kung University Hospital, College of Medicine, National Cheng Kung University, Tainan 704, Taiwan
9   Department of Medical Research, China Medical University Hospital, China Medical University, Taichung 404, Taiwan
*   Correspondence: wuyh@mail.ncku.edu.tw (Y.-H.W.); yjwang@mail.ncku.edu.tw (Y.-J.W.); Tel.: +886-6-235-3535 (ext. 2441) (Y.-H.W.); +886-6-235-3535 (ext. 5804) (Y.-J.W.); Fax: +886-6-235-9333 (Y.-H.W.); +886-6-275-2484 (Y.-J.W.)
†   These authors contributed equally to this work.

Received: 21 February 2020; Accepted: 28 March 2020; Published: 30 March 2020

Abstract: Nanotechnology has rapidly promoted the development of a new generation of industrial and commercial products; however, it has also raised some concerns about human health and safety. To evaluate the toxicity of the great diversity of nanomaterials (NMs) in the traditional manner, a tremendous number of safety assessments and a very large number of animals would be required. For this reason, it is necessary to consider the use of alternative testing strategies or methods that reduce, refine, or replace (3Rs) the use of animals for assessing the toxicity of NMs. Autophagy is considered an early indicator of NM interactions with cells and has been recently recognized as an important form of cell death in nanoparticle-induced toxicity. Impairment of autophagy is related to the accelerated pathogenesis of diseases. By using mechanism-based high-throughput screening in vitro, we can predict the NMs that may lead to the generation of disease outcomes in vivo. Thus, a tiered testing strategy is suggested that includes a set of standardized assays in relevant human cell lines followed by critical validation studies carried out in animals or whole organism models such as *C. elegans* (Caenorhabditis elegans), zebrafish (Danio rerio), and Drosophila (Drosophila melanogaster)for improved screening of NM safety. A thorough understanding of the mechanisms by which NMs perturb biological systems, including autophagy induction, is critical for a more comprehensive elucidation of nanotoxicity. A more profound understanding of toxicity mechanisms will also facilitate the development of prevention and intervention policies against adverse outcomes induced by NMs. The development of a tiered testing strategy for NM hazard assessment not only promotes a more widespread adoption of non-rodent or 3R principles but also makes nanotoxicology testing more ethical, relevant, and cost- and time-efficient.

Keywords: nanomaterials; autophagy; alternative testing strategy; high throughput screening; tiered testing strategy; *C. elegans*; zebrafish and Drosophila models

1. Introduction

Nanomaterials (NMs) are defined as having at least one dimension that is 1–100 nm in diameter [1] and unique properties; for example, they can change reactivity, optical characteristics, or conductivity, thereby enabling novel applications. Furthermore, particle properties can be modified to promote different applications, resulting in consumer benefits, particularly in medical and industrial applications [2]. In recent years, nanotechnology has rapidly been promoted in the development of a new generation of industrial and commercial products. It has been estimated that the nanoproduct demands in medicine and pharmaceuticals, and especially the cosmetics industry, are expected to rise by over 17% each year and at a much higher rate in the food industry [3,4]. However, the application of nanotechnology has also raised some concerns about human health and safety. In some cases, nanomaterials present unexpected risks to both humans and the environment. Regulatory authorities in the European Union, United States, and Asian countries carefully observe developments in nanotechnology, trying to find a balance between consumer safety and the interests of the industry [5]. In addition, several international planning activities have been proposed or performed with the expectation that significant advances will be made in understanding the potential hazards triggered by nanomaterial exposure in both occupational and consumer environments [2].

Assessments of the potential hazards associated with nanotechnology have been emerging, but substantial challenges remain because all of the different nanoparticle (NP) types cannot be effectively evaluated for safety and environmental effects in a timely manner [2]. Identification of the physicochemical properties of nanomaterials that confer toxicity is a core component of toxicity studies. To evaluate the toxicity of the great diversity of NMs, a tremendous number of safety assessments would need to be conducted. It was estimated that, in 2009, a complete toxicity evaluation of all the nanomaterial on the market using traditional animal approaches would cost more than 1 billion US dollars, take at least 50 years, and require a very large number of animals [6,7]. Whether animals can be used to predict human response to toxicant exposure is still under debate, attributing to data gap between human and animal studies. Thus, there is a need for developing and using human-cell-based methods that generate human-relevant mechanistic data that are not necessarily obtainable from traditional animal studies conducted by vertebrate animals [8–10]. Furthermore, there are government regulations that have resulted in an enhanced need for alternative methods, such as the E.U. Cosmetics Directive that prohibits the testing of cosmetics products on animals in the European Union (EU Regulation 1223/2009). For all these reasons, it is necessary to consider the use of alternative testing strategies or methods that reduce, refine, or replace (3Rs) the use of animals for assessing the toxicity of nanomaterials [10].

Alternative testing strategies are commonly used to assess the safety of chemicals, and many of these strategies have been evaluated for their applicability to the testing of nanomaterials. A single alternative testing method may contribute to basic mechanistic or toxicity knowledge but may not be sufficient for use in hazard assessment. However, incorporating multiple alternative testing methods into alternative testing strategies will provide an understanding of the behavior and toxicity of nanomaterials in humans and the environment [10,11]. In vitro testing was proposed as the principal approach with the support of in vivo assays to fill knowledge gaps, including tests conducted in non-mammalian species such as *C. elegans*, Drosophila, and zebrafish, or genetically engineered animal models. These tools are being used to identify responses in cells exposed to chemicals expected to result in toxic effects [9,12]. Well-designed alternative testing strategies will not only allow for the prioritization of nanomaterials for further testing but can also assist in the prediction of risk to human beings and the environment.

Autophagy is a catabolic mechanism that is evolutionarily conserved from yeast to mammals. The autophagy pathway first described by Christian De Duve in 1963 [13] is a ubiquitous process that involves the degradation of cytoplasmic components and cytoplasm organelles, that degrade through the lysosomal pathway, and is distinct from other degradative pathways, such as proteasomal degradation [14]. When energy is limiting (ATP shortage), AMP kinase (AMPK) is activated, which can drive autophagy. Similarly, deprivation of growth factors or amino acids leads to the inhibition of TORC1, which is a repressor of conventional autophagy [15]. The inability to regulate autophagy is associated with aging, neurodegeneration, and a variety of diseases, including cancer, type 2 diabetes, and atherosclerosis [16]. Autophagy was recently recognized as an important form of cell death in various types of nanoparticle-induced toxicity, but the details of the underlying mechanisms are still unclear. A thorough understanding of the cellular and molecular mechanisms of nanoparticle-triggered toxicity is critical for a more comprehensive elucidation of nanotoxicity [17]. Our recent work provides the first demonstration that autophagy activated by silver nanoparticles (AgNPs) in normal cells fails to trigger lysosomal degradation pathway and led to a toxicity phenomenon called defective autophagic flux or autophagy dysfunction, which is relevant to the accelerated cellular pathogenesis of diseases [18–20]. The toxic effects induced by AgNPs and some of the metal oxide NPs, such as ZnONPs, have been shown, either in vitro or in vivo, to be quite similar in terms of cytotoxicity, genotoxicity, hematotoxicity, immunotoxicity, hepatotoxicity, and embryotoxicity [19,21]. A more profound understanding of these toxicity mechanisms will facilitate the development of prevention and intervention policies against adverse outcomes induced by metal and metal oxide nanomaterials. Knowledge derived from the cellular and molecular processes underlying nanomaterial-induced toxic effects may also facilitate the establishment of the scientific foundations of nanomaterial risk assessment. Therefore, an overview of current findings regarding the mediation of autophagy triggered by NPs both in vitro and in vivo will shed light on the pivotal role of autophagy in nanomaterial toxicity and the useful implementation of autophagy in safety assessments conducted through alternative testing strategies.

2. Alternative Testing Strategy for Nanomaterial Safety Assessments

2.1. Approach towards the Use of Alternatives to Testing on Animals

The "3 Rs", standing for reduction, refinement, or replacement, is a strategy being applied to the use of laboratory animals through implementation of different methods and alternative organisms to provide integrated approaches that could provide insight into the minimal use of animals in scientific experiments [22]. In 2007, the US National Research Council (NRC) published a report entitled "Toxicity Testing in the 21st Century (TT21C): A Vision and a Strategy", putting forward a long-term strategy taking advantage of newly developed technologies to enhance the efficiency of the toxicity testing of chemicals to which human beings may be exposed. The important parts of this strategy are the increased use of high-throughput in vitro test systems and methods in computational toxicology for the purpose of reducing the reliance on time-consuming and costly toxicological studies using experimental animals. This vision has received international support and has provided a blueprint for implementing change in toxicological science [9,23]. In 2011, the European Union launched SEURAT-1, the first execution phase of "Safety Evaluation Ultimately Replacing Animal Testing (SEURAT)", with the ultimate goals of the future implementation of mechanism-based, integrated toxicity testing strategies into modern safety assessment approaches [24]. In addition, promotion of non-animal approaches is also among the objectives of the REACH (Registration, Evaluation, Authorization, and Restriction of Chemicals),

CLP (Classification, Labeling, and Packaging) and BPR (Biocidal Products Regulations) initiatives, which are based on the 3Rs principle of animal use for testing [25]. Through the abovementioned efforts, many government agencies within the United States, European Union, and other international bodies are beginning to incorporate the new approach methodologies envisioned in the original TT21C vision into regulatory practice [23].

Current experimental toxicology approaches are being promoted rapidly by the incorporation of novel techniques and methods that provide a much more in-depth view into the mechanisms of potential adverse effects of chemical exposure to human health [26]. For example, the basal cytotoxicity level determined by in vitro cytotoxicity assays is considered a key factor in many prevalent toxicological modes of action associated with the mechanisms of organ failure, including disruption of cell membrane structure and/or function, disturbance of protein turnover, inhibition of mitochondrial function, and disruption of metabolism and energy production [27–30]. In 2017, Vinken and Blaauboer proposed three consecutive steps, including initial cell injury, mitochondrial dysfunction, and cell death, as the adverse outcome pathway (AOP) framework for measuring basal cytotoxicity. The outcome of basal cytotoxicity assessment could serve as the first step of a tiered strategy aimed at evaluating the toxicity of new chemicals, and then, more specific types of toxicity could be evaluated in a second step [29]. The data from various in vitro assays are useful for both increasing confidence in hazard and risk decisions and enabling better, faster, and less expensive assessments of a large number of chemicals, mixtures, and complex products.

2.2. Alternative Testing Strategy for Nanomaterials

The lack of availability of regulatory guidelines for the safety assessment of nanomaterials is a major problem. In general, the Organization for Economic Cooperation and Development (OECD) safety assessment of traditional chemicals is suitable for the nanomaterial safety assessment but needs adaptation. Owing to the unique physical/chemical properties of nanomaterials, the original OECD guidelines need to be further adjusted and improved [7]. Our understanding of the mechanisms of nanomaterial-induced toxicity is insufficient for drawing a general consensus and/or conclusion on the toxicity of nanomaterials [31]. To assess nanomaterial hazards, reliable screening approaches are required to test the basic materials as well as the nano-enabled products. The European Union launched the FP7 NanoTEST project (www.nanotest-fp7.eu) to provide testing strategies for the hazard identification and risk assessment of nanomaterials and to propose recommendations for evaluating the potential risks of newly designed nanomaterials [11]. However, the knowledge gaps of nanomaterial behavior, such as its transformation and fate in biological systems, make it difficult to perform adequate hazard assessment. Thus, a better understanding of nanomaterials with cells, tissues and organs for addressing critical issues related to toxicity testing, especially with respect to alternatives to testing animals, is needed [11]. How nanomaterials interact with biological systems has become an important and complex issue in terms of both the research and regulatory options. When nanomaterials encounter biomolecules or cells, their physicochemical properties have a major impact on the degree to which the material adversely perturbs biological systems [32,33]. Nano-bio interactions may also be affected by the properties of different cell types, the biological environment, and the assay methods applied, making the issues more complicated. A thorough understanding of the mechanisms regarding nanomaterials-induced perturbation in biological systems such as autophagy induction is critical for a more comprehensive elucidation of nanotoxicity [33,34].

In vitro studies are mainly performed on cell lines, which are transformed and immortalized cells to escape normal cellular senescence. These well-established cell lines are cheap, readily available, and easy to passage because of enhanced proliferation ability [35]. On the contrary, primary cells are isolated from tissue without any modification and have similar characteristics as the original donor tissue, so more and more people are considering the use of primary cells in in vitro studies to establish more biologically representative models [35]. Several research groups have compared the effects of NPs exposure between primary cells and immortalized cell lines representing the same tissue, and claimed that immortalized cell lines were more sensitive to NPs toxicity than primary cells [35]. However, in some cases, since immortalized cell lines are de-differentiated in culture, they may not respond to certain cell reactions [36,37]. Therefore, the nanotoxicity results of in vitro studies are often not very relevant to the results of in vivo studies. However, Drasler et al. suggest that immortalized cell lines are suitable for the first stage of nanosafety assessment, because they can provide greater comparability and reproducibility for interlaboratory comparisons of the same NM type or between different NMs. In terms of higher tier evaluation, using primary cells to better understand the NM mechanism in the human body is preferred, as they can more closely mimic in vivo conditions [36].

A strategy for in vitro toxicity testing requires a series of tests addressing and covering different mechanisms important toxicity endpoints. Thus, to identify relevant short-term hazard models, several different outcomes, such as cell viability, oxidative stress, genotoxicity, the proinflammatory response, immunotoxicity, cell uptake, and transport, are conducted [38]. By using mechanism-based high-throughput screening in vitro, [7] we can predict the nanomaterials that may lead to the generation of target organ toxicity in vivo. Additional in vivo studies are used to validate and improve the in vitro high-throughput screening process and to establish structure–function relationships that enable hazard ranking and modeling by an appropriate combination of in vitro and in vivo testing [38,39]. Thus, a tiered testing strategy was suggested that includes a set of standardized cytotoxicity assays in relevant human cell lines followed by critical validation studies carried out in animals or whole organism models such as C. *elegans*, zebrafish, or Drosophila for improved screening of nanomaterial safety [31]. Zebrafish have been deemed acceptable by regulatory agencies for use in chemical safety assessments for evaluating developmental toxicity and are now regularly accepted models in biomedical research, providing strong foundations for their use in nanotoxicology [39]. The development of a tiered testing strategy for nanomaterial hazard assessment not only promotes the widespread adoption of non-rodent and the 3R principles but also makes nanotoxicology testing more ethical, relevant, and cost- and time-efficient [7].

3. Updated Nanotoxicology Knowledge Regarding Autophagy Dysfunction

3.1. Autophagy and Autophagy-Induced Cell Death (ACD)

Dysregulated cell death is a common feature of many human diseases, and the complex mechanisms and pathways that control cell death are becoming increasingly understood. It is now clear that different cell death pathways have a critical role in multiple diseases (such as Crohn's, Parkinson's, and Alzheimer's diseases) [16]. Currently, there are currently three common well known cell death pathways, including apoptosis (type I programmed cell death), necrosis, and the so call autophagic cell death (also referred to as type II programmed cell death) [40]. The association between autophagy and cell death has been known for many years. Originally, based on morphology, autophagic structures were observed in dying cells and distinguished autophagic from apoptotic cell death. Under stress conditions, autophagy is initially induced as an early pro-survival response in the cell, but accumulated autophagy-related substances contributes to autophagic cell death (ACD) [16]. ACD is characterized by the large-scale autophagosomes sequestration in the cytoplasm, giving the cell a characteristic vacuolated shape [16]. Autophagosomes can be identified by transmission electron microscopy as double-membraned vesicles that contain cytosol or cytoplasmic organelles such as mitochondria or the endoplasmic reticulum. Numerous reports, particularly from model systems, provide support for a direct role of autophagy in cell death in context-dependent settings [16,40]. Despite this evidence, our understanding of the mechanism by which autophagy contributes to cell death is not clear. The cross talk between autophagy and other cell death pathway components suggests that the role of autophagy may be context-specific, and understanding the molecular nature of these relationships will aid in understanding the role of ACD.

3.2. Autophagy Dysfunction as a Cell Toxicity Mechanism

Autophagy dysfunction is defined as massive autophagy induction or a blockade of autophagic flux. It is recognized as a potential mechanism of cell death, resulting in either apoptosis or ACD [40]. Previous report proposed that autophagy proteins LC3-II and ATG5 may directly activate caspase, causing cell death through its interactions with Fas and Fas-associated proteins with a death domain (FADD) [41]. In addition, defective autophagy can lead to cancer development, possibly by accumulation of damaged organelles, such as mitochondria, that can induce oxidative stress, inflammation, and DNA damage. The pro-autophagy gene beclin-1 is commonly deleted in several types of cancer, such like breast, ovarian, and prostate cancer, suggesting a tumor suppressor function in the autophagy pathway. Furthermore, beclin-1-knockout mice exhibit enhanced susceptibility to cancer development [42]. Defective autophagy has also been associated with many diseases, such as Crohn's, Parkinson's, and Alzheimer's diseases and may play a role in disease development [43]. In the case of Crohn's disease, disruption of autophagy during immune and inflammation responses may be involved in disease progression. In Parkinson's and Alzheimer's disease, blockade of autophagy-mediated elimination of amyloid beta and alpha synuclein proteins or damaged mitochondria may be involved [43].

Disruption of lysosomal trafficking is a major mechanism for blocking autophagic flux, which results in the accumulation of autophagic and lysosomal vacuoles. There are several possible mechanisms by which NPs might disrupt autophagy and lysosomal trafficking. For example, lysosomal overload by particulates has been proposed as a mechanism by which cigarette smoke blocks autophagic flux in alveolar macrophages [44]. One common form of lysosomal dysfunction that has been associated with nanomaterial treatments is increased lysosomal membrane permeabilization (LMP). Proton pump inhibitors, such as bafilomycin A1, that block autophagic flux predispose cells to LMP, triggering apoptosis through the LMP-induced release of pro-apoptosis mediators such as cathepsin [45]. LMP is a recognized cell death mechanism that can result in mitochondrial membrane permeabilization through several mechanisms, including lysosomal-iron mediated oxidative stress and the release of cathepsin or other lysosomal associated hydrolases [45]. As lysosomal dysfunction has been involved in disease pathogenesis, the association of nanoparticle exposure and lysosomal dysfunction may have relevance to nanomaterial-induced toxicity levels, especially chronic toxicity. Since lysosomal degradation pathways play vital roles in cellular homeostasis, lysosomal dysfunction has been associated with several diseases, termed lysosomal storage disorders [43,46]. Many types of lysosomal perturbations are also associated with autophagy dysfunction, blocking autophagosomes and lysosome fusion and promoting the accumulation of autophagosomes and other autophagy-related substrates (e.g., ubiquitinated protein aggregates) [45]. Autophagy dysfunction can result from lysosomal overload, which prevents autophagosome–lysosome fusion. Similar to those causing lysosomal dysfunction, perturbations in the autophagy pathway have been linked to a variety of diseases [47].

3.3. Autophagy Dysfunction Induced by Nanomaterials

Autophagy disturbances have been reported consistently across several types of nanomaterials and biological models. There are several plausible pathways of nanomaterials-induced autophagy dysfunction (Figure 1). Studies have attempted to illuminate the importance of cellular nanoparticle internalization pathways in nanotoxicity. For example, researchers found that intracellular AgNPs possess limited or no cytotoxic effects when intracellularized AgNPs were shown to exhibit free random Brownian motions within the cytosol rather than accumulate in lysosomes [19]. By contrast, they identified that AgNPs actively internalized via endocytosis were predominantly trafficked within the endo-lysosomal compartments and were obviously toxic to cells [19]. For instance, nanomaterials may induce autophagy through an oxidative stress mechanism [48], such as by the accumulation of reactive oxygen species (ROS), damaged proteins and the endoplasmic reticulum Stress (ER stress) or mitochondrial damage [48]. The involvement of oxidative stress in the induction of autophagy by nanomaterials is supported by a study in which silica nanoparticles induced autophagy in human endothelial cells via reactive oxygen species-mediated MAPK/Bcl-2 and PI3K/Akt/mTOR signaling, which was suppressed by cell exposure to the antioxidant N-acetyl-L-cysteine [49]. Alternatively, nanomaterials may directly affect autophagy-dependent signaling pathways or gene/protein expression. It is also likely that autophagy induction by nanomaterials is the result of an attempt by the cell to degrade what is perceived as foreign or aberrant, similar to cellular action against bacteria and other pathogens. As discussed above, nanoparticles are commonly observed within the autophagosomes compartment, suggesting that the activation of autophagy is triggered by the attempt to sequester and degrade materials that enter into the cytoplasm [50,51]. Cytoplasmic nanoparticles undergo ubiquitination and are co-localized with polyubiquitin complexes that are then translocated to the autophagosomes by p62 [52]. The induction of autophagy has also been observed following treatment with different types of nanoparticles, such as silica nanoparticles, silver nanoparticles, and others [50,53]. Autophagy is activated in human cancer cells by zinc oxide nanoparticles, which are currently under development for use in enhancing tumor chemotherapy and overcoming drug resistance [54]. Quantum dots are also currently under development for use in a broad range of biomedical imaging applications and have been shown to induce lysosome-dependent autophagy activation, ROS production, and toxicity in human hepatocytes [55]. As explained above, there is little evidence of autophagy as an actual effector of cell death, and the cytotoxicity resulting from blocking autophagic pro-survival mechanisms appears to be the more likely effect of nanomaterial exposure. Since blocking autophagic flux and autophagy induction can both lead to autophagosome accumulation [16], the mechanism by which nanomaterials induce autophagosome accumulation is unclear in many cases. Nonetheless, the disruption or blockade of autophagic flux is often observed in cells exposed to nanomaterials.

Many studies have revealed the connection between nanomaterial-induced autophagy dysfunction and mitochondrial damage [56,57]. Disruption of the autophagy pathway by gene knockout has also been associated with the accumulation of dysfunctional mitochondria and ROS, thus providing a potential link between nanomaterial-induced autophagy blockade and oxidative stress (Figure 1) [58]. Nanomaterial-induced autophagy blockade may also be a mechanism by which nanomaterials induce inflammation, due to the fact that autophagy plays an important role in negatively regulating the NLRP3 inflammasome [53]. Blockade of autophagic flux may result in mitochondrial dysfunction through prevention of the removal of the damaged mitochondria, which are normally degraded in the normal autophagy pathway [57]. Consistent with autophagy being assumed to be a mechanism to diminish damaged mitochondria, there is also evidence that mitochondrial depolarization actually led to autophagy induction [57]. Thus, it is conceivable that nanoparticles might be expected to result from a combination of autophagy induction and autophagy blockade, which may be triggered with an increased number of depolarized, dysfunctional mitochondria that cannot be cleared because of impaired autophagic flux. Another major mechanism of nanoparticle-induced autophagy pathway dysfunction is lysosomal dysfunction. Many studies have observed nanomaterial-induced lysosomal dysfunction (Figure 1) [59,60]. There are many plausible explanations for nanoparticle-induced lysosomal dysfunction, including inhibited enzyme ability and bio-persistence [55]. The "proton sponge" hypothesis for cationic nanoparticles is a well-known theory of nanoparticle-induced lysosomal dysfunction, which involves osmotic swelling and membrane rupture [61]. Another direct mechanism that might account for nanoparticle-induced lysosomal dysfunction is the generation of ROS [62]. As many nanoparticles can induce ROS, the oxidative stress model is by far the most accepted theory of nanoparticle-induced toxicity [48].

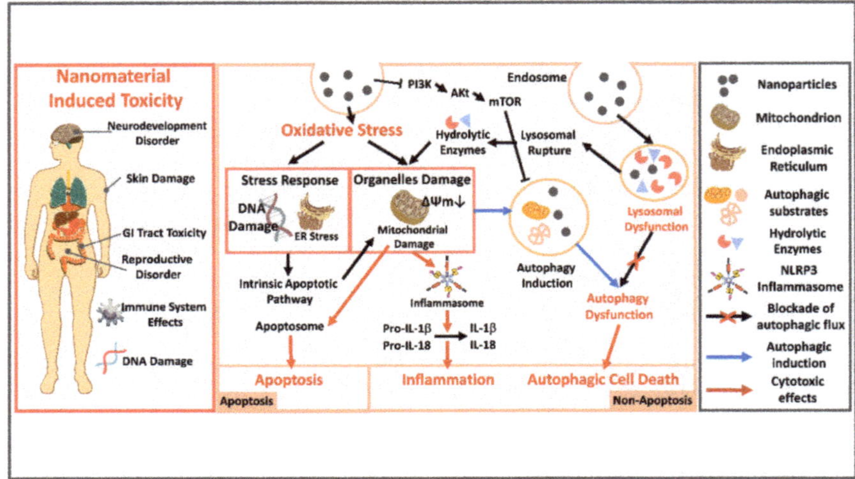

Figure 1. An overview of the mechanism of nanomaterial-induced autophagy-related toxicity and subsequent systematic potential toxic effects. The overall toxicity caused by nanomaterials includes neurodevelopmental disorders, skin damage, gastrointestinal tract toxicity, reproductive disorders, immune system effects, DNA damage, and so on. One of the major forms of toxicity caused by nanomaterials is the oxidative stress induced by ROS and resulting in ER stress and mitochondria and DNA damage. The stress response and organelle damage can eventually induce apoptotic cell death. ROS production results in mitochondrial damage that activates the NLRP3 inflammasome and cellular inflammation. Another major implication for nanomaterial-induced toxicity is autophagy dysfunction. Nanomaterial-induced autophagy and lysosomal dysfunction are displayed as blue arrows in the figure. The initiation step of autophagy is induced by the accumulation of nanomaterials in autophagosomes and blocked vesicle trafficking or by the inhibition of the PI3K/Akt/mTOR pathways. The second step of autophagic toxicity can be induced by overloading of nanomaterials in the lysosomes, leading to damage to the organelle compartments, lysosomal membrane permeabilization (LMP), and release of hydrolytic enzymes. The damaged lysosomes also cause blocked autophagosome-lysosome fusion, eventually leading to autophagic cell death. Altogether, stress responses and organelle damage may synergistically promote cell death, including that caused by apoptosis activation, NLRP3 inflammasome activation, or autophagy.

4. In Vivo Model Systems for the Detection of Nanomaterial Toxicity and Autophagy

4.1. C. elegans Model

Caenorhabditis elegans (*C. elegans*) is a well-established small nematode model organism that has been used since the 1970s [63]. Unlike the traditional toxicity cell culture testing systems, *C. elegans* provide data from a whole animal with complete and metabolically active digestive, reproductive, endocrine, sensory, and neuromuscular systems [64]. Indeed, *C. elegans* research has been proved to be essential in the clarification of several basic aspects of biology, including apoptosis, autophagy, RNA interference, and miRNA function. It has also been demonstrated that the results conducted in *C. elegans* have consistently shown good correlation with rodent oral LD50 ranking [65]. Due to the rapid needs of nanotechnology assessment, especially at the environmental exposure and risk assessment, *C. elegans*, as a complete model organism, has become an important in vivo alternative assay system to assess the risk of NPs [66]. The interaction between NPs and *C. elegans* can be used for providing the toxicity outcome of NPs in a multicellular organism. Recently, *C. elegans* has been used in acute, prolonged, and chronic exposure by using oral treatment, topical applications, or microinjection to particular organs. In addition, *C. elegans* as a whole organism, is able to provide different toxicity endpoints, such as immunotoxicity, neurotoxicity, reproductive toxicity, and genotoxicity [66]. The assessment of the interaction between NPs and *C. elegans* provides information of the in vivo behavior and biocompatibility in a multicellular organism of some NPs for evaluating their fate and toxicity [67].

Regarding the nanomaterial-induced autophagy, *C. elegans* has been employed to identify and characterize the autophagy-regulating mechanism, and 139 conserved genes that regulate autophagy activity are identified, offering a framework for thorough dissection of the autophagy process [68]. Studies show that treatment of quantum dots (QDs), gold nanoparticles (AuNPs), and carbon dots (CDs) are able to induce massive autophagosome formation, autophagy related gene upregulation, and autophagy substrate degradation in cultured HeLa cells and in live *C. elegans* [64]. Due to the small size of *C. elegans*, it is very easy to track the autophagosome formation in real-time by simply using the fluorescence microscopy [69], making *C. elegans* a suitable model organism for the alternative nanotoxicity approach and providing great connection between in vitro and in vivo toxicity [69].

4.2. Zebrafish Model

The extensive applications of nanoparticles in various aspects of daily life, such as the healthcare and industrial sectors, have increased the concern of their impact on human health and the environment [70,71]. Several models have been applied to investigate the toxicity of nanomaterials, including rodent, cell culture, zebrafish, and Drosophila system models [72–74]. Although the higher-vertebrate platform is an important model for evaluating complicated physiological situations, vertebrates present various disadvantages that make them ill-suited for use in exploring nanotoxicity [75]. The vertebrate animals are costly to obtain, time-consuming to maintain, and may not align with animal welfare concerns. Therefore, cell culture, zebrafish, and Drosophila models have become attractive alternative approaches due to their high throughput and cost efficiency [73,76].

Zebrafish constitute a well-established model that is often applied to study issues of development, disease, and environmental contamination [77–80]. They exhibit various advantages that make them suitable for studying toxicology [73,81]. Zebrafish are highly efficient models due to their high reproduction rates [81]. Moreover, the maintenance of zebrafish is relatively inexpensive, with small tank requirements, rapid development, and transparent embryos. In addition, the genes of this model organism share 70% similarity with human genes [82,83], and their critical organ systems, such as the nervous system, intestinal system, and cardiovascular systems, are similar those of humans [76,84]. Furthermore, the results of acute toxicity via inhalation or injection in zebrafish have been demonstrated to exhibit a high correlation with zebrafish embryos and rodents. It is worth mentioning that the zebrafish model can help us quickly and efficiently understand cellular and molecular mechanisms. Due to the advancement of genetic tools, mechanisms, such as those of ROS and autophagy triggered by toxicants, can be easily observed via fluorescent reporters and transgenic lines [85–88].

According to several studies, zebrafish and their embryos have great potential as models to evaluate nanotoxicity, serving as alternatives in the approach for testing nanomaterials [73,76]. Moreover, the zebrafish model provides various means to measure nanotoxicity, such as the quick assessment of productive toxicity, teratogenicity, and developmental toxicity, as well as for evaluating immunotoxicity, genotoxicity, and neurotoxicity [7,89–92]. For example, zebrafish were employed to evaluate nanoparticle-induced adverse effects. ZnONPs increased the mortality rate of zebrafish embryos and induced malformation phenotypes, such as pericardial edema and yolk-sac edema. Moreover, ZnO affected the expression of inflammatory and immune response genes, including *aicda*, *cyb5d1*, *edar*, *intl2*, *ogfrl2*, and *tnfsf13b* [93]. In addition, silica NPs cause lower blood flow and blood velocity in zebrafish embryos. Silica NPs trigger inflammatory responses via neutrophils and damage vascular endothelial cells [94]. Another study revealed that AgNPs influence the richness and diversity of the microbiota in zebrafish, particularly in males. Therefore, the zebrafish model provides an ideal platform for a relatively quick, high-throughput screening of hazardous nanomaterials and for determining nanomaterial-triggered toxic mechanisms.

As described above, autophagy is an important cellular response induced by nanomaterials. Over the past decade, the zebrafish model gained attention in the autophagy field. Zebrafish, as a tractable animal model, are suitable for use in manipulating the molecular and cellular mechanisms of autophagy [88,95]. Several transgene techniques are applicable to zebrafish models, such as tissue-specific genomic manipulation and the CRISPR system for genome editing. Moreover, several transgenic reporter lines have been used to monitor the autophagy process, including Tg (CMV:EGFP-Mapllc3b) [30], Tg (TαCP:YFP-2XFYVE) [96], and Tg (TαCP:mCherry-GFP-map1 lc3b) [88,95,97]. These techniques promote the understanding of the physiological functions of autophagy in zebrafish. Similarly, the application of these technologies in a zebrafish system to detect nanomaterial-induced autophagy may be an ideal strategy. Indeed, exposure to high doses of TiO_2NPs has been reported to lead to abnormal testicular morphology and spermatocyte necrosis in zebrafish. The application of TiO_2NPs resulted in mitochondria being swallowed and autophagic vacuoles accumulating in zebrafish testes [98].

4.3. Drosophila Model

Drosophila melanogaster has a long history of significant contributions to biomedical research, including in the research of nine Nobel Laureates in Physiology or Medicine. This model organism has orthologs for approximately 75% of human genes and can be profoundly manipulated with genetic and molecular tools. Drosophila also complies with the recommendations of the European Center for the Validation of Alternative Methods (ECVAM), because they present minor practical and ethical obstacles. Further advantages of the fruit fly system includes easy maintenance, low cost, a relatively short life cycle, and much-reduced genetic redundancy compared to mammals, making Drosophila an efficient system for high-throughput screenings and assays [99].

It has been shown that exposure to AgNPs in the diet in an effective Drosophila in vivo platform led to the generation of ROS and high-level autophagy activation, providing strong in vivo evidence that dietary AgNPs activate a series of cytotoxic pathways, including autophagy [45]. In addition, many established Drosophila autophagy transgenic lines are readily available to further study of autophagy induction/activation/dysfunction; for example, UAS-GFP-mCherry-tagged Atg8a is an autophagy reporter transgenic fly line used to monitor the progression of autophagic flux in vivo. These transgenic lines can greatly facilitate the use and development of Drosophila as an in vivo animal model for alternative strategies for testing nanomaterials. Therefore, the Drosophila autophagy system may provide an excellent system for the in vivo assessment of nanoparticle toxicity [100,101]. Furthermore, AgNPs shortened the life span and reduced the stress resistance capacity of the adult flies [45]. Interestingly, other nanomaterials, such as copper oxide nanoparticles (CuONPs), also induced toxicity in Drosophila via ROS; whether CuONPs and other nanomaterials also induce autophagy in Drosophila remains to be explored [102]. As basal autophagy provides a protection mechanism and thus is generally considered a promoter of longevity [103], it is a possibility that AgNPs-induced autophagy activation plays an important role in aging and longevity. Thus, inducing autophagy activation in Drosophila can serve as an alternative testing strategy for determining nanomaterial toxicity. More importantly, autophagy genes, encoding Atg proteins, are structurally, functionally, and mechanistically conserved between Drosophila and humans [104].

Nevertheless, the molecular mechanisms by which nanomaterials induce autophagy and shorten the life span of Drosophila remain to be explored. Given the potential risks associated with nanomaterial-induced autophagy in longevity and diseases, it is important to further study autophagy using an in vivo model and to develop systematic alternative strategies for testing nanomaterials, particularly regarding autophagy activation in Drosophila. In other words, Drosophila can serve as a practical and ethical in vivo animal model for alternative strategies for testing nanomaterials and make an important contribution to preventing/treating nanomaterial-induced toxicity in humans. Moreover, it is critical to establish Drosophila autophagy as a systematic and effective alternative strategy for testing nanomaterials for improving nanomaterial risk management and human safety regulations.

4.4. Rodent Model

The classic animal models most frequently used in toxicology are mammals, including rodents, dogs, pigs, and non-human primates. However, major species used in nanotoxicology are mainly rodents, such as mice and rats. Laboratory mice are popular research models because of their size, availability, ease of handling, and high genetic similarity to humans. Many mice and rat models are well-established and widely used in pharmacological and toxicological studies of NMs [105,106]. Using mouse models for precisely controlled exposure to NMs, we can perform NMs toxicity assessment include LD50, biodistribution, clearance, hematology, serum chemistry, and histopathology. Biodistribution can evaluate the NMs circulation route and the target organs by measuring fluorescent or radioactive labeled nanoparticles in animals [107]. Examining the excretion and metabolism of nanoparticles at regular time points after exposure can determine the clearance of NMs [108]. Monitoring changes in serum chemistry, cell type, and organ histopathology of mice after nanoparticle exposure is also another method for nanotoxicity assessment [109–111]. Coupling with the development of transgenic mice and disease models, we can also take various host factors, such as genetic defects and pre-existing pathology, into consideration in nanosafety [105]. The REACH Guidance states that acute, subchronic, and chronic toxicity, skin and eye irritation, or corrosion and skin sensitization, genetic toxicity, reproductive toxicity, carcinogenicity, and toxicokinetics of NPs should be evaluated when conducting in vivo nanotoxicity assessments. In some cases, REACH requests collection of urine and blood at specified time points, recording the weight of the mice or rats and their behavior, measuring the consumption of food and water [35]. NMs is widely used in multiple medical applications, so the potential side effect of these NMs has become an important issue. As the in vitro and in vivo effects of NPs are not fully matched, side effects cannot be accurately estimated by in vitro tests, it is recognized that animal tests are essential for safety assessment of these medical NMs [106]. For example, AgNPs is one of the most common NMs in medical use because of its antimicrobial activity. Of course, many studies have evaluated the safety and biocompatibility of AgNPs in rodent models. Kim et al. have reported the results of a 28-day oral exposure study of 60 nm AgNPs in rats [112]. Data showed dose-dependent changes in serum cholesterol and ALP and liver damage after 300 mg AgNPs treatment. In addition, eosinophil infiltration of the hepatic lobules and portal tract and bile duct hyperplasia was also found. Another subchronic dermal toxicity assessment analyzed biodistribution of AgNPs in guinea pigs, and the results showed that the tissue level of AgNPs and dermal exposure show a dose-dependent correlation with the following ranking: kidney > muscle > bone > skin > liver > heart > spleen [113]. Histopathological data further reported toxicities in kidney, bone, and cardiocytes after dermal exposure to AgNPs. Chuang's group also established allergen-provocation mice models to investigate the effects of inhaled AgNPs in healthy and allergic individuals [114].

5. Autophagy Detection as a Toxicity Biomarker-Like Indicator for Medical, Food, and Cosmetic NPs Safety Assessments in Future

5.1. Silver Nanoparticles-Induced Toxicity and the Possible Role of Autophagy

Silver has been used in our lives throughout history and recently for many medical applications due to its effectiveness in arresting the growth of microorganisms [115]. Recent studies have shown the enormous therapeutic potential of AgNPs against numerous cancer cells by modulating autophagy action as cytotoxic agents or as nanocarriers that, combined with other treatments, deliver therapeutic molecules [116,117]. The toxicity of AgNPs has been suggested as the result of lysosome-dependent silver ion release that leads to massive ROS production. These ROS cause disruption of the lysosomal membrane integrity and enables the escape of AgNPs into the cytosolic space, through which they subsequently target other subcellular compartments [118]. In addition, AgNPs-induced lysosomal dysfunction, including loss of membrane integrity or internal acidity, is also related to an impaired autophagosome–lysosome fusion process that critically interferes with the functionality of the autophagy machinery [119]. AgNPs have high affinity for thiol groups, which are important for protein folding and function as ROS scavengers. Therefore, AgNPs cause the protein misfolding that induces the ER stress and glutathione depletion that leads to ROS metabolism imbalance. All of these results enhance the autophagy process and ultimately cause cell death.

It has been documented that AgNPs are potential sources of oxidative stress, leading to ROS production and subsequent autophagy induction in the NIH3T3 mouse embryonic fibroblast cells to which they were exposed [18]. In another study, administration of AgNPs upregulated LC3-II protein expression and accumulated in liver tissue [120]. Under oxidative stress conditions, autophagy can be induced to suppress cellular ROS levels in a cell survival mechanism of normal cells. While it has also been reported that chronic low-dose AgNPs exposure resulted in to HaCaT noncancerous cell transformation, and despite the activation of EGF receptors and the related gene expression that enhances cell proliferation, cells treated with a high dose of AgNPs within a short time showed inhibited proliferation [121]. Ag-NPs have been observed to have a higher cytotoxic effect on PANC1 pancreatic cancer cells than on non-tumor cells of the same tissue [122]. Furthermore, combining AgNPs with drugs synergistically enhanced the cytotoxicity to cancer cells [116]. In addition to autophagy induction, AgNPs have also been demonstrated to block autophagic flux to induce autophagosome accumulation, resulting in the impedance of monocyte–macrophage differentiation [123]. These findings provide new perspectives on anticancer therapy strategies using nanomaterials.

5.2. ZnO Nanoparticles-Induced Toxicity and the Possible Role of Autophagy

ZnONPs have been employed in biomedical and cancer applications due to their unique properties [124]. It has been reported that ZnONPs induced significant cytotoxicity with increased intracellular ROS and oxidative stress that led to apoptosis and autophagy in SKOV3 ovarian cancer cells [125]. ZnONPs have also been observed to induce toxicity by activating PINK1/Parkin-mediated mitophagy in CAL27 oral cancer cell lines [126]. In addition, some studies reported that ZnONPs exhibited preferential cytotoxicity to highly proliferative tumor cells through a lysosome-mediated zinc ion release mechanism that subsequently led to ROS-mediated cell death [124,127]. These reports strongly suggest the potential of ZnONPs as anticancer agents.

In the cosmetics industry, ZnONPs are present in daily supplies, such as shampoos, conditioners, soaps, deodorants, sunscreens, and skin care products, and makeup, in general, to function as antibacterial agents, UV-filters, and pigments and for deeper skin penetration, anti-wrinkling, or moisturizing [128–130]. For example, sunscreens containing ZnONPs and TiO_2NPs are effective barriers against ultraviolet light (UV-light) damage to skin and do not leave white or other residues on the skin [131,132]. AgNPs are used in toothpaste and soap and other cleaning products to achieve an antibacterial effect, and gold nanoparticles (AuNPs) are commonly used as carriers that easily penetrate the skin [132–134]. However, the diversity of NMs applied in cosmetics has raised concerns about their potential risks. Studies have indicated that NMs can be translocated to main organs, such as the brain, kidney, or heart, from different exposure routes; the transdermal penetration and translocation of NMs through the skin are still controversial [135–137]. Zvyagin et al. reported that ZnONPs stayed in the stratum corneum (SC) and accumulated into skin folds and/or hair follicle roots when applied topically in excised and in vivo human skin. [137]. In contrast, skin exposure to ZnONPs and TiO_2NPs led to the incorporation of nanoparticles in the SC and induced phototoxicity and genotoxicity [138]. Numerous studies have reported the autophagy-inducing activities of ZnONPs. For example, abnormal autophagosome accumulation and mitochondrial dysfunction was observed in normal ZnONPs-treated skin cells, and ZnONPs toxicity was found to be related to the induction of ROS in a concentration- and time-dependent manner [139]. In another study, ZnONPs induced ROS generation in immune cells and activated autophagy through PI3K/Akt/mTOR signaling pathway inhibition [140]. All these results show the ROS-related autophagy-inducing and cytotoxicity-inducing abilities of ZnONPs. Therefore, the toxicity of ZnONPs to cells has attracted researchers' attention.

5.3. TiO$_2$ Nanoparticles-Induced Toxicity and the Possible Role of Autophagy

TiO$_2$NPs, which also serve as common ingredients in sunscreens and cosmetics, through which they absorb ultraviolet radiation, have also been investigated as autophagy modulators [141]. TiO$_2$NPs induced autophagy in primary human keratinocytes in a dose-dependent manner, which played a vital role in determining keratinocyte survival [142]. Another study also revealed that TiO$_2$NPs induced autophagy at a low dose while blocking autophagic flux at a high dose, which is caused by a large amount of TiO$_2$ NP accumulation-mediated overload of the degradative capacity of human keratinocytes [143]. Generally, TiO$_2$NPs induce lower toxicity in cells than ZnONPs because of their resistance to lysosome degradation and low metal ion release [139]. These findings suggest that a pro-survival mode of autophagy induction by TiO$_2$NPs provides further insights into the debate of the NPs for use in consumer products.

Recently, the unique super-photocatalytic properties of TiO$_2$NPs showed potential application in photodynamic therapy (PDT) upon irradiation [144]. It has been reported that TiO$_2$NPs were successfully used in PDT for many different types of cancers [145,146]. Under UV light illumination, the excited valence band electrons in TiO$_2$ jump to the conduction band, resulting in electron holes that have the ability to generate various ROS, including hydroxyl radicals (OH·), hydrogen peroxide (H$_2$O$_2$), and superoxide (O$_2$$^-$) [145]. Excess ROS can further trigger autophagy-associated apoptotic cell death, making TiO$_2$ much more efficient at killing cancer cells. These findings provide another application potential of TiO$_2$NPs as anticancer agents.

In food industry, most of the nanoparticle in the products are designed "out-of-food" that are not directly added to human food but some of these products have been used as food pigments and colorants [147]. For instance, inorganic oxide chemicals such as SiO$_2$ (E551), MgO (E530), and TiO$_2$ (E171) are used as anti-caking agents, food flavor carriers, food pigments, and colorants that are permitted by the U.S. FDA [147]. Food-grade TiO$_2$NPs are the most widely used NPs in food, as additives in gum, white sauce, cake icing, candy, and pudding, with approximately 40% at concentrations in the nanometer range [148]. TiO$_2$NPs, AgNPs, and ZnONPs have also been used in food packaging because each can be easily and effectively incorporated into nanocomposites to inhibit bacterial growth and extend the shelf life of food products [149]. As mentioned above, NPs might increase the levels of intracellular ROS, in turn damaging mitochondria and the ER, leading to apoptosis, DNA damage, an impaired cell cycle, and autophagy [148,150]. However, ROS production may not be the sole mechanism for the toxicity found in vitro with NPs. Previous studies have indicated that TiO$_2$NPs may interact with DNA directly, since particles were detected by transmission electron microscopy inside the nucleus in various cells, including blood lymphocytes and nasal, pulmonary, and dermal cells [151]. TiO$_2$NPs-triggered DNA damage can also be induced through indirect mechanisms, such as dysregulated cell division, DNA replication, transcription, and repair [151]. Zhang et al. indicated that TiO$_2$NPs entered trophoblast HTR-8/SVneo cells and were distributed primarily to lysosomes, where they ultimately induced autophagy in the cells [152]. Thus, food NPs could trigger autophagy blockage and lead to the accumulation of damaged ER and mitochondria and the production of ROS, resulting in further cellular damage such as NLRP3 inflammasome activation [51]. These findings provide insight into autophagy and may account for the early and specific toxicity-inducing mechanisms of NPs.

6. Conclusions and Perspectives

Nanotechnology has potential to be widely utilized in different fields, including the pharmaceutical, food, and cosmetics industries. The biosafety issue of nanoparticles has drawn great attention because many NPs induce various levels of cytotoxicity that eventually lead to cell death, cell cycle arrest, or differentiation disruption. It should also be noted that the toxicity induced by the long-term exposure to NPs has elicited significant concerns, especially the toxic effects to fertility, carcinogenesis, neuron, skin, and the gastrointestinal tract [138,148,153–155]. All these concerns suggest that more research and optimized evaluation systems are needed to define the exact mode of toxicity of NPs. Additionally, uncovering the underlying mechanisms that regulate toxicity will contribute greatly to adequate hazard assessments of NPs and regulation and legislation development for the management of NPs. Although several in vitro toxicity tests or in silica analyses covering important toxicity endpoints have been established as high-throughput methods for the evaluation of chemical toxicity [8,22,24,26], universally accepted protocols and well-designed alternative testing strategies for NP toxicity are still relatively scarce. A primary cause of these limited protocols and strategies is the complexity of the physicochemical properties of NPs and of NM interactions with biological systems that dictate the diverse fates of exposed cells. To fill the gaps of understanding on nano–bio interactions, more systematic research approaches using high-throughput in vitro models are needed to provide toxicity results of NM use (Figure 2). Of importance, a better understanding of the interaction of NPs with cells, tissues, and organs, for addressing critical issues related to toxicity testing, especially with respect to alternatives to tests on animals, is needed. While nanotoxicity is often the major concern when toxicologists discuss the novel nanomaterials' safe assessment, there are many nanomaterials that have been applied for therapeutic or industry purposes. Since the introduction of the first FDA-approved nano-drug in 1995, nanomedicine has constantly revolutionized medical therapeutics and diagnostics [156]. Manipulating molecules and atoms in the nanoscale has empowered researchers to come up with novel particles and formulations that possess more beneficiary characteristics and less unwanted features [156].

Adverse outcome pathways (AOPs) are an important tool to organize data and facilitate the understanding of the specific bioactivity of NPs. With respect to AOPs, this review demonstrated that autophagy and ROS production elicited by NPs appeared to be critical responses to toxicity, as autophagy is a basic stress response and a potential regulator of toxicity. Undoubtedly, the autophagic effects of NPs are highly dependent on their physicochemical characterization [51]. Some NPs can induce both autophagy blockage and autophagic flux in different testing systems. Autophagy dysfunction can lead to the accumulation of damaged DNA, proteins, and organelles that in turn increase the risk of cancer, neurodegenerative diseases, and reproductive dysfunction [51]. Effective risk assessment of NPs depends on in vitro testing strategies and relevant non-mammalian models with sufficient sensitivity to these substances. The integrated approach applying autophagy as an early sensitivity marker combined with the appropriate AOPs would enable the determination of the possible toxicity of NPs. In this review, we focused on NPs that are widely used in several industries, describing their applications, toxic effects, and autophagy-inducing potency. We also discussed several alternative methods for nanoparticle toxicity evaluation and suggested the potential application of autophagy as a tier I early toxicity endpoint in the testing framework. These results will enable the development of more-relevant testing strategies to predict the possible long-term toxicity of NPs. In addition, these strategies can be applied in the future for regulatory decision making and risk assessment of NP uses (Figure 2).

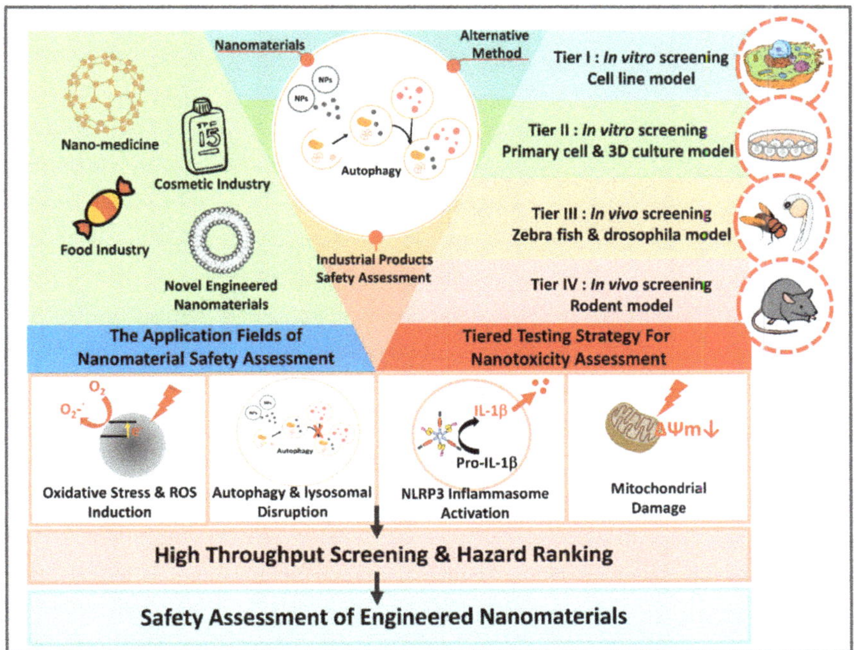

Figure 2. The proposed predictive, tiered toxicological testing strategy for nanomaterial hazard testing. The tiered testing strategy we have suggested for the evaluation of the toxicity of nanomaterials is based on screening with in vitro cell lines and high-throughput systems (Tier I). The next testing step is performed using primary cells and a 3D cell culture system to increase confidence in the data obtained from the cell lines (Tier II). Then, the zebrafish and/or Drosophila models (Tier III) are used to fill the gap in the in vitro, and the potential effects are then detected in rodents (Tier IV). When a significant potential hazard is identified in these test system steps, the rodent toxicity testing is needed. More importantly, we focused on the assessment of autophagy-related effects (autophagy and lysosomal dysfunction) and oxidative stress-related responses (ROS, mitochondrial damage, and DNA damage) as primary sensitive markers for evaluating the toxicity of the nanomaterials. As autophagy is a significant and sensitive effect induced by nanomaterials, evaluating autophagy-related pathways in the first step would improve the testing efficiency of the nanomaterials. The development of a tiered testing strategy for nanomaterial (NM) hazard assessment not only promotes the widespread adoption of non-rodent models and/or the 3Rs (reduces, refines, or replaces) principle but also makes nanotoxicology testing more ethical, relevant, and cost- and time-efficient.

Author Contributions: Y.-J.W. and Y.-H.W. wrote the "Abstract" and "Introduction". Y.-L.L. wrote "Alternative testing strategy for nanomaterial safety assessments". S.-J.Y. and Z.-Y.C. wrote "In vivo model systems for the detection of nanomaterial toxicity and autophagy". Y.-Y.C. wrote "Updated nanotoxicology knowledge regarding autophagy dysfunction" and prepared figures. M.-Y.L., L.-X.Y. and Y.-H.L. wrote "Autophagy detection as a toxicity biomarker-like indicator for medical, food and cosmetic NPSs safety assessments in future". R.-J.C. wrote "Conclusions and Perspectives", organized and revised the manuscript. Y.-L.Y. organized the References. All authors have read and agreed to the published version of the manuscript.

Acknowledgments: This work was supported by the Ministry of Science and Technology, Taiwan (MOST 108-2314-B-006-057, MOST 106-2314-B-006-029-MY3, MOST 107-2311-B-006-004-MY3, and MOST 108-2113-M-153-001) and Toxic and Chemical Substance Bureau, Taiwan (107A024).

Conflicts of Interest: The authors declare no conflict of interest.

Abbreviations

3R's	Reduces, Refines, or Replaces
ACD	Autophagic Cell Death
AgNPs	Silver Nanoparticles
AMPK	AMP Kinase
AOP	Adverse Outcome Pathway
AuNPs	Gold Nanoparticles
BBB	Blood–Brain Barrier
BPR	Biocidal Products Regulations
CLP	Classification, Labelling and Packaging
CuONPs	Copper Oxide Nanoparticles
DMH	Dimethylhydrazine
ECVAM	European Center for the Validation of Alternative Methods
ER	Endoplasmic Reticulum
FADD	Fas-Associated Protein with Death Domain
GIT	Gastrointestinal Tract
INRA	French National Institute for Agricultural Research
ISO	International Organization for Standardization
LMP	Lysosomal Membrane Permeabilization
NCI	National Cancer Institute
NCL	Nanotechnology Characterization Lab
NMs	Nanomaterials
NRC	National Research Council
OECD	Organization for Economic Co-operation and Development
PDT	Photodynamic Therapy
REACH	Registration, Evaluation, Authorization and Restriction of Chemicals
ROS	Reactive Oxygen Species
SC	Stratum Corneum
SCCS	Scientific Committee on Consumer Safety
SEURAT	Safety Evaluation Ultimately Replacing Animal Testing
TEM	Transmission Electron Microscopy
TiO_2NPs	Titanium Dioxide Nanoparticles
TT21C	Toxicity Testing in the 21st Century
UV	Ultraviolet
UV-light	Ultraviolet Light
ZnONPs	Zinc Oxide Nanoparticles

References

1. European Commission. Commission Recommendation of 18 October 2011 on the definition of nanomaterial. *J. Off. Eur. Union* **2011**, *275*, 38–40.
2. Warheit, D.B. Hazard and risk assessment strategies for nanoparticle exposures: How far have we come in the past 10 years? *F1000Research* **2018**, *7*, 376. [CrossRef]
3. Jones, C.F.; Grainger, D.W. In vitro assessments of nanomaterial toxicity. *Adv. Drug Deliv. Rev.* **2009**, *61*, 438–456. [CrossRef]
4. Jain, A.; Ranjan, S.; Dasgupta, N.; Ramalingam, C. Nanomaterials in food and agriculture: An overview on their safety concerns and regulatory issues. *Crit. Rev. Food Sci. Nutr.* **2018**, *58*, 297–317. [CrossRef] [PubMed]
5. Wacker, M.G.; Proykova, A.; Santos, G.M.L. Dealing with nanosafety around the globe—Regulation vs. innovation. *Int. J. Pharm.* **2016**, *509*, 95–106. [CrossRef] [PubMed]
6. Choi, J.-Y.; Ramachandran, G.; Kandlikar, M. The impact of toxicity testing costs on nanomaterial regulation. *Environ. Sci. Technol.* **2009**, *43*, 3030–3034. [CrossRef] [PubMed]

7.   Johnston, H.J.; Verdon, R.; Gillies, S.; Brown, D.M.; Fernandes, T.F.; Henry, T.B.; Rossi, A.G.; Tran, L.; Tucker, C.; Tyler, C.R.; et al. Adoption of in vitro systems and zebrafish embryos as alternative models for reducing rodent use in assessments of immunological and oxidative stress responses to nanomaterials. *Crit. Rev. Toxicol.* **2018**, *48*, 252–271. [CrossRef] [PubMed]

8.   Krewski, D.; Acosta, D., Jr.; Andersen, M.; Anderson, H.; Bailar, J.C., III; Boekelheide, K.; Brent, R.; Charnley, G.; Cheung, V.G.; Green, S., Jr.; et al. Toxicity testing in the 21st century: A vision and a strategy. *J. Toxicol. Environ. Health Part B* **2010**, *13*, 51–138. [CrossRef] [PubMed]

9.   National Research Council. *Toxicity Testing in the 21st Century: A Vision and a Strategy*; National Academies Press: Washington, DC, USA, 2007; ISBN 978-030-910-992-5.

10. Shatkin, J.; Ong, K. Alternative testing strategies for nanomaterials: State of the science and considerations for risk analysis. *Risk Anal.* **2016**, *36*, 1564–1580. [CrossRef]

11. Dusinska, M.; Boland, S.; Saunders, M.; Juillerat-Jeanneret, L.; Tran, L.; Pojana, G.; Marcomini, A.; Volkovova, K.; Tulinska, J.; Knudsen, L.E.; et al. Towards an alternative testing strategy for nanomaterials used in nanomedicine: Lessons from NanoTEST. *Nanotoxicology* **2015**, *9* (Suppl. 1), 118–132. [CrossRef]

12. Collins, F.S.; Gray, G.M.; Bucher, J.R. Transforming environmental health protection. *Science* **2008**, *319*, 906–907. [CrossRef] [PubMed]

13. De Duve, C. The lysosome. *Sci. Am.* **1963**, *208*, 64–73. [CrossRef] [PubMed]

14. Mizushima, N. Autophagy in protein and organelle turnover. *Cold Spring Harbor Symp. Quant. Biol.* **2011**, *76*, 397–402. [CrossRef] [PubMed]

15. Green, D.R.; Levine, B. To be or not to be? How selective autophagy and cell death govern cell fate. *Cell* **2014**, *157*, 65–75. [CrossRef] [PubMed]

16. Denton, D.; Kumar, S. Autophagy-dependent cell death. *Cell Death Differ.* **2019**, *26*, 605–616. [CrossRef] [PubMed]

17. Li, Y.; Ju, D. The role of autophagy in nanoparticles-induced toxicity and its related cellular and molecular mechanisms. *Adv. Exp. Med. Biol.* **2018**, *1048*, 71–84.

18. Lee, Y.-H.; Cheng, F.-Y.; Chiu, H.-W.; Tsai, J.-C.; Fang, C.-Y.; Chen, C.-W.; Wang, Y.-J. Cytotoxicity, oxidative stress, apoptosis and the autophagic effects of silver nanoparticles in mouse embryonic fibroblasts. *Biomaterials* **2014**, *35*, 4706–4715. [CrossRef]

19. Mao, B.H.; Tsai, J.C.; Chen, C.W.; Yan, S.J.; Wang, Y.J. Mechanisms of silver nanoparticle-induced toxicity and important role of autophagy. *Nanotoxicology* **2016**, *10*, 1021–1040. [CrossRef]

20. Lee, Y.-H.; Fang, C.-Y.; Chiu, H.-W.; Cheng, F.-Y.; Tsai, J.-C.; Chen, C.-W.; Wang, Y.-J. Endoplasmic reticulum stress-triggered autophagy and lysosomal dysfunction contribute to the cytotoxicity of amine-modified silver nanoparticles in NIH 3T3 cells. *J. Biomed. Nanotechnol.* **2017**, *13*, 778–794. [CrossRef]

21. Ding, L.; Liu, Z.; Okweesi Aggrey, M.; Li, C.; Chen, J.; Tong, L. Nanotoxicity: The toxicity research progress of metal and metal-containing nanoparticles. *Mini Rev. Med. Chem.* **2015**, *15*, 529–542. [CrossRef]

22. Doke, S.K.; Dhawale, S.C. Alternatives to animal testing: A review. *Saudi Pharm. J.* **2015**, *23*, 223–229. [CrossRef] [PubMed]

23. Krewski, D.; Andersen, M.; Tyshenko, M.; Krishnan, K.; Hartung, T.; Boekelheide, K.; Wambaugh, J.; Jones, D.; Whelan, M.; Thomas, R.; et al. Toxicity testing in the 21st century: Progress in the past decade and future perspectives. *Arch. Toxicol.* **2019**, *94*, 1–58. [CrossRef] [PubMed]

24. Gocht, T.; Berggren, E.; Ahr, H.J.; Cotgreave, I.; Cronin, M.T.; Daston, G.; Hardy, B.; Heinzle, E.; Hescheler, J.; Knight, D.J. The SEURAT-1 approach towards animal free human safety assessment. *ALTEX Altern. Anim. Exp.* **2015**, *32*, 9–24. [CrossRef] [PubMed]

25. European Chemicals Agency. *Non-Animal Approaches-Current Status of Regulatory Applicability under the REACH, CLP and Biocidal Products Regulations*; European Chemicals Agency: Helsinki, Finland, 2017; ISBN 978-92-9020-208-0.

26. Rusyn, I.; Greene, N. The impact of novel assessment methodologies in toxicology on green chemistry and chemical alternatives. *Toxicol. Sci.* **2018**, *161*, 276–284. [CrossRef] [PubMed]

27. Gennari, A.; Van Den Berghe, C.; Casati, S.; Castell, J.; Clemedson, C.; Coecke, S.; Colombo, A.; Curren, R.; Negro, G.D.; Goldberg, A.; et al. Strategies to replace in vivo acute systemic toxicity testing: The report and recommendations of ECVAM workshop 50. *ATLA Altern. Lab. Anim.* **2004**, *32*, 437–459. [CrossRef]

28. Andrew, D.J. Acute systemic toxicity: Oral, dermal and inhalation exposures. In *Reducing, Refining and Replacing the Use of Animals in Toxicity Testing*; Allen, D., Waters, M.D., Eds.; Royal Society of Chemistry: London, UK, 2013; pp. 183–214.

29. Prieto, P.; Graepel, R.; Gerloff, K.; Lamon, L.; Sachana, M.; Pistollato, F.; Gribaldo, L.; Bal-Price, A.; Worth, A. Investigating cell type specific mechanisms contributing to acute oral toxicity. *ALTEX Altern. Anim. Exp.* **2019**, *36*, 39–64. [CrossRef]

30. National Institutes of Health. *Report on the ICCVAM-NICEATM/ECVAM/JaCVAM Scientific Workshop on Acute Chemical Safety Testing: Advancing In Vitro Approaches and Humane Endpoints for Systemic Toxicity Evaluations*; National Toxicology Program; National Institutes of Health: Bethesda, MD, USA, 2009.

31. Saifi, M.A.; Khan, W.; Godugu, C. Cytotoxicity of nanomaterials: Using nanotoxicology to address the safety concerns of nanoparticles. *Pharm. Nanotechnol.* **2018**, *6*, 3–16. [CrossRef]

32. Braakhuis, H.M.; Park, M.V.; Gosens, I.; De Jong, W.H.; Cassee, F.R. Physicochemical characteristics of nanomaterials that affect pulmonary inflammation. *Part. Fibre Toxicol.* **2014**, *11*, 18. [CrossRef]

33. Bai, X.; Liu, F.; Liu, Y.; Li, C.; Wang, S.; Zhou, H.; Wang, W.; Zhu, H.; Winkler, D.A.; Yan, B. Toward a systematic exploration of nano-bio interactions. *Toxicol. Appl. Pharmacol.* **2017**, *323*, 66–73. [CrossRef]

34. Wu, L.; Zhang, Y.; Zhang, C.; Cui, X.; Zhai, S.; Liu, Y.; Li, C.; Zhu, H.; Qu, G.; Jiang, G.; et al. Tuning cell autophagy by diversifying carbon nanotube surface chemistry. *ACS Nano* **2014**, *8*, 2087–2099. [CrossRef]

35. Joris, F.; Manshian, B.B.; Peynshaert, K.; De Smedt, S.C.; Braeckmans, K.; Soenen, S.J. Assessing nanoparticle toxicity in cell-based assays: Influence of cell culture parameters and optimized models for bridging the in vitro-in vivo gap. *Chem. Soc. Rev.* **2013**, *42*, 8339–8359. [CrossRef] [PubMed]

36. Drasler, B.; Sayre, P.; Steinhauser, K.G.; Petri-Fink, A.; Rothen-Rutishauser, B. In vitro approaches to assess the hazard of nanomaterials. *Nanoimpact* **2017**, *8*, 99–116. [CrossRef]

37. Hanley, C.; Layne, J.; Punnoose, A.; Reddy, K.M.; Coombs, I.; Coombs, A.; Feris, K.; Wingett, D. Preferential killing of cancer cells and activated human T cells using ZnO nanoparticles. *Nanotechnology* **2008**, *19*, 295103. [CrossRef] [PubMed]

38. Nel, A.; Xia, T.; Meng, H.; Wang, X.; Lin, S.; Ji, Z.; Zhang, H. Nanomaterial toxicity testing in the 21st century: Use of a predictive toxicological approach and high-throughput screening. *Acc. Chem. Res.* **2013**, *46*, 607–621. [CrossRef] [PubMed]

39. Pereira, A.C.; Gomes, T.; Machado, M.R.F.; Rocha, T.L. The zebrafish embryotoxicity test (ZET) for nanotoxicity assessment: From morphological to molecular approach. *Environ. Pollut.* **2019**, *252*, 1841–1853. [CrossRef] [PubMed]

40. Doherty, J.; Baehrecke, E.H. Life, death and autophagy. *Nat. Cell Biol.* **2018**, *20*, 1110–1117. [CrossRef]

41. Chen, Z.-H.; Lam, H.C.; Jin, Y.; Kim, H.-P.; Cao, J.; Lee, S.-J.; Ifedigbo, E.; Parameswaran, H.; Ryter, S.W.; Choi, A.M. Autophagy protein microtubule-associated protein 1 light chain-3B (LC3B) activates extrinsic apoptosis during cigarette smoke-induced emphysema. *Proc. Natl. Acad. Sci. USA* **2010**, *107*, 18880–18885. [CrossRef]

42. White, E.; DiPaola, R.S. The double-edged sword of autophagy modulation in cancer. *Clin. Cancer Res.* **2009**, *15*, 5308–5316. [CrossRef]

43. Ravikumar, B.; Sarkar, S.; Davies, J.E.; Futter, M.; Garcia-Arencibia, M.; Green-Thompson, Z.W.; Jimenez-Sanchez, M.; Korolchuk, V.I.; Lichtenberg, M.; Luo, S.; et al. Regulation of mammalian autophagy in physiology and pathophysiology. *Physiol. Rev.* **2010**, *90*, 1383–1435. [CrossRef]

44. Monick, M.M.; Powers, L.S.; Walters, K.; Lovan, N.; Zhang, M.; Gerke, A.; Hansdottir, S.; Hunninghake, G.W. Identification of an autophagy defect in smokers' alveolar macrophages. *J. Immunol.* **2010**, *185*, 5425–5435. [CrossRef]

45. Mao, B.H.; Chen, Z.Y.; Wang, Y.J.; Yan, S.J. Silver nanoparticles have lethal and sublethal adverse effects on development and longevity by inducing ROS-mediated stress responses. *Sci. Rep.* **2018**, *8*, 2445. [CrossRef] [PubMed]

46. Nair, V.; Belanger, E.C.; Veinot, J.P. Lysosomal storage disorders affecting the heart: A review. *Cardiovasc. Pathol.* **2019**, *39*, 12–24. [CrossRef] [PubMed]

47. Saha, S.; Panigrahi, D.P.; Patil, S.; Bhutia, S.K. Autophagy in health and disease: A comprehensive review. *Biomed. Pharmacother.* **2018**, *104*, 485–495. [CrossRef] [PubMed]

48. Flores-López, L.Z.; Espinoza-Gómez, H.; Somanathan, R. Silver nanoparticles: Electron transfer, reactive oxygen species, oxidative stress, beneficial and toxicological effects. Mini review. *J. Appl. Toxicol.* **2019**, *39*, 16–26. [CrossRef] [PubMed]

49. Guo, C.; Yang, M.; Jing, L.; Wang, J.; Yu, Y.; Li, Y.; Duan, J.; Zhou, X.; Li, Y.; Sun, Z. Amorphous silica nanoparticles trigger vascular endothelial cell injury through apoptosis and autophagy via reactive oxygen species-mediated MAPK/Bcl-2 and PI3K/Akt/mTOR signaling. *Int. J. Nanomed.* **2016**, *11*, 5257–5276. [CrossRef]

50. Wang, J.; Yu, Y.; Lu, K.; Yang, M.; Li, Y.; Zhou, X.; Sun, Z. Silica nanoparticles induce autophagy dysfunction via lysosomal impairment and inhibition of autophagosome degradation in hepatocytes. *Int. J. Nanomed.* **2017**, *12*, 809–825. [CrossRef]

51. Mohammadinejad, R.; Moosavi, M.A.; Tavakol, S.; Vardar, D.Ö.; Hosseini, A.; Rahmati, M.; Dini, L.; Hussain, S.; Mandegary, A.; Klionsky, D.J. Necrotic, apoptotic and autophagic cell fates triggered by nanoparticles. *Autophagy* **2019**, *15*, 4–33. [CrossRef]

52. Zheng, Y.T.; Shahnazari, S.; Brech, A.; Lamark, T.; Johansen, T.; Brumell, J.H. The adaptor protein p62/SQSTM1 targets invading bacteria to the autophagy pathway. *J. Immunol.* **2009**, *183*, 5909–5916. [CrossRef] [PubMed]

53. Mishra, A.R.; Zheng, J.; Tang, X.; Goering, P.L. Silver nanoparticle-induced autophagic-lysosomal disruption and NLRP3-inflammasome activation in HepG2 cells is size-dependent. *Toxicol. Sci.* **2016**, *150*, 473–487. [CrossRef]

54. Hu, Y.; Zhang, H.-R.; Dong, L.; Xu, M.-R.; Zhang, L.; Ding, W.-P.; Zhang, J.-Q.; Lin, J.; Zhang, Y.-J.; Qiu, B.-S.; et al. Enhancing tumor chemotherapy and overcoming drug resistance through autophagy-mediated intracellular dissolution of zinc oxide nanoparticles. *Nanoscale* **2019**, *11*, 11789–11807. [CrossRef]

55. Fan, J.; Wang, S.; Zhang, X.; Chen, W.; Li, Y.; Yang, P.; Cao, Z.; Wang, Y.; Lu, W.; Ju, D. Quantum dots elicit hepatotoxicity through lysosome-dependent autophagy activation and reactive oxygen species production. *ACS Biomater. Sci. Eng.* **2018**, *4*, 1418–1427. [CrossRef]

56. Eid, N.; Ito, Y.; Horibe, A.; Otsuki, Y.; Kondo, Y. Ethanol-induced mitochondrial damage in sertoli cells is associated with parkin overexpression and activation of mitophagy. *Cells* **2019**, *8*, 283. [CrossRef] [PubMed]

57. Kruppa, A.J.; Buss, F. Actin cages isolate damaged mitochondria during mitophagy. *Autophagy* **2018**, *14*, 1644–1645. [CrossRef] [PubMed]

58. Lee, J.; Giordano, S.; Zhang, J. Autophagy, mitochondria and oxidative stress: Cross-talk and redox signalling. *Biochem. J.* **2012**, *441*, 523–540. [CrossRef] [PubMed]

59. Liu, X.; Tu, B.; Jiang, X.; Xu, G.; Bai, L.; Zhang, L.; Meng, P.; Qin, X.; Chen, C.; Zou, Z. Lysosomal dysfunction is associated with persistent lung injury in dams caused by pregnancy exposure to carbon black nanoparticles. *Life Sci.* **2019**, *233*, 116741. [CrossRef]

60. Zhou, H.; Gong, X.; Lin, H.; Chen, H.; Huang, D.; Li, D.; Shan, H.; Gao, J. Gold nanoparticles impair autophagy flux through shape-dependent endocytosis and lysosomal dysfunction. *J. Mater. Chem. B* **2018**, *6*, 8127–8136. [CrossRef]

61. Vermeulen, L.M.; De Smedt, S.C.; Remaut, K.; Braeckmans, K. The proton sponge hypothesis: Fable or fact? *Eur. J. Pharm. Biopharm.* **2018**, *129*, 184–190. [CrossRef]

62. Anozie, U.C.; Dalhaimer, P. Molecular links among non-biodegradable nanoparticles, reactive oxygen species, and autophagy. *Adv. Drug Deliv. Rev.* **2017**, *122*, 65–73. [CrossRef]

63. Brenner, S. The genetics of *Caenorhabditis elegans*. *Genetics* **1974**, *77*, 71–94.

64. Wang, Q.; Zhou, Y.; Fu, R.; Zhu, Y.; Song, B.; Zhong, Y.; Wu, S.; Shi, Y.; Wu, Y.; Su, Y.; et al. Distinct autophagy-inducing abilities of similar-sized nanoparticles in cell culture and live *C. elegans*. *Nanoscale* **2018**, *10*, 23059–23069. [CrossRef]

65. Hunt, P.R. The *C. elegans* model in toxicity testing. *J. Appl. Toxicol.* **2017**, *37*, 50–59. [CrossRef] [PubMed]

66. Wu, T.; Xu, H.; Liang, X.; Tang, M. *Caenorhabditis elegans* as a complete model organism for biosafety assessments of nanoparticles. *Chemosphere* **2019**, *221*, 708–726. [CrossRef] [PubMed]

67. Gonzalez-Moragas, L.; Roig, A.; Laromaine, A. *C. elegans* as a tool for in vivo nanoparticle assessment. *Adv. Colloid Interface Sci.* **2015**, *219*, 10–26. [CrossRef] [PubMed]

68. Guo, B.; Huang, X.; Zhang, P.; Qi, L.; Liang, Q.; Zhang, X.; Huang, J.; Fang, B.; Hou, W.; Han, J.; et al. Genome-wide screen identifies signaling pathways that regulate autophagy during *Caenorhabditis elegans* development. *EMBO Rep.* **2014**, *15*, 705–713.

69. Kim, J.H.; Lee, S.H.; Cha, Y.J.; Hong, S.J.; Chung, S.K.; Park, T.H.; Choi, S.S. *C. elegans*-on-a-chip for in situ and in vivo Ag nanoparticles' uptake and toxicity assay. *Sci. Rep.* **2017**, *7*, 40225. [CrossRef]

70. Khan, I.; Saeed, K.; Khan, I. Nanoparticles: Properties, applications and toxicities. *Arab. J. Chem.* **2019**, *12*, 908–931. [CrossRef]

71. Bundschuh, M.; Filser, J.; Lüderwald, S.; McKee, M.S.; Metreveli, G.; Schaumann, G.E.; Schulz, R.; Wagner, S. Nanoparticles in the environment: Where do we come from, where do we go to? *Environ. Sci. Eur.* **2018**, *30*, 6. [CrossRef]

72. Savage, D.T.; Hilt, J.Z.; Dziubla, T.D. In vitro methods for assessing nanoparticle toxicity. *Methods Mol. Biol.* **2019**, *1894*, 1–29.

73. Haque, E.; Ward, A.C. Zebrafish as a model to evaluate nanoparticle toxicity. *Nanomaterials (Basel)* **2018**, *8*, 561. [CrossRef]

74. Pappus, S.A.; Mishra, M. A drosophila model to decipher the toxicity of nanoparticles taken through oral routes. *Adv. Exp. Med. Biol.* **2018**, *1048*, 311–322.

75. Bahadar, H.; Maqbool, F.; Niaz, K.; Abdollahi, M. Toxicity of nanoparticles and an overview of current experimental models. *Iran Biomed. J.* **2016**, *20*, 1–11. [PubMed]

76. Chakraborty, C.; Sharma, A.R.; Sharma, G.; Lee, S.-S. Zebrafish: A complete animal model to enumerate the nanoparticle toxicity. *J. Nanobiotechnol.* **2016**, *14*, 65. [CrossRef] [PubMed]

77. Tanguay, R.L. The rise of zebrafish as a model for toxicology. *Toxicol. Sci.* **2018**, *163*, 3–4. [CrossRef]

78. Dooley, K.; Zon, L.I. Zebrafish: A model system for the study of human disease. *Curr. Opin. Genet. Dev.* **2000**, *10*, 252–256. [CrossRef]

79. Haffter, P.; Granato, M.; Brand, M.; Mullins, M.C.; Hammerschmidt, M.; Kane, D.A.; Odenthal, J.; Van Eeden, F.J.; Jiang, Y.J.; Heisenberg, C.P.; et al. The identification of genes with unique and essential functions in the development of the zebrafish, *Danio rerio*. *Development* **1996**, *123*, 1–36. [PubMed]

80. Stainier, D.Y.; Fishman, M.C. The zebrafish as a model system to study cardiovascular development. *Trends Cardiovasc. Med.* **1994**, *4*, 207–212. [CrossRef]

81. Bambino, K.; Chu, J. Zebrafish in toxicology and environmental health. *Curr. Top Dev. Biol.* **2017**, *124*, 331–367.

82. Howe, K.; Clark, M.D.; Torroja, C.F.; Torrance, J.; Berthelot, C.; Muffato, M.; Collins, J.E.; Humphray, S.; McLaren, K.; Matthews, L.; et al. The zebrafish reference genome sequence and its relationship to the human genome. *Nature* **2013**, *496*, 498–503. [CrossRef]

83. Kettleborough, R.N.; Busch-Nentwich, E.M.; Harvey, S.A.; Dooley, C.M.; De Bruijn, E.; Van Eeden, F.; Sealy, I.; White, R.J.; Herd, C.; Nijman, I.J.; et al. A systematic genome-wide analysis of zebrafish protein-coding gene function. *Nature* **2013**, *496*, 494–497. [CrossRef]

84. Hsu, C.H.; Wen, Z.H.; Lin, C.S.; Chakraborty, C. The zebrafish model: Use in studying cellular mechanisms for a spectrum of clinical disease entities. *Curr. Neurovasc. Res.* **2007**, *4*, 111–120. [CrossRef]

85. Liu, H.; Gooneratne, R.; Huang, X.; Lai, R.; Wei, J.; Wang, W. A rapid in vivo zebrafish model to elucidate oxidative stress-mediated PCB126-induced apoptosis and developmental toxicity. *Free Radic. Biol. Med.* **2015**, *84*, 91–102. [CrossRef] [PubMed]

86. Dai, Y.J.; Jia, Y.F.; Chen, N.; Bian, W.P.; Li, Q.K.; Ma, Y.B.; Chen, Y.L.; Pei, D.S. Zebrafish as a model system to study toxicology. *Environ. Toxicol. Chem.* **2014**, *33*, 11–17. [CrossRef] [PubMed]

87. Clark, E.M.; Nonarath, H.J.T.; Bostrom, J.R.; Link, B.A. Establishment and validation of an endoplasmic reticulum stress reporter to monitor zebrafish ATF6 activity in development and disease. *Dis. Model Mech.* **2020**, *13*. [CrossRef]

88. Mathai, B.J.; Meijer, A.H.; Simonsen, A. Studying autophagy in zebrafish. *Cells* **2017**, *6*, 21. [CrossRef] [PubMed]

89. Christen, V.; Capelle, M.; Fent, K. Silver nanoparticles induce endoplasmatic reticulum stress response in zebrafish. *Toxicol. Appl. Pharmacol.* **2013**, *272*, 519–528. [CrossRef] [PubMed]

90. Truong, L.; Tilton, S.C.; Zaikova, T.; Richman, E.; Waters, K.M.; Hutchison, J.E.; Tanguay, R.L. Surface functionalities of gold nanoparticles impact embryonic gene expression responses. *Nanotoxicology* **2013**, *7*, 192–201. [CrossRef] [PubMed]

91. Rocco, L.; Santonastaso, M.; Mottola, F.; Costagliola, D.; Suero, T.; Pacifico, S.; Stingo, V. Genotoxicity assessment of TiO_2 nanoparticles in the teleost *Danio rerio*. *Ecotoxicol. Environ. Saf.* **2015**, *113*, 223–230. [CrossRef]

92. Villacis, R.A.R.; Filho, J.S.; Pina, B.; Azevedo, R.B.; Pic-Taylor, A.; Mazzeu, J.F.; Grisolia, C.K. Integrated assessment of toxic effects of maghemite (gamma-Fe_2O_3) nanoparticles in zebrafish. *Aquat. Toxicol.* **2017**, *191*, 219–225. [CrossRef]

93. Choi, J.S.; Kim, R.-O.; Yoon, S.; Kim, W.-K. Developmental toxicity of zinc oxide nanoparticles to zebrafish (*Danio rerio*): A transcriptomic analysis. *PLoS ONE* **2016**, *11*, e0160763. [CrossRef]

94. Duan, J.; Liang, S.; Yu, Y.; Li, Y.; Wang, L.; Wu, Z.; Chen, Y.; Miller, M.R.; Sun, Z. Inflammation-coagulation response and thrombotic effects induced by silica nanoparticles in zebrafish embryos. *Nanotoxicology* **2018**, *12*, 470–484. [CrossRef]

95. Varga, M.; Fodor, E.; Vellai, T. Autophagy in zebrafish. *Methods* **2015**, *75*, 172–180. [CrossRef]

96. Simonsen, A.; Tooze, S.A. Coordination of membrane events during autophagy by multiple class III PI3-kinase complexes. *J. Cell Biol.* **2009**, *186*, 773–782. [CrossRef]

97. George, A.A.; Hayden, S.; Stanton, G.R.; Brockerhoff, S.E. Arf6 and the 5′phosphatase of synaptojanin 1 regulate autophagy in cone photoreceptors. *BioEssays* **2016**, *38* (Suppl. 1), S119–S135. [CrossRef]

98. Kotil, T.; Akbulut, C.; Yon, N.D. The effects of titanium dioxide nanoparticles on ultrastructure of zebrafish testis (*Danio rerio*). *Micron* **2017**, *100*, 38–44. [CrossRef]

99. Yamaguchi, M.; Yoshida, H. Drosophila as a model organism. *Adv. Exp. Med. Biol.* **2018**, *1076*, 1–10.

100. Lorincz, P.; Mauvezin, C.; Juhasz, G. Exploring autophagy in drosophila. *Cells* **2017**, *6*, 22. [CrossRef]

101. Ong, C.; Yung, L.Y.; Cai, Y.; Bay, B.H.; Baeg, G.H. Drosophila melanogaster as a model organism to study nanotoxicity. *Nanotoxicology* **2015**, *9*, 396–403. [CrossRef]

102. Baeg, E.; Sooklert, K.; Sereemaspun, A. Copper oxide nanoparticles cause a dose-dependent toxicity via inducing reactive oxygen species in drosophila. *Nanomaterials (Basel)* **2018**, *8*, 824. [CrossRef]

103. Hansen, M.; Rubinsztein, D.C.; Walker, D.W. Autophagy as a promoter of longevity: Insights from model organisms. *Nat. Rev. Mol. Cell Biol.* **2018**, *19*, 579–593. [CrossRef]

104. Mizushima, N. Autophagy: Process and function. *Genes Dev.* **2007**, *21*, 2861–2873. [CrossRef]

105. He, X. In vivo nanotoxicity assays in animal models. In *Toxicology of Nanomaterials*; Zhao, Y., Zhang, Z., Eds.; Wiley-VCH: Weinheim, Germany, 2016; pp. 151–197.

106. Ajdary, M.; Moosavi, M.A.; Rahmati, M.; Falahati, M.; Mahboubi, M.; Mandegary, A.; Jangjoo, S.; Mohammadinejad, R.; Varma, R.S. Health concerns of various nanoparticles: A review of their in vitro and in vivo toxicity. *Nanomaterials (Basel)* **2018**, *8*, 634. [CrossRef]

107. Kim, S.C.; Kim, D.W.; Shim, Y.H.; Bang, J.S.; Oh, H.S.; Kim, S.W.; Seo, M.H. In vivo evaluation of polymeric micellar paclitaxel formulation: Toxicity and efficacy. *J. Control. Release* **2001**, *72*, 191–202. [CrossRef]

108. Li, Y.P.; Pei, Y.Y.; Zhang, X.Y.; Gu, Z.H.; Zhou, Z.H.; Yuan, W.F.; Zhou, J.J.; Zhu, J.H.; Gao, X.J. PEGylated PLGA nanoparticles as protein carriers: Synthesis, preparation and biodistribution in rats. *J. Control. Release* **2001**, *71*, 203–211. [CrossRef]

109. Baker, G.L.; Gupta, A.; Clark, M.L.; Valenzuela, B.R.; Staska, L.M.; Harbo, S.J.; Pierce, J.T.; Dill, J.A. Inhalation toxicity and lung toxicokinetics of C-60 fullerene nanoparticles and microparticles. *Toxicol. Sci.* **2008**, *101*, 122–131. [CrossRef]

110. Lei, R.; Wu, C.; Yang, B.; Ma, H.; Shi, C.; Wang, Q.; Wang, Q.; Yuan, Y.; Liao, M. Integrated metabolomic analysis of the nano-sized copper particle-induced hepatotoxicity and nephrotoxicity in rats: A rapid in vivo screening method for nanotoxicity. *Toxicol. Appl. Pharmacol.* **2008**, *232*, 292–301. [CrossRef]

111. Carraro, T.C.M.M.; Altmeyer, C.; Khalil, N.M.; Mainardes, R.M. Assessment of in vitro antifungal efficacy and in vivo toxicity of Amphotericin B-loaded PLGA and PLGA-PEG blend nanoparticles. *J. Mycol. Med.* **2017**, *27*, 519–529. [CrossRef]

112. Kim, Y.S.; Kim, J.S.; Cho, H.S.; Rha, D.S.; Kim, J.M.; Park, J.D.; Choi, B.S.; Lim, R.; Chang, H.K.; Chung, Y.H.; et al. Twenty-eight-day oral toxicity, genotoxicity, and gender-related tissue distribution of silver nanoparticles in Sprague-Dawley rats. *Inhal. Toxicol.* **2008**, *20*, 575–583. [CrossRef]

113. Korani, M.; Rezayat, S.M.; Arbabi Bidgoli, S. Sub-chronic dermal toxicity of silver nanoparticles in guinea pig: Special emphasis to heart, bone and kidney toxicities. *Iran. J. Pharm. Res.* **2013**, *12*, 511–519.

114. Chuang, H.C.; Hsiao, T.C.; Wu, C.K.; Chang, H.H.; Lee, C.H.; Chang, C.C.; Cheng, T.J.; Taiwan CardioPulmonary Research Group. Allergenicity and toxicology of inhaled silver nanoparticles in allergen-provocation mice models. *Int. J. Nanomed.* **2013**, *8*, 4495–4506. [CrossRef]

115. Rai, M.; Yadav, A.; Gade, A. Silver nanoparticles as a new generation of antimicrobials. *Biotechnol. Adv.* **2009**, *27*, 76–83. [CrossRef]

116. Yuan, Y.G.; Gurunathan, S. Combination of graphene oxide-silver nanoparticle nanocomposites and cisplatin enhances apoptosis and autophagy in human cervical cancer cells. *Int. J. Nanomed.* **2017**, *12*, 6537–6558. [CrossRef]

117. Zhu, L.Y.; Guo, D.W.; Sun, L.L.; Huang, Z.H.; Zhang, X.Y.; Ma, W.J.; Wu, J.; Xiao, L.; Zhao, Y.; Gu, N. Activation of autophagy by elevated reactive oxygen species rather than released silver ions promotes cytotoxicity of polyvinylpyrrolidone-coated silver nanoparticles in hematopoietic cells. *Nanoscale* **2017**, *9*, 5489–5498. [CrossRef]

118. Setyawati, M.I.; Yuan, X.; Xie, J.P.; Leong, D.T. The influence of lysosomal stability of silver nanomaterials on their toxicity to human cells. *Biomaterials* **2014**, *35*, 6707–6715. [CrossRef]

119. Kroemer, G.; Jaattela, M. Lysosomes and autophagy in cell death control. *Nat. Rev. Cancer* **2005**, *5*, 886–897. [CrossRef]

120. Lee, T.Y.; Liu, M.S.; Huang, L.J.; Lue, S.I.; Lin, L.C.; Kwan, A.L.; Yang, R.C. Bioenergetic failure correlates with autophagy and apoptosis in rat liver following silver nanoparticle intraperitoneal administration. *Part. Fibre Toxicol.* **2013**, *10*, 40. [CrossRef]

121. Comfort, K.K.; Braydich-Stolle, L.K.; Maurer, E.I.; Hussain, S.M. Less is more: Long-term in vitro exposure to low levels of silver nanoparticles provides new insights for nanomaterial evaluation. *ACS Nano* **2014**, *8*, 3260–3271. [CrossRef]

122. Zielinska, E.; Zauszkiewicz-Pawlak, A.; Wojcik, M.; Inkielewicz-Stepniak, I. Silver nanoparticles of different sizes induce a mixed type of programmed cell death in human pancreatic ductal adenocarcinoma. *Oncotarget* **2018**, *9*, 4675–4697. [CrossRef]

123. Xu, Y.Y.; Wang, L.M.; Bai, R.; Zhang, T.L.; Chen, C.Y. Silver nanoparticles impede phorbol myristate acetate-induced monocyte-macrophage differentiation and autophagy. *Nanoscale* **2015**, *7*, 16100–16109. [CrossRef]

124. Rasmussen, J.W.; Martinez, E.; Louka, P.; Wingett, D.G. Zinc oxide nanoparticles for selective destruction of tumor cells and potential for drug delivery applications. *Expert Opin. Drug Deliv.* **2010**, *7*, 1063–1077. [CrossRef]

125. Bai, D.P.; Zhang, X.F.; Zhang, G.L.; Huang, Y.F.; Gurunathan, S. Zinc oxide nanoparticles induce apoptosis and autophagy in human ovarian cancer cells. *Int. J. Nanomed.* **2017**, *12*, 6521–6535. [CrossRef]

126. Wang, J.; Gao, S.; Wang, S.; Xu, Z.; Wei, L. Zinc oxide nanoparticles induce toxicity in CAL 27 oral cancer cell lines by activating PINK1/Parkin-mediated mitophagy. *Int. J. Nanomed.* **2018**, *13*, 3441–3450. [CrossRef] [PubMed]

127. Taccola, L.; Raffa, V.; Riggio, C.; Vittorio, O.; Iorio, M.C.; Vanacore, R.; Pietrabissa, A.; Cuschieri, A. Zinc oxide nanoparticles as selective killers of proliferating cells. *Int. J. Nanomed.* **2011**, *6*, 1129–1140.

128. Mu, L.; Sprando, R.L. Application of nanotechnology in cosmetics. *Pharm. Res.* **2010**, *27*, 1746–1749. [CrossRef] [PubMed]

129. Fakhravar, Z.; Ebrahimnejad, P.; Daraee, H.; Akbarzadeh, A. Nanoliposomes: Synthesis methods and applications in cosmetics. *J. Cosmet. Laser Ther.* **2016**, *18*, 174–181. [CrossRef]

130. Subramaniam, V.D.; Prasad, S.V.; Banerjee, A.; Gopinath, M.; Murugesan, R.; Marotta, F.; Sun, X.-F.; Pathak, S. Health hazards of nanoparticles: Understanding the toxicity mechanism of nanosized ZnO in cosmetic products. *Drug Chem. Toxicol.* **2019**, *42*, 84–93. [CrossRef]

131. Monteiro-Riviere, N.A.; Wiench, K.; Landsiedel, R.; Schulte, S.; Inman, A.O.; Riviere, J.E. Safety evaluation of sunscreen formulations containing titanium dioxide and zinc oxide nanoparticles in UVB sunburned skin: An in vitro and in vivo study. *Toxicol. Sci.* **2011**, *123*, 264–280. [CrossRef]

132. Nikolic, S.; Keck, C.M.; Anselmi, C.; Müller, R.H. Skin photoprotection improvement: Synergistic interaction between lipid nanoparticles and organic UV filters. *Int. J. Pharm.* **2011**, *414*, 276–284. [CrossRef]

133. Gupta, R.; Rai, B. Effect of size and surface charge of gold nanoparticles on their skin permeability: A molecular dynamics study. *Sci. Rep.* **2017**, *7*, 45292. [CrossRef]

134. Dorier, M.; Brun, E.; Veronesi, G.; Barreau, F.; Pernet-Gallay, K.; Desvergne, C.; Rabilloud, T.; Carapito, C.; Herlin-Boime, N.; Carriere, M. Impact of anatase and rutile titanium dioxide nanoparticles on uptake carriers and efflux pumps in Caco-2 gut epithelial cells. *Nanoscale* **2015**, *7*, 7352–7360. [CrossRef]

135. Raj, S.; Jose, S.; Sumod, U.S.; Sabitha, M. Nanotechnology in cosmetics: Opportunities and challenges. *J. Pharm. Bioallied Sci.* **2012**, *4*, 186–193. [CrossRef]

136. Hathaway, Q.A.; Nichols, C.E.; Shepherd, D.L.; Stapleton, P.A.; McLaughlin, S.L.; Stricker, J.C.; Rellick, S.L.; Pinti, M.V.; Abukabda, A.B.; McBride, C.R.; et al. Maternal-engineered nanomaterial exposure disrupts progeny cardiac function and bioenergetics. *Am. J. Physiol. Heart Circ. Physiol.* **2017**, *312*, H446–H458. [CrossRef] [PubMed]

137. Zvyagin, A.V.; Zhao, X.; Gierden, A.; Sanchez, W.; Ross, J.A.; Roberts, M.S. Imaging of zinc oxide nanoparticle penetration in human skin in vitro and in vivo. *J. Biomed. Opt.* **2008**, *13*, 064031. [CrossRef] [PubMed]

138. Smijs, T.G.; Pavel, S. Titanium dioxide and zinc oxide nanoparticles in sunscreens: Focus on their safety and effectiveness. *Nanotechnol. Sci. Appl.* **2011**, *4*, 95–112. [CrossRef] [PubMed]

139. Gautam, A.; Rakshit, M.; Nguyen, K.T.; Kathawala, M.H.; Nguyen, L.T.H.; Tay, C.Y.; Wong, E.; Ng, K.W. Understanding the implications of engineered nanoparticle induced autophagy in human epidermal keratinocytes in vitro. *NanoImpact* **2019**, *15*, 100177. [CrossRef]

140. Johnson, B.M.; Fraietta, J.A.; Gracias, D.T.; Hope, J.L.; Stairiker, C.J.; Patel, P.R.; Mueller, Y.M.; McHugh, M.D.; Jablonowski, L.J.; Wheatley, M.A.; et al. Acute exposure to ZnO nanoparticles induces autophagic immune cell death. *Nanotoxicology* **2015**, *9*, 737–748. [CrossRef] [PubMed]

141. DeLouise, L.A. Applications of nanotechnology in dermatology. *J. Investig. Dermatol.* **2012**, *132*, 964–975. [CrossRef]

142. Zhao, Y.; Howe, J.L.; Yu, Z.; Leong, D.T.; Chu, J.J.; Loo, J.S.; Ng, K.W. Exposure to titanium dioxide nanoparticles induces autophagy in primary human keratinocytes. *Small* **2013**, *9*, 387–392. [CrossRef]

143. Lopes, V.R.; Loitto, V.; Audinot, J.N.; Bayat, N.; Gutleb, A.C.; Cristobal, S. Dose-dependent autophagic effect of titanium dioxide nanoparticles in human HaCaT cells at non-cytotoxic levels. *J. Nanobiotechnol.* **2016**, *14*, 22. [CrossRef]

144. Vinardell, M.P.; Mitjans, M. Antitumor activities of metal oxide nanoparticles. *Nanomaterials (Basel).* **2015**, *5*, 1004–1021. [CrossRef]

145. Cai, R.; Kubota, Y.; Shuin, T.; Sakai, H.; Hashimoto, K.; Fujishima, A. Induction of cytotoxicity by photoexcited TiO_2 particles. *Cancer Res.* **1992**, *52*, 2346–2348.

146. Moosavi, M.A.; Sharifi, M.; Ghafary, S.M.; Mohammadalipour, Z.; Khataee, A.; Rahmati, M.; Hajjaran, S.; Los, M.J.; Klonisch, T.; Ghavami, S. Photodynamic N-TiO_2 nanoparticle treatment induces controlled ROS-mediated autophagy and terminal differentiation of leukemia cells. *Sci. Rep.* **2016**, *6*, 34413. [CrossRef] [PubMed]

147. He, X.; Deng, H.; Hwang, H.M. The current application of nanotechnology in food and agriculture. *J. Food Drug Anal.* **2019**, *27*, 1–21. [CrossRef] [PubMed]

148. Baranowska-Wojcik, E.; Szwajgier, D.; Oleszczuk, P.; Winiarska-Mieczan, A. Effects of titanium dioxide nanoparticles exposure on human health—A review. *Biol. Trace Elem. Res.* **2020**, *193*, 118–129. [CrossRef] [PubMed]

149. Mao, X.; Nguyen, T.H.; Lin, M.; Mustapha, A. Engineered nanoparticles as potential food contaminants and their toxicity to Caco-2 cells. *J. Food Sci.* **2016**, *81*, T2107–T2113. [CrossRef]

150. Yu, S.; Mu, Y.; Zhang, X.; Li, J.; Lee, C.; Wang, H. Molecular mechanisms underlying titanium dioxide nanoparticles (TiO_2NP) induced autophagy in mesenchymal stem cells (MSC). *J. Toxicol. Environ. Health* **2019**, *82*, 997–1008. [CrossRef]

151. Charles, S.; Jomini, S.; Fessard, V.; Bigorgne-Vizade, E.; Rousselle, C.; Michel, C. Assessment of the in vitro genotoxicity of TiO_2 nanoparticles in a regulatory context. *Nanotoxicology* **2018**, *12*, 357–374. [CrossRef]

152. Zhang, Y.; Xu, B.; Yao, M.; Dong, T.; Mao, Z.; Hang, B.; Han, X.; Lin, Z.; Bian, Q.; Li, M.; et al. Titanium dioxide nanoparticles induce proteostasis disruption and autophagy in human trophoblast cells. *Chem. Biol. Interact.* **2018**, *296*, 124–133. [CrossRef]

153. Bettini, S.; Boutet-Robinet, E.; Cartier, C.; Comera, C.; Gaultier, E.; Dupuy, J.; Naud, N.; Tache, S.; Grysan, P.; Reguer, S.; et al. Food-grade TiO_2 impairs intestinal and systemic immune homeostasis, initiates preneoplastic lesions and promotes aberrant crypt development in the rat colon. *Sci. Rep.* **2017**, *7*, 40373. [CrossRef]

154. Guo, Z.; Martucci, N.J.; Moreno-Olivas, F.; Tako, E.; Mahler, G.J. Titanium dioxide nanoparticle ingestion alters nutrient absorption in an in vitro model of the small intestine. *NanoImpact* **2017**, *5*, 70–82. [CrossRef]

155. Hong, F.; Si, W.; Zhao, X.; Wang, L.; Zhou, Y.; Chen, M.; Ge, Y.; Zhang, Q.; Wang, Y.; Zhang, J. TiO_2 nanoparticle exposure decreases spermatogenesis via biochemical dysfunctions in the testis of male mice. *J. Agric. Food Chem.* **2015**, *63*, 7084–7092. [CrossRef]
156. Mousavi, S.Z.; Nafisi, S.; Maibach, H.I. Fullerene nanoparticle in dermatological and cosmetic applications. *Nanomed. Nanotechnol. Biol. Med.* **2017**, *13*, 1071–1087. [CrossRef] [PubMed]

International Journal of
Molecular Sciences

Article

Evaluation of the Genotoxic and Oxidative Damage Potential of Silver Nanoparticles in Human NCM460 and HCT116 Cells

Mingxi Jia [1,2,†], Wenjing Zhang [1], Taojin He [1], Meng Shu [1], Jing Deng [1,2,†], Jianhui Wang [3,†], Wen Li [1,2,*], Jie Bai [1,*], Qinlu Lin [1], Feijun Luo [1], Wenhua Zhou [1] and Xiaoxi Zeng [2]

1   Hunan Key Laboratory of Processed Food for Special Medical Purpose, Hunan Key Laboratory of Grain-Oil Deep Process and Quality Control, College of Food Science and Engineering, National Engineering Laboratory for Deep Process of Rice and Byproducts, Central South University of Forestry and Technology, Changsha 410004, China; minxijia@outlook.com (M.J.); zswenjing@foxmail.com (W.Z.); hetaojin@outlook.com (T.H.); Shu-Meng@outlook.com (M.S.); dengjing0102@yahoo.com (J.D.); LinQL0403@126.com (Q.L.); luofeijun@outlook.com (F.L.); zhouwenhua@outlook.com (W.Z.)
2   Key Laboratory of Biological Nanomaterials and Devices, College of life sciences and chemistry, Hunan University of Technology, Zhuzhou 412007, China; zengxiaoxi2003@hut.edu.cn
3   School of Chemistry and Food Engineering, Changsha University of Science and Technology, Changsha 410114, China; wangjh0909@csust.edu.cn
*   Correspondence: liwen@hut.edu.cn (W.L.); baijie2811@outlook.com (J.B.)
†   These authors contributed equally to this work.

Received: 14 January 2020; Accepted: 24 February 2020; Published: 27 February 2020

Abstract: Nano Ag has excellent antibacterial properties and is widely used in various antibacterial materials, such as antibacterial medicine and medical devices, food packaging materials and antibacterial textiles. Despite the many benefits of nano-Ag, more and more research indicates that it may have potential biotoxic effects. Studies have shown that people who ingest nanoparticles by mouth have the highest uptake in the intestinal tract, and that the colon area is the most vulnerable to damage and causes the disease. In this study, we examined the toxic effects of different concentrations of Ag-NPs on normal human colon cells (NCM460) and human colon cancer cells (HCT116). As the concentration of nanoparticles increased, the activity of the two colon cells decreased and intracellular reactive oxygen species (ROS) increased. RT-qPCR and Western-blot analyses showed that Ag NPs can promote the increase in P38 protein phosphorylation levels in two colon cells and promote the expression of P53 and Bax. The analysis also showed that Ag NPs can promote the down-regulation of Bcl-2, leading to an increased Bax/Bcl-2 ratio and activation of P21, further accelerating cell death. This study showed that a low concentration of nano Ag has no obvious toxic effect on colon cells, while nano Ag with concentrations higher than 15 µg/mL will cause oxidative damage to colon cells.

Keywords: nano-Ag; colon cells; biological toxicity; oxidative damage

---

1. Introduction

Nanomaterials have unique physical and chemical properties and are widely used in medical equipment, industrial fields, biomedical fields, electronics and environmental research [1]. Especially in recent years, their use in medicine, health products and food has grown exponentially [2]. Human exposure to nanomaterials in daily life is increasing rapidly, so any potential harmful health effects need special attention.

Nano-Ag is one of the most widely used materials in the world, used in everyday consumer products [3]. Silver nanoparticles (Ag NPs) have highly effective antibacterial properties and have been widely used in medicine, medical devices, coatings, textiles, food and cosmetics to inhibit bacterial growth [4–7]. In contrast to the widespread use of nanomaterials, their biological and toxicological information is still relatively lacking. It is well known that the physical and chemical characteristics of NPs are the key factors affecting their biological effects [8]. Compared with conventional materials, the highly specific surface area of Ag-NPs enhances their interaction with components such as serum, saliva, mucus and lung lining fluid, which may cause negative effects on biological systems [9]. On the other hand, Ag-NPs can also cause strong oxidative activity by releasing Ag^+, and can then induce cytotoxicity, genotoxicity, immune responses and even cell death [10–12].

The transport and distribution of nanoparticles in the body depends mainly on the blood circulation system. Nanoparticles can destroy lung epithelial cells, intestinal epithelial cells and vascular endothelial cells, resulting in increased cell membrane permeability. This allows nanoparticles to easily enter blood vessels and be transported throughout the body [13]. Although no significant human disease has been attributed to food-grade nanoparticles to date, existing research suggests that NPs may cause potentially adverse biological reactions. Nanoparticles can interact with various components of the coagulation system in blood vessels and regulate their activity [14]. For example, Ag-NPs can induce platelet aggregation and enhance thrombus formation in rats [15], and silver colloids can cause platelet aggregation and fibrin polymerization [16]. Further, nanoparticles are distributed to tissues and organs such as the liver, kidneys, brain, testis and ovaries through blood circulation. This distribution can induce the production of reactive oxygen species (ROS), causing cellular oxidative damage, immune response and genetic damage [17–19].

At present, little is known about the multiple mechanisms of action of Ag-NPs biological toxicity and their physiological effects on humans during short-term exposure [20,21]. The toxicity of other nanoparticles in different organisms has been reported in many studies, but the toxicity of Ag-NPs has not been widely explored, especially the colonic cytotoxicity of Ag-NPs. Studies have shown that people who ingest nanoparticles by mouth have the highest uptake in the intestinal tract, and that the colon area is the most vulnerable to damage and causes the disease. In this study, we used normal human colon cells (NCM460) and human colon cancer cells (HCT116) as models to detect cytotoxicity, oxidative stress and apoptosis after exposure to Ag-NPs and to investigate the molecular mechanisms involved.

2. Results

2.1. Characterization of Ag-NPs

The characteristics of the Ag-NPs used in this study are shown in Figure 1. The TEM pattern shows that the nanoparticles have good dispersibility, a uniform size and a complete morphology. A particle size analysis shows that the size of the nanoparticles is mainly distributed around 20 nm. The XRD pattern of nano-Ag is basically the same as that of the standard JCPDS card (JCPDS NO14-0781). The characteristic peaks are shown in Figure 1b. The figure shows five obvious peaks, indicating that the nano-Ag has five crystal planes, and the corresponding indices are (111), (200), (220), (311), (322) from inside to outside. This is because there are a large number of finely oriented crystal particles in the sample area irradiated by the electron incident beam, which belongs to a polycrystalline structure.

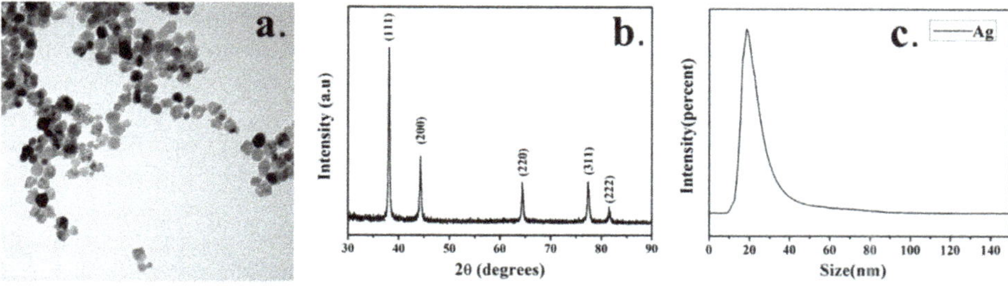

Figure 1. Characterization of nano-Ag. (**a**) Transmission electron micrograph; (**b**) XRD pattern. (**c**) Granularity analysis.

2.2. Morphology Changes of Ag-NPs Treated Cells

After the cells were exposed to different concentrations of Ag-NPs for 24 h, the cell morphology was observed through an inverted microscope, as shown in Figure 2. When the Ag-NPs concentration was 3–6 µg/mL, there was no significant change in cell morphology. When the Ag-NPs concentration was higher than 15 µg/mL, the morphology of the two colon cells changed, the cells became round and the intercellular space increased.

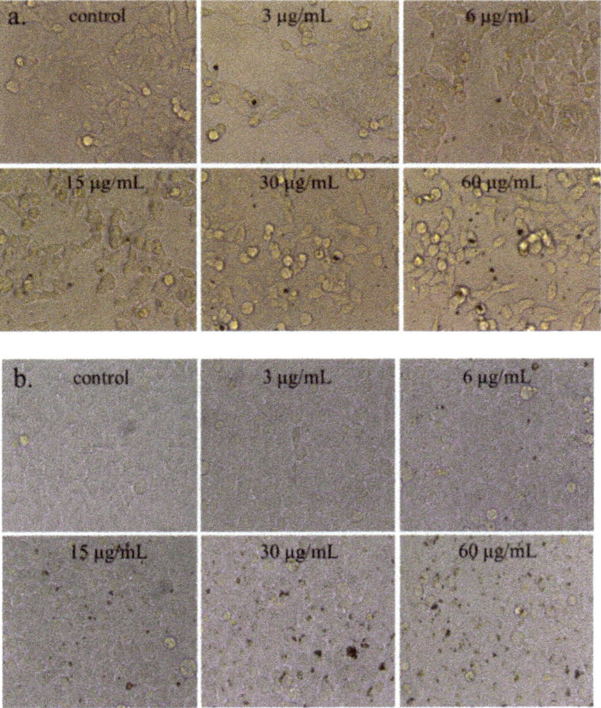

Figure 2. The morphology of two cells was observed after exposure to nano-Ag. (**a**) HCT116 cells; (**b**) NCM460 cells.

2.3. Cytotoxicity Analysis

The cellular activity of the two colon cells after exposure to different concentrations of nanoparticles is shown in Figure 3a. After 24 h of exposure to Ag-NPs at a concentration of 3–6 µg/mL, the activity of the two cells did not change significantly. As the nanoparticle concentration increased, the viability of the two cells began to decrease when it was higher than 15 µg/mL. When the nanoparticle concentration was 60 µg/mL, the viability of NCM460 cells decreased to 60%, about 15% lower than that of the HCT116 cells. Ag-NPs showed toxicity to colon cells and inhibited cell viability at 15–60 µg/mL. A flow cytometry analysis of HCT116 cells showed that with the increase in nano Ag concentration, the proportion of apoptosis increased (Supplementary Figure S1). We also examined the levels of ROS in cells exposed to different concentrations of Ag NPs (Figure 4). When the nanoparticle concentration was 15 µg/mL, the ROS content increased significantly by about 50%. When the nanoparticle concentration reached 60 µg/mL, the intracellular ROS content increased by a factor of 2 relative to the control group (Figure 3b). However, within the tested nanoparticle concentration, there was no significant difference in the lactate dehydrogenase (LDH) concentration in the medium of each experimental group, indicating that the integrity of the cell membrane was not damaged (Figure 3c).

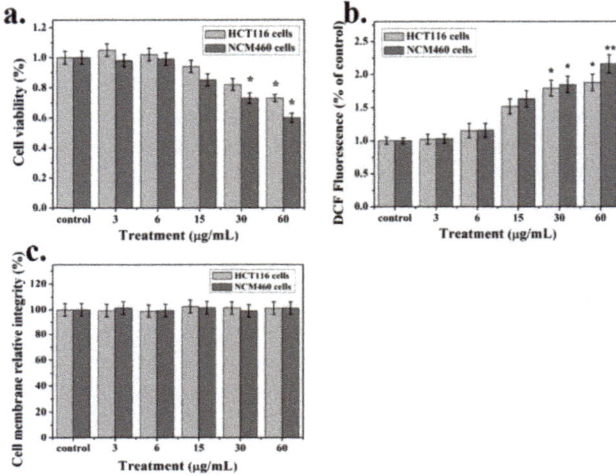

Figure 3. Nano-Ag cytotoxicity test. (**a**) MTT assay for cell viability; (**b**) Intracellular ROS content after exposure to nano Ag; (**c**) The lactate dehydrogenase (LDH) content in the medium was measured to assess the integrity of the cell membrane.

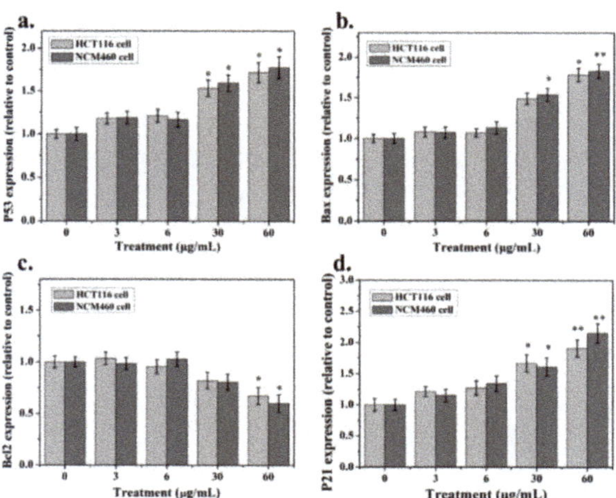

Figure 4. RT-PCR Analysis of the Expression Changes in mRNA Level of P53, Bax, Bcl-2, P21 after exposure to Ag NPs for HCT116 and NCM460 cells. The experiment was in triplicate and data represent mean. * means $p < 0.05$; ** means $p < 0.01$.

2.4. Apoptosis-Associated mRNA and Protein Expression

The mRNA levels of the cell apoptosis markers (P53, Bax, Bcl-2, P21 and GAPDH) were detected by RT-PCR after exposure to Ag-NPs and are shown in Figure 4. After exposure to 3 and 6 μg/mL Ag NPs, the expression of P53, Bax, Bcl-2 and P21 did not change significantly in the two colon cells. When the nanoparticle concentration reached 30–60 μg/mL, P53, Bax and P21 significantly up-regulated expression, while Bcl-2 down-regulated expression.

Further, we performed Western-blot on the protein expression of related genes (Figure 5). In HCT116 cells, when the nano Ag concentration was 3–6 μg/mL, there was no significant change in the expression of each related protein. When the nano-Ag concentration was increased to 30 μg/mL, the phosphorylation level of the P38 protein increased, but the phosphorylation level of the ERK protein did not change significantly. At the same time, the expression of the P53, Bax and P21 proteins increased and the expression of Bcl-2 decreased. In the NCM460 cells, the expression of the P38, P53, Bax and P21 proteins was the same as that of the HCT116 cells, with the difference being that the phosphorylation level of the ERK protein was reduced. The results of Western-blot were consistent with mRNA expression.

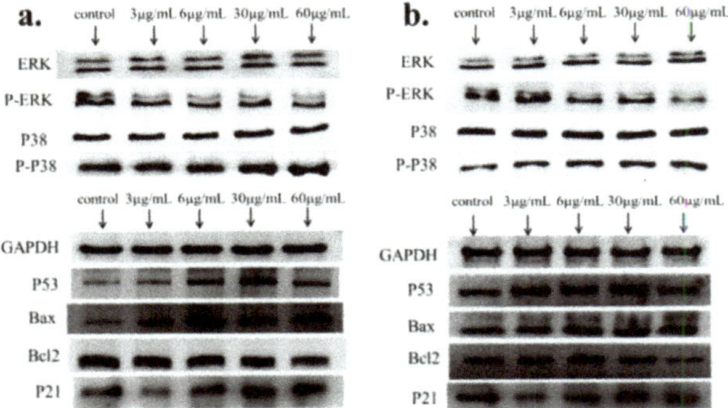

Figure 5. Western-blot analysis of the expression changes in protein level of P53, Bax, Bcl-2, P21 after exposure to Ag NPs for HCT116 and NCM460 cells. (**a**) HCT116 cell protein expression result; (**b**) NCM460 cell protein expression result.

3. Discussion

Currently, Ag NPs are widely used as antibacterial agents in various foods, cosmetics, healthcare products and medical devices. With the increasing application of Ag NPs in daily life, their toxicology and environmental impact have caused widespread concern. No matter how the nanoparticles enter the human body, they can reach all parts of the body through the blood system. Due to the special size of the nanoparticles, they may remain in human organs and cause toxic effects. In vivo studies have shown that the accumulation of Ag NPs in organs may cause pathological changes in organs such as the liver, kidneys and lungs [22,23]. In recent years, research on the biological safety of nanoparticles has become a hot topic. The cytotoxicity of many nanoparticles related to food, cosmetics and medicine, such as Ag, TiO_2 and ZnO [24–26], has been studied. Studies have shown that Ag NPs can induce cytotoxicity and genotoxicity in human cells [27]. At present, the interaction between Ag NPs and mammalian systems is not fully understood and the underlying mechanisms of Ag NPs toxicity are poorly understood.

In this study, HCT116 and NCM460 cells were selected to evaluate the colonic cytotoxicity of nano Ag. Morphological observation is the easiest and most direct way to identify cells, because the morphology of most cells is related to the action of drugs or poisons. After treatment with Ag NPs, the shape of the two colon cells changed, while the controls remained unchanged. Cell viability testing is a vital step in toxicology testing which shows the response of cells to drugs and provides information about cell death and survival. In this experiment, Ag NPs did not significantly inhibit the growth of the two cells in 0–6 μg/mL. As the nanoparticle concentration increased, the viability of the two cells began to decrease when it was higher than 15 μg/mL. When the nanoparticle concentration was 60 μg/mL, the viability of NCM460 cells decreased to 60%, which was about 15% lower than that of HCT116 cells. These results indicate that the cytotoxicity of Ag NP is positively correlated with its concentration. In addition, previous studies have shown that Ag NPs can release Ag $^+$ in biological fluids, while cytotoxicity is largely caused by Ag$^+$ [28]. This will be further studied in our next work.

The ROS mechanism is one of the most commonly accepted conclusions for assessing the toxicity of nanomaterials [29]. Studies have found that mitochondria are the main source of ROS. If the oxidant is imbalanced and expresses antioxidant enzymes and proteins, it will produce oxidative stress [30–32]. This condition leads to the formation of peroxides and free radicals, thereby destroying proteins, lipids and DNA, causing a series of cascades such as apoptosis, inflammation and the promotion of tissue and organ damage [33,34]. In our study, ROS content was positively correlated with Ag NPs concentration. When the Ag NPs concentration reached 60 μg/mL, the intracellular ROS content was about twice that of the control group. This indicates that the excessive accumulation of ROS induced by Ag NPs may be an important cause of cytotoxicity.

For detecting the impact of apoptosis-regulating genes such as P53, Bax, Bcl-2 and P21 on the survival or cytotoxicity of HCT116 and NCM460 cells exposed to Ag NPs, we determined the mRNA expression in the cells. In the HCT116 cell line, Ag NPs can promote the increase in P38 protein phosphorylation levels without significant effects on ERK. By promoting the expression of P53 and Bax and the down-regulation of Bcl-2, this leads to an increase in the Bax/Bcl-2 ratio and the activation of P21, which further increases cell death. The difference is that ERK protein phosphorylation is reduced in NCM460 cells. Bcl2 can prevent the release of apoptosis-forming factors (such as cytochrome C) in mitochondria and plays an anti-apoptotic role. Moreover, Bax can act on voltage-dependent ion channels on mitochondria, mediate the release of cytochrome C and promote apoptosis. P53 promotes apoptosis by up-regulating Bax and down-regulating Bcl-2 [35,36]. There have been reports of colorectal cancer cells treated with Ag NPs, and the down-regulation of Bcl-2 expression, which inhibited colon cancer proliferation, has also been recorded [37]. At the same time, the P53 protein activates the transcription and high expression of the P21 gene, stopping the cell cycle at the G1, G2 or S phase and hindering DNA replication and mitosis thereby inhibiting cell proliferation [38,39]. Our flow cytometry results of HCT116 cells can confirm this conclusion (Supplementary Figure S1). Compared with the control, significant concentration-dependent cytotoxicity of the nano Ag in HCT116 cells was observed, in the concentration range of 15 to 60 μg/mL. It has also been found in other related studies that 20 nm Ag particles increase ROS content in LoVo intestinal cell lines, leading to mitochondrial dysfunction and subsequent induced apoptosis [40]. This is similar to our results. However, studies on Ag nano-human liver HepG2 cells and colon Caco2 cytotoxicity studies have shown that the differences in the mechanisms of toxicity induced by nanosilver may be largely a consequence of the type of cells used [41]. This differential, rather than universal, response of different cell types exposed to nanoparticles may play an important role in the mechanism of their toxicity. Similarly, our study also showed that nano Ag has toxic differences on HCT116 and NCM460 cells. In addition, there is no doubt that supplementing the results of studies with molecules such as caspase3 and other cell lines can better prove our experimental conclusions. However, due to the recent severe pneumonia epidemic in China, experimental supplementation of other related genes and cell lines cannot be performed in a short period of time and should be explored further in the subsequent research.

In conclusion, this study evaluated the effect of Ag NPs exposure on colon cells. Ag NPs may induce oxidative stress in cells, increase ROS, reduce cell viability and induce apoptosis. Furthermore, the cytotoxicity is dependent on the concentration of Ag nanoparticles. In summary, these findings provide very useful toxicological information and help to better evaluate the biological safety of nanomaterials.

4. Materials and Methods

4.1. Nanoparticles

Nano Ag powder was purchased from Beijing Deke Daojin Technology Co., Ltd. (Deke Daojin, Beijing, China). The structure of Ag-NPs was studied by transmission electron microscopy (TEM) JEM-2010 (JEOL Ltd., Tokyo, Japan) and the observation samples were fixed on a copper mesh according to standard procedures. The crystal structure of Ag-NPs was determined by X-ray diffraction (XRD) (BRUKER AXS GMBH, Karlsruhe, Germany) and the particle size was detected by a nanometer particle size analyzer Zetasizer Nano S90 (Malvern Panalytical, Malvern, United Kingdom). Ag-NPs were added to a serum-free 1640 medium before each use and then sonicated for 30 min to fully disperse the nanoparticles.

4.2. Cell Culture and Nanoparticle Treatment

A stock solution of Ag NPs (1.0 mg/mL) was prepared in a serum-free RPMI-1640 medium and stored at −20 °C (pay attention to avoid repeated freezing-thawing). The desired concentration (3, 6, 15, 30, and 60 μg/mL) was obtained by diluting the nanoparticle stock solution with a complete cell culture medium each time. The HCT116 and NCM460 cells were provided by the Molecular Nutrition Laboratory of Central South University of Forestry and Technology (Changsha, China). The cells were dispersed in a RPMI-1640 (Invitrogen Gibco, New York, NY, USA) complete medium (containing 10% calf serum and 1% penicillin-streptomycin) and cultured in a 37 °C, 5% CO_2 incubator for 8 h. The old medium was removed, a fresh medium with different concentrations of nanoparticles was added and incubation was continued for 24 h. The changes to cell morphology under an inverted microscope were observed (Olympus CK 40, Tokyo, Japan) and saved.

4.3. Cell Viability Assay

MTT cell proliferation and cytotoxicity assay kits (Sangon, Shanghai, China) were used to detect colon cell viability after exposure to different concentrations of Ag NPs. This method is based on succinate dehydrogenase activity and determines the number of viable cells. The cells (1.0×10^5 mL^{-1}) were seeded into a 96-well plate and grown at 37 °C, 5% CO_2 for adherent growth. The Ag NPs stock solution was added to complete the medium in order to obtain the desired concentration of Ag NPs (3, 6, 15, 30 and 60 μg/mL). The old medium was carefully removed from the 96-well plate and a new medium containing different concentrations of Ag NPs was added and placed in an incubator for further culture. For each concentration, four well replicates were prepared. After 24 h of exposure, 10 μL of MTT solution was added to each cell culture well according to the manufacturer's instructions. Gently shake the culture plate to evenly disperse the MTT reagent, and continue incubating for 4 h in a 37 °C, 5% CO_2 incubator. The MTT solution was then discarded, and 100 μL of formazan solution was added to each well, and the plate was gently shaken at room temperature to mix them for 10 min. The optical density (OD) of each well was measured by a multi-functional microplate reader SpectraMax i3x (Molecular Devices, San Jose, California, USA) at a wavelength of 570 nm. Cell viability was calculated as the ratio of the average OD obtained at each nanoparticle concentration to the average OD under control (without Ag nanoparticles).

4.4. Intracellular Reactive Oxygen Species (ROS) Assay

Intracellular ROS levels were measured using a fluorescent probe 2′,7′-dichlorodihydrofluorescein diacetate (DCFH-DA) according to the kit manufacturer's instructions (Beyctime, Shanghai, China). In short, HCT116 cells and NCM460 cells were seeded into 6-well plates with 5×10^6 cells/well and allowed to grow overnight. The cells were then treated with Ag NPs (3, 6, 15, 30 and 60 μg/mL) for 24 h. The cell culture medium was removed, 1 mL of serum-free medium diluted DCFH-DA (10 μM) was added and incubation was continued for 20 min. Finally, the cell culture medium was washed 3 times with PBS to completely remove free DCFH-DA. The fluorescence value of the sample was immediately measured by a multifunctional microplate reader (Molecular Devices, California, USA). The percentage of active oxygen in the cell was calculated by the following formula: test sample fluorescence value/control sample fluorescence value × 100%.

4.5. LDH Release

In this study, we used a lactate dehydrogenase (LDH) cytotoxicity assay kit (Beyotime, Shanghai, China) to detect nano-Ag cytotoxicity. LDH is a cytoplasmic enzyme found in all living cells. If cell membranes are damaged and integrity is compromised, intracellular LDH will be released into external media. Therefore, we evaluated the cytotoxicity of nanoparticles by measuring the amount of LDH in the medium after the cells were exposed to nano-Ag. Cells were seeded into 6-well plates (5.0×10^6 cells/well), cultured to adherent growth and then exposed to different concentrations (0, 3, 6, 15, 30, and 60 µg/mL) of Ag nanoparticles. After 24 h, the cell culture plates were centrifuged with a multi-well plate centrifuge at 400 g for 5 min, the supernatant (120 µL) was removed and 120 µL of diluted LDH detection reagent was added. The culture was then mixed and shaken appropriately, and incubation was continued for 1 h. The absorbance of each well was measured using a multifunctional microplate reader (Molecular Devices) at 490 nm. The calculation formula is as follows: cell membrane integrity (%) = absorbance of treated samples/absorbance of sample control wells × 100%.

4.6. RNA Isolation and Reverse-Transcriptase PCR Analysis

After cells were exposed to different concentrations of Ag-NP (3, 6, 15, 30, and 60 µg/mL) for 24 h, RNA was extracted using the "Column Total RNA Purification Kit" (Sangon Biotech, Shanghai, China) according to the manufacturer's instructions. cDNA was synthesized using a M-MuLV First Strand cDNA Synthesis Kit (Sangon Biotech, Shanghai, China) and further detected by RT-PCR using specific primers [42,43]. The PCR conditions include 1 cycle at 95 °C for 3 min, then 40 cycles at 95 °C for 30 s, 60 °C for 30 s and 72 °C for 30 s and, finally, at 72 °C for the extended time of 5 min. The sequences of the specific primers used in this study are shown in Supplementary Table S1. Glyceraldehyde-3-phosphate dehydrogenase (GAPDH) was used as an internal reference gene to quantify the expression level of the target gene.

4.7. Western-Blotting Analysis

The cells were seeded in a 10 cm petri dish at a density of 1×10^7 and cultured in a 37 °C, 5% CO_2 incubator to adhere the cells. The old culture medium was discarded and replaced with a fresh medium containing different concentrations of Ag-NP (3, 6, 15, 30, and 60 µg/mL). Incubation was then continued for 24 h. After the treatment was completed, the cells were washed twice with PBS. The total protein was extracted using a whole protein extraction kit (Sangon Biotech, Shanghai, China) according to the manufacturer's instructions. 12% sodium lauryl sulfate - polyacrylamide gel (SDS-PAGE) was used to separate the target protein fragment, and then the protein was transferred to a polyvinylidene fluoride (PVDF) membrane. The PVDF membrane was blocked with TBST containing 5% skimmed milk for 30 min at room temperature and then incubated with a specific primary antibody at 4 °C overnight. The PVDF membranes were incubated with horseradish peroxidase (HRP)-conjugated (Sangon Biotech, Shanghai, China) secondary antibodies at room temperature and the blots were developed by a chemiluminescence visualization substrate system. GAPDH were used as a normalization control for equal loading.

4.8. Statistical Analysis

The results of at least three independent experiments are presented as mean ± SD. The comparison between the treated and control groups was carried out by ANOVA analysis using the SPSS 23.0 software package (SPSS Company, Chicago, IL, USA) [44]. * $p < 0.05$ was considered as the significance level for all analyses performed.

Supplementary Materials: Supplementary materials can be found at http://www.mdpi.com/1422-0067/21/5/1618/s1.

Author Contributions: Conceptualization, J.D., W.L. and Q.L.; Data curation, M.J., W.Z. and F.L.; Formal analysis, T.H., J.D., J.W., W.L., W.Z. and X.Z.; Funding acquisition, J.D. and X.Z.; Investigation, F.L.; Methodology, M.J., M.S. and F.L.; Project administration, J.W., W.L., J.B. and Q.L.; Software, M.S., J.W. and J.B.; Writing—original draft, M.J. and W.L. All authors have read and agreed to the published version of the manuscript.

Funding: This research was funded by the National Key R&D Program of China (2018YFE0110200 and 2015BAD05B02), the National Natural Science Foundation of China (51774128), the Natural Science Foundation of Hunan Province of China (2018JJ4061, 2018JJ2672, 2018JJ2090 and 2018JJ4009), the key program of Hunan Provincial Department of science and technology (2016NK2096, 2019NK2111 and 2019SK2121), Program for Science & Technology Innovation Talents of Hunan Province (2017TP1021, kc1704007), the Hunan Scientific Talents Promotion Program of Hunan Association for Science and Technology (2019TJ-Q01), the Scientific Research Fund of Hunan Provincial Education Department (17A055 and 18B290) and the Zhuzhou Key Science & Technology Program of Hunan Province (2017, 2018 and 2019).

Conflicts of Interest: The authors declare no conflict of interest.

References

1. Hamzeh, M.; Sunahara, G.I. In vitro cytotoxicity and genotoxicity studies of titanium dioxide (TiO_2) nanoparticles in Chinese hamster lung fibroblast cells. *Toxicol. In Vitro* **2013**, *27*, 864–873. [CrossRef] [PubMed]

2. Coto-García, A.M.; Sotelo-González, E.; Fernández-Argüelles, M.T.; Pereiro, R.; Costa-Fernández, J.M.; Sanz-Medel, A. Nanoparticles as fluorescent labels for optical imaging and sensing in genomics and proteomics. *Anal. Bioanal. Chem.* **2011**, *399*, 29–42. [CrossRef] [PubMed]

3. Edwards-Jones, V. The benefits of silver in hygiene, personal care and healthcare. *Lett. Appl. Microbiol.* **2009**, *49*, 147–152. [CrossRef] [PubMed]

4. Chen, X.; Schluesener, H.J. Nanosilver: A nanoproduct in medical application. *Toxicol. Lett.* **2008**, *176*, 1–12. [CrossRef] [PubMed]

5. Gaiser, B.K.; Hirn, S.; Kermanizadeh, A.; Kanase, N.; Fytianos, K.; Wenk, A.; Stone, V. Effects of silver nanoparticles on the liver and hepatocytes in vitro. *Toxicol. Sci.* **2012**, *131*, 537–547. [CrossRef] [PubMed]

6. Deng, J.; Ding, Q.M.; Li, W. Preparation of Nano-Silver-Containing Polyethylene Composite Film and Ag Ion Migration into Food-Simulants. *J. Nanosci. Nanotechnol.* **2020**, *20*, 1613–1621. [CrossRef]

7. Deng, J.; Chen, Q.J.; Peng, Z.Y. Nano-silver-containing polyvinyl alcohol composite film for grape fresh-keeping. *Mater. Express* **2019**, *9*, 985–992. [CrossRef]

8. Sayes, C.M.; Warheit, D.B. Characterization of nanomaterials for toxicity assessment. *Wiley Interdiscip. Rev. Nanomed. Nanobiotechnol.* **2009**, *1*, 660–670. [CrossRef]

9. Beer, C.; Foldbjerg, R.; Hayashi, Y.; Sutherland, D.S.; Autrup, H. Toxicity of silver nanoparticles—Nanoparticle or silver ion? *Toxicol. Lett.* **2012**, *208*, 286–292. [CrossRef]

10. Chernousova, S.; Epple, M. Silver as antibacterial agent: Ion, nanoparticle, and metal. *Angew. Chem. Int. Ed.* **2013**, *52*, 1636–1653. [CrossRef]

11. Simon-Deckers, A.; Gouget, B.; Mayne-L'Hermite, M.; Herlin-Boime, N.; Reynaud, C.; Carriere, M. In vitro investigation of oxide nanoparticle and carbon nanotube toxicity and intracellular accumulation in A549 human pneumocytes. *Toxicology* **2008**, *253*, 137–146. [CrossRef] [PubMed]

12. Cho, J.G.; Kim, K.T.; Ryu, T.K. Stepwise embryonic toxicity of silver nanoparticles on Oryzias latipes. *Biomed Res. Int.* **2013**, *2013*. [CrossRef] [PubMed]

13. Rezaei, Z.R. Effect of Titanium Dioxide Nanoparticles on The Amount of Blood Cells and Liver Enzymes in Wistar Rats. *J. Shahid Sadoughi Univ. Med. Sci.* **2011**, *19*, 618–626.

14. Ruckerl, R.; Ibald-Mulli, A. Air pollution and markers of inflammation and coagulation in patients with coronary heart disease. *Am. J. Respir. Crit. Care Med.* **2006**, *173*, 432–441. [CrossRef]

15. Veronesi, B.; Makwana, O.; Pooler, M.; Chen, L.C. Effects of Subchronic Exposures to Concentrated Ambient Particles: VII. Degeneration of Dopaminergic Neurons in Apo E−/− Mice. *Inhal. Toxicol.* **2005**, *17*, 235–241. [CrossRef]

16. Guildford, A.L.; Poletti, T.; Osbourne, L.H.; Di Cerbo, A.; Gatti, A.M.; Santin, M. Nanoparticles of a different source induce different patterns of activation in key biochemical and cellular components of the host response. *J. R. Soc. Interface* **2009**, *6*, 1213–1221. [CrossRef]

17. Sarhan, O.M.M.; Hussein, R.M. Effects of intraperitoneally injected silver nanoparticles on histological structures and blood parameters in the albino rat. *Int. J. Nanomed.* **2014**, *9*, 1505. [CrossRef]

18. Lee, C.M.; Jeong, H.J.; Yun, K.N. Optical imaging to trace near infrared fluorescent zinc oxide nanoparticles following oral exposure. *Int. J. Nanomed.* **2012**, *7*, 3203. [CrossRef]

19. Wang, Y.; Chen, Z.; Ba, T. Susceptibility of young and adult rats to the oral toxicity of titanium dioxide nanoparticles. *Small* **2013**, *9*, 1742–1752. [CrossRef]

20. Nel, A.; Xia, T.; Mädler, L.; Li, N. Toxic potential of materials at the nanolevel. *Science* **2006**, *311*, 622–627. [CrossRef]

21. Mishra, A.R.; Zheng, J.; Tang, X.; Goering, P.L. Silver nanoparticle-induced autophagic-lysosomal disruption and NLRP3-inflammasome activation in HepG2 cells is size-dependent. *Toxicol. Sci.* **2016**, *150*, 473–487. [CrossRef] [PubMed]

22. Dziendzikowska, K.; Gromadzka-Ostrowska, J.; Lankoff, A. Time-dependent biodistribution and excretion of silver nanoparticles in male Wistar rats. *J. Appl. Toxicol.* **2012**, *32*, 920–928. [CrossRef] [PubMed]

23. Xue, Y.; Zhang, S.; Huang, Y. Acute toxic effects and gender-related biokinetics of silver nanoparticles following an intravenous injection in mice. *J. Appl. Toxicol.* **2012**, *32*, 890–899. [CrossRef] [PubMed]

24. Wei, L.; Lu, J.; Xu, H.; Patel, A.; Chen, Z.S.; Chen, G. Silver nanoparticles: Synthesis, properties, and therapeutic applications. *Drug Discov. Today* **2015**, *20*, 595–601. [CrossRef] [PubMed]

25. Ozkaleli, M.; Erdem, A. Biotoxicity of TiO$_2$ nanoparticles on *Raphidocelis subcapitata* microalgae exemplified by membrane deformation. *Int. J. Env. Res. Public Health* **2018**, *15*, 416. [CrossRef] [PubMed]

26. Subramaniam, V.D.; Prasad, S.V.; Banerjee, A. Health hazards of nanoparticles: Understanding the toxicity mechanism of nanosized ZnO in cosmetic products. *Drug Chem. Toxicol.* **2019**, *42*, 84–93. [CrossRef]

27. Sahu, S.C.; Njoroge, J.; Bryce, S.M.; Yourick, J.J.; Sprando, R.L. Comparative genotoxicity of nanosilver in human liver HepG2 and colon Caco2 cells evaluated by a flow cytometric in vitro micronucleus assay. *J. Appl. Toxicol.* **2014**, *34*, 1226–1234. [CrossRef]

28. Xue, Y.; Zhang, T.; Zhang, B.; Gong, F.; Huang, Y.; Tang, M. Cytotoxicity and apoptosis induced by silver nanoparticles in human liver HepG2 cells in different dispersion media. *J. Appl. Toxicol.* **2016**, *36*, 352–360. [CrossRef]

29. Wilson, C.L.; Natarajan, V.; Hayward, S.L.; Khalimonchuk, O.; Kidambi, S. Mitochondrial dysfunction and loss of glutamate uptake in primary astrocytes exposed to titanium dioxide nanoparticles. *Nanoscale* **2015**, *7*, 18477–18488. [CrossRef]

30. Liang, Y.; Lin, Q.; Huang, P.; Wang, Y.; Li, J.; Zhang, L.; Cao, J. Rice Bioactive Peptide Binding with TLR4 To Overcome H$_2$O$_2$-Induced Injury in Human Umbilical Vein Endothelial Cells through NF-κB Signaling. *J. Agric. Food Chem.* **2018**, *66*, 440–448. [CrossRef]

31. Natarajan, V.; Wilson, C.L.; Hayward, S.L.; Kidambi, S. Titanium dioxide nanoparticles trigger loss of function and perturbation of mitochondrial dynamics in primary hepatocytes. *PLoS ONE* **2015**, *10*, e0134541. [CrossRef] [PubMed]

32. Shvedova, A.A.; Pietroiusti, A.; Fadeel, B.; Kagan, V.E. Mechanisms of carbon nanotube-induced toxicity: Focus on oxidative stress. *Toxicol. Appl. Pharmacol.* **2012**, *261*, 121–133. [CrossRef]

33. Foldbjerg, R.; Dang, D.A.; Autrup, H. Cytotoxicity and genotoxicity of silver nanoparticles in the human lung cancer cell line, A549. *Arch. Toxicol.* **2011**, *85*, 743–750. [CrossRef] [PubMed]

34. Madl, A.K.; Plummer, L.E.; Carosino, C.; Pinkerton, K.E. Nanoparticles, lung injury, and the role of oxidant stress. *Annu. Rev. Physiol.* **2014**, *76*, 447–465. [CrossRef] [PubMed]

35. Zhang, L.; Ma, L.; Yan, T.; Han, X.; Xu, J.; Xu, J.; Xu, X. Activated mitochondrial apoptosis in hESCs after dissociation involving the PKA/p-p53/Bax signaling pathway. *Exp. cell Res.* **2018**, *369*, 226–233. [CrossRef] [PubMed]

36. Beberok, A.; Wrześniok, D.; Rzepka, Z.; Respondek, M.; Buszman, E. Ciprofloxacin triggers the apoptosis of human triple-negative breast cancer MDA-MB-231 cells via the p53/Bax/Bcl-2 signaling pathway. *Int. J. Oncol.* **2018**, *52*, 1727–1737. [CrossRef]

37. El-Deeb, N.M.; El-Sherbiny, I.M.; El-Aassara, M.R.; Hafez, E.E. Novel trend in colon cancer therapy using silver nanoparticles synthesized by honey bee. *J. Nanomed. Nanotechnol.* **2015**, *6*, 265. [CrossRef]

38. Karimian, A.; Ahmadi, Y.; Yousefi, B. Multiple functions of p21 in cell cycle, apoptosis and transcriptional regulation after DNA damage. *DNA Repair* **2016**, *42*, 63–71. [CrossRef]

39. Georgakilas, A.G.; Martin, O.A.; Bonner, W.M. p21: A two-faced genome guardian. *Trends Mol. Med.* **2017**, *23*, 310–319. [CrossRef]
40. Miethling-Graff, R.; Rumpker, R.; Richter, M.; Verano-Braga, T.; Kjeldsen, F.; Brewer, J. Exposure to silver nanoparticles induces size-and dose-dependent oxidative stress and cytotoxicity in human colon carcinoma cells. *Toxicol. Vitr.* **2014**, *28*, 1280–1289. [CrossRef]
41. Sahu, S.C.; Zheng, J.; Graham, L.; Chen, L.; Ihrie, J.; Yourick, J.J.; Sprando, R.L. Comparative cytotoxicity of nano silver in human liver HepG2 and colon Caco2 cells in culture. *J. Appl. Toxicol.* **2014**, *34*, 1155–1166. [CrossRef] [PubMed]
42. Li, W.; Jia, M.X.; Wang, J.H.; Lu, J.L.; Deng, J.; Tang, J.X.; Liu, C. Association of MMP9-1562C/T and MMP13-77A/G Polymorphisms with Non-Small Cell Lung Cancer in Southern Chinese Population. *Biomolecules* **2019**, *9*, 107. [CrossRef] [PubMed]
43. Li, W.; Jia, M.X.; Deng, J.; Wang, J.H.; Lin, Q.L.; Tang, J.X.; Liu, X.Y. Down-regulation of microRNA-200b is a potential prognostic marker of lung cancer in southern-central Chinese population. *Saudi J. Biol. Sci.* **2019**, *26*, 173–177. [CrossRef] [PubMed]
44. Li, W.; Jia, M.X.; Deng, J.; Wang, J.H.; Lin, Q.L.; Liu, C.; Su, W. Isolation, genetic identification and degradation characteristics of COD-degrading bacterial strain in slaughter wastewater. *Saudi J. Biol. Sci.* **2018**, *25*, 1800–1805. [CrossRef] [PubMed]

International Journal of
Molecular Sciences

Review

Health Impact of Silver Nanoparticles: A Review of the Biodistribution and Toxicity Following Various Routes of Exposure

Zannatul Ferdous [1] and Abderrahim Nemmar [1,2,*]

[1] Department of Physiology, College of Medicine and Health Sciences, United Arab Emirates University, P.O. Box 17666 Al Ain, UAE; 201370004@uaeu.ac.ae

[2] Zayed Center for Health Sciences, United Arab Emirates University, P.O. Box 17666 Al Ain, UAE

* Correspondence: anemmar@uaeu.ac.ae; Tel.: +971-37137533; Fax: +971-37671966

Received: 14 October 2019; Accepted: 18 December 2019; Published: 30 March 2020

Abstract: Engineered nanomaterials (ENMs) have gained huge importance in technological advancements over the past few years. Among the various ENMs, silver nanoparticles (AgNPs) have become one of the most explored nanotechnology-derived nanostructures and have been intensively investigated for their unique physicochemical properties. The widespread commercial and biomedical application of nanosilver include its use as a catalyst and an optical receptor in cosmetics, electronics and textile engineering, as a bactericidal agent, and in wound dressings, surgical instruments, and disinfectants. This, in turn, has increased the potential for interactions of AgNPs with terrestrial and aquatic environments, as well as potential exposure and toxicity to human health. In the present review, after giving an overview of ENMs, we discuss the current advances on the physiochemical properties of AgNPs with specific emphasis on biodistribution and both in vitro and in vivo toxicity following various routes of exposure. Most in vitro studies have demonstrated the size-, dose- and coating-dependent cellular uptake of AgNPs. Following NPs exposure, in vivo biodistribution studies have reported Ag accumulation and toxicity to local as well as distant organs. Though there has been an increase in the number of studies in this area, more investigations are required to understand the mechanisms of toxicity following various modes of exposure to AgNPs.

Keywords: silver nanoparticles; cytotoxicity; routes of exposure; biodistribution

---

1. Introduction

Nanotechnology

The concept of novel nanoscale technology was introduced in 1959 in a lecture of physicist Richard Feynman entitled "There is plenty of room at the bottom," which discussed the importance of manipulating and controlling things at the atomic scale [1]. The term nanotechnology was first introduced by Professor Norio Taniguchi in 1974, and the American engineer Kim Eric Drexler popularized the concept of molecular nanotechnology by using it in his 1986 book Engines of Creation; The Coming Era of Nanotechnology [1]. However, it was only after the inventions of instruments for imaging surfaces at atomic level—the scanning tunneling microscope in 1981, which was followed by the atomic force microscope—that the growth of nanotechnology was sparked in the modern era [1]. The emerging and exponential nanotechnology involves the manipulation, design and precision placement of atoms and molecules at the nanoscale level [1]. Over the past two decades, nanotechnology has witnessed breakthroughs in the fields of medicine, environment, therapeutics, drug development, and biotechnology.

A major element of nanotechnology comprises engineered nanomaterials (ENMs). According to European commission, NMs are "a natural, incidental or manufactured material containing particles in an unbound state or as an aggregate or as an agglomerate and where, for 50% or more of the particles in the number size distribution, one or more external dimensions is in the size range of 1–100 nm" [2]. NMs-containing consumer products include cosmetics, electronics, kitchenware, textiles and sporting goods [3]. This wide utilization is due to unique properties of NMs, such as their small size, large surface area to volume ratio, high reactivity, high carrier capacity, and easy variation of surface properties [3]. The same unique properties that led to their widespread applications raise questions about potential environmental and health effects that might result from occupational exposures during the manufacture and use of NMs at the consumer end [4]. The toxic effect of NMs on humans has recently gained much attention in the health industry [5,6]. Most exposure to airborne NMs occurs in the workplace during the manufacture of these materials, the formulation of them into products, their transport, or their handling in the storage facilities [7]. Additionally, widespread consumer exposure via oral inhalation and direct contact with ENM-containing products is likely to occur [7]. Subsequently, the small size of this type of particle facilitates its translocation from natural barriers such as the gastrointestinal tract (GIT), lungs, or skin, and this translocation can induce acute and chronic toxic effects [8].

In spite of several advantages of nanoscale materials, their potential health hazards cannot be overlooked due to their uncontrollable use, discharge to the natural environment and potential toxic effects. Hence, nanotoxicology warrants intensive research studies in make the use of NMs more convenient and environment friendly. Some of the most commonly studied NMs include fullerenes, carbon nanotubes (CNTs), silver nanoparticles (AgNPs), gold nanoparticles (AuNPs), titanium oxide nanoparticles (TiO_2), zinc oxide nanoparticles, iron oxide (FeO), and silica nanoparticles [9]. Among these, AgNPs have gained strong popularity among researchers over the past few decades [9]. Initial investigations on AgNPs were more focused on synthesizing and characterizing them by using chemical approaches [10]. However, current works have been more concentrated on their biological effects and applications for several purposes [11]. AgNPs are also known to have unique properties in terms of toxicity, surface plasmon resonance, and electrical resistance [11]. Based on these, intensive works have been conducted to investigate their properties and potential applications for several purposes such as antimicrobial agents in wound dressings, water disinfectants, electronic devices, and anticancer agents [11,12].

Previous review papers have addressed the toxicological properties of AgNPs during their use as antimicrobial agents for textiles, dental biomaterials, and bio-detectors, as well as during their syntheses [13–18]. Recent reviews have covered topics such as their biosynthesis by using plant extracts for antimicrobial applications, biocidal properties, and cytotoxicity based on their physiochemical properties such as size, concentration and coating [19–21]. Moreover, the chemical and toxicological interactions of AgNPs with other metal NPs such as Ag, Fe, and TiO_2 were also discussed by Sharma et al. [22]. An important factor regarding the potential health impact of AgNPs is their various routes of exposure. In this regard, the potential toxic effects of AgNPs after oral exposure were also thoroughly discussed by Gaillet et al. [23]. However, other major routes including respiratory, dermal and intravenous exposure have not been covered in previous review articles. In our present paper, we summarize the existing physical, chemical, and biological synthetic approaches of AgNPs, followed by a description on their characterization techniques and applications. Notably, our paper aims to report the most recent update and important studies on the toxic effect of AgNPs following various routes of exposure including oral, inhalation, dermal, and intravenous administration. Moreover, important in vitro and in vivo pathophysiological effects at the site of exposure and in remote and distant organs following exposure are highlighted. To our knowledge, no other previous reviews have structurally presented these topics so far. We conducted an extensive literature search of bibliographic databases (e.g., PubMed, Google Scholar, Medline, and Web of Science) by using different keywords and combinations of keywords (AgNPs, biodistribution, in vitro toxicity, in vivo toxicity, pulmonary exposure, inhalation, instillation, oral exposure, dermal exposure and organ toxicity) to retrieve the relevant information.

2. Silver Nanoparticles (AgNPs)

Of all the ENMs developed and characterized so far, AgNPs has assumed a significant position due to their potential uses in commercial applications [12,24]. Silver, symbolized as Ag, is a lustrous, soft, ductile and malleable metal that has the highest electrical conductivity of all metals and is widely used in electrical appliances [25]. This precious metal is chemically inactive, stable in water, and does not oxidize in air—hence, it is used in the manufacturing of coins, ornaments and jewelry [25]. Silver can be obtained from pure deposits as well as from silver ores such as horn silver and argentite. Most silver is derived as a by-product along with deposits of ores containing gold, copper, and lead [25]. It is estimated that nearly 320 tons of nanoparticulate form of Ag are manufactured every year and used in nanomedical imaging, biosensing and food products [21]. AgNPs exhibit special properties relative to their bulk components due to their unique physicochemical properties including their small size, greater surface area, surface chemistry, shape, particle morphology, particle composition, coating/capping, agglomeration, rate of particle dissolution, particle reactivity in solution, efficiency of ion release, and type of reducing agents used for the synthesis of AgNPs [12,26]. In addition, AgNPs are also well known for their antimicrobial, optical, electrical, and catalytic properties [11]. Owing to their unique properties, AgNPs have been extensively used in household utensils, food storage, the health care industry, environmental applications, and biomedical applications such as wound dressings, surgical instruments, and disinfectants [26]. Furthermore, due to their optical activities, these NPs have been used in catalysis, electronics and biosensors [26].

2.1. AgNPs Synthesis and Characterization

Numerous methods have been adopted for the synthesis of AgNPs in order to meet these increasing requirements. The conventional physical method of synthesis includes spark discharge and pyrolysis [27,28]. A chemical method that can be a top–down or bottom–up approach involves three main components: metal precursors, reducing agents and stabilizing/capping agents [12]. The common approach is usually chemical reduction by organic or inorganic reducing agents, such as sodium citrate, ascorbate, sodium borohydride, elemental hydrogen, polyol process, Tollens reagent, N, N-dimethylformamide, and polyethylene glycol-block copolymer [12]. Other procedures include cryochemical synthesis, laser ablation, lithography, electrochemical reduction, laser irradiation, sono-decomposition and thermal decomposition [12]. The major advantage of the chemical method is its high yield unlike the physical method, which has a low yield [10]. However, contrary to the physical method, the chemical method is extremely expensive, toxic and hazardous [10]. In order to overcome the later limitations, the biologically-mediated synthesis of NPs has emerged as a better option. This simple, cost effective and environment friendly approach uses biological systems including bacteria, fungi, plant extracts, and small biomolecules like vitamins, amino acids and enzymes for the synthesis of AgNPs [29]. The green approach is widely accepted due to the availability of a vast array of biological resources, a decreased time requirement, high density, stability, and the ready solubility of prepared NPs in water [29].

The characterization of AgNPs is a very crucial step to evaluate the functional effect of synthesized particles [12]. It has been documented in various studies that the biological activity of AgNPs depends on morphology, structure, size, shape, charge and coating/capping, chemical composition, redox potential, particle dissolution, ion release, and degree of aggregation [21,30–33]. Like all other NPs, these parameters can be determined by using various analytical techniques, such as dynamic light scattering (DLS), zeta potential, and advanced microscopic techniques such as atomic force microscopy, scanning electron microscopy (SEM) and transmission electron microscopy (TEM), UV-vis spectroscopy, X-ray diffractometry (XRD), Fourier transform infrared spectroscopy (FTIR), and X-ray photoelectron spectroscopy (XPS) [12,34]. To capture the concept of importance of AgNPs characterization, it must be noted that from a toxicological perspective, studies that used similar AgNPs provided by the same manufacturer demonstrated different results. For example, a repeated exposure study by Vandebriel RJ et al. [35] in rats showed that AgNPs are cytotoxic to various cells. On the other hand, Boudreau

M.D et al. [36], using similar particles, showed the dose-dependent accumulation of AgNPs in various tissues of rats, without causing significant cytotoxicity. Moreover, some published nanotoxicity studies have reported characteristics of the particles by using only manufacturer's data without investigators to confirm their found characteristics or by using single analytical tool that provides limited information about the particle-type being studied [35,37–39]. This brings in the importance of the adequate physicochemical characterization of AgNPs prior to undertaking toxicity assessment studies. In addition, a standardized measurement approach like the application of validated methods and the use of reference materials that are specific to AgNPs needs to be more developed in order to assure the comparability of results among toxicity studies that used similar AgNPs.

2.2. AgNPs Physicochemical Properties

Size is an important property that influence the NPs uptake and effect. In this regard, a review study reported that the common sizes of AgNPs used in general applications ranging from 1 to 10 nm [40]. The reason behind this is that smaller particles have been found to display better antimicrobial and cytotoxic activity, as demonstrated in various studies [38,41,42]. Furthermore, as the particle size gets smaller, the specific surface area increases and, hence, a greater proportion of its atoms is displayed on the surface [4] This implies that for the same mass of an NP, biological interactions and toxicity are more dependent on particle number and surface area than on particle mass. The properties of AgNPs can be further enhanced by the functionalization of NPs by using various coatings that in turn influence NPs' surface charge, solubility, and/or hydrophobicity. There is considerable literature that has suggested that the fate and toxicity of AgNPs are determined by their types of coating [43,44]. Various types of surface coatings applied to AgNPs, particularly to improve their biocompatibility and stability against agglomeration are trisodium citrate (CT-AgNP), sodium bis(2-ethylhexyl) sulfosuccinate, cetyltrimethylammonium bromide, polyvinylpyrrolidone (PVP-AgNP), poly(L-lysine), bovine serum albumin, Brij 35 and Tween 20 [45]. Among the latter, CT and PVP coatings are the most commonly used as stabilizing agents [40]. Furthermore, the coating also modifies their charge, which again influences their toxic effect in cells. For example, positively charged NPs are considered more suitable drug delivery tools for anticancer drugs because they can stay for a long time in the blood stream compared to negatively charged particles [12]. Shape-dependent effects were also reported in studies by using varying sized AgNPs. For instance, silver nanocubes showed greater antibacterial effect against *Escherichia coli* compared to spheres and wires in a study that used 55 nm AgNPs [46]. Pal. S et al. [47] also successfully demonstrated a shape-dependent interaction with *E. coli* where truncated a triangular-shaped AgNP had stronger biocidal action than spherical and rod shaped AgNPs. Contrarily, Actis et al. [48] reported no biocidal effect on *Staphylococcus aureus* after using spherical, triangular and cuboid AgNPs. Cellular uptake and biological responses are also defined by the agglomeration state of NPs, and there is sufficient evidence that interaction of the AgNPs with biological media and biomolecules can lead to particle agglomeration and aggregation [49,50]. Though the easy penetration of agglomerated AgNPs in mesenchymal stems cells and nuclei have been reported in several studies, reduced cytotoxicity has also been evident with agglomerated particles compared to free AgNPs [49,51]. A good amount of research has also been conducted on various types of surface corona resulting from interfacial interactions between AgNPs and biological fluids [20]. This has included studies involving both single and complex molecule protein coronas like bovine and human serum albumin, tubulin, ubiquitin, and fetal bovine serum [52]. The formation of a corona, depending on composition, has been shown to interfere with AgNPs' dissolution to Ag ions and, thus, their toxicity [52]. Researchers have also successfully established the importance of various AgNP formulation during synthesis with respect to biomedical applications [53]. For example, the loading of AgNPs inside multiwalled carbon nanotubes has demonstrated an improved targeting of AgNPs to sperm cells and, hence, the potential for development as diagnostic tools for infertility management [54]. Similarly, Bilal et al. [55] synthesized an AgNPs-loaded chitosan-alginate construct that interestingly showed excellent biocompatibility with normal cell line (L929) and cytotoxicity to cancer cells (HeLa cells). Azizi et al. [56] formulated albumin-coated AgNPs with the aim of developing new anticancer agents and showed that the latter was taken specifically by tumor cells and induced apoptosis.

2.3. AgNPs Application and Mechanism of Action

Among various metal salts and NMs that are known to be effective in inhibiting the growth of many bacteria, AgNPs are noteworthy for their strong inhibitory and bactericidal effects [57,58]. The use of AgNPs as well as Ag salts in catheters, cuts, burns and wounds to protect them against infection has been well established [59–62]. However, the exact mechanism underlying the antimicrobial effects of AgNPs is still unresolved, though the literature has suggested that these particles can interact with the membranes of bacteria [15,63]. A potential proposed pathway is that AgNPs, upon interaction with bacteria, induce reactive oxygen species and free radicals, thus damaging the intracellular organelles and modulating the intracellular signaling pathways towards apoptosis [64]. Another widely accepted mechanism of bacterial cytotoxicity is the adhesion of AgNPs to the bacterial wall, followed by the infiltration of the particles, with bacterial cell membrane damage leading to the leakage of cellular contents and death [63,65]. In this context, the antimicrobial activity assessment of small sized AgNPs (12 nm) by Das et al. [66] demonstrated these NPs to be excellent inhibitors against both Gram-positive and Gram-negative bacteria, including *Staphylococcus bacillus*, *Staphylococcus aureus*, and *Pseudomonas aeruginosa*. This indicates that both the membrane thickness and surface charge facilitates particle attachment onto the cell membrane [67]. Finally, the large surface area of AgNPs releasing Ag^+ ions is another crucial factor that contributes to the cytotoxic activity. As it is well established, smaller AgNPs have a faster rate of silver ion (Ag^+) dissolution in the surrounding microenvironment due to their larger surface area to volume ratio and, hence, an increased bioavailability, enhanced distribution, and toxicity of Ag compared with larger NPs [68,69]. Furthermore, Ag^+ ions' release rate is dependent on a number of factors including the size, shape, concentration, capping agent and colloidal state of NPs [70,71]. In particular, the rate of Ag^+ ion release has been shown to be associated with the presence of chlorine, thiols, sulfur, and oxygen [14]. Released Ag^+ ions are suggested to interact with respiratory chain proteins on the membrane, interrupt intracellular O_2 reduction, and induce reactive oxygen species (ROS) production, thus causing cellular oxidative stress in microbes and death [72]. AgNPs are also familiar for their antifungal, antiviral and anti-inflammatory activity [71]. Several studies have reported the potent anti-fungal activity of AgNPs against several phytopathogenic fungi (e.g., *Alternaria alternate*, *Sclerotinia sclerotiorum*, *Macrophomina phaseolina*, *Rhizoctonia solani*, *Botrytis cinereal* and *Curvularia lunata*) as well as human pathogenic fungi (e.g., *Candida* and *Trichoderma sp*) [73,74]. Likewise, AgNPs have demonstrated efficient inhibitory activities against several viruses including human immunodeficiency virus, hepatitis B virus, herpes simplex virus, human parainfluenza virus, peste des petits ruminants virus, and bean yellow mosaic virus, a plant pathogenic virus [75–79]. Inflammation is an early immunological response against foreign particles by tissue, and, interestingly, AgNPs have been recently recognized to play important roles as anti-inflammatory agents. Studies evaluating the anti-inflammatory effect of AgNPs have shown a significant reduction in wound inflammation, a modulation of fibrogenic cytokines, a down regulation of pro-inflammatory cytokines, and apoptosis in inflammatory cells [61,80,81].

3. Routes of Exposure and Biodistribution

The major routes of entry of NPs are ingestion, inhalation, dermal contact and, directly in systemic circulation via intraperitoneal (i.p.) or intravenous (i.v.) injection [7]. The various modes of exposure to AgNPs, their biodistribution, and their mechanisms underlying the effects are illustrated in Figure 1. As AgNPs are extensively used in various household and biomedical products, the following section discusses the various potential routes of entry of these NPs.

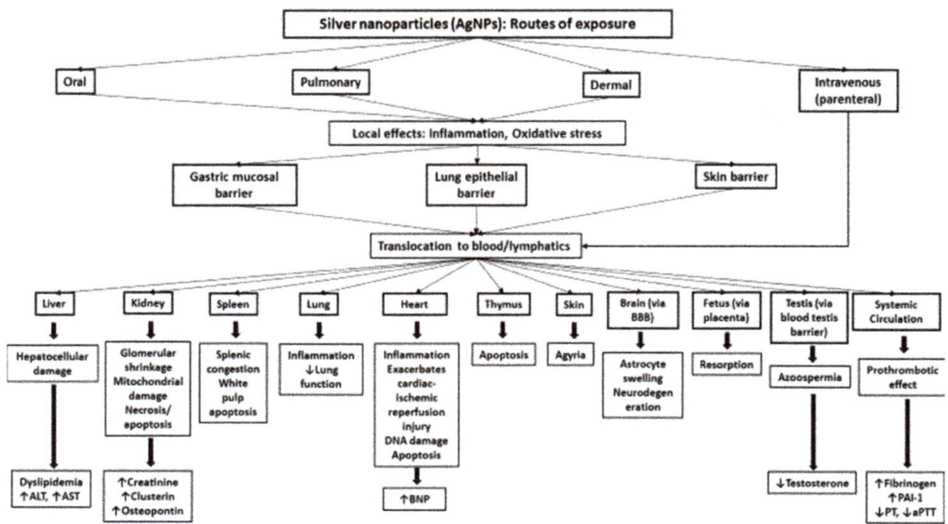

Figure 1. Schematic representation of the biodistribution and toxicity of silver nanoparticles (AgNPs) following various routes of exposure.

After their exposure, AgNPs are able to induce inflammation and oxidative stress at the site of exposure. Moreover, they can cross various biological barriers and enter the systemic circulation. Intravenously-administered AgNPs are directly available in circulation. From then onwards, AgNPs are distributed to various organs and cause organ-specific pathophysiological effects. It remains to be seen whether the effects observed in the distant organs are due to the direct impact of the translocated AgNPs and/or particle-induced inflammatory and oxidative stress responses at the site of exposure. Some abbreviation are as follows: alanine aminotransferase (ALT), aspartate aminotransferase (AST), brain natriuretic peptide, plasminogen activator inhibitor-1, prothrombin time, activated partial thromboplastin time, blood–brain barrier.

3.1. Respiratory Exposure

The release of AgNPs in the environment during the manufacturing, washing or disposal of products enables the NPs to enter the human respiratory system through inhalation [82]. An exposure assessment in an NMs manufacturing facility showed a significant release of AgNPs during processing as soon as the reactor, dryer and grinder were opened, leading to potential occupational exposure even for wet production processes [83]. Similar studies evaluating workplace exposure and health hazard have reported that concentrations of AgNPs in the processes of manufacturing and integration of AgNPs into various consumer products can reach up to 1.35 µg/m^3 [84,85]. Several healthcare, hygiene and antibacterial spray products containing AgNPs have now entered in our daily use. Nazarenko et al. [86] and Lorenz et al. [87] reported that the use of nanotechnology-based consumer sprays containing AgNPs can lead to the generation of nanosized aerosols and the release of NPs near the human breathing zone. Furthermore, Ag-treated textiles can be a source of AgNPs in washing solutions when laundering fabrics, regardless of either conventional Ag or nano Ag treatment [88]. A recent study that evaluated the effluent from a commercially available silver nanowashing machine showed that AgNPs, at an average concentration of 11 µg/L, were released in the environment [89]. AgNPs with a maximum concentration of 145 µg/L were also reported to be released from the outdoor paints during initial runoff events [90]. An occupational study in a silver manufacturing plant revealed that AgNPs in the air increases during production, and a peak area concentration of more than 290 µg/m^3 could be been detected [84]. The authors suggested the possibility for workers to be exposed to airborne AgNPs at concentrations ranging from 0.005 to 0.289 mg/m^3. In spite of considerable studies that have evaluated pulmonary exposure and toxicity, there are still a lack of long-term toxicity data, consumer exposure data, and human health effect data on AgNPs information. Nevertheless, an occupational exposure limit of 0.19 μg/m^3 for AgNPs has recently been proposed based on a subchronic rat inhalation toxicity study and by taking the human equivalent concentration with kinetics into consideration [91]. Following inhalation, the transport and deposition of NPs is not uniform and is influenced by several factors including flow rate, the structure of the airway, pulmonary function, age, and, most importantly, particle size [82]. Particles smaller than 0.1 μm have been shown to penetrate deeply into the alveolar region, mainly by diffusion [7,92]. Consequently, due to deeper particle deposition, the clearance mechanism takes longer and leads to prolonged particle–tissue interactions and more pathophysiological effects [7]. In addition, translocation in blood capillaries is relatively easy for particles with a diameter lower than 0.1 μm [7]. The alveolar–capillary barrier consists of a very thin monolayer of epithelial cells, the endothelial cells of the capillaries, and the basement membrane between the two cells, and this barrier maintains the homeostasis of the lung [7]. NP penetration has been demonstrated following damage to the epithelial layer of the alveolar capillary membrane [7,93].

3.2. Oral Exposure

In the food industry, AgNPs are used in packaging and storage in order to increase the shelf life and quality of food [23]. Moreover, urban and industrial effluents enter the aquatic ecosystem and accumulate along trophic chains [94]. Thus, the presence of AgNPs in dietary supplements, water contamination, or food fish and other aquatic organisms provides the potential sources of oral exposure [23]. Recent studies have also demonstrated that AgNPs incorporated in food packaging can migrate from packaging into food under several usage conditions [95,96]. Inhalation exposure during manufacturing also ultimately leads to oral exposure, since particles cleared via the mucociliary escalator are swallowed and cleared through the GIT. It is estimated that the amount of daily consumption of silver in humans by ingestion is around 20–80 μg [7]. After ingestion, the GIT serves as a mucosal barrier that selectively promotes the degradation and uptake of nutrients such as carbohydrates, peptides, and fats. NPs can act on the mucus layer, translocate to the blood stream and consequently access each organ upon crossing the epithelium. It has been reported that the uptake of NPs with a diameter lower than 100 nm occurs mainly by endocytosis in epithelial cells [97]. Within enterocytes, AgNPs can trigger oxidative stress, DNA damage, and inflammation [7].

3.3. Skin and Parenteral Exposure

Human exposure to AgNPs may also take place through the skin, the largest organ of the body and the first line of defense between the external environment and the internal environment. The potential of solid NPs to penetrate healthy and breached human skin, as well as their ability to diffuse into underlying structures, has been well demonstrated [98,99]. In this context, the use of AgNPs in cosmetics production has been estimated to reach up to 20%. In addition to cosmetics, dermal contact to wound dressings and antibacterial textiles has also shown large diffusion of AgNPs [100]. In a laboratory set up, i.v., i.p., and subcutaneous injection enables AgNPs to directly gain access into systemic circulation. Furthermore, the development of AgNP-based drugs or drug carriers could enable the direct entry of these particles into the human circulatory system.

Following exposure, the distribution and toxicity of AgNPs is further discussed broadly under the in vivo toxicity section of this review article. In general, exposure and gender-related differences in the target tissue AgNP accumulation have been evident in previous research [101–104]. Next to biodistribution, the assessment of the clearance behavior of NPs is an important indicator of cumulative toxicity. In this context, there are several studies that have investigated the post exposure clearance kinetics following subacute inhalation, i.v, and, oral exposure to various sizes of AgNPs and Ag$^+$ ions [104–107]. These studies have revealed silver clearance from most organs after the recovery period, which is generally 17 days to four months. However, tissues with biological barriers like the brain and testes have exhibited a persistence of silver in long term oral exposure studies, suggesting the difficulty of the silver to be cleared from these organs [106,107]. The persistence of silver in these organs also enhances chances of increased toxicity.

4. Pathophysiological Effects of AgNPs

The increasing concern about the possible impact of AgNPs on the environment and human health has directed researchers to focus on the in vitro and in vivo toxicity induced by these particles.

4.1. In Vitro Effects

In vitro cytotoxicity studies are often used to characterize the biological response to AgNPs, and the results of these studies may be used to identify hazards associated with exposure to AgNPs. Some important studies that have shown the toxic effects of AgNPs on different cell lines, including macrophages (RAW 264.7), including bronchial epithelial cells (BEAS-2B), alveolar epithelial cells (A549), hepatocytes (C3A, HepG2), colon cells (Caco2), skin keratinocytes (HaCaT), human epidermal keratinocytes (HEKs), erythrocytes, neuroblastoma cells, embryonic kidney cells (HEK293T), porcine kidney cells (Pk 15), monocytic cells (THP-1), and stem cells [20,108–117], are discussed below.

The exposure of A549 cells to increasing concentrations of AgNPs for 24 h has been to shown cause morphological changes including cell shrinkage, few cellular extensions, a restricted spreading pattern, and cell death in a dose-dependent manner [118]. In another study that used the same type of cells, treatment with 20 nm AgNPs induced DNA damage and the overexpression of metallothioneins at a concentration of 0.6 nM up to 48 h [119]. Size-dependent changes in cellular morphology were observed in a rat alveolar macrophage cell line incubated with hydrocarbon-coated AgNPs of different sizes (15, 30 and 55 nm) [120]. Gliga et al. [31] explored the mechanism of toxicity in BEAS-2B cells exposed to CT-AgNPs of different particle sizes (10, 40 and 75 nm) as well as to 10 nm PVP-coated and 50 nm uncoated AgNPs. In the latter study, cytotoxicity was evaluated with Alamar Blue and lactate dehydrogenase (LDH) assay. The Alamar Blue reagent assessed cell viability and proliferation based on the reduction potential of metabolically active cells. Their results showed cytotoxicity only of the 10 nm particles, independently of surface coating, and toxicity observed was associated with the rate of intracellular Ag release, a 'Trojan Horse' effect. Nguyen et al. [44] exposed a macrophage cell line to uncoated (20, 40, 60, and 80 nm) and PVP-AgNPs (10, 50, and 75 nm) and found a cell shrinkage effect due to uncoated particles, whereas cell elongation was evident after treatment with PVP-coated particles. The exposure of BEAS-2B and RAW 264.7 cell lines to 20 and 110 nm PVP- and CT-Ag-AuNPs (AgNPs with a gold core) showed that 20 nm Ag-AuNPs induced a significant reduction in cell viability in the dose range of 6.25–50 µg/mL for 24 h [121]. In addition, significant ROS generation, intracellular calcium influx, and a decline in mitochondrial membrane potential were also demonstrated in 20 nm CT- and PVP-AgNPs and 110 nm CT-AgNP-treated cells. Bastos V. et al. [122] also evaluated the cytotoxicity of 30 nm CT-AgNPs on RAW 264.7 cells by using parameters including viability, oxidative stress, and cytostaticity at 24 and 48 h of exposure. Their findings revealed decreased cell proliferation and viability at a concentration of only 75 µg/mL, thereby suggesting the low sensitivity of RAW 264.7 cells to lower doses of AgNPs. Recently, Gliga et al. [116], using a combination of RNA sequence and functional assays, showed that repeated, low doses (1 µg/mL) and long term exposure (six weeks) of BEAS-2B cells to 10 and 75 nm CT-AgNPs is profibrotic, indicated by the upregulation of *TGFβ1* and induce epithelial–mesenchymal transition and cell transformation. This evidence suggests that the observed cellular effects are dose-, size-, coating- and duration of exposure-dependent.

The exposure of 20 nm AgNPs to C3A cells at sublethal concentrations (1.95 µg/10^6 cells) revealed size-dependent cytotoxicity, as indicated by elevated LDH levels, an increased release of inflammatory proteins (interleukin (IL) 8 and tumor necrosis factor (TNF) α), oxidative stress, and a decrease in albumin synthesis [123]. Cell viability was also evaluated by a 3-(4,5-dimethylthiazol-2-yl)-2,5-diphenyl tetrazolium bromide (MTT) assay, a colorimetric assay measuring cell metabolic activity based on nicotinamide adenine dinucleotide phosphate -dependent cellular oxidoreductase enzymes, in human hepatoblastoma HepG2 and mice primary liver cells. Interestingly, AgNPs caused a concentration-dependent decrease of cell viability in both cell types [124]. A study by Xue et al. [125] in HepG2 cells demonstrated that AgNPs are able to cause time- (24 and 48 h) and dose-dependent (40, 80, 160 µg/mL) decreases in cell viability, and they are induce cell-cycle arrest in the gap/mitotic phase, significantly increasing the apoptosis rate and ROS generation. A similar study that used PVP and CT-AgNPs at concentrations of 1–100 mg/L also showed coating- (with CT causing more effects than PVP) and dose-dependent reductions in cell viability along with the inhibition of albumin synthesis, as well as a decrease in alanine transaminase activity and apoptosis in HepG2 cells, thus indicating the therapeutic potential of the AgNPs against hepatic cancer [126]. Intestinal cells treated with Ag showed an induction of cytokine release and a higher genotoxicity compared to other inorganic metallic NPs (TiO2 and silicon dioxide) [127]. Böhmert et al. [128] analyzed the toxicity of AgNPs with a primary size of 7.02 ± 0.68 nm in Caco-2 cells by using NP concentrations between 1 and 100 µg/ml. A partial aggregation between digested and not-digested particles was observed by field fractionation (A4F) combined with DLS and X-ray dispersion at small angles. The authors concluded that AgNPs entered the GIT barrier without forming large aggregates in digestive fluids. These results confirmed the importance of body fluids on NP behavior and toxicity.

Samberg et al. [129] assayed the potential cytotoxicity of AgNPs in HEKs cells following 24 h of exposure and reported that unwashed and uncoated AgNPs caused a significant dose-dependent decrease of HEK cell viability and an increase in inflammatory cytokines, whereas washed and carbon-coated AgNPs did not induce any effect. Moreover, an in vitro percutaneous penetration of Ag study revealed that the accumulation of Ag and silver chloride aggregates of smaller than 1 µm, both in the epidermis and dermis [130].

NPs readily enter systemic circulation and may interact with circulatory components like blood cells, the heart, and blood vessels [131–133]. The potential impacts of human exposure to AgNPs on hemolysis, platelet activity and coagulation have recently gained interest. Studies that used human erythrocytes have investigated the effects of AgNPs on hemolysis, morphology, and their uptake [132,134]. In our recent in vitro study, we assessed the effect of the coating and dose of AgNPs on oxidative damage and eryptosis on mice erythrocytes [111]. Both PVP and CT-AgNPs induced oxidative stress and increased cytosolic calcium, annexin V binding, and calpain activity. The latter data may explain the mechanism of hemolysis and eryptosis induced by AgNPs. These NPs could also prevent platelet responses, as evidenced by the inhibitory effects of AgNPs of different sizes (13–15, 30–35, and 40–45 nm) on platelet aggregation [135]. Conversely, Bian et al. [136] recently compared the AgNPs (<100 nm) with Ag macro particles (5–8 µm) and showed that the former can promote phosphatidylserine (PS) exposure and microvesicle generation in freshly isolated human erythrocytes, mainly through ROS generation and intracellular calcium increases, hence suggesting that AgNPs may have prothrombotic risks by promoting the procoagulant activity of red blood cells, more importantly at non-hemolytic concentrations (≤100 µg/mL). These discrepancies in results, though not fully understood, could be related to the different cell types used in studies. The various blood biological effects of AgNPs, such as hemolysis, the interference of plasma coagulation, the enhancement of platelet aggregation, and the inhibition of lymphocyte proliferation, are size-, coating- and concentration-dependent [111,137,138]. Lin et al. [139] studied the potential toxicity of AgNPs on cardiac electrophysiology. The particles caused the concentration-dependent (10^{-9}–10^{-6} g/mL) depolarization of resting membrane potential and diminished action potential, subsequently leading to a loss of excitability in mice cardiac papillary muscle cells in vitro. Milic M et al. [113] investigated the interaction of CT-AgNPs (13–61 nm) with porcine kidney (Pk15) cells, and compared the effect of the particles in their ionic form. For both forms of silver, concentration (1–75 mg/L) dependently decreased the viability of Pk15 cells after 24 h.

Furthermore, AgNPs exhibited an increased toxicity in stem cells, and this was attributed to their properties such as size, concentration and coating. The biocompatibility of 100 nm AgNPs was tested in human mesenchymal stem cells (hMSCs) by Greulich C et al. [140], and there was a dose-dependent (0.5–50 µg/mL) effect on cytotoxicity exhibited by decreased cell proliferation and chemotaxis. In addition, He W et al. [141] showed an increased LDH release and ROS production and reduction in both cell viability and mitochondrial membrane potential in hMSCs exposed to 30 nm AgNPs. Murine spermatogonial stem cells showed less cell viability, LDH leakage, and prolonged apoptosis after exposure to 15 nm AgNPs at concentrations of 5, 10, 25, 50, and 100 µg/mL [142]. Similarly, neural stem cells (NSCs) showed an increase in cell death and LDH leakage, an induction of ROS, an upregulation of the pro-apoptotic Bax protein, and increased in apoptosis when exposed to various concentrations (0.01–80 µg/mL) of PVP-Ag-NPs [143]. In the latter study, AgNPs induced neurotoxicity was compared to Ag^+ ions, and the authors demonstrated that AgNPs caused cell apoptosis by inducing intracellular ROS generation coupled with c-Jun N-terminal kinases phosphorylation, while Ag ions caused cell necrosis via the alteration of cell membrane integrity and direct binding with cellular thiol groups.

Results of in vitro studies have indicated that AgNPs are toxic to the mammalian cells that are derived from the skin, the liver, the lung, the brain, the vascular system and reproductive organs [144]. The cytotoxicity of AgNPs depends on their size, shape, surface charge, coating/capping agent, dosage, oxidation state, agglomeration and type of pathogens against which their toxicity is investigated [42,108,145,146]. Despite these studies, the toxicological of AgNPs mechanism is still unclear. Several studies have reported DNA damage and apoptosis induced by NPs. In this context, AgNPs have been shown to cause apoptosis in mouse embryonic stem cells [147]. A follow up by the same group also demonstrated the involvement of AgNPs in the activation of apoptotic markers, caspase 3 and caspase 9, at concentrations of 50 and 100 µg/mL [148]. DNA damage at a concentration of 0.1 µg/mL of AgNPs was also reported in a study that investigated chromosomal aberrations in human mesenchymal cells [149]. Furthermore, the potential of AgNPs to induce genes that are associated with cell cycle progression, cause chromosomal damage, cell cycle arrest, and cell death in human BEAS-2B cells, umbilical vein endothelial cells, and hepatocellular liver carcinoma cells at various concentrations was also reported [144]. In spite of these numerous studies, a major limitation of the in vitro study of hazard identification with respect to human health is related to the doses used in in vitro studies, as these doses may not be comparable to realistic exposure doses in human. Hence, this necessitates in vivo toxicity research, a review of which is presented in the following section based on potential routes of exposure.

4.2. In Vivo Toxicity

In vivo biodistribution and toxicity studies in rats and mice have demonstrated that AgNPs that are administered by inhalation, ingestion or i.v./i.p. injection are subsequently detected in blood and cause toxicity in several organs including the lung, the liver, the kidney, the intestine and the brain.

Inhalation is a proposed major route of exposure, not only during manufacturing of Ag-containing materials but also during the use of aerosolized products. Tables 1 and 2 summarize the important toxicity and biodistribution studies of AgNPs in rodents following pulmonary exposure via inhalation and intratracheal (i.t.) instillation, respectively. The data from these studies showed diverse outcomes related to biodistribution and remote organ toxicity. Some studies showed no induction of adverse effects [150,151], while other studies reported adverse effects varying from a minimal inflammatory response to the presence of inflammatory lesions in the lungs [103,104,152,153]. For instance, a 28-day inhalation toxicity study on rats showed no significant changes in the hematology and blood biochemistry in either the male or female rats following exposure to 11–14 nm AgNPs at concentrations of $1.73 \times 10^4/cm^3$, $1.27 \times 10^5/cm^3$ and 1.32×10^6 particles/cm^3 [151]. Hyun et al. [154] also exposed rats to 12–15 nm AgNPs for similar durations and doses and showed no remarkable histopathological changes in the nasal cavity and the lung in the NPs exposed group compared to the control group. Nevertheless, Lee et al. [155] found that a short term (14 days) nose-only exposure of mice to 20 nm AgNPs at concentration of 1.91×10^7 particles/cm^3 led to alterations in brain gene expression. Sub-chronic

(90 days) inhalation studies showed mild, dose-dependent pulmonary inflammation and alterations in pulmonary function in rats exposed to 18 nm AgNPs [153]. In addition, inhaled AgNPs may also enter systemic circulation to become distributed to extra-pulmonary organs such as the liver and the brain, as demonstrated in studies that used ~15 nm NPs at concentrations of 1–3 × 10^6 particles/cm^3 [151,156]. A 90 days, an inhalation study by Sung et al. showed alterations in lung function and inflammatory responses in rats exposed to 18 nm AgNPs [153]. Additionally, the accumulation of Ag in the lungs and the liver were more evident in rats after 90 days of inhalation [103]. Silver accumulation has been also observed in the brain, the olfactory bulb, the kidney and the spleen [103,157,158]. Moreover, AgNPs (18–20 nm) were also shown to reach and cross mouse placenta in an inhalation study, where pregnant females were exposed to freshly produced aerosols for either 1 or 4 h/day during the first 15 days of gestation at a particle number concentration of 3.80 × 10^7 part/cm^3 [159].

Table 1. Toxicity and distribution of AgNPs following pulmonary exposure in rodents via inhalation.

Size	Dose	Model	End-Point Measurement	Effect	Tissue Accumulation	References
11–14 nm	1.73 × 10^4/cm^3 (low dose), 1.27 × 10^5/cm^3 (middle dose), (1.32 × 10^6 particles/cm^3 (high dose)	Sprague–Dawley rats	Inhalation 6 h/day, 5 days/week, for 4 weeks, sacrificed 1 day post last exposure	AgNPs concentration below the American Conference of Governmental Industrial Hygienists silver dust limit (100 µg/m^3 did not produce significant toxic effects.	Lu, Li, Br, Ob	[151]
18 nm	0.7 × 10^6 particles/cm^3 (low dose), 1.4 × 10^6 particles /cm^3 (middle dose), and 2.9 × 10^6 particles/cm^3 (high dose)	Sprague–Dawley rats	Inhalation 6 h/day, 5 days/week, for 90 days, sacrificed 1 day post last exposure	Subchronic exposure to AgNPs compromised the lung function.	N/A	[153]
18 nm	0.6 × 10^6 particle/cm^3, 49 µg/m^3 (low dose), 1.4 × 10^6 particle/cm^3, 133 µg/m^3 (middle dose) and 3.0 × 10^6 particle/cm^3, 515 µg/m^3 (high dose)	Sprague–Dawley rats	Inhalation 6 h/day, 5 days/week, for 90 days, sacrificed 1 day post last exposure	Silver accumulation in kidney was gender-dependent. Dose-dependent increase of bile duct hyperplasia in AgNP-exposed liver.	Lu, Li, Br, Ob, Ki	[103]
10 nm (PVP-coated)	3.3 ± 0.5 mg/m^3 or 31 µg/g lung	Male C57Bl/6 mice	Inhalation 4 h/day, 5 days a week, for 10 days, sacrifice at 1 hr and 21 days post last exposure	Subacute inhalation of nanosilver induced minimal pulmonary toxicity.	Lu	[152]
15 nm; 410 nm	179 µg/m^3 and 167 µg/m^3 or 7.9 × 10^6 particles/mm^3 and 118 particles/mm^3 for 15 and 410, respectively	Male Fischer rats	Inhalation 6 h/day, 4 consecutive days, sacrifice at 1 and 7 days post exposure	Size-dependent effect on pulmonary toxicity after inhalation of similar mass concentration of 15 and 410 nm AgNPs.	Lu, Li	[158]
15 nm	8, 28 µg	BrownNorway and Sprague–Dawley rats	Inhalation 3 h/1 day and 3 h/4 days. Sacrifice at 1 and 7 days post last exposure	AgNPs induced an acute pulmonary neutrophilic inflammation with the production of proinflammatory and pro-neutrophilic cytokines.	Lu	[160]

Table 1. *Cont.*

Size	Dose	Model	End-Point Measurement	Effect	Tissue Accumulation	References
20 nm; 110 nm (CT-coated)	7.2 ± 0.8 mg/m^3 and 5.3 ± 1.0 mg/m^3 or 86 and 53 µg/rat for C20 and C110, respectively	Male Sprague–Dawley rats	Inhalation 6 h/1 day, sacrifice at 1, 7, 21, and 56 days postexposure	Delayed peak and short-lived inflammatory and cytotoxic effects in lungs with greater response due to smaller sized nanoparticles.	N/A	[161]
18–20 nm	3.80×10^7 part./cm^{-3}	Female C57BL/6 mice	Inhalation for 1 h/day or 4 h/day, for first 15 days of gestation, sacrificed 4 h post last exposure	Increased number of resorbed fetuses associated with reduced estrogen plasma levels, in the 4 h/day exposed mothers.	Lu, Sp, Li, Pl	[159]

Abbreviations: lung (Lu), liver (Li), spleen (Sp), kidney (Ki), olfactory bulb (Ob), brain (Br), placenta (Pl), polyvinylpyrrolidone (PVP), and citrate (CT).

Table 2. Toxicity and distribution of AgNPs following pulmonary exposure in rodents via intratracheal instillation.

Size	Dose	Model	End-Point Measurement	Effect	Tissue Accumulation	References
20 nm; 110 nm (PVP- and CT-coated)	0.5, 1 mg/kg	Male Sprague–Dawley rats	Single i.t. instillation, sacrifice at 1, 7 and 21 days post exposure	Coating- and size-dependent AgNPs retention in lungs. PVP-coated AgNPs had less retention over time and larger particles were more rapidly cleared from large airways than smaller particles.	Lu	[162]
20 nm; 110 nm (PVP- and CT-coated)	0.1, 0.5, 1 mg/kg	Male Sprague–Dawley rats	Single i.t. instillation, sacrifice at 1, 7 and 21 days post exposure	Smaller sized AgNPs produced more inflammatory and cytotoxic response. Larger particles produce lasting effects post 21 days instillation.	N/A	[163]
20 nm (CT-capped)	1 mg/kg	Male Sprague–Dawley rats	Single i.t. instillation, sacrifice at 1 and 7 days post exposure	AgNP resulted in exacerbation of cardiac ischemic-reperfusion injury.	N/A	[164]
20 nm; 110 nm	1 mg/kg	Male Sprague–Dawley rats	Single i.t. instillation, sacrifice at 1 and 7 days post exposure	Both sizes of AgNP resulted in exacerbation cardiac I/R injury 1 day following instillation independent of capping agent. Persistence of injury was greater for 110 nm PVP-capped AgNP following 7 days instillation.	N/A	[165]
50 nm; 200 nm (PVP-coated)	0.1875, 0.375, 0.75, 1.5, 3 mg/kg	Female Wistar rats	Single i.t. instillation, sacrifice at 3 and 21 days post exposure	Focal accumulation of Ag in peripheral organs along with transient inflammation in lung.	Li, Sp, Ki	[157]
50 nm; 200 nm (PVP- and CT-coated)	0.05, 0.5, 2.5 mg/kg	Female BALB/C mice	Single i.t. instillation, sacrifice 1 day post instillation	Size-, dose- and coating-dependent pro-inflammatory effects in healthy and sensitized lungs following pulmonary exposure to AgNPs.	N/A	[100]
10 nm	0.05, 0.5, 5 mg/kg	BALB/C mice	Single i.t. instillation, sacrifice at 1 and 7 days post exposure	Oxidative stress, DNA damage, apoptosis in heart. Induced prothrombotic events and altered coagulation markers.	N/A	[166]

Abbreviations: lung (Lu), liver (Li), spleen (Sp), kidney (Ki), intratracheal (i.t.), polyvinylpyrrolidone (PVP), and citrate (CT).

In several cases, a gender-dependent difference for AgNPs accumulation in kidneys has been reported, with females exhibiting a higher concentration than males [101,153,167]. One possible explanation for the sex differences in the distribution of Ag may be hormonal regulation in the rat kidney [36]. A gender-dependent difference was also reported in terms of the persistence of pulmonary inflammation and a decrease in lung function in male rats following the termination of exposure, while females showed a gradual improvement in lung inflammation following the cessation of exposure [104]. However, the exact mechanism of sex-related differences is still not clear.

The potential mechanisms of the cardiovascular effects of lung-deposited particles were previously discussed by Nemmar et al. [168]. In this context, Holland et al. [164] investigated the effect of 20 nm AgNPs on cardiovascular injury and showed the exacerbation of cardiac ischemic-reperfusion injury following a single i.t. instillation in rats. The authors further evaluated the impact of the size (20 and 110 nm) and coating (PVP and CT) of AgNPs, and they demonstrated that the acute effect was size- and coating-independent, whereas the persistence of injury was greater for 110 nm PVP-AgNPs [165]. A significant dose-dependent effect of pulmonary-exposed PVP- and CT-AgNPs on cardiovascular homeostasis was also demonstrated in our recent study [166]. The mechanism through which lung injury occurs and how the physicochemical properties of inhaled AgNPs affect their interactions with the lung have recently begun to be investigated in vivo [100,158,162,163]. Along with in vitro studies, in vivo results have suggested that size, coating and dose affect pulmonary inflammation and cellular toxicity.

As mentioned above, besides respiratory exposure, consumer exposure to AgNPs via ingestion can also occur due to the incorporation of AgNP into products such as food containers and dietary supplements. Table 3 summarizes the important toxicity and biodistribution studies of AgNPs in rodents via oral exposure. The deposition of orally-exposed AgNPs in the GIT has been widely demonstrated in previous studies [36,169]. Jeong et al. [170] showed an increase of goblet cells in the intestine, together with a high mucus granule release in mice orally-treated with AgNPs (60 nm) at a concentration of 30 mg/kg bw/day for 28 days. In addition, AgNPs (5–20 nm) that were orally administrated for 21 days in mice (20 mg/kg of body weight) disrupted epithelial cell microvilli and intestinal glands [106]. The distribution of PVP-AgNPs (14 nm) to multiple organs including the intestine, the liver, the kidney, the lung and the brain following oral administration have been reported [169]. Several other investigators also showed that oral exposure to AgNPs may lead to liver, intestinal and neuronal damage [171–173]. Cases of argyria (a condition characterized by an irreversible gray or bluish gray pigmentation of the skin), irreversible neurologic toxicity, and death have been reported upon the long-term ingestion of colloidal silver [174]. The liver appears to be a major accumulation site of circulatory AgNPs [175]. In fact, PVP-AgNPs (20–30 nm) have been shown to increase oxidative stress, enhance autophagy, and deplete insulin signaling pathways following oral exposure for 90 days in the liver of male rats [173]. Changes in blood parameters indicated by a significant elevation of ALT, AST, and hepatoxicity, shown by histological damages (necrosis, hepatocytic inflammation, and the resultant aggregation of lymphocytes in liver tissue) were also observed in a study that evaluated the toxic effect of 14 days of oral exposure to AgNPs (40 nm) at doses 20 and 50 ppm in BALB/C mice [176]. Moreover, Tiwari et al. [177] determined the effect of 60 days AgNPs (10–40 nm) treatment on the kidneys of female Wistar rats at doses of 50 and 200 ppm, and they demonstrated significant mitochondrial damage, increased levels of serum creatinine, and early toxicity markers such as KIM-1, clusterin and osteopontin.

Table 3. Toxicity and biodistribution of AgNPs in rodents via oral exposure.

Size	Dose mg/kg	Model	End-Point Measurement	Effect	Tissue Accumulation	References
56 nm	30, 125, 500	F344 rats	Daily exposure for 90 days, sacrifice 24 h post last exposure	Accumulation of silver in kidneys was gender-dependent, with a 2-fold increase in female kidneys. Liver is the target of silver toxicity for both male and female rats.	Li, Ki, Br, Lu, Bl	[102]
15 nm; 20 nm	90	Male Sprague Dawley rats	Daily exposure 28 days, sacrifice 24 h, 1 week, 8 weeks post last exposure	Main target organ for AgNPs and $AgNO_3$ upon oral exposure are the liver and spleen. Silver was cleared from all organs after 8 weeks post dosing except brain and testes.	Li, Sp, Te, Ki, Br, Lu, Bl, Blr, Ht	[106]
20 nm	820	Male Sprague Dawley rats	Daily exposure for 81 days, sacrifice 24 h post last exposure	AgNPs induces liver and cardiac oxidative stress and mild inflammatory response in liver.	N/A	[178]
20, nm; 110 nm (PVP- and CT-coated)	0.1, 1, 10	Male C57BL/6NCrl mice	3 days exposure, sacrifice at 1 and 7 days post exposure	Acutely ingested AgNP, irrespective of size or coating, are well-tolerated in rodents even in markedly high doses and associated with predominant fecal accumulation.	Absence of tissue accumulation	[179]
10 nm; 75 nm; 110 nm	9, 18, 36	Sprague Dawley rats	Daily exposure for 13 weeks. Sacrifice 24 h post last exposure	Silver accumulation in tissues showed a size-dependent relationship 10>75>110. Statistically significant difference in distribution and accumulation of silver in male and female rats. No toxic effect on blood, reproductive and genetic system tested was observed.	Ki, Sp, Li, Ht, Ut	[36]
20–30 nm (PVP-coated)	50, 100, 200	Male Sprague Dawley rats	Daily exposure for 90 days, sacrifice 24 h post last exposure	Though AgNPs accumulated in hepatic and ileum cells, no harmful effects in liver and kidney, as well as no histopathological, hematological and biochemical markers changes was observed.	Il, Li, Ki, Br, Th, Spl	[180]
20–30 nm (PVP-coated)	50, 100, 200	Male Sprague Dawley rats	Daily exposure for 90 days, sacrifice 24 h post last exposure	High dose showed an increase of sperm morphology abnormalities.	N/A	[181]
10 ± 4 nm (CT-capped)	0.2	Male Wistar rats	Daily exposure for 14 days, sacrifice 24 h post exposure	Prolonged low dose AgNPs exposure induced oxidative stress in brain but not in liver.	N/A	[182]
91.71 ± 1.6 nm	0.5	Male Wistar rats	Daily exposure for 45 days, sacrificed 24 h post exposure	AgNPs caused significant oxidative stress compared to TiO_2NP with similar dose concentration.	Bl, Li, Ki	[183]
20 nm; 110 nm	10	Pregnant Sprague Dawley rats	Single exposure, sacrificed at 24 h and 48 h post exposure	Silver enters and crosses placenta regardless of route and form of silver.	Ce, Ll, Pl, Ki, Bl	[184]
20–30 nm (PVP-coated)	50, 100, 200	Male Sprague Dawley rats	Daily exposure for 90 days, sacrifice 24 h post last exposure	High dose of AgNPs induced hepatocellular damage by increased ROS production, enhanced autophagy and depleted insulin signaling pathway.	N/A	[173]

Abbreviations: lung (Lu), liver (Li), spleen (Sp), kidney (Ki), brain (Br), heart (Ht), bladder (Bl), uterus (Ut), thymus (Th), cecum (Ce), placenta (Pl), polyvinylpyrrolidone (PVP), and citrate (CT).

The use of AgNPs in wound dressings and other applications designed to regulate skin microbiome composition is an established strategy that has been shown to inhibit a broad range of bacteria, including *Pseudomonas aeruginosa*, *Escherichia coli* and *Staphylococcus aureus* [57,81]. Numerous reports have indicated that AgNPs promote wound healing by decreasing the inflammatory response [185,186]. Nevertheless, an aspect that remains sparsely researched is the possibility of the sensitization of the skin by these NPs. It is well known that patients with wounds are at particular risk of developing either an allergic or irritant contact dermatitis, and silver compounds are widely used in wound care. However, so far, there have been very few confirmed cases of contact dermatitis secondary to silver-containing wound products like silver sulfadiazine and skin markers that contain silver nitrate [187,188]. Contrary, there is considerable evidence for the significant transdermal penetration of AgNPs into capillaries during the use of surgical dressings, textiles, and cosmetics [189,190]. In order to elucidate the mechanism of cytotoxicity, Samberg et al. [129] evaluated the potential ability of AgNPs to penetrate porcine skin and showed the existence of the focal inflammation and localization of AgNPs on the surface and in the upper stratum corneum layers of porcine skin. Acute and sub-acute dermal studies conducted by Korani et al. [6] suggested a correlation between dermal exposure, tissue accumulation of AgNPs (100 nm), and dose-dependent histopathological abnormalities in the skin, all of which was exhibited by a reduced thickness of the papillary layer and the epidermis. Compared to animals treated with a single dose, animals that were subjected to sub-chronic exposure showed a considerable accumulation of AgNPs, as well as a dose-dependent toxic response in several organs, including the spleen, liver and skin [191].

Another potential route of AgNPs entry in the case of biomedical applications includes parenteral administration. A summary of important toxicity and biodistribution studies of AgNP in rodents via exposure of i.v. injection is given in Table 4. In a recent study, comparing the biodistribution and toxicological examinations after repeated i.v. administration of AgNPs and AuNPs in mice showed a higher deposition of AgNPs in the heart, the lung, and the kidney than that of AuNPs [175]. Moreover, the AgNPs induced adverse effects in a dose-dependent manner (the concentrations tested were 4, 10, 20, and 40 mg/kg) [175]. Another study, following the subcutaneous administration of AgNPs of different sizes in rats also revealed that the particles translocated to the blood circulation and were distributed throughout the main organs, especially in the kidney, the liver, the spleen, the brain and the lung [131]. The results also suggested the potential of AgNPs to cross the blood–brain barrier and to induce astrocyte swelling and neuronal degeneration [131]. A few studies have reported the transfer of AgNPs across the placenta in rats and mice [192,193]. Following the i.v. administration of 10 nm AgNPs at a dose of 66 µg Ag/mouse to pregnant animals on gestational days 7, 8 and 9, Ag accumulation was revealed in all examined organs, with the highest accumulation being in the maternal liver, spleen and visceral yolk sac and the lowest concentrations being in the embryos [192]. Another study comparing administration methods (i.v. versus i.p.) showed a similar localization of Ag in the liver and the spleen for both methods [193]. However, Ag was more quickly excreted from the body with i.v. administration, as compared to i.p. administration. The latter study also showed that the AgNPs could cross the placental and the blood–testes barriers, thus resulting in an accumulation in the fetus and the testes, respectively [193].

Table 4. Toxicity and biodistribution of AgNPs in rodents via exposure through intravenous injection.

Size	Dose	Model	End-Point Measurement	Effect	Tissue Accumulation	References
15–40 nm	4, 10, 20, 40 mg/kg	Male Wistar rats	32 days i.v, sacrifice 24 h post last i.v administration	AgNPs in doses <10mg/kg is safe, while >20mg/kg is toxic.	Li, Ki	[194]
21.8 nm	7.5, 30, 120 mg/kg	ICR mice	Single i.v, parameters measured at 1,7,14 days post injection	Inflammatory reactions in lung and liver cells were induced in mice treated at the high dose of AgNPs. Gender-related differences in distribution and elimination of AgNPs, elimination in female is longer than the males.	Sp, Li, Lu, Ki	[195]
20 nm; 200 nm	5 mg/kg	Male Wistar rats	Single i.v, sacrifice at 1, 7, 28 days post i.v administration	High tissue concentration of silver in tissues of 20 nm group as compared to 200 nm groups	Li, Sp, Ki, Lu, Br, Ur, Fe	[196]
7.2 ± 3.3 nm	5, 10, 45 mg/kg	Male Sprague–Dawley rats	Daily tail vein injection for 3 consecutive days, parameters measured at 1 and 3rd day	Decrease in body weight and locomotor activity.	N/A	[197]
20, nm; 100 nm	0.0082, 0.0025, 0.074, 0.22, 0.67, 2, 6 mg/kg	Wistar rats	28 days, repeated i.v, sacrifice at 24 h post last injection	Immune system is the most sensitive parameter affected by AgNPs; reduced thymus weight, increased spleen weight and spleen cell number, strongly reduced NK cell activity, and reduced IFN-γ production were observed.	N/A	[37]
20 nm	0.0082, 0.0025, 0.074, 0.22, 0.67, 2, 6 mg/kg	Male Wistar rats	28 days, repeated i.v, sacrifice at 24 h post last injection	AgNPs suppress the functional immune system.	N/A	[35]
10 nm (CT-coated)	1 mg/kg	CD1 mice	IV administration, once every 3 days, sacrifice at 15, 60 120 days post initial exposure	AgNPs induced toxicity to male reproduction, altered Leydig cell function, increased testosterone level.	Te	[198]
10 nm; 75 nm; 110 nm (CT-coated)	0.1 mg/kg	Female BALB/C mice	Single dose injection sacrifice at 4h, 1, 3 or 7 days post injection. Multi dose: i.v injection on day 1, 4 and 10, sacrifice at 7 days post last injection	Injection of a single dose of AgNPs induced a less toxicity in liver and lung tissues than that induced by multi-dose injections. The toxic effect of AgNPs was time-dependent.	N/A	[199]
10 nm; 40 nm, 100 nm (PVP- or CT-coated)	10 mg/kg	Male CD-1 (ICR) mice	Single i.v, sacrifice at 24 h post injection	10 nm AgNP group showed increased silver distribution and overt hepatobiliary toxicity compared to larger ones.	Sp, Li, Lu, Ki, Bl, Br	[68]
3 ± 1.57 nm	11.4–13.3 mg/kg	Male KunMing mice	i.v. injection 2 times/week for 4 weeks, sacrificed at 1, 28, 56 days post injection	AgNPs preferentially accumulated in all major organs compared to other metallic NPs.	Li, Sp, Ki, He, Lu, Te, Fe, Bl, In, St, Br, SV	[175]

Table 4. *Cont.*

Size	Dose	Model	End-Point Measurement	Effect	Tissue Accumulation	References
20 nm; 110 nm	1 mg/kg	Pregnant Sprague Dawley rats	Single exposure, sacrificed at 24 and 48 h post exposure	Silver crosses the placenta and is transferred to the fetus regardless of the form of silver.	Sp, Pl, Li, Lu, Ce, Bl, Ki	[184]
6.3–629 nm	5 mg/kg	Female Sprague Dawley rats	Single exposure, sacrificed at 24 h	Histopathologically, AgNPs caused mild irritation in thymus and spleen and significantly increased chromosome breakage and polyploidy cell rates.	Lu, Sp. Li, Ki, Th, Ht	[200]
20 nm; 110 nm (PVP or CT)	701.75 µg/kg	Pregnant female Sprague Dawley rats	Single i.v administration	Exposure to CT-AgNPs is associated with changes in fetal growth and increased contractile force in both uterine and aortic vessels.	N/A	[201]

Abbreviations: lung (Lu), liver (Li), spleen (Sp), kidney (Ki), brain (Br), heart (Ht), bladder (Bl), uterus (Ut), thymus (Th), cecum (Ce), placenta (Pl), intestine (It), stomach (St), feces (Fe), urine (Ur), intravenous (i.v.), polyvinylpyrrolidone (PVP), and citrate (CT).

5. Knowledge Gaps in Human and Environmental Risk Assessment

For all NPs studies, a crucial issue remains the composition, particle surface area, surface chemistry, and the careful, accurate characterization of particle size and morphologic features, especially in the physiological environment [202]. Moreover, equally important to the latter is the control of assays and assay conditions [202]. It is only with the complete characterization of NPs and the appropriate control of assays that the results of reported studies can be comparable with those of other studies conducted with similar NMs [68,203]. Unfortunately, the characterization of materials, especially following in vivo applications, is still inadequate for many published studies. In this regard, though several characterizing tools have been developed, each has its own limitations. For instance, DLS is the most commonly used tool, especially in studies that adopt limited characterization steps, due to its accessibility, low cost, and easy handling [204]. However, its disadvantages include low resolution, multiple light scattering, sedimentation, a lack of selectivity, and a relatively low signal strength, particularly in complex biological media such as in plasma [204]. Likewise, zeta potential is affected not only by the properties of NPs but also by the nature of the solution, such as pH and ionic strength [204]. Moreover, the understanding of operating principles, as well as dealings with critical issues like sample preparation and data interpretation, proposes challenges to the application of these characterization techniques. As the outcome of particle effects is largely governed by the NP's physicochemical properties, the thorough characterization of AgNPs is extremely important, especially when investigating in vivo effects following various routes of contact.

Toxicokinetic studies of NPs including absorption, distribution, metabolism and elimination have provided important data related to their *in vivo* behavior and risk assessment. In this context, there are very limited data on AgNPs' toxicokinetic properties, particularly metabolism and clearance. Though several of the studies discussed in this review have demonstrated organ distribution following the translocation of AgNPs, it is not yet clear whether the distribution and effects observed were due to their particulate forms, ionic forms, or a combination of both forms. In order to find out whether the distribution of AgNPs were in ionic or particulate forms, Lee JH et al. [205] recently investigated the toxicokinetic of i.v.-administered AgNPs (10 nm) and AuNPs (14 nm), either separately or in combination, and evaluated NP clearance after a four-week recovery period. Interestingly, their data revealed that the co-administration of AgNPs with AuNPs of a similar size distribution not only decreased NPs' distribution to organs, thus indicating a competitive cellular uptake, but also confirmed the particulate form of NP tissue distribution rather than ionic form [205]. Another potential form suggested is the distribution in secondary particle form, because NPs can interact with proteins like thiol after dissolution in ionic form. The latter was demonstrated by Liu et al. [206], who suggested that the newly formed secondary AgNPs circulated in systemic circulation and photoreduced to metallic silver, eventually contributing to agyria silver deposits in light-affected skin areas. Another study that used silver nitrates or AgNPs also demonstrated the formation of similar secondary AgNPs [106]. This in fact gives another dimension to research, as the detected AgNPs in tissues could not only be the product of exposed nanotechnology but could also have been due to any chemical forms of silver exposure that eventually transformed into secondary AgNPs. A parallel controversy pertains to the tissue clearance of accumulated silver. In spite of studies that have evaluated silver clearance with different exposure routes, exposure periods, and recovery periods, there not yet a clear understanding of whether the ionic or particulate form is eliminated from tissues.

Another aspect that has not been studied much is the impeding effects of AgNPs on susceptible populations like pulmonary disorders, obesity, hypertension, and diabetes. It is well established that the impact of air pollution is aggravated in patients with pre-existing cardiorespiratory diseases, such as asthma, and chronic respiratory diseases, such as chronic obstructive pulmonary disease, pneumonia, cystic fibrosis, and ischemic heart diseases [207]. A recent study also reported the exacerbation of autoimmune diseases to short term exposure to particulate matter (PM_{10} and $PM_{2.5}$) [208]. However, information in regard to the effect of AgNP exposure on susceptible populations is very much limited. The pathophysiological effects of the latter could be well studied by using animal models of increased susceptibility, e.g., hypertension and diabetes, changes in blood biochemistry, acute phase response, and hepatic pathology. In this context, Ramirez-Lee et al. [209] recently evaluated the cardiovascular effect of 15 nm AgNPs by using isolated perfused hearts from male, spontaneously hypertensive rats. The authors concluded that hypertension intensified AgNP-induced cardiotoxicity due to an observed reduction in NO and an increase in oxidative stress, leading to increased vasoconstriction and myocardial contractility. Jia et al. [210] studied the effect of orally-exposed PVP-AgNPs (30 nm) in overweight mice and showed the progression of fatty liver disease from steatosis to steatohepatitis. The mechanisms proposed in the latter study were the activation of Kupffer cells, the enhancement of hepatic inflammation, and the suppression of fatty acid oxidation. Kermanizadeh, A. et al. [211] investigated AgNP-induced hepatic pathology in models representative of pre-existing alcoholic liver disease. Their data showed that following oral exposure, AgNP-induced hepatic effects were aggravated in the alcohol-pretreated mice in comparison to controls with regards to an organ-specific inflammatory response.

The evaluation of trans-generational impact is also an important point to understand the long-term effect of NPs on human health and the environment. In this context, Hartmann et al. [212], assessed the impact of pristine and waste water-borne AgNPs on the aquatic invertebrate *Daphnia magna* in a multi generation approach that covered six generations. The authors showed that while pristine AgNPs caused a significant reduction in the mean number of offspring compared with the control, the waste water-borne AgNPs had no effects on reproduction in any generation. Raj et al. [213] investigated impact of ingested AgNPs (20–100 nm) on the adult and larval stages of *Drosophila*. Their results demonstrated a significant reduction of survival, longevity, ovary size, and egg laying capability in flies fed with AgNPs compared with a control [213]. The latter effects persisted in the next generation without AgNP feeding, thereby suggesting the transgenerational effects of AgNPs. Despite these findings, the in vivo transgenerational effects of AgNPs involving higher mammalian systems or humans still remains the least explored area of NP research.

Another aspect that lacks detailed research is the effect of AgNPs on humans via various routes of exposure. In this regard, very few studies have attempted to investigate whether AgNPs can penetrate physically and functionally intact human skin [99,214]. George et al. [99] demonstrated the in vivo penetration of AgNPs (10–40 nm) by using healthy human participants with normal skin. Their data suggested that AgNPs, applied as nanocrystalline silver dressing for four-to-six days, can penetrate beyond the stratum corneum and reach as deep as the reticular dermis. A controlled, cross over time exposure (three, seven, and 14 days) study of orally dosed (10 ppm) commercial AgNPs (5–10 nm) demonstrated the absence of any changes in human metabolic, hematologic, urine, and physical findings or imaging morphology [215]. However, AgNP toxicology research with respect to susceptible individual and human exposure thus far remains understudied, and these areas are particularly important with regard to NP risk assessment.

The increasing concern about the possible impact of AgNPs on the environment and, subsequently, human health has directed researchers to focus on the in vitro and in vivo toxicity induced by these particles.

6. Conclusions and Recommendations for Future Studies

This paper critically reviewed and structurally presented the toxicity and biodistribution studies of AgNPs following various routes of exposure. Our conclusions drawn from these studies are listed below:

- The cytotoxic effects of AgNPs, documented in in vitro studies in various cell lines, are governed by factors such as size, shape, coating, dose and cell type.
- Toxicity and biodistribution studies, in vivo, following various routes of exposure, like inhalation, instillation, oral, dermal and intravenous, have established Ag translocation, accumulation, and toxicity to various organs.
- Both the local and distant organ effects are influenced by particle size, coating, route and duration of exposure, doses, and end point measurement time.
- There is lack of adequate and standard characterization techniques that could be adapted for studies that evaluate the toxicity of AgNPs in order to make the results of one study comparable to another by using similar NPs.
- The mechanisms of action of AgNPs are still not well understood, and there is lack of information on the potential effects of AgNP exposure on animal models of enhanced susceptibility, such as hypertension, diabetes, and asthma.

Owing to the evidence provided in this review, there are still gaps in the risk assessment of the Ag in the form of NPs both for humans and the environment. For example, it is still not clear to what extent the intact AgNPs themselves can enter the human body, whether the AgNPs undergo changes in the physiological environment, if the Ag^+ ions released from the NPs absorbed, or if the effect observed is due to tge AgNP-induced inflammatory response or due to the ions released or due to the nanoparticulate from itself. Since no clear pathway has been proven to be the most important mechanism of AgNP-induced pathophysiological effects, we have some recommendation for future research listed below.

- In order to overcome the limitation of a single method of particle characterization and to efficiently evaluate the functional effect of synthesized particles, the characterization of AgNPs should be done by using multiple relevant techniques.
- AgNPs' characteristics should be evaluated in an appropriate medium because interactions with a biological fluid can alter NPs' properties, intake, and cellular effects.
- There is a need for extensive data on the biodistribution and accumulation of AgNPs, and these data should take AgNPs' various physicochemical properties into consideration in order to get a concrete idea on the local and distant tissue toxicity of AgNPs, as well as the mechanisms behind the toxicity.
- Appropriate techniques and methodologies have to be constructed in order to estimate Ag^+ ions originating from AgNPs in vivo and to calculate AgNPs' surface ionization fraction in various tissues.
- AgNPs effect on animal models with pre-existing diseases like asthma, obesity, hypertension, and diabetes needs to be carried out, as toxicological consequences might be aggravated in animal models of enhanced susceptibility.

- Multi-generation studies assessing the transgenerational impact of AgNPs in higher mammalian systems needs to be carried out in order to identify the potential long-term effects of AgNPs in a more realistic scenario.
- Multidisciplinary investigations taking in account long term exposure, variable routes of exposure, and the dosing of AgNPs should be conducted in humans in order to ascertain the human toxicity threshold.

Author Contributions: Z.F. and A.N. drafted the manuscript. All authors have read and agreed to the published version of the manuscript.

Funding: This work was supported by funds of the Sheikh Hamdan Foundation for Medical Research and the College of Medicine and Health Sciences grants.

Acknowledgments: All contributors who provided help during the research have been listed.

Conflicts of Interest: The authors declare no conflict of interest.

Abbreviations

AuNPs	Gold nanoparticles
AgNPs	Silver nanoparticles
Bl	Bladder
Br	Brain
Ce	Caecum
CNTs	Carbon nanotubes
CT	Citrate
ENMs	Engineered nanomaterials
Fe	Feces
FeO	Iron Oxide
Ht	Heart
It	Intestine
i.p.	Intra-peritoneal
i.t.	Intra-tracheal
i.v.	Intra-venous
Ki	Kidney
LDH	Lactate Dehydrogenase
Lu	Lung
Li	Liver
NM	Nanomaterials
Pl	Placenta
PVP	Polyvinylpyrrolidone
ROS	Reactive Oxygen Species
Sp	Spleen
St	Stomach
Th	Thymus
TiO_2	Titanium Oxide
Ur	Urine
Ut	Uterus

References

1. Keiper, A. The nanotechnology revolution. *New Atlantis* **2003**, *2*, 17–34.
2. Boverhof, D.R.; Bramante, C.M.; Butala, J.H.; Clancy, S.F.; Lafranconi, M.; West, J.; Gordon, S.C. Comparative assessment of nanomaterial definitions and safety evaluation considerations. *Regul. Toxicol. Pharmacol.* **2015**, *73*, 137–150. [CrossRef] [PubMed]
3. Khan, I.; Saeed, K.; Khan, I. Nanoparticles: Properties, applications and toxicities. *Arab. J. Chem.* **2017**, *12*, 908–931. [CrossRef]

4. Bakand, S.; Hayes, A. Toxicological considerations, toxicity assessment, and risk management of inhaled nanoparticles. *Int. J. Mol. Sci.* **2016**, *17*, 929. [CrossRef] [PubMed]

5. Li, Z.; Cong, H.; Yan, Z.; Liu, A.; Yu, B. The Potential Human Health and Environmental Issues of Nanomaterials. In *Handbook of Nanomaterials for Industrial Applications*; Elsevier: Amsterdam, The Netherlands, 2018; pp. 1049–1054.

6. Korani, M.; Ghazizadeh, E.; Korani, S.; Hami, Z.; Mohammadi-Bardbori, A. Effects of silver nanoparticles on human health. *Eur. J. Nanomed.* **2015**, *7*, 51–62. [CrossRef]

7. De Matteis, V. Exposure to inorganic nanoparticles: Routes of entry, immune response, biodistribution and in vitro/in vivo toxicity evaluation. *Toxics* **2017**, *5*, 29. [CrossRef]

8. Jeevanandam, J.; Barhoum, A.; Chan, Y.S.; Dufresne, A.; Danquah, M.K. Review on nanoparticles and nanostructured materials: History, sources, toxicity and regulations. *Beilstein J. Nanotechnol.* **2018**, *9*, 1050–1074. [CrossRef]

9. Vance, M.E.; Kuiken, T.; Vejerano, E.P.; McGinnis, S.P.; Hochella, M.F., Jr.; Rejeski, D.; Hull, M.S. Nanotechnology in the real world: Redeveloping the nanomaterial consumer products inventory. *Beilstein J. Nanotechnol.* **2015**, *6*, 1769–1780. [CrossRef]

10. Iravani, S.; Korbekandi, H.; Mirmohammadi, S.V.; Zolfaghari, B. Synthesis of silver nanoparticles: Chemical, physical and biological methods. *Res. Pharm. Sci.* **2014**, *9*, 385.

11. Syafiuddin, A.; Salim, M.R.; Beng Hong Kueh, A.; Hadibarata, T.; Nur, H. A review of silver nanoparticles: Research trends, global consumption, synthesis, properties, and future challenges. *J. Chin. Chem. Soc.* **2017**, *64*, 732–756. [CrossRef]

12. Zhang, X.-F.; Liu, Z.-G.; Shen, W.; Gurunathan, S. Silver nanoparticles: Synthesis, characterization, properties, applications, and therapeutic approaches. *Int. J. Mol. Sci.* **2016**, *17*, 1534. [CrossRef] [PubMed]

13. Ahmed, K.B.R.; Nagy, A.M.; Brown, R.P.; Zhang, Q.; Malghan, S.G.; Goering, P.L. Silver nanoparticles: Significance of physicochemical properties and assay interference on the interpretation of in vitro cytotoxicity studies. *Toxicol. In Vitro* **2017**, *38*, 179–192. [CrossRef] [PubMed]

14. Maurer, L.L.; Meyer, J.N. A systematic review of evidence for silver nanoparticle-induced mitochondrial toxicity. *Environ. Sci.* **2016**, *3*, 311–322. [CrossRef]

15. Le Ouay, B.; Stellacci, F. Antibacterial activity of silver nanoparticles: A surface science insight. *Nano Today* **2015**, *10*, 339–354. [CrossRef]

16. Singh, R.; Shedbalkar, U.U.; Wadhwani, S.A.; Chopade, B.A. Bacteriagenic silver nanoparticles: Synthesis, mechanism, and applications. *Appl. Microbiol. Biotechnol.* **2015**, *99*, 4579–4593. [CrossRef]

17. Simončič, B.; Klemenčič, D. Preparation and performance of silver as an antimicrobial agent for textiles: A review. *Text. Res. J.* **2016**, *86*, 210–223. [CrossRef]

18. Corrêa, J.M.; Mori, M.; Sanches, H.L.; Cruz, A.D.D.; Poiate, E.; Poiate, I.A.V.P. Silver nanoparticles in dental biomaterials. *Int. J. Biomater.* **2015**, *2015*, 9. [CrossRef]

19. Ahmed, S.; Ahmad, M.; Swami, B.L.; Ikram, S. A review on plants extract mediated synthesis of silver nanoparticles for antimicrobial applications: A green expertise. *J. Adv. Res.* **2016**, *7*, 17–28. [CrossRef]

20. Akter, M.; Sikder, M.T.; Rahman, M.M.; Ullah, A.K.M.A.; Hossain, K.F.B.; Banik, S.; Hosokawa, T.; Saito, T.; Kurasaki, M. A systematic review on silver nanoparticles-induced cytotoxicity: Physicochemical properties and perspectives. *J. Adv. Res.* **2018**, *9*, 1–16. [CrossRef]

21. Siddiqi, K.S.; Husen, A.; Rao, R.A.K. A review on biosynthesis of silver nanoparticles and their biocidal properties. *J. Nanobiotechnol.* **2018**, *16*, 14. [CrossRef]

22. Sharma, V.K.; Sayes, C.M.; Guo, B.; Pillai, S.; Parsons, J.G.; Wang, C.; Yan, B.; Ma, X. Interactions between silver nanoparticles and other metal nanoparticles under environmentally relevant conditions: A review. *Sci. Total Environ.* **2019**, *653*, 1042–1051. [CrossRef] [PubMed]

23. Gaillet, S.; Rouanet, J.-M. Silver nanoparticles: Their potential toxic effects after oral exposure and underlying mechanisms—A review. *Food Chem. Toxicol.* **2015**, *77*, 58–63. [CrossRef] [PubMed]

24. Natsuki, J.; Natsuki, T.; Hashimoto, Y. A review of silver nanoparticles: Synthesis methods, properties and applications. *Int. J. Mater. Sci. Appl.* **2015**, *4*, 325–332. [CrossRef]

25. Rumble, J. *CRC Handbook of Chemistry and Physics*; CRC Press: Boca Raton, FL, USA, 2017.

26. Zivic, F.; Grujovic, N.; Mitrovic, S.; Ahad, I.U.; Brabazon, D. Characteristics and applications of silver nanoparticles. In *Commercialization of Nanotechnologies–A Case Study Approach*; Springer: Berlin/Heidelberg, Germany, 2018; pp. 227–273.

27. Tien, D.C.; Liao, C.Y.; Huang, J.C.; Tseng, K.H.; Lung, J.K.; Tsung, T.T.; Kao, W.S.; Tsai, T.H.; Cheng, T.W.; Yu, B.S. Novel technique for preparing a nano-silver water suspension by the arc-discharge method. *Rev. Adv. Mater. Sci.* **2008**, *18*, 750–756.

28. Shih, S.-J.; Chien, I.C. Preparation and characterization of nanostructured silver particles by one-step spray pyrolysis. *Powder Technol.* **2013**, *237*, 436–441. [CrossRef]

29. Gudikandula, K.; Charya Maringanti, S. Synthesis of silver nanoparticles by chemical and biological methods and their antimicrobial properties. *J. Exp. Nanosci.* **2016**, *11*, 714–721. [CrossRef]

30. Kim, D.H.; Park, J.C.; Jeon, G.E.; Kim, C.S.; Seo, J.H. Effect of the size and shape of silver nanoparticles on bacterial growth and metabolism by monitoring optical density and fluorescence intensity. *Biotechnol. Bioprocess Eng.* **2017**, *22*, 210–217. [CrossRef]

31. Gliga, A.R.; Skoglund, S.; Wallinder, I.O.; Fadeel, B.; Karlsson, H.L. Size-dependent cytotoxicity of silver nanoparticles in human lung cells: The role of cellular uptake, agglomeration and Ag release. *Part. Fibre Toxicol.* **2014**, *11*, 11. [CrossRef]

32. Loza, K.; Diendorf, J.; Sengstock, C.; Ruiz-Gonzalez, L.; Gonzalez-Calbet, J.M.; Vallet-Regi, M.; Köller, M.; Epple, M. The dissolution and biological effects of silver nanoparticles in biological media. *J. Mater. Chem. B* **2014**, *2*, 1634–1643. [CrossRef]

33. Wei, L.; Lu, J.; Xu, H.; Patel, A.; Chen, Z.-S.; Chen, G. Silver nanoparticles: Synthesis, properties, and therapeutic applications. *Drug Discov. Today* **2015**, *20*, 595–601. [CrossRef]

34. Lin, P.-C.; Lin, S.; Wang, P.C.; Sridhar, R. Techniques for physicochemical characterization of nanomaterials. *Biotechnol. Adv.* **2014**, *32*, 711–726. [CrossRef] [PubMed]

35. Vandebriel, R.J.; Tonk, E.C.M.; de la Fonteyne-Blankestijn, L.J.; Gremmer, E.R.; Verharen, H.W.; van der Ven, L.T.; van Loveren, H.; de Jong, W.H. Immunotoxicity of silver nanoparticles in an intravenous 28-day repeated-dose toxicity study in rats. *Part. Fibre Toxicol.* **2014**, *11*, 21. [CrossRef] [PubMed]

36. Boudreau, M.D.; Imam, M.S.; Paredes, A.M.; Bryant, M.S.; Cunningham, C.K.; Felton, R.P.; Jones, M.Y.; Davis, K.J.; Olson, G.R. Differential effects of silver nanoparticles and silver ions on tissue accumulation, distribution, and toxicity in the Sprague Dawley rat following daily oral gavage administration for 13 weeks. *Toxicol. Sci.* **2016**, *150*, 131–160. [CrossRef] [PubMed]

37. De Jong, W.H.; Van Der Ven, L.T.; Sleijffers, A.; Park, M.V.; Jansen, E.H.; Van Loveren, H.; Vandebriel, R.J. Systemic and immunotoxicity of silver nanoparticles in an intravenous 28 days repeated dose toxicity study in rats. *Biomaterials* **2013**, *34*, 8333–8343. [CrossRef] [PubMed]

38. Cho, Y.-M.; Mizuta, Y.; Akagi, J.-I.; Toyoda, T.; Sone, M.; Ogawa, K. Size-dependent acute toxicity of silver nanoparticles in mice. *J. Toxicol. Pathol.* **2018**, *31*, 73–80. [CrossRef]

39. Gherkhbolagh, M.H.; Alizadeh, Z.; Asari, M.J.; Sohrabi, M. In Vivo Induced Nephrotoxicity of Silver Nanoparticles in Rat after Oral Administration. *J. Res. Med. Dent. Sci.* **2018**, *6*, 43–51.

40. Tolaymat, T.M.; El Badawy, A.M.; Genaidy, A.; Scheckel, K.G.; Luxton, T.P.; Suidan, M. An evidence-based environmental perspective of manufactured silver nanoparticle in syntheses and applications: A systematic review and critical appraisal of peer-reviewed scientific papers. *Sci. Total Environ.* **2010**, *408*, 999–1006. [CrossRef]

41. Lu, Z.; Rong, K.; Li, J.; Yang, H.; Chen, R. Size-dependent antibacterial activities of silver nanoparticles against oral anaerobic pathogenic bacteria. *J. Mater. Sci.* **2013**, *24*, 1465–1471. [CrossRef]

42. Jeong, Y.; Lim, D.W.; Choi, J. Assessment of size-dependent antimicrobial and cytotoxic properties of silver nanoparticles. *Adv. Mater. Sci. Eng.* **2014**, *2014*, 6. [CrossRef]

43. Sharma, V.K.; Siskova, K.M.; Zboril, R.; Gardea-Torresdey, J.L. Organic-coated silver nanoparticles in biological and environmental conditions: Fate, stability and toxicity. *Adv. Colloid Interface Sci.* **2014**, *204*, 15–34. [CrossRef]

44. Nguyen, K.C.; Seligy, V.L.; Massarsky, A.; Moon, T.W.; Rippstein, P.; Tan, J.; Tayabali, A.F. Comparison of toxicity of uncoated and coated silver nanoparticles. *J. Phys.* **2013**, *429*, 012025. [CrossRef]

45. Jurašin, D.D.; Ćurlin, M.; Capjak, I.; Crnković, T.; Lovrić, M.; Babič, M.; Horák, D.; Vrček, I.V.; Gajović, S. Surface coating affects behavior of metallic nanoparticles in a biological environment. *Beilstein J. Nanotechnol.* **2016**, *7*, 246–262. [CrossRef] [PubMed]

46. Hong, X.; Wen, J.; Xiong, X.; Hu, Y. Shape effect on the antibacterial activity of silver nanoparticles synthesized via a microwave-assisted method. *Environ. Sci. Pollut. Res.* **2016**, *23*, 4489–4497. [CrossRef] [PubMed]

47. Pal, S.; Tak, Y.K.; Song, J.M. Does the antibacterial activity of silver nanoparticles depend on the shape of the nanoparticle? A study of the gram-negative bacterium Escherichia coli. *Appl. Environ. Microbiol.* **2007**, *73*, 1712–1720. [CrossRef]

48. Actis, L.; Srinivasan, A.; Lopez-Ribot, J.L.; Ramasubramanian, A.K.; Ong, J.L. Effect of silver nanoparticle geometry on methicillin susceptible and resistant *Staphylococcus aureus*, and osteoblast viability. *J. Mater. Sci.* **2015**, *26*, 215. [CrossRef]

49. Lankoff, A.; Sandberg, W.J.; Wegierek-Ciuk, A.; Lisowska, H.; Refsnes, M.; Sartowska, B.; Schwarze, P.E.; Meczynska-Wielgosz, S.; Wojewodzka, M.; Kruszewski, M. The effect of agglomeration state of silver and titanium dioxide nanoparticles on cellular response of HepG2, A549 and THP-1 cells. *Toxicol. Lett.* **2012**, *208*, 197–213. [CrossRef]

50. Argentiere, S.; Cella, C.; Cesaria, M.; Milani, P.; Lenardi, C. Silver nanoparticles in complex biological media: Assessment of colloidal stability and protein corona formation. *J. Nanopart. Res.* **2016**, *18*, 253. [CrossRef]

51. Bae, E.; Lee, B.-C.; Kim, Y.; Choi, K.; Yi, J. Effect of agglomeration of silver nanoparticle on nanotoxicity depression. *Korean J. Chem. Eng.* **2013**, *30*, 364–368. [CrossRef]

52. Durán, N.; Silveira, C.P.; Durán, M.; Martinez, D.S.T. Silver nanoparticle protein corona and toxicity: A mini-review. *J. Nanobiotechnol.* **2015**, *13*, 55. [CrossRef]

53. Mathur, P.; Jha, S.; Ramteke, S.; Jain, N.K. Pharmaceutical aspects of silver nanoparticles. *Artif. Cells Nanomed. Biotechnol.* **2018**, *46*, 115–126. [CrossRef]

54. Jha, P.K.; Jha, R.K.; Rout, D.; Gnanasekar, S.; Rana, S.V.S.; Hossain, M. Potential targetability of multi-walled carbon nanotube loaded with silver nanoparticles photosynthesized from Ocimum tenuiflorum (tulsi extract) in fertility diagnosis. *J. Drug Target.* **2017**, *25*, 616–625. [CrossRef] [PubMed]

55. Bilal, M.; Rasheed, T.; Iqbal, H.M.N.; Li, C.; Hu, H.; Zhang, X. Development of silver nanoparticles loaded chitosan-alginate constructs with biomedical potentialities. *Int. J. Biol. Macromol.* **2017**, *105*, 393–400. [CrossRef] [PubMed]

56. Azizi, M.; Ghourchian, H.; Yazdian, F.; Bagherifam, S.; Bekhradnia, S.; Nyström, B. Anti-cancerous effect of albumin coated silver nanoparticles on MDA-MB 231 human breast cancer cell line. *Sci. Rep.* **2017**, *7*, 5178. [CrossRef] [PubMed]

57. Nam, G.; Purushothaman, B.; Rangasamy, S.; Song, J.M. Investigating the versatility of multifunctional silver nanoparticles: Preparation and inspection of their potential as wound treatment agents. *Int. Nano Lett.* **2016**, *6*, 51–63. [CrossRef]

58. Firdhouse, M.J.; Lalitha, P. Biosynthesis of silver nanoparticles and its applications. *J. Nanotechnol.* **2015**, *2015*, 18. [CrossRef]

59. Catauro, M.; Raucci, M.G.; De Gaetano, F.; Marotta, A. Antibacterial and bioactive silver-containing Na$_2$O·CaO·2SiO$_2$ glass prepared by sol–gel method. *J. Mater. Sci.* **2004**, *15*, 831–837.

60. Crabtree, J.H.; Burchette, R.J.; Siddiqi, R.A.; Huen, I.T.; Hadnott, L.L.; Fishman, A. The efficacy of silver-ion implanted catheters in reducing peritoneal dialysis-related infections. *Perit. Dial. Int.* **2003**, *23*, 368–374.

61. Tian, J.; Wong, K.K.Y.; Ho, C.M.; Lok, C.N.; Yu, W.Y.; Che, C.M.; Chiu, J.F.; Tam, P.K.H. Topical delivery of silver nanoparticles promotes wound healing. *ChemMedChem* **2007**, *2*, 129–136. [CrossRef]

62. Roe, D.; Karandikar, B.; Bonn-Savage, N.; Gibbins, B.; Roullet, J.-B. Antimicrobial surface functionalization of plastic catheters by silver nanoparticles. *J. Antimicrob. Chemother.* **2008**, *61*, 869–876. [CrossRef]

63. Qing, Y.; Cheng, L.; Li, R.; Liu, G.; Zhang, Y.; Tang, X.; Wang, J.; Liu, H.; Qin, Y. Potential antibacterial mechanism of silver nanoparticles and the optimization of orthopedic implants by advanced modification technologies. *Int. J. Nanomed.* **2018**, *13*, 3311. [CrossRef]

64. Lee, S.H.; Jun, B.-H. Silver nanoparticles: Synthesis and application for nanomedicine. *Int. J. Mol. Sci.* **2019**, *20*, 865. [CrossRef] [PubMed]

65. Seong, M.; Lee, D.G. Silver nanoparticles against Salmonella enterica serotype typhimurium: Role of inner membrane dysfunction. *Curr. Microbiol.* **2017**, *74*, 661–670. [CrossRef] [PubMed]

66. Das, R.; Gang, S.; Nath, S.S. Preparation and antibacterial activity of silver nanoparticles. *J. Biomater. Nanobiotechnol.* **2011**, *2*, 472. [CrossRef]

67. Abbaszadegan, A.; Ghahramani, Y.; Gholami, A.; Hemmateenejad, B.; Dorostkar, S.; Nabavizadeh, M.; Sharghi, H. The effect of charge at the surface of silver nanoparticles on antimicrobial activity against gram-positive and gram-negative bacteria: A preliminary study. *J. Nanomater.* **2015**, *16*, 53. [CrossRef]

68. Recordati, C.; De Maglie, M.; Bianchessi, S.; Argentiere, S.; Cella, C.; Mattiello, S.; Cubadda, F.; Aureli, F.; D'Amato, M.; Raggi, A. Tissue distribution and acute toxicity of silver after single intravenous administration in mice: Nano-specific and size-dependent effects. *Part. Fibre Toxicol.* **2016**, *13*, 12. [CrossRef]
69. Shang, L.; Nienhaus, K.; Nienhaus, G.U. Engineered nanoparticles interacting with cells: Size matters. *J. Nanobiotechnol.* **2014**, *12*, 5. [CrossRef]
70. Raza, M.; Kanwal, Z.; Rauf, A.; Sabri, A.; Riaz, S.; Naseem, S. Size-and shape-dependent antibacterial studies of silver nanoparticles synthesized by wet chemical routes. *Nanomaterials* **2016**, *6*, 74. [CrossRef]
71. Kailasa, S.K.; Park, T.-J.; Rohit, J.V.; Koduru, J.R. Antimicrobial activity of silver nanoparticles. In *Nanoparticles in Pharmacotherapy*; Elsevier: Amsterdam, The Netherlands, 2019; pp. 461–484.
72. Long, Y.-M.; Hu, L.-G.; Yan, X.-T.; Zhao, X.-C.; Zhou, Q.-F.; Cai, Y.; Jiang, G.-B. Surface ligand controls silver ion release of nanosilver and its antibacterial activity against Escherichia coli. *Int. J. Nanomed.* **2017**, *12*, 3193. [CrossRef]
73. Kim, S.W.; Jung, J.H.; Lamsal, K.; Kim, Y.S.; Min, J.S.; Lee, Y.S. Antifungal effects of silver nanoparticles (AgNPs) against various plant pathogenic fungi. *Mycobiology* **2012**, *40*, 53–58. [CrossRef]
74. Elgorban, A.M.; El-Samawaty, A.E.-R.M.; Yassin, M.A.; Sayed, S.R.; Adil, S.F.; Elhindi, K.M.; Bakri, M.; Khan, M. Antifungal silver nanoparticles: Synthesis, characterization and biological evaluation. *Biotechnol. Biotechnol. Equip.* **2016**, *30*, 56–62. [CrossRef]
75. Lara, H.H.; Ayala-Nuñez, N.V.; Ixtepan-Turrent, L.; Rodriguez-Padilla, C. Mode of antiviral action of silver nanoparticles against HIV-1. *J. Nanobiotechnol.* **2010**, *8*, 1. [CrossRef] [PubMed]
76. Gaikwad, S.; Ingle, A.; Gade, A.; Rai, M.; Falanga, A.; Incoronato, N.; Russo, L.; Galdiero, S.; Galdiero, M. Antiviral activity of mycosynthesized silver nanoparticles against herpes simplex virus and human parainfluenza virus type 3. *Int. J. Nanomed.* **2013**, *8*, 4303.
77. Khandelwal, N.; Kaur, G.; Chaubey, K.K.; Singh, P.; Sharma, S.; Tiwari, A.; Singh, S.V.; Kumar, N. Silver nanoparticles impair Peste des petits ruminants virus replication. *Virus Res.* **2014**, *190*, 1–7. [CrossRef] [PubMed]
78. Elbeshehy, E.K.F.; Elazzazy, A.M.; Aggelis, G. Silver nanoparticles synthesis mediated by new isolates of Bacillus spp., nanoparticle characterization and their activity against Bean Yellow Mosaic Virus and human pathogens. *Front. Microbiol.* **2015**, *6*, 453. [CrossRef]
79. Lu, L.; Sun, R.W.; Chen, R.; Hui, C.-K.; Ho, C.-M.; Luk, J.M.; Lau, G.K.; Che, C.-M. Silver nanoparticles inhibit hepatitis B virus replication. *Antivir. Ther.* **2008**, *13*, 253.
80. David, L.; Moldovan, B.; Vulcu, A.; Olenic, L.; Perde-Schrepler, M.; Fischer-Fodor, E.; Florea, A.; Crisan, M.; Chiorean, I.; Clichici, S. Green synthesis, characterization and anti-inflammatory activity of silver nanoparticles using European black elderberry fruits extract. *Colloids Surf. B* **2014**, *122*, 767–777. [CrossRef]
81. Hebeish, A.; El-Rafie, M.H.; El-Sheikh, M.A.; Seleem, A.A.; El-Naggar, M.E. Antimicrobial wound dressing and anti-inflammatory efficacy of silver nanoparticles. *Int. J. Biol. Macromol.* **2014**, *65*, 509–515. [CrossRef]
82. Qiao, H.; Liu, W.; Gu, H.; Wang, D.; Wang, Y. The transport and deposition of nanoparticles in respiratory system by inhalation. *J. Nanomater.* **2015**, *2015*, 2. [CrossRef]
83. Park, J.; Kwak, B.K.; Bae, E.; Lee, J.; Kim, Y.; Choi, K.; Yi, J. Characterization of exposure to silver nanoparticles in a manufacturing facility. *J. Nanopart. Res.* **2009**, *11*, 1705–1712. [CrossRef]
84. Lee, J.H.; Mun, J.; Park, J.D.; Yu, I.J. A health surveillance case study on workers who manufacture silver nanomaterials. *Nanotoxicology* **2012**, *6*, 667–669. [CrossRef]
85. Lee, J.H.; Kwon, M.; Ji, J.H.; Kang, C.S.; Ahn, K.H.; Han, J.H.; Yu, I.J. Exposure assessment of workplaces manufacturing nanosized TiO_2 and silver. *Inhal. Toxicol.* **2011**, *23*, 226–236. [CrossRef] [PubMed]
86. Nazarenko, Y.; Han, T.W.; Lioy, P.J.; Mainelis, G. Potential for exposure to engineered nanoparticles from nanotechnology-based consumer spray products. *J. Expo. Sci. Environ. Epidemiol.* **2011**, *21*, 515. [CrossRef] [PubMed]
87. Lorenz, C.; Hagendorfer, H.; von Goetz, N.; Kaegi, R.; Gehrig, R.; Ulrich, A.; Scheringer, M.; Hungerbühler, K. Nanosized aerosols from consumer sprays: Experimental analysis and exposure modeling for four commercial products. *J. Nanopart. Res.* **2011**, *13*, 3377–3391. [CrossRef]
88. Mitrano, D.M.; Rimmele, E.; Wichser, A.; Erni, R.; Height, M.; Nowack, B. Presence of nanoparticles in wash water from conventional silver and nano-silver textiles. *ACS Nano* **2014**, *8*, 7208–7219. [CrossRef] [PubMed]

89. Farkas, J.; Peter, H.; Christian, P.; Urrea, J.A.G.; Hassellöv, M.; Tuoriniemi, J.; Gustafsson, S.; Olsson, E.; Hylland, K.; Thomas, K.V. Characterization of the effluent from a nanosilver producing washing machine. *Environ. Int.* **2011**, *37*, 1057–1062. [CrossRef] [PubMed]

90. Kaegi, R.; Sinnet, B.; Zuleeg, S.; Hagendorfer, H.; Mueller, E.; Vonbank, R.; Boller, M.; Burkhardt, M. Release of silver nanoparticles from outdoor facades. *Environ. Pollut.* **2010**, *158*, 2900–2905. [CrossRef]

91. Weldon, B.A.; Faustman, E.; Oberdörster, G.; Workman, T.; Griffith, W.C.; Kneuer, C.; Yu, I.J. Occupational exposure limit for silver nanoparticles: Considerations on the derivation of a general health-based value. *Nanotoxicology* **2016**, *10*, 945–956. [CrossRef]

92. Deng, Q.; Deng, L.; Miao, Y.; Guo, X.; Li, Y. Particle deposition in the human lung: Health implications of particulate matter from different sources. *Environ. Res.* **2019**, *169*, 237–245. [CrossRef]

93. Theodorou, I.G.; Ryan, M.P.; Tetley, T.D.; Porter, A.E. Inhalation of silver nanomaterials—Seeing the risks. *Int. J. Mol. Sci.* **2014**, *15*, 23936–23974. [CrossRef]

94. Gambardella, C.; Costa, E.; Piazza, V.; Fabbrocini, A.; Magi, E.; Faimali, M.; Garaventa, F. Effect of silver nanoparticles on marine organisms belonging to different trophic levels. *Mar. Environ. Res.* **2015**, *111*, 41–49. [CrossRef]

95. Choi, J.I.; Chae, S.J.; Kim, J.M.; Choi, J.C.; Park, S.J.; Choi, H.J.; Bae, H.; Park, H.J. Potential silver nanoparticles migration from commercially available polymeric baby products into food simulants. *Food Addit. Contam.* **2018**, *35*, 996–1005. [CrossRef]

96. Mackevica, A.; Olsson, M.E.; Hansen, S.F. Silver nanoparticle release from commercially available plastic food containers into food simulants. *J. Nanopart. Res.* **2016**, *18*, 5. [CrossRef]

97. Axson, J.L.; Stark, D.I.; Bondy, A.L.; Capracotta, S.S.; Maynard, A.D.; Philbert, M.A.; Bergin, I.L.; Ault, A.P. Rapid kinetics of size and pH-dependent dissolution and aggregation of silver nanoparticles in simulated gastric fluid. *J. Phys. Chem. C* **2015**, *119*, 20632–20641. [CrossRef] [PubMed]

98. Wang, L.-P.; Wang, J.-Y. Skin penetration of inorganic and metallic nanoparticles. *J. Shanghai Jiaotong Univ.* **2014**, *19*, 691–697. [CrossRef]

99. George, R.; Merten, S.; Wang, T.T.; Kennedy, P.; Maitz, P. In vivo analysis of dermal and systemic absorption of silver nanoparticles through healthy human skin. *Australas. J. Dermatol.* **2014**, *55*, 185–190. [CrossRef]

100. Alessandrini, F.; Vennemann, A.; Gschwendtner, S.; Neumann, A.; Rothballer, M.; Seher, T.; Wimmer, M.; Kublik, S.; Traidl-Hoffmann, C.; Schloter, M. Pro-inflammatory versus immunomodulatory effects of silver nanoparticles in the lung: The critical role of dose, size and surface modification. *Nanomaterials* **2017**, *7*, 300. [CrossRef]

101. Kim, Y.S.; Kim, J.S.; Cho, H.S.; Rha, D.S.; Kim, J.M.; Park, J.D.; Choi, B.S.; Lim, R.; Chang, H.K.; Chung, Y.H. Twenty-eight-day oral toxicity, genotoxicity, and gender-related tissue distribution of silver nanoparticles in Sprague-Dawley rats. *Inhal. Toxicol.* **2008**, *20*, 575–583. [CrossRef]

102. Kim, Y.S.; Song, M.Y.; Park, J.D.; Song, K.S.; Ryu, H.R.; Chung, Y.H.; Chang, H.K.; Lee, J.H.; Oh, K.H.; Kelman, B.J. Subchronic oral toxicity of silver nanoparticles. *Part. Fibre Toxicol.* **2010**, *7*, 20. [CrossRef]

103. Sung, J.H.; Ji, J.H.; Park, J.D.; Yoon, J.U.; Kim, D.S.; Jeon, K.S.; Song, M.Y.; Jeong, J.; Han, B.S.; Han, J.H. Subchronic inhalation toxicity of silver nanoparticles. *Toxicol. Sci.* **2008**, *108*, 452–461. [CrossRef]

104. Song, K.S.; Sung, J.H.; Ji, J.H.; Lee, J.H.; Lee, J.S.; Ryu, H.R.; Lee, J.K.; Chung, Y.H.; Park, H.M.; Shin, B.S. Recovery from silver-nanoparticle-exposure-induced lung inflammation and lung function changes in Sprague Dawley rats. *Nanotoxicology* **2013**, *7*, 169–180. [CrossRef]

105. Lankveld, D.P.K.; Oomen, A.G.; Krystek, P.; Neigh, A.; Troost–de Jong, A.; Noorlander, C.W.; Van Eijkeren, J.C.H.; Geertsma, R.E.; De Jong, W.H. The kinetics of the tissue distribution of silver nanoparticles of different sizes. *Biomaterials* **2010**, *31*, 8350–8361. [CrossRef]

106. van der Zande, M.; Vandebriel, R.J.; Van Doren, E.; Kramer, E.; Herrera Rivera, Z.; Serrano-Rojero, C.S.; Gremmer, E.R.; Mast, J.; Peters, R.J.B.; Hollman, P.C.H. Distribution, elimination, and toxicity of silver nanoparticles and silver ions in rats after 28-day oral exposure. *ACS Nano* **2012**, *6*, 7427–7442. [CrossRef] [PubMed]

107. Lee, J.H.; Kim, Y.S.; Song, K.S.; Ryu, H.R.; Sung, J.H.; Park, J.D.; Park, H.M.; Song, N.W.; Shin, B.S.; Marshak, D. Biopersistence of silver nanoparticles in tissues from Sprague–Dawley rats. *Part. Fibre Toxicol.* **2013**, *10*, 36. [CrossRef] [PubMed]

108. Jiang, X.; Lu, C.; Tang, M.; Yang, Z.; Jia, W.; Ma, Y.; Jia, P.; Pei, D.; Wang, H. Nanotoxicity of silver nanoparticles on HEK293T cells: A combined study using biomechanical and biological techniques. *ACS Omega* **2018**, *3*, 6770–6778. [CrossRef] [PubMed]

109. Carrola, J.; Bastos, V.; Jarak, I.; Oliveira-Silva, R.; Malheiro, E.; Daniel-da-Silva, A.L.; Oliveira, H.; Santos, C.; Gil, A.M.; Duarte, I.F. Metabolomics of silver nanoparticles toxicity in HaCaT cells: Structure–activity relationships and role of ionic silver and oxidative stress. *Nanotoxicology* **2016**, *10*, 1105–1117. [CrossRef] [PubMed]

110. Sahu, S.C.; Zheng, J.; Graham, L.; Chen, L.; Ihrie, J.; Yourick, J.J.; Sprando, R.L. Comparative cytotoxicity of nanosilver in human liver HepG2 and colon Caco2 cells in culture. *J. Appl. Toxicol.* **2014**, *34*, 1155–1166. [CrossRef]

111. Ferdous, Z.; Beegam, S.; Tariq, S.; Ali, B.H.; Nemmar, A. The in Vitro Effect of Polyvinylpyrrolidone and Citrate Coated Silver Nanoparticles on Erythrocytic Oxidative Damage and Eryptosis. *Cell. Physiol. Biochem.* **2018**, *49*, 1577–1588. [CrossRef]

112. Xin, L.; Wang, J.; Fan, G.; Che, B.; Wu, Y.; Guo, S.; Tong, J. Oxidative stress and mitochondrial injury-mediated cytotoxicity induced by silver nanoparticles in human A549 and HepG2 cells. *Environ. Toxicol.* **2016**, *31*, 1691–1699. [CrossRef]

113. Milić, M.; Leitinger, G.; Pavičić, I.; Zebić Avdičević, M.; Dobrović, S.; Goessler, W.; Vinković Vrček, I. Cellular uptake and toxicity effects of silver nanoparticles in mammalian kidney cells. *J. Appl. Toxicol.* **2015**, *35*, 581–592. [CrossRef]

114. Lategan, K.L.; Walters, C.R.; Pool, E.J. The effects of silver nanoparticles on RAW 264.7. Macrophages and human whole blood cell cultures. *Front. Biosci.* **2019**, *24*, 347–365.

115. Galbiati, V.; Cornaghi, L.; Gianazza, E.; Potenza, M.A.; Donetti, E.; Marinovich, M.; Corsini, E. In vitro assessment of silver nanoparticles immunotoxicity. *Food Chem. Toxicol.* **2018**, *112*, 363–374. [CrossRef] [PubMed]

116. Gliga, A.R.; Di Bucchianico, S.; Lindvall, J.; Fadeel, B.; Karlsson, H.L. RNA-sequencing reveals long-term effects of silver nanoparticles on human lung cells. *Sci. Rep.* **2018**, *8*, 6668. [CrossRef] [PubMed]

117. Abdal Dayem, A.; Lee, S.; Choi, H.; Cho, S.-G. Silver Nanoparticles: Two-Faced Neuronal Differentiation-Inducing Material in Neuroblastoma (SH-SY5Y) Cells. *Int. J. Mol. Sci.* **2018**, *19*, 1470. [CrossRef] [PubMed]

118. Lee, Y.S.; Kim, D.W.; Lee, Y.H.; Oh, J.H.; Yoon, S.; Choi, M.S.; Lee, S.K.; Kim, J.W.; Lee, K.; Song, C.-W. Silver nanoparticles induce apoptosis and G2/M arrest via PKCζ-dependent signaling in A549 lung cells. *Arch. Toxicol.* **2011**, *85*, 1529–1540. [CrossRef]

119. De Matteis, V.; Malvindi, M.A.; Galeone, A.; Brunetti, V.; De Luca, E.; Kote, S.; Kshirsagar, P.; Sabella, S.; Bardi, G.; Pompa, P.P. Negligible particle-specific toxicity mechanism of silver nanoparticles: The role of Ag+ ion release in the cytosol. *Nanomed. Nanotechnol. Biol. Med.* **2015**, *11*, 731–739. [CrossRef]

120. Carlson, C.; Hussain, S.M.; Schrand, A.M.; Braydich-Stolle, L.K.; Hess, K.L.; Jones, R.L.; Schlager, J.J. Unique cellular interaction of silver nanoparticles: Size-dependent generation of reactive oxygen species. *J. Phys. Chem. B* **2008**, *112*, 13608–13619. [CrossRef]

121. Wang, X.; Ji, Z.; Chang, C.H.; Zhang, H.; Wang, M.; Liao, Y.P.; Lin, S.; Meng, H.; Li, R.; Sun, B. Use of coated silver nanoparticles to understand the relationship of particle dissolution and bioavailability to cell and lung toxicological potential. *Small* **2014**, *10*, 385–398. [CrossRef]

122. Bastos, V.; Duarte, I.F.; Santos, C.; Oliveira, H. A study of the effects of citrate-coated silver nanoparticles on RAW 264.7 cells using a toolbox of cytotoxic endpoints. *J. Nanopart. Res.* **2017**, *19*, 163. [CrossRef]

123. Gaiser, B.K.; Hirn, S.; Kermanizadeh, A.; Kanase, N.; Fytianos, K.; Wenk, A.; Haberl, N.; Brunelli, A.; Kreyling, W.G.; Stone, V. Effects of silver nanoparticles on the liver and hepatocytes in vitro. *Toxicol. Sci.* **2012**, *131*, 537–547. [CrossRef]

124. Faedmaleki, F.; Shirazi, F.H.; Salarian, A.-A.; Ashtiani, H.A.; Rastegar, H. Toxicity effect of silver nanoparticles on mice liver primary cell culture and HepG2 cell line. *Iran. J. Pharm. Res.* **2014**, *13*, 235.

125. Xue, Y.; Zhang, T.; Zhang, B.; Gong, F.; Huang, Y.; Tang, M. Cytotoxicity and apoptosis induced by silver nanoparticles in human liver HepG2 cells in different dispersion media. *J. Appl. Toxicol.* **2016**, *36*, 352–360. [CrossRef] [PubMed]

126. Vrček, I.V.; Žuntar, I.; Petlevski, R.; Pavičić, I.; Dutour Sikirić, M.; Ćurlin, M.; Goessler, W. Comparison of in vitro toxicity of silver ions and silver nanoparticles on human hepatoma cells. *Environ. Toxicol.* **2016**, *31*, 679–692. [CrossRef] [PubMed]

127. Kaiser, J.-P.; Roesslein, M.; Diener, L.; Wick, P. Human health risk of ingested nanoparticles that are added as multifunctional agents to paints: An in vitro study. *PLoS ONE* **2013**, *8*, e83215. [CrossRef] [PubMed]

128. Böhmert, L.; Girod, M.; Hansen, U.; Maul, R.; Knappe, P.; Niemann, B.; Weidner, S.M.; Thünemann, A.F.; Lampen, A. Analytically monitored digestion of silver nanoparticles and their toxicity on human intestinal cells. *Nanotoxicology* **2014**, *8*, 631–642. [CrossRef]

129. Samberg, M.E.; Oldenburg, S.J.; Monteiro-Riviere, N.A. Evaluation of silver nanoparticle toxicity in skin in vivo and keratinocytes in vitro. *Environ. Health Perspect.* **2009**, *118*, 407–413. [CrossRef]

130. Bianco, C.; Kezic, S.; Crosera, M.; Svetličić, V.; Šegota, S.; Maina, G.; Romano, C.; Larese, F.; Adami, G. In vitro percutaneous penetration and characterization of silver from silver-containing textiles. *Int. J. Nanomed.* **2015**, *10*, 1899. [CrossRef]

131. Tang, J.; Xiong, L.; Wang, S.; Wang, J.; Liu, L.; Li, J.; Yuan, F.; Xi, T. Distribution, translocation and accumulation of silver nanoparticles in rats. *J. Nanosci. Nanotechnol.* **2009**, *9*, 4924–4932. [CrossRef]

132. Kim, M.J.; Shin, S. Toxic effects of silver nanoparticles and nanowires on erythrocyte rheology. *Food Chem. Toxicol.* **2014**, *67*, 80–86. [CrossRef]

133. Gupta, R.; Xie, H. Nanoparticles in Daily Life: Applications, Toxicity and Regulations. *J. Environ. Pathol. Toxicol. Oncol.* **2018**, *37*, 209–230. [CrossRef]

134. Chen, L.Q.; Fang, L.; Ling, J.; Ding, C.Z.; Kang, B.; Huang, C.Z. Nanotoxicity of silver nanoparticles to red blood cells: Size dependent adsorption, uptake, and hemolytic activity. *Chem. Res. Toxicol.* **2015**, *28*, 501–509. [CrossRef]

135. Shrivastava, S.; Bera, T.; Singh, S.K.; Singh, G.; Ramachandrarao, P.; Dash, D. Characterization of antiplatelet properties of silver nanoparticles. *ACS Nano* **2009**, *3*, 1357–1364. [CrossRef] [PubMed]

136. Bian, Y.; Kim, K.; Ngo, T.; Kim, I.; Bae, O.-N.; Lim, K.-M.; Chung, J.-H. Silver nanoparticles promote procoagulant activity of red blood cells: A potential risk of thrombosis in susceptible population. *Part. Fibre Toxicol.* **2019**, *16*, 9. [CrossRef] [PubMed]

137. Jun, E.-A.; Lim, K.-M.; Kim, K.; Bae, O.-N.; Noh, J.-Y.; Chung, K.-H.; Chung, J.-H. Silver nanoparticles enhance thrombus formation through increased platelet aggregation and procoagulant activity. *Nanotoxicology* **2011**, *5*, 157–167. [CrossRef] [PubMed]

138. Huang, H.; Lai, W.; Cui, M.; Liang, L.; Lin, Y.; Fang, Q.; Liu, Y.; Xie, L. An evaluation of blood compatibility of silver nanoparticles. *Sci. Rep.* **2016**, *6*, 25518. [CrossRef] [PubMed]

139. Lin, C.-X.; Yang, S.-Y.; Gu, J.-L.; Meng, J.; Xu, H.-Y.; Cao, J.-M. The acute toxic effects of silver nanoparticles on myocardial transmembrane potential, I Na and I K1 channels and heart rhythm in mice. *Nanotoxicology* **2017**, *11*, 827–837.

140. Greulich, C.; Kittler, S.; Epple, M.; Muhr, G.; Köller, M. Studies on the biocompatibility and the interaction of silver nanoparticles with human mesenchymal stem cells (hMSCs). *Langenbeck's Arch. Surg.* **2009**, *394*, 495–502. [CrossRef]

141. He, W.; Liu, X.; Kienzle, A.; Müller, W.E.G.; Feng, Q. In vitro uptake of silver nanoparticles and their toxicity in human mesenchymal stem cells derived from bone marrow. *J. Nanosci. Nanotechnol.* **2016**, *16*, 219–228. [CrossRef]

142. Braydich-Stolle, L.; Hussain, S.; Schlager, J.J.; Hofmann, M.-C. In vitro cytotoxicity of nanoparticles in mammalian germline stem cells. *Toxicol. Sci.* **2005**, *88*, 412–419. [CrossRef]

143. Sun, C.; Yin, N.; Wen, R.; Liu, W.; Jia, Y.; Hu, L.; Zhou, Q.; Jiang, G. Silver nanoparticles induced neurotoxicity through oxidative stress in rat cerebral astrocytes is distinct from the effects of silver ions. *Neurotoxicology* **2016**, *52*, 210–221. [CrossRef]

144. Liao, C.; Li, Y.; Tjong, S.C. Bactericidal and Cytotoxic Properties of Silver Nanoparticles. *Int. J. Mol. Sci.* **2019**, *20*, 449. [CrossRef]

145. Butler, K.S.; Peeler, D.J.; Casey, B.J.; Dair, B.J.; Elespuru, R.K. Silver nanoparticles: Correlating nanoparticle size and cellular uptake with genotoxicity. *Mutagenesis* **2015**, *30*, 577–591. [CrossRef] [PubMed]

146. Abdal Dayem, A.; Hossain, M.; Lee, S.; Kim, K.; Saha, S.; Yang, G.-M.; Choi, H.; Cho, S.-G. The role of reactive oxygen species (ROS) in the biological activities of metallic nanoparticles. *Int. J. Mol. Sci.* **2017**, *18*, 120. [CrossRef] [PubMed]

147. Ahamed, M.; Karns, M.; Goodson, M.; Rowe, J.; Hussain, S.M.; Schlager, J.J.; Hong, Y. DNA damage response to different surface chemistry of silver nanoparticles in mammalian cells. *Toxicol. Appl. Pharmacol.* **2008**, *233*, 404–410. [CrossRef] [PubMed]

148. Ahamed, M.; Posgai, R.; Gorey, T.J.; Nielsen, M.; Hussain, S.M.; Rowe, J.J. Silver nanoparticles induced heat shock protein 70, oxidative stress and apoptosis in Drosophila melanogaster. *Toxicol. Appl. Pharmacol.* **2010**, *242*, 263–269. [CrossRef] [PubMed]

149. Hackenberg, S.; Scherzed, A.; Kessler, M.; Hummel, S.; Technau, A.; Froelich, K.; Ginzkey, C.; Koehler, C.; Hagen, R.; Kleinsasser, N. Silver nanoparticles: Evaluation of DNA damage, toxicity and functional impairment in human mesenchymal stem cells. *Toxicol. Lett.* **2011**, *201*, 27–33. [CrossRef] [PubMed]

150. Sung, J.H.; Ji, J.H.; Song, K.S.; Lee, J.H.; Choi, K.H.; Lee, S.H.; Yu, I.J. Acute inhalation toxicity of silver nanoparticles. *Toxicol. Ind. Health* **2011**, *27*, 149–154. [CrossRef]

151. Ji, J.H.; Jung, J.H.; Kim, S.S.; Yoon, J.-U.; Park, J.D.; Choi, B.S.; Chung, Y.H.; Kwon, I.H.; Jeong, J.; Han, B.S. Twenty-eight-day inhalation toxicity study of silver nanoparticles in Sprague-Dawley rats. *Inhal. Toxicol.* **2007**, *19*, 857–871. [CrossRef]

152. Stebounova, L.V.; Adamcakova-Dodd, A.; Kim, J.S.; Park, H.; O'Shaughnessy, P.T.; Grassian, V.H.; Thorne, P.S. Nanosilver induces minimal lung toxicity or inflammation in a subacute murine inhalation model. *Part. Fibre Toxicol.* **2011**, *8*, 5. [CrossRef]

153. Sung, J.H.; Ji, J.H.; Yoon, J.U.; Kim, D.S.; Song, M.Y.; Jeong, J.; Han, B.S.; Han, J.H.; Chung, Y.H.; Kim, J. Lung function changes in Sprague-Dawley rats after prolonged inhalation exposure to silver nanoparticles. *Inhal. Toxicol.* **2008**, *20*, 567–574. [CrossRef]

154. Hyun, J.-S.; Lee, B.S.; Ryu, H.Y.; Sung, J.H.; Chung, K.H.; Yu, I.J. Effects of repeated silver nanoparticles exposure on the histological structure and mucins of nasal respiratory mucosa in rats. *Toxicol. Lett.* **2008**, *182*, 24–28. [CrossRef]

155. Lee, H.-Y.; Choi, Y.-J.; Jung, E.-J.; Yin, H.-Q.; Kwon, J.-T.; Kim, J.-E.; Im, H.-T.; Cho, M.-H.; Kim, J.-H.; Kim, H.-Y. Genomics-based screening of differentially expressed genes in the brains of mice exposed to silver nanoparticles via inhalation. *J. Nanopart. Res.* **2010**, *12*, 1567–1578.

156. Takenaka, S.; Karg, E.; Roth, C.; Schulz, H.; Ziesenis, A.; Heinzmann, U.; Schramel, P.; Heyder, J. Pulmonary and systemic distribution of inhaled ultrafine silver particles in rats. *Environ. Health Perspect.* **2001**, *109*, 547. [PubMed]

157. Wiemann, M.; Vennemann, A.; Blaske, F.; Sperling, M.; Karst, U. Silver Nanoparticles in the Lung: Toxic Effects and Focal Accumulation of Silver in Remote Organs. *Nanomaterials* **2017**, *7*, 441. [CrossRef] [PubMed]

158. Braakhuis, H.M.; Gosens, I.; Krystek, P.; Boere, J.A.F.; Cassee, F.R.; Fokkens, P.H.B.; Post, J.A.; Van Loveren, H.; Park, M.V.D.Z. Particle size dependent deposition and pulmonary inflammation after short-term inhalation of silver nanoparticles. *Part. Fibre Toxicol.* **2014**, *11*, 49. [PubMed]

159. Campagnolo, L.; Massimiani, M.; Vecchione, L.; Piccirilli, D.; Toschi, N.; Magrini, A.; Bonanno, E.; Scimeca, M.; Castagnozzi, L.; Buonanno, G. Silver nanoparticles inhaled during pregnancy reach and affect the placenta and the foetus. *Nanotoxicology* **2017**, *11*, 687–698. [CrossRef]

160. Seiffert, J.; Buckley, A.; Leo, B.; Martin, N.G.; Zhu, J.; Dai, R.; Hussain, F.; Guo, C.; Warren, J.; Hodgson, A. Pulmonary effects of inhalation of spark-generated silver nanoparticles in Brown-Norway and Sprague–Dawley rats. *Respir. Res.* **2016**, *17*, 85. [CrossRef]

161. Silva, R.M.; Anderson, D.S.; Peake, J.; Edwards, P.C.; Patchin, E.S.; Guo, T.; Gordon, T.; Chen, L.C.; Sun, X.; Van Winkle, L.S. Aerosolized silver nanoparticles in the rat lung and pulmonary responses over time. *Toxicol. Pathol.* **2016**, *44*, 673–686.

162. Anderson, D.S.; Silva, R.M.; Lee, D.; Edwards, P.C.; Sharmah, A.; Guo, T.; Pinkerton, K.E.; Van Winkle, L.S. Persistence of silver nanoparticles in the rat lung: Influence of dose, size, and chemical composition. *Nanotoxicology* **2015**, *9*, 591–602. [CrossRef]

163. Silva, R.M.; Anderson, D.S.; Franzi, L.M.; Peake, J.L.; Edwards, P.C.; Van Winkle, L.S.; Pinkerton, K.E. Pulmonary effects of silver nanoparticle size, coating, and dose over time upon intratracheal instillation. *Toxicol. Sci.* **2015**, *144*, 151–162. [CrossRef]

164. Holland, N.A.; Becak, D.P.; Shannahan, J.H.; Brown, J.M.; Carratt, S.A.; Winkle, L.S.V.; Pinkerton, K.E.; Wang, C.M.; Munusamy, P.; Baer, D.R. Cardiac ischemia reperfusion injury following instillation of 20 nm citrate-capped nanosilver. *J. Nanomed. Nanotechnol.* **2015**, *6*. [CrossRef]

165. Holland, N.A.; Thompson, L.C.; Vidanapathirana, A.K.; Urankar, R.N.; Lust, R.M.; Fennell, T.R.; Wingard, C.J. Impact of pulmonary exposure to gold core silver nanoparticles of different size and capping agents on cardiovascular injury. *Part. Fibre Toxicol.* **2016**, *13*, 48. [CrossRef] [PubMed]

166. Ferdous, Z.; Al-Salam, S.; Greish, Y.E.; Ali, B.H.; Nemmar, A. Pulmonary exposure to silver nanoparticles impairs cardiovascular homeostasis: Effects of coating, dose and time. *Toxicol. Appl. Pharmacol.* **2019**, *367*, 36–50. [CrossRef] [PubMed]

167. Kim, W.-Y.; Kim, J.; Park, J.D.; Ryu, H.Y.; Yu, I.J. Histological study of gender differences in accumulation of silver nanoparticles in kidneys of Fischer 344 rats. *J. Toxicol. Environ. Health Part A* **2009**, *72*, 1279–1284. [CrossRef] [PubMed]

168. Nemmar, A.; Holme, J.A.; Rosas, I.; Schwarze, P.E.; Alfaro-Moreno, E. Recent advances in particulate matter and nanoparticle toxicology: A review of the in vivo and in vitro studies. *Biomed Res. Int.* **2013**, *2013*, 22. [CrossRef] [PubMed]

169. Loeschner, K.; Hadrup, N.; Qvortrup, K.; Larsen, A.; Gao, X.; Vogel, U.; Mortensen, A.; Lam, H.R.; Larsen, E.H. Distribution of silver in rats following 28 days of repeated oral exposure to silver nanoparticles or silver acetate. *Part. Fibre Toxicol.* **2011**, *8*, 18. [CrossRef]

170. Jeong, G.N.; Jo, U.B.; Ryu, H.Y.; Kim, Y.S.; Song, K.S.; Yu, I.J. Histochemical study of intestinal mucins after administration of silver nanoparticles in Sprague–Dawley rats. *Arch. Toxicol.* **2010**, *84*, 63. [CrossRef]

171. Shahare, B.; Yashpal, M.; Singh, G. Toxic effects of repeated oral exposure of silver nanoparticles on small intestine mucosa of mice. *Toxicol. Mech. Methods* **2013**, *23*, 161–167. [CrossRef]

172. Skalska, J.; Frontczak-Baniewicz, M.; Strużyńska, L. Synaptic degeneration in rat brain after prolonged oral exposure to silver nanoparticles. *Neurotoxicology* **2015**, *46*, 145–154. [CrossRef]

173. Blanco, J.; Tomás-Hernández, S.; García, T.; Mulero, M.; Gómez, M.; Domingc, J.L.; Sánchez, D.J. Oral exposure to silver nanoparticles increases oxidative stress markers in the liver of male rats and deregulates the insulin signalling pathway and p53 and cleaved caspase 3 protein expression. *Food Chem. Toxicol.* **2018**, *115*, 398–404. [CrossRef]

174. Kim, J.J.; Konkel, K.; McCulley, L.; Diak, I.-L. Cases of Argyria Associated With Colloidal Silver Use. *Ann. Pharmacother.* **2019**. [CrossRef]

175. Yang, L.; Kuang, H.; Zhang, W.; Aguilar, Z.P.; Wei, H.; Xu, H. Comparisons of the biodistribution and toxicological examinations after repeated intravenous administration of silver and gold nanoparticles in mice. *Sci. Rep.* **2017**, *7*, 3303. [CrossRef] [PubMed]

176. Heydrnejad, M.S.; Samani, R.J.; Aghaeivanda, S. Toxic effects of silver nanoparticles on liver and some hematological parameters in male and female mice (Mus musculus). *Biol. Trace Elem. Res.* **2015**, *165*, 153–158. [CrossRef] [PubMed]

177. Tiwari, R.; Singh, R.D.; Khan, H.; Gangopadhyay, S.; Mittal, S.; Singh, V.; Arjaria, N.; Shankar, J.; Roy, S.K.; Singh, D. Oral subchronic exposure to silver nanoparticles causes renal damage through apoptotic impairment and necrotic cell death. *Nanotoxicology* **2017**, *11*, 671–686.

178. Elle, R.E.; Gaillet, S.; Vide, J.; Romain, C.; Lauret, C.; Rugani, N.; Cristol, J.P.; Rouanet, J.M. Dietary exposure to silver nanoparticles in Sprague–Dawley rats: Effects on oxidative stress and inflammation. *Food Chem. Toxicol.* **2013**, *60*, 297–301.

179. Bergin, I.L.; Wilding, L.A.; Morishita, M.; Walacavage, K.; Ault, A.P.; Axson, J.L.; Stark, D.I.; Hashway, S.A.; Capracotta, S.S.; Leroueil, P.R. Effects of particle size and coating on toxicologic parameters, fecal elimination kinetics and tissue distribution of acutely ingested silver nanoparticles in a mouse model. *Nanotoxicology* **2016**, *10*, 352–360. [PubMed]

180. Garcia, T.; Lafuente, D.; Blanco, J.; Sánchez, D.J.; Sirvent, J.J.; Domingo, J.L.; Gómez, M. Oral subchronic exposure to silver nanoparticles in rats. *Food Chem. Toxicol.* **2016**, *92*, 177–187. [CrossRef]

181. Lafuente, D.; Garcia, T.; Blanco, J.; Sánchez, D.J.; Sirvent, J.J.; Domingo, J.L.; Gómez, M. Effects of oral exposure to silver nanoparticles on the sperm of rats. *Reprod. Toxicol.* **2016**, *60*, 133–139.

182. Skalska, J.; Dąbrowska-Bouta, B.; Strużyńska, L. Oxidative stress in rat brain but not in liver following oral administration of a low dose of nanoparticulate silver. *Food Chem. Toxicol.* **2016**, *97*, 307–315.

183. Martins, A.D.C., Jr.; Azevedo, L.F.; de Souza Rocha, C.C.; Carneiro, M.F.H.; Venancio, V.P.; de Almeida, M.R.; Antunes, L.M.G.; de Carvalho Hott, R.; Rodrigues, J.L.; Ogunjimi, A.T. Evaluation of distribution, redox parameters, and genotoxicity in Wistar rats co-exposed to silver and titanium dioxide nanoparticles. *J. Toxicol. Environ. Health Part A* **2017**, *80*, 1156–1165. [CrossRef]

184. Fennell, T.R.; Mortensen, N.P.; Black, S.R.; Snyder, R.W.; Levine, K.E.; Poitras, E.; Harrington, J.M.; Wingard, C.J.; Holland, N.A.; Pathmasiri, W. Disposition of intravenously or orally administered silver nanoparticles in pregnant rats and the effect on the biochemical profile in urine. *J. Appl. Toxicol.* **2017**, *37*, 530–544. [CrossRef]
185. Orlowski, P.; Zmigrodzka, M.; Tomaszewska, E.; Ranoszek-Soliwoda, K.; Czupryn, M.; Antos-Bielska, M.; Szemraj, J.; Celichowski, G.; Grobelny, J.; Krzyzowska, M. Tannic acid-modified silver nanoparticles for wound healing: The importance of size. *Int. J. Nanomed.* **2018**, *13*, 991. [CrossRef] [PubMed]
186. Zhang, S.; Liu, X.; Wang, H.; Peng, J.; Wong, K.K.Y. Silver nanoparticle-coated suture effectively reduces inflammation and improves mechanical strength at intestinal anastomosis in mice. *J. Pediatr. Surg.* **2014**, *49*, 606–613. [CrossRef] [PubMed]
187. García, A.A.; Rodríguez Martín, A.M.; Serra Baldrich, E.; Manubens Mercade, E.; Puig Sanz, L. Allergic contact dermatitis to silver in a patient treated with silver sulphadiazine after a burn. *J. Eur. Acad. Dermatol. Venereol.* **2016**, *30*, 365–366. [CrossRef] [PubMed]
188. Özkaya, E. A rare case of allergic contact dermatitis from silver nitrate in a widely used special patch test marker. *Contact Dermat.* **2009**, *61*, 120–122. [CrossRef] [PubMed]
189. Wang, M.; Lai, X.; Shao, L.; Li, L. Evaluation of immunoresponses and cytotoxicity from skin exposure to metallic nanoparticles. *Int. J. Nanomed.* **2018**, *13*, 4445. [CrossRef] [PubMed]
190. Zhu, Y.; Choe, C.-S.; Ahlberg, S.; Meinke, M.C.; Ulrike, A.; Lademann, J.M.; Darvin, M.E. Penetration of silver nanoparticles into porcine skin ex vivo using fluorescence lifetime imaging microscopy, Raman microscopy, and surface-enhanced Raman scattering microscopy. *J. Biomed. Opt.* **2014**, *20*, 051006. [CrossRef]
191. Korani, M.; Rezayat, S.M.; Bidgoli, S.A. Sub-chronic dermal toxicity of silver nanoparticles in guinea pig: Special emphasis to heart, bone and kidney toxicities. *Iran. J. Pharm. Res.* **2013**, *12*, 511.
192. Austin, C.A.; Hinkley, G.K.; Mishra, A.R.; Zhang, Q.; Umbreit, T.H.; Betz, M.W.; Wildt, B.E.; Casey, B.J.; Francke-Carroll, S.; Hussain, S.M. Distribution and accumulation of 10 nm silver nanoparticles in maternal tissues and visceral yolk sac of pregnant mice, and a potential effect on embryo growth. *Nanotoxicology* **2016**, *10*, 654–661. [CrossRef]
193. Wang, Z.; Qu, G.; Su, L.; Wang, L.; Yang, Z.; Jiang, J.; Liu, S.; Jiang, G. Evaluation of the biological fate and the transport through biological barriers of nanosilver in mice. *Curr. Pharm. Des.* **2013**, *19*, 6691–6697. [CrossRef]
194. Tiwari, D.K.; Jin, T.; Behari, J. Dose-dependent in-vivo toxicity assessment of silver nanoparticle in Wistar rats. *Toxicol. Mech. Methods* **2011**, *21*, 13–24. [CrossRef]
195. Xue, Y.; Zhang, S.; Huang, Y.; Zhang, T.; Liu, X.; Hu, Y.; Zhang, Z.; Tang, M. Acute toxic effects and gender-related biokinetics of silver nanoparticles following an intravenous injection in mice. *J. Appl. Toxicol.* **2012**, *32*, 890–899. [CrossRef] [PubMed]
196. Dziendzikowska, K.; Gromadzka-Ostrowska, J.; Lankoff, A.; Oczkowski, M.; Krawczyńska, A.; Chwastowska, J.; Sadowska-Bratek, M.; Chajduk, E.; Wojewódzka, M.; Dušinská, M. Time-dependent biodistribution and excretion of silver nanoparticles in male Wistar rats. *J. Appl. Toxicol.* **2012**, *32*, 920–928. [CrossRef] [PubMed]
197. Zhang, Y.; Ferguson, S.A.; Watanabe, F.; Jones, Y.; Xu, Y.; Biris, A.S.; Hussain, S.; Ali, S.F. Silver nanoparticles decrease body weight and locomotor activity in adult male rats. *Small* **2013**, *9*, 1715–1720. [CrossRef] [PubMed]
198. Garcia, T.X.; Costa, G.M.J.; Franca, L.R.; Hofmann, M.-C. Sub-acute intravenous administration of silver nanoparticles in male mice alters Leydig cell function and testosterone levels. *Reprod. Toxicol.* **2014**, *45*, 59–70. [CrossRef]
199. Guo, H.; Zhang, J.; Boudreau, M.; Meng, J.; Yin, J.J.; Liu, J.; Xu, H. Intravenous administration of silver nanoparticles causes organ toxicity through intracellular ROS-related loss of inter-endothelial junction. *Part. Fibre Toxicol.* **2016**, *13*, 21. [CrossRef]
200. Wen, H.; Dan, M.; Yang, Y.; Lyu, J.; Shao, A.; Cheng, X.; Chen, L.; Xu, L. Acute toxicity and genotoxicity of silver nanoparticle in rats. *PLoS ONE* **2017**, *12*, e0185554. [CrossRef]
201. Vidanapathirana, A.K.; Thompson, L.C.; Herco, M.; Odom, J.; Sumner, S.J.; Fennell, T.R.; Brown, J.M.; Wingard, C.J. Acute intravenous exposure to silver nanoparticles during pregnancy induces particle size and vehicle dependent changes in vascular tissue contractility in Sprague Dawley rats. *Reprod. Toxicol.* **2018**, *75*, 10–22. [CrossRef]

202. Singh, A.V.; Laux, P.; Luch, A.; Sudrik, C.; Wiehr, S.; Wild, A.-M.; Santomauro, G.; Bill, J.; Sitti, M. Review of emerging concepts in nanotoxicology: Opportunities and challenges for safer nanomaterial design. *Toxicol. Mech. Methods* **2019**, *29*, 378–387. [CrossRef]

203. Holland, N.A. Intratracheal Instillation of Silver Nanoparticles Exacerbates Cardiac Ischemia/Reperfusion Injury in Male Sprague-Dawley Rats. Ph.D. Thesis, East Carolina University, Greenville, NC, USA, 2014.

204. Carvalho, P.M.; Felício, M.R.; Santos, N.C.; Gonçalves, S.; Domingues, M. Application of light scattering techniques to nanoparticle characterization and development. *Front. Chem.* **2018**, *6*, 237. [CrossRef]

205. Lee, J.H.; Sung, J.H.; Ryu, H.R.; Song, K.S.; Song, N.W.; Park, H.M.; Shin, B.S.; Ahn, K.; Gulumian, M.; Faustman, E.M. Tissue distribution of gold and silver after subacute intravenous injection of co-administered gold and silver nanoparticles of similar sizes. *Arch. Toxicol.* **2018**, *92*, 1393–1405. [CrossRef]

206. Liu, J.; Wang, Z.; Liu, F.D.; Kane, A.B.; Hurt, R.H. Chemical transformations of nanosilver in biological environments. *ACS Nano* **2012**, *6*, 9887–9899. [CrossRef] [PubMed]

207. Hooper, L.G.; Kaufman, J.D. Ambient air pollution and clinical implications for susceptible populations. *Ann. Am. Thorac. Soc.* **2018**, *15*, S64–S68. [CrossRef] [PubMed]

208. Faustini, A.; Renzi, M.; Kirchmayer, U.; Balducci, M.; Davoli, M.; Forastiere, F. Short-term exposure to air pollution might exacerbate autoimmune diseases. *Environ. Epidemiol.* **2018**, *2*, e025. [CrossRef]

209. Ramirez-Lee, M.A.; Aguirre-Bañuelos, P.; Martinez-Cuevas, P.P.; Espinosa-Tanguma, R.; Chi-Ahumada, E.; Martinez-Castañon, G.A.; Gonzalez, C. Evaluation of cardiovascular responses to silver nanoparticles (AgNPs) in spontaneously hypertensive rats. *Nanomedicine* **2018**, *14*, 385–395. [CrossRef] [PubMed]

210. Jia, J.; Li, F.; Zhou, H.; Bai, Y.; Liu, S.; Jiang, Y.; Jiang, G.; Yan, B. Oral exposure to silver nanoparticles or silver ions may aggravate fatty liver disease in overweight mice. *Environ. Sci. Technol.* **2017**, *51*, 9334–9343. [CrossRef]

211. Kermanizadeh, A.; Jacobsen, N.R.; Roursgaard, M.; Loft, S.; Møller, P. Hepatic hazard assessment of silver nanoparticle exposure in healthy and chronically alcohol fed mice. *Toxicol. Sci.* **2017**, *158*, 176–187. [CrossRef]

212. Hartmann, S.; Louch, R.; Zeumer, R.; Steinhoff, B.; Mozhayeva, D.; Engelhard, C.; Schönherr, H.; Schlechtriem, C.; Witte, K. Comparative multi-generation study on long-term effects of pristine and wastewater-borne silver and titanium dioxide nanoparticles on key lifecycle parameters in Daphnia magna. *NanoImpact* **2019**, *14*, 100163. [CrossRef]

213. Raj, A.; Shah, P.; Agrawal, N. Dose-dependent effect of silver nanoparticles (AgNPs) on fertility and survival of Drosophila: An In-Vivo study. *PLoS ONE* **2017**, *12*, e0178051. [CrossRef]

214. Larese, F.F.; D'Agostin, F.; Crosera, M.; Adami, G.; Renzi, N.; Bovenzi, M.; Maina, G. Human skin penetration of silver nanoparticles through intact and damaged skin. *Toxicology* **2009**, *255*, 33–37. [CrossRef]

215. Munger, M.A.; Radwanski, P.; Hadlock, G.C.; Stoddard, G.; Shaaban, A.; Falconer, J.; Grainger, D.W.; Deering-Rice, C.E. In vivo human time-exposure study of orally dosed commercial silver nanoparticles. *Nanomedicine* **2014**, *10*, 1–9. [CrossRef]

International Journal of
Molecular Sciences

Article

Role of Autophagy in Zinc Oxide Nanoparticles-Induced Apoptosis of Mouse LEYDIG Cells

Jingcao Shen [1,†], Dan Yang [1,†], Xingfan Zhou [2], Yuqian Wang [2], Shichuan Tang [2], Hong Yin [3], Jinglei Wang [1], Rui Chen [2,*] and Jiaxiang Chen [1,4,*]

1   Department of Physiology, Medical College of Nanchang University, Nanchang 330006, China
2   Key Laboratory of Occupational Health and Safety, Beijing Municipal Institute of Labor Protection, Beijing 100054, China
3   School of Aerospace, Mechanical and Manufacturing Engineering, RMIT University, Bundoora, VIC 3083, Australia
4   Jiangxi Provincial Key Laboratory of Reproductive Physiology and Pathology, Nanchang 330006, China
*   Correspondence: chenxiaorui@hotmail.com (R.C.); chenjiaxiang@ncu.edu.cn (J.C.); Tel.: +86-10-6352-3998 (R.C.); Tel./Fax: +86-791-8636-0556 (J.C.); Fax: +86-10-6352-3877 (R.C.)
†   These authors contributed equally to this work.

Received: 28 July 2019; Accepted: 16 August 2019; Published: 19 August 2019

Abstract: Zinc oxide nanoparticles (ZnO NPs) have shown adverse health impact on the human male reproductive system, with evidence of inducing apoptosis. However, whether or not ZnO NPs could promote autophagy, and the possible role of autophagy in the progress of apoptosis, remain unclear. In the current study, in vitro and in vivo toxicological responses of ZnO NPs were explored by using a mouse model and mouse Leydig cell line. It was found that intragastrical exposure of ZnO NPs to mice for 28 days at the concentrations of 100, 200, and 400 mg/kg/day disrupted the seminiferous epithelium of the testis and decreased the sperm density in the epididymis. Furthermore, serum testosterone levels were markedly reduced. The induction of apoptosis and autophagy in the testis tissues was disclosed by up-regulating the protein levels of cleaved Caspase-8, cleaved Caspase-3, Bax, LC3-II, Atg 5, and Beclin 1, accompanied by down-regulation of Bcl 2. In vitro tests showed that ZnO NPs could induce apoptosis and autophagy with the generation of oxidative stress. Specific inhibition of autophagy pathway significantly decreased the cell viability and up-regulated the apoptosis level in mouse Leydig TM3 cells. In summary, ZnO NPs can induce apoptosis and autophagy via oxidative stress, and autophagy might play a protective role in ZnO NPs-induced apoptosis of mouse Leydig cells.

Keywords: ZnO NPs; Leydig cells; apoptosis; autophagy; oxidative stress

1. Introduction

Nanotechnology manipulates matters at the atomic, molecular, and supramolecular scales and has grown rapidly worldwide in the past decades. With the development of nanotechnology, environmental exposure to nanoparticles (NPs) is increasing dramatically [1,2]. Metal oxide nanoparticles are the most abundantly produced types of engineered nanomaterials in industry [3]. Among them, zinc oxide nanoparticles (ZnO NPs) are used in various applications, such as cosmetics, rubber manufacture, pigments, food additives, biosensors, chemical fibers, bioimaging, and antibacterial agents, due to their low production cost and unique physicochemical properties [4]. ZnO NPs may enter human bodies by various routes, including inhalation, dermal penetration, injection, and ingestion [5]. These NPs can then accumulate in various organs, such as the liver, spleen, lungs, kidney, and heart via circulation, and may produce adverse consequences, such as edema and degeneration of hepatocytes, inflammation of the pancreas, or damage to the stomach and spleen [6,7]. Consequently, toxicity research and health risk assessments of ZnO NPs have attracted tremendous attention recently [8].

Tissue damage due to NPs exposure arises from direct cell–NPs interaction and is associated with local concentrations of exogenous substances in the tissues, i.e., the nanoparticle itself or solubilized ions [9,10]. Previous researches show that zinc ion dissolved from the surface of ZnO NPs is a primary reason for its cytotoxicity [11,12]. Reducing the ion release from the surface, such as by pre-coating with a protein corona, could greatly decrease their cytotoxicity [13,14]. Furthermore, our previous research illustrates that ZnO NPs and their soluble ions can induce significant cellular endoplasmic reticulum (ER) stress responses before triggering ER-related apoptosis [12]. Generation of reactive oxygen species (ROS) is generally involved in cellular damage from the exposure to ZnO NPs [12,13,15]. ROS are chemically-reactive molecules containing oxygen, which are generated as by-products of biological oxidation during mitochondrial respiration under physiological conditions. ROS include both free radicals, such as nitric oxide (NO), superoxide ($O_2^{\bullet-}$), and hydroxyl radical ($^{\bullet}OH$), and peroxides [16]. Reduced glutathione (GSH) and antioxidant enzymes such as glutathione peroxidase (GSH-PX) and superoxide dismutase (SOD) are normally used to scavenge ROS. Oxidative stress occurs when there is an imbalance between ROS production and the cellular antioxidant defence system [17,18].

Spermatogenesis consists of highly organized and sequential steps of undifferentiated spermatogonial stem cell proliferation and differentiation, which generates functional sperms in the testis [19–21]. ZnO NPs can cause vacuolization of germinal epithelium and sloughing of germ, and even decrease the sperm number and motility in the epididymis [22]. In addition to Sertoli cells, Leydig cells play an important role in maintaining spermatogenesis and are prone to being affected by various chemicals [23,24]. ZnO NPs have been reported to exert cytotoxic effects on mouse Leydig cells [25,26]. Similar toxic effects were revealed in the testis of six-month-old common carp Cyprinus carpio after exposure to 10, 50, and 100 µg/L ZnO NPs for 21 days [27]. Furthermore, increasing evidence suggests that the toxicity of ZnO NPs may result from ROS production [9,28,29].

Autophagy is an evolutionarily-conserved, highly-regulated lysosomal degradative pathway involving the delivery of cytoplasmic cargo to the lysosome, which occurs at low basal levels to perform protein and organelle turnover in normal situations [30,31]. Autophagy can be induced during starvation or growth factor withdrawal in order to generate more intracellular nutrients and energy [32]. Autophagy can also be induced under stressful conditions such as neurodegenerative diseases, pathogen infections, chemotherapy, and chemical exposure [33–37]. Increasing evidence has shown that ZnO NPs can induce autophagy in immune cells, normal skin cells, gastrointestinal tract cells, and kidney tissue [7,38–40]. Until now, there has been no evidence that ZnO NPs exposure could induce autophagy in testis tissue.

The aims of the present study were to investigate whether oxidative stress was involved in ZnO NPs-induced apoptosis and autophagy of mouse Leydig cells, and to determine the role of autophagy in ZnO NPs-induced apoptosis. These results will provide fundamental understanding of ZnO NPs-induced spermatogenesis failure.

2. Results

2.1. Characteristics and Morphology of ZnO NPs

Transmission electron microscopy (TEM) test shows the primary size of ZnO NPs is about 30 nm with a propensity to agglomerate (Figure S2). These characteristics are comparable to previous publications using the same nanoparticles [41,42]. The hydrodynamic sizes and zeta potentials of ZnO NPs suspended in water are 66.36 ± 0.39 nm (PDI = 0.167, n = 3) and 38.25 ± 1.06 mV (n = 3), respectively.

2.2. ZnO NPs Cause Testis Damage to Male Mice

As shown in Figure 1A, the testes of vehicle-treated mice showed normal seminiferous tubules lined with both spermatogenic cells and Sertoli cells. No detached germ cells were found in the tubular lumen. In the 100 mg/kg/day ZnO NPs exposure group, no significant morphologic changes were observed at the seminiferous epithelium. However, the seminiferous tubule demonstrated mildly disorganized histo-architecture in the 200 mg/kg/day group. In the 400 mg/kg/day group, seminiferous tubules exhibited disintegration of the germinal epithelium, germ cell depletion, and a reduction in round sperm. There was a significant decrease in sperm density of the epididymis after exposure to 100, 200, or 400 mg ZnO NPs/kg/day compared to the vehicle control group (Figure 1B), indicating that ZnO NPs exposure significantly inhibited spermatogenesis.

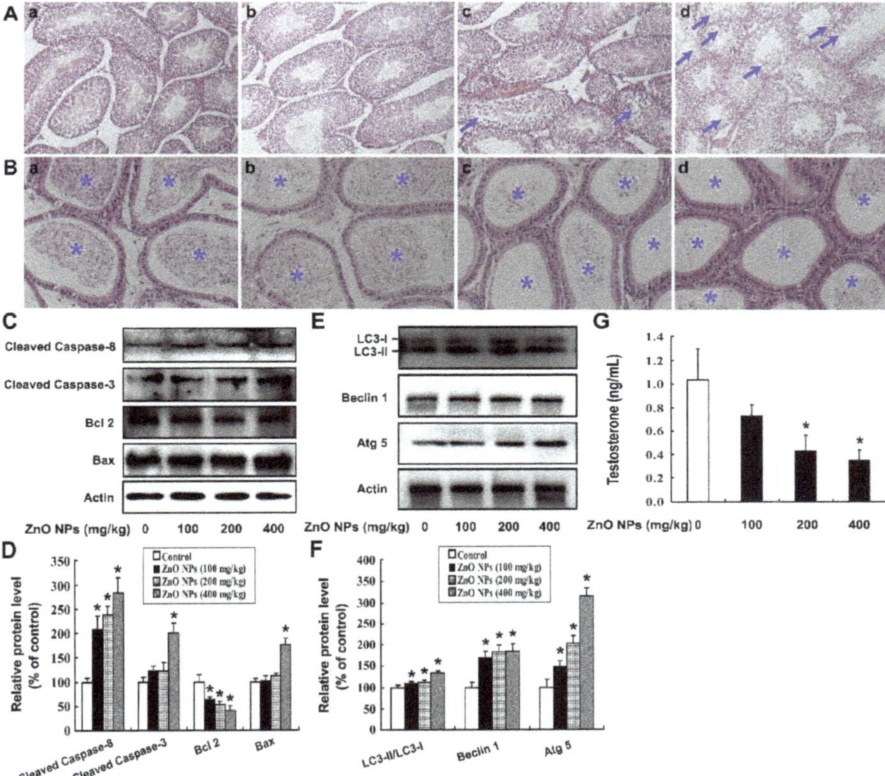

Figure 1. Intragastrical exposure of zinc oxide nanoparticles (ZnO NPs) cause toxic damage to the mouse male reproductive system. (**A**) Testes were obtained from male mice treated with 0 (**a**), 100 (**b**), 200 (**c**), or 400 (**d**) mg ZnO NPs/kg/day for 28 days. The testes were stained with hematoxylin and eosin (HE) and then were visualized under an IX51 Olympus microscope. The disruption of the seminiferous epithelium in the testis is indicated by arrows. Magnification: 100×. (**B**) Epididymides were obtained from male mice treated with 0 (**a**), 100 (**b**), 200 (**c**), or 400 (**d**) mg ZnO NPs/kg/day for 28 days, and stained with HE. The sperm in the epididymis are indicated by an asterisk. Magnification: 200×. (**C**) The protein levels of cleaved Caspase-3, cleaved Caspase-8, Bax, and Bcl 2 and (**E**) the levels of LC3, Beclin 1, and Atg 5 were detected by Western blot; Actin was used as an internal control. (**D,F**) The relative protein levels were quantified by densitometry. (**G**) The serum testosterone concentration. The experiment was done in triplicate and repeated three times ($n = 9$). Data were analyzed by one-way ANOVA. * $p < 0.05$.

To further investigate the potential mechanism of ZnO NPs-induced spermatogenesis failure, the apoptosis level in the mouse testis tissues was assessed. As can be seen from Figure 1C,D, ZnO NPs significantly increased the levels of apoptosis-related proteins, including cleaved Caspase-8, cleaved Caspase-3 and Bax, along with a decreased protein level of Bcl 2 in the testis tissue, which indicates that ZnO NPs induced apoptosis of the testis tissue. Additionally, ZnO NPs markedly increased the ratio of LC3-II/LC3-I, as well as the levels of autophagy proteins Atg 5 and Beclin 1, indicating that ZnO NPs induced autophagy of the testis tissue (Figure 1E,F). Furthermore, ZnO NPs decreased the serum testosterone concentration in a dose-dependent manner ($p < 0.05$), which implies that ZnO NPs disrupted the physiological function of the male reproductive system by targeting the Leydig cells (Figure 1G).

2.3. ZnO NPs Induce Apoptosis of Mouse Leydig TM3 Cells

The content of testosterone dramatically decreased in the ZnO NPs-treated groups, which implies that ZnO NPs might cause damage to Leydig cells. To further verify the hypothesis, mouse Leydig TM3 cell line was utilized as an in vitro model. As shown in Figure 2A, ZnO NPs at concentrations of 3, 4, and 8 µg/mL significantly inhibited cell viability. Further tests showed that the cell viability was further suppressed at time points of 24, 48, and 72 h post-exposure to 4 µg/mL ZnO NPs (Figure 2B). To determine whether the anti-proliferative effect of ZnO NPs resulted from apoptosis, the apoptosis-related proteins were investigated, including cleaved Caspase-8, cleaved Caspase-3, Bcl 2 and Bax, after the cells were incubated with 0, 2, 3, and 4 µg/mL ZnO NPs for 24 h. It was shown that ZnO NPs dramatically increased the protein levels of cleaved Caspase-8, cleaved Caspase-3, and Bax, as well as decreased Bcl 2 protein level (Figure 2C,D). Furthermore, ZnO NPs increased the numbers of AnnexinV-FITC positive staining cells (Figure 2E). These results indicate that ZnO NPs induced apoptosis of mouse Leydig TM3 cells.

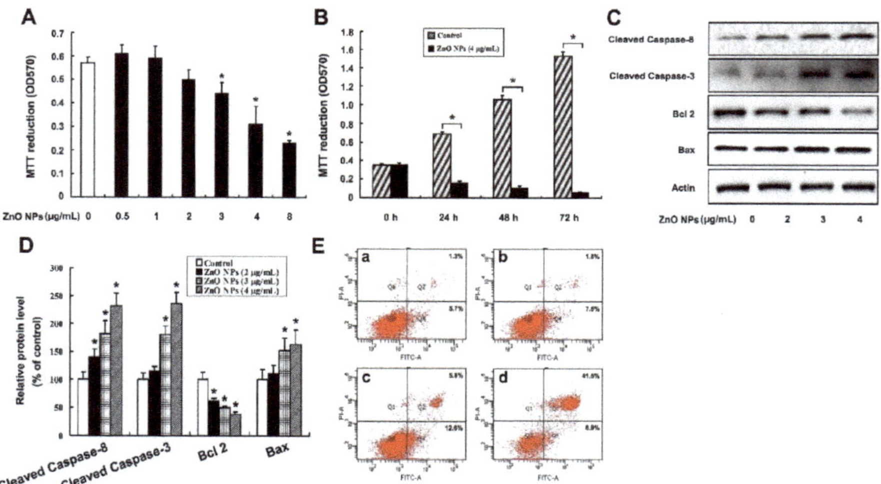

Figure 2. ZnO NPs induce apoptosis in mouse Leydig TM3 cells. 3-(4,5-dimethylthiazol-2-yl)-2,5-diphenyltetrazolium bromide (MTT) assay results of mouse Leydig TM3 cells treated with 0–8 µg/mL ZnO NPs for 24 h (**A**) or treated with 4 µg/mL ZnO NPs for 24~72 h (**B**). (**C**) The cells were treated with 0–4 µg/mL ZnO NPs for 24 h; then, the protein levels of cleaved Caspase-3, cleaved Caspase-8, Bcl 2, and Bax were investigated by Western blot; Actin was used as an internal control. (**D**) The relative protein levels were quantified by densitometry. (**E**) The cells were treated with 0 (**a**), 2 (**b**), 3 (**c**), 4 (**d**) µg/mL ZnO NPs for 24 h, then the AnnexinV-FITC positive staining cells were counted by flow cytometry. The experiment was done in triplicate and repeated three times. Data were analyzed by one-way ANOVA. * $p < 0.05$.

2.4. ZnO NPs Induce Apoptosis through Activation of Oxidative Stress

In order to investigate whether oxidative stress was involved in ZnO NPs-induced apoptosis of mouse Leydig TM3 cells, the contents of malondialdehyde (MDA) and GSH and the enzyme activities of SOD and GSH-PX were determined after the cells were treated with ZnO NPs for 24 h. As shown in Figure 3, ZnO NPs significantly increased MDA level in the cells in a dose-dependent manner, whereas the content of GSH and the activities of the antioxidant enzymes SOD and GSH-PX were decreased in the ZnO NPs-treated cells, which implies that ZnO NPs induced oxidative stress in mouse Leydig TM3 cells. The same markers were detected after the cells were treated with H_2O_2 and the cell viability was significantly inhibited with the induction of apoptosis, suggesting that oxidative stress could induce apoptosis of mouse Leydig TM3 cells (Figure S3). The mouse Leydig TM3 cells were treated with 4 µg/mL ZnO NPs for 24 h in the presence or absence of 5 mM NAC, an inhibitor of ROS, to further confirm the role of oxidative stress in ZnO NPs-induced apoptosis. As shown in Figure 4, inhibition of viability and induction of apoptosis by ZnO NPs was significantly rescued by NAC. These results illustrate that oxidative stress was involved in ZnO NPs-induced apoptosis of mouse Leydig TM3 cells.

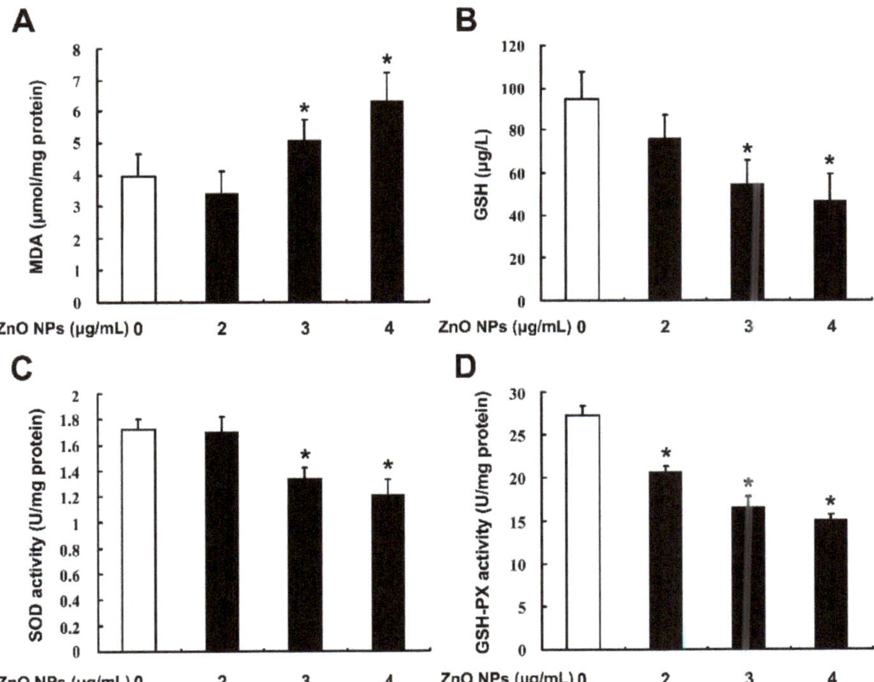

Figure 3. ZnO NPs induce oxidative stress in mouse Leydig TM3 cells. Mouse Leydig TM3 cells were treated with 0–4 µg/mL ZnO NPs for 24 h; then the contents of MDA (**A**) and GSH (**B**) and the enzyme activities of SOD (**C**) and GSH-PX (**D**) were determined. The experiment was done in triplicate and repeated three times. Data were analyzed by one-way ANOVA. * $p < 0.05$.

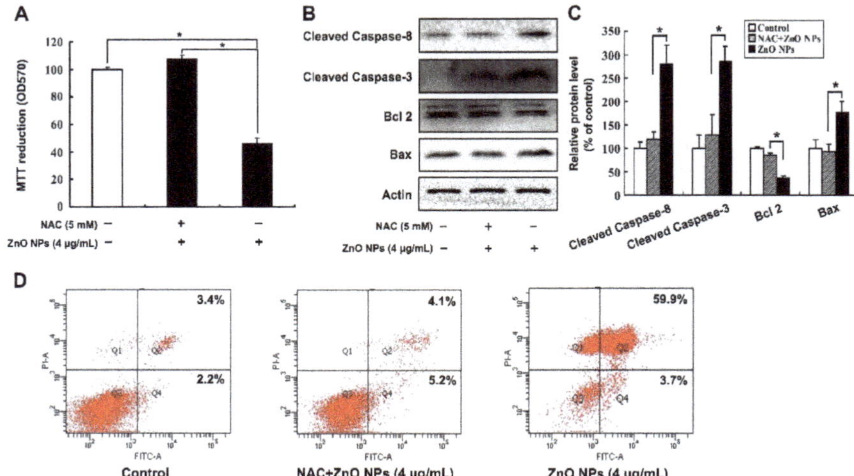

Figure 4. Oxidative stress is involved in ZnO NPs-induced apoptosis of mouse Leydig TM3 cells. Mouse Leydig TM3 cells were treated with 4 µg/mL ZnO NPs for 24 h in the absence or presence of 5 mM NAC, then cell viability (**A**), the protein levels of cleaved Caspase-8, cleaved Caspase-3, Bcl 2, and Bax (**B**) and the AnnexinV-FITC positive staining cells (**D**) were detected by MTT assay, Western blot, and flow cytometry, respectively. (**C**) The relative protein levels of cleaved Caspase-8, cleaved Caspase-3, Bcl 2, and Bax were quantified by densitometry. The experiment was done in triplicate and repeated three times. Data were analyzed by one-way ANOVA. * $p < 0.05$.

2.5. Oxidative Stress is Involved in ZnO NPs-Induced Autophagy

As shown in Figure 5A,B, ZnO NPs increased the ratio of LC3-II to LC3-I, as well as the protein levels of Atg 5 and Beclin 1. Similarly, H_2O_2 markedly increased the ratio of LC3-II to LC3-I and the contents of Atg 5 and Beclin 1, indicating that oxidative stress could induce autophagy of mouse Leydig TM3 cells (Figure S4). Furthermore, inhibition of oxidative stress could rescue the induction of autophagy by ZnO NPs (Figure 5C,D). ZnO NPs-induced autophagy was further investigated by TEM. As shown in Figure 5E, there were relatively few autophagosomes in the cytoplasm of the control cells, while autophagic vacuoles containing extensively-degraded organelles (such as mitochondria and endoplasmic reticulum) significantly increased in both ZnO NPs-treated cells and starvation-treated cells. Interestingly, inhibition of oxidative stress decreased the number of autophagosomes. These results suggest that oxidative stress played an important role in ZnO NPs-induced autophagy. It is worthy to note that ZnO NPs-induced autophagy and apoptosis of mouse Leydig TM3 cells might be closely related to the soluble zinc ions, as similar bio-effects were observed after the cells were treated with 0–1 µg/mL $ZnCl_2$ (Figure S5).

2.6. Inhibition of Autophagy Increases ZnO NP-Induced Apoptosis

As apoptosis and autophagy were both induced by ZnO NPs, the effects of autophagy on ZnO NPs-induced apoptosis were studied. Cell viability was measured after the cells were treated with 4 µg/mL ZnO NPs for 24 h in the absence or presence of an autophagy inhibitor, either 10 mM 3-Methyladenine (3-MA) or 1 µM Wortmannin (Wort). Compared with the ZnO NPs-treated cells, inhibition of autophagy further decreased viability of mouse Leydig TM3 cells (Figure 6A) and up-regulated the protein levels of cleaved Caspase-8, cleaved Caspase-3, and Bax, accompanied by the down-regulation of Bcl 2 protein (Figure 6B,C). The number of AnnexinV-FITC positive staining cells were also markedly increased when autophagy was inhibited (Figure 6D). These results indicate that autophagy might play a protective role in ZnO NPs-induced apoptosis of mouse Leydig TM3 cells.

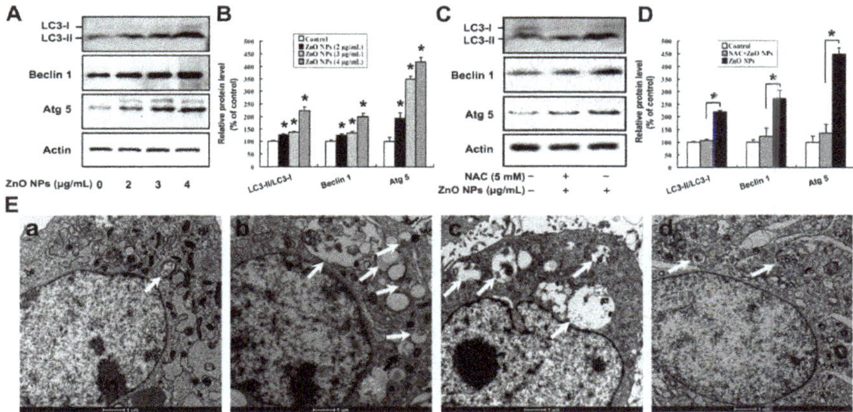

Figure 5. Oxidative stress is involved in ZnO NPs-induced autophagy of mouse Leydig TM3 cells. Mouse Leydig TM3 cells were treated with 0–4 µg/mL ZnO NPs for 24 h (**A**) or treated with 4 µg/mL ZnO NPs for 24 h in absence or presence of 5 mM NAC (**C**); then, the protein levels of LC 3, Atg 5, and Beclin 1 were quantified by Western blot. (**B,D**) The relative protein levels of LC 3, Atg 5, and Beclin 1 were quantified by densitometry. (**E**) The cells were treated with ddH$_2$O, bars: 1 µm, (**a**), 4 µg/mL ZnO NPs (**b**), or 5 mM N-acetyl-L-cysteine (NAC) plus 4 µg/mL ZnO NPs for 24 h (**d**). Then, autophagic vacuoles in the cells were visualized by transmission electron microscopy (TEM), with starvation-treated cells as a positive control (**c**). The autophagic vacuoles are indicated by white arrows. The experiment was done in triplicate and repeated three times. Data were analyzed by one-way ANOVA. * $p < 0.05$.

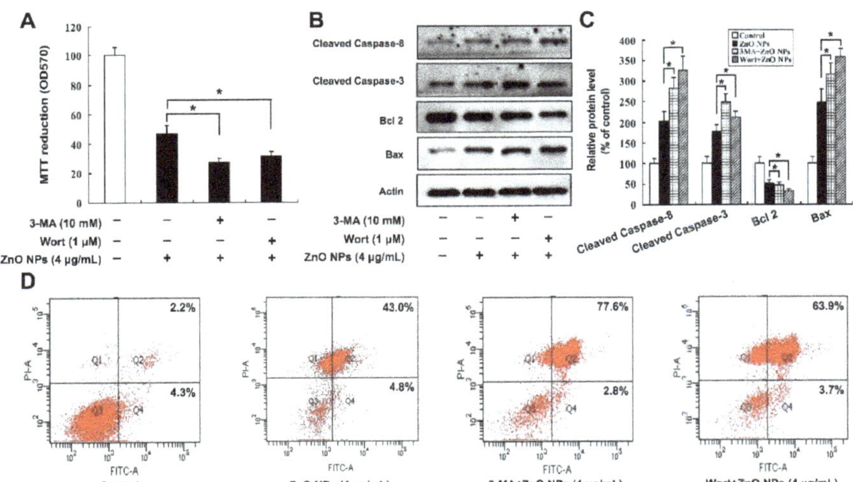

Figure 6. Inhibition of autophagy increases the ZnO NPs-induced apoptosis level in mouse Leydig TM3 cells. Mouse Leydig TM3 cells were treated with 4 µg/mL ZnO NPs for 24 h in the absence or presence of 10 mM 3-MA or 1 µM Wortmannin (Wort), then cell viability (**A**), the protein levels of cleaved Caspase-8, cleaved Caspase-3, Bcl 2, and Bax (**B**) and the AnnexinV-FITC positive staining cells (**D**) were tested by MTT assay, Western blot, and flow cytometry, respectively. (**C**) The relative protein levels of cleaved Caspase-8, cleaved Caspase-3, Bcl 2, and Bax were quantified by densitometry. The experiment was done in triplicate and repeated three times. Data were analyzed by one-way ANOVA. * $p < 0.05$.

3. Discussion

Exposure to ZnO NPs for humans is inevitable due to their wide applications in commercial and industrial products. Thus, the adverse effects from the exposure of ZnO NPs need clear definition. Nanoparticles can pass through the blood–brain barrier (BBB), blood–testis barrier (BTB), and blood–air barrier (BAB), with the ability to accumulate in the brain, the testis, or peripheral organs [43–46]. Recently, Qian et al. showed that ZnO NPs could cause adverse effects throughout the male reproductive system by impairing the BTB [47]. Oral dose toxicity has been reported in SD mice after repetitive exposure to positively-charged 100 nm ZnO NPs over 14 or 90 days, and the target organs were found to be the spleen, stomach, and pancreas, with a no-observed-adverse-effect dose level of about 125 mg/kg (b.w.) [48,49]. In our research, it was shown that 28-day gavage exposure of ZnO NPs (30 nm positively charged) at the concentrations of 100, 200, and 400 mg/kg/day caused disruption and atrophy of the seminiferous epithelium in the testis of mice. Furthermore, the sperm density in the epididymis significantly decreased in the ZnO NPs-treated groups, which was in good agreement with some previous work [22,27]. This toxic dosage range is also similar to the research of Hong et al., in which they tested the toxicity on embryo-fetal development in rats from 15 days of repeated oral doses of 20 nm negatively-charged ZnO NPs [50]. Therefore, ZnO NPs, with the high chance of daily contact and exposure, may pose a high risk of reproductive toxicity after long-term accumulation in the human body.

ZnO NPs have been shown to induce apoptosis in many cells such as human epidermal keratinocytes, human aortic endothelial cells, human liver and kidney podocytes [51–54]. Han et al. showed that ZnO NPs took cytotoxic effects on mouse testicular cells and induced apoptosis in Leydig cells [25]. In this research, we confirmed that ZnO NPs up-regulated the protein levels of Bax, cleaved Caspase-3, and cleaved Caspase-8 in the testis tissue, as well as decreased the protein level of Bcl 2, which indicates that ZnO NPs could induce apoptosis in the testis.

Autophagy protein LC3, a widely used marker of mammalian autophagy, has two forms, i.e., a cytosolic form (LC3-I) and an autophagic vesicle-associated form (LC3-II). During induction of autophagy, LC3-I covalently conjugates with phosphatidylethanolamine and develops LC3-II, which is recruited and bound to the autophagosome membrane [33,55]. The conversion of LC3-I to LC3-II is considered to be a crucial step in initiating autophagy [56], with the amount of LC3-II related to the extent of autophagosome formation [55]. In the present study, ZnO NPs exposure significantly increased the ratio of LC3-II to LC3-I in the testis tissue, along with similar up-regulation of autophagy proteins Atg 5 and Beclin 1. These results implied that ZnO NPs could induce autophagy in the testis tissue.

The primary function of Leydig cells is the synthesis and secretion of androgen, which plays an important role in spermatogenesis [57]. In our study, ZnO NPs could decrease serum testosterone level, indicating that Leydig cells might be the target for ZnO NPs-induced spermatogenesis failure. To further verify this hypothesis, the mouse Leydig TM3 cell line was utilized as an in vitro research model. In agreement with the in vivo findings, ZnO NPs exposure inhibited viability and induced apoptosis of mouse Leydig TM3 cells. Thus, it is reasonable to speculate that the inhibition of cell viability upon ZnO NPs exposure might result from the induction of apoptosis.

Oxidative stress has been identified as a critical pathophysiological mechanism of reproductive toxicity from environmental chemicals or organophosphorus compounds [58]. Asani et al. showed that ZnO NPs could induce oxidative stress in pancreatic β-cells [59]. Similar to the toxicity effects of H_2O_2, exposure to ZnO NPs significantly increased the MDA in the cells, along with a marked decrease in both the GSH levels and the enzyme activities of SOD and GSH-PX. Further evidence demonstrates that apoptosis could be distinctly reduced when oxidative stress was inhibited, which confirmed that oxidative stress was involved in ZnO NPs-induced apoptosis of mouse Leydig TM3 cells. Oxidative stress has been shown to induce autophagy and plays an important role in chemical-induced autophagy [21,60]. In the current study, ZnO NPs exposure induced autophagy of mouse Leydig TM3 cells, which could be inhibited by NAC, a scavenger of ROS. Collectively, these results provide clear evidence that oxidative stress was critical in ZnO NPs-induced autophagy in mouse Leydig TM3 cells.

Both cell survival and death can be related to autophagy when the cells are subject to stressful conditions. In most circumstances, autophagy will promote cell survival [35,61]. However, autophagy is also considered to be a form of non-apoptotic programmed cell death—"type II" or "autophagic" cell death [62,63]. To investigate the role of autophagy in ZnO NPs-induced apoptosis, apoptosis was measured after the treatment of ZnO NPs in the absence or presence of autophagy inhibitor. Surprisingly, inhibition of autophagy could further induce apoptosis of mouse Leydig TM3 cells (Figure 7). These results illustrate that autophagy plays a cytoprotective role in ZnO NPs-induced apoptosis of mouse Leydig TM3 cells.

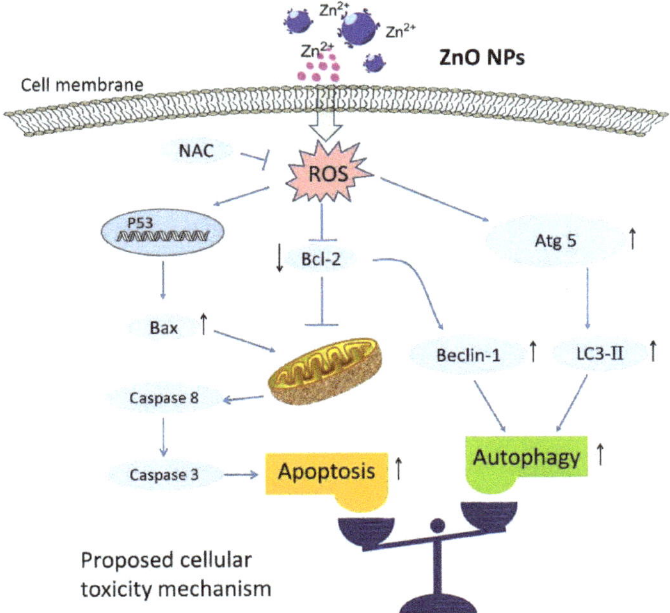

Figure 7. Schematic representation of the activation mechanism of apoptosis and cytoprotective autophagy in mouse Leydig cells after ZnO NP exposure. The up-regulation expression of protein is indicated by up arrow (↑), and down-regulation expression is indicated by down arrow (↓) in schematic illustration.

4. Materials and Methods

4.1. Reagents

ZnO NPs (No. 721077), N-acetyl-L-cysteine (A7250), 3-Methyladenine (M9281), and Wortmannin (12-338) were obtained from Sigma (St. Louis, MO, USA). Mouse Leydig cell line (TM3) was obtained from the Cell Culture Center of the Institute of Basic Medical Science, Chinese Academy of Medical Sciences (Beijing, China). Anti-Caspase-3 (sc-7148), anti-Caspase-8 (sc-7890), anti-Bax (sc-493), anti-Bcl-2 (sc-492), and anti-β-actin (sc-69879) were purchased from Santa Cruz Biotechnology (Santa Cruz, CA, USA). Anti-LC3 (PD014), anti-Atg5 (PM050), and anti-Beclin-1 (PD017) were gained from MBL Co. Ltd. (Nagoya, Japan). The AnnexinV-FITC/PI Apoptosis Kit (V13242) was purchased from Invitrogen Life Technologies (Waltham, MA, USA). Oxidation-antioxidation assay kits of malondialdehyde (MDA) (A003-1), glutathione (GSH) (A006-1), superoxide dismutase (SOD) (A001-1-1) and glutathione peroxidase (GSH-PX) (A005), testosterone Assay Kit (H090), and protease inhibitor cocktail (W060) were bought from Nanjing Jiancheng Bioengineering Institute (Nanjing, China).

4.2. Nanoparticles and Characterization

In characterization tests, ZnO NPs dispersed in sterile Milli-Q water (final concentration 1 mg/mL, Milford, MA, USA) were put into an ultrasound bath (100 W, Shanghai, China) to break up the aggregates before transmission electron microscopy (TEM, JEM-200CX, JEOL, Japan) and dynamic light scattering (DLS) analyses (Malvern Zeta sizer Nano ZS, Malvern Instruments, U.K.).

4.3. Animal Administration

Adult male Kunming mice (8 weeks old, 20–25 g) were obtained from the Shanghai Laboratory Animal Center, Chinese Academy of Sciences (CAS, Shanghai, China). The mice were housed in an isolated and air-conditioned animal room with water and rodent food supplement. All animals were acclimated to this environment for at least one week prior to the experiment. Experiments were approved by the Animal Ethics Committee of Nanchang University, China, SYKX2015-0001, 12 October 2015. The mice were intragastrically (i.g.) administered with ZnO NPs (0, 100, 200, 400 mg/kg/day, diluted in water) for 28 days and were then anesthetized with carbon dioxide inhalation, followed by cervical dislocation. The serum samples were collected following standard operation procedures. Then, the testes and the epididymis were quickly dissected free of fat, decapsulated, and frozen in liquid nitrogen.

4.4. Histology

Male mouse testis and epididymis tissues were stained with hematoxylin and eosin (HE) according to the method described by Chen et al. [43].

4.5. Western Blotting Analysis

The homogenized testis tissue and mouse Leydig TM3 cells were harvested in lysis buffer (50 mM Tris pH 7.5, 0.3 M NaCl, 5 mM EGTA, 1 mM EDTA, 0.5% Triton X-100, 0.5% NP40) containing protease inhibitor cocktail. Then, the supernatants were collected for Western blot after centrifuge for 10 min at 12 000× g. All primary antibodies and their recommended secondary antibodies were diluted 1:1000 and 1:5000, respectively.

4.6. Detection of Testosterone Content

The testosterone level in the serum was determined by ELISA kit according to the manufacturer's instructions (Nanjing Jiancheng Bioengineering Institute, Nanjing, China).

4.7. Cell Culture and ZnO NP Treatment

Mouse Leydig TM3 cells were cultured at 37 °C in a 5% CO_2 atmosphere in Dulbecco's modified Eagle's medium (DMEM, Gibco, Langley, OK, USA), supplemented with 5% horse serum (Gibco, Langley, OK, USA) and 2.5% fetal bovine serum (FBS, Gibco, Langley, OK, USA). The cultured cells were seeded and incubated for 24 h before exposure to varying concentrations of ZnO NPs or $ZnCl_2$ as per experimental designs. The summary of the research design was illustrated in Figure S1.

4.8. Cell Viability Assay

The cells were seeded at a density of 1 × 104 cells per well in medium in 96-well plates and incubated for 24 h. The medium was then replaced with ZnO NPs of indicated concentrations in the presence or absence of 5 mM NAC, 10 mM 3-MA, or 1 µM Wortmannin for 24 h. Cell viability was determined by measuring the absorbance at 570 nm after the cells were incubated with 0.5 mg/mL MTT in medium for 4 h.

4.9. AnnexinV-FITC/PI Apoptosis Assay

Apoptosis was determined by using an AnnexinV-FITC/PI Apoptosis Kit from Invitrogen Life Technologies (Waltham, MA, USA) as described previously [64].

4.10. Oxidative Stress Measurement

The resultant supernatants of homogenized mouse Leydig TM3 cells were utilized to determine the activities of GSH-PX and SOD and the levels of GSH and MDA by using the commercial kits following the manufacturer's instructions. The protein concentration was detected by the Bradford assay.

4.11. Transmission Electron Microscopy (TEM) Analysis

Mouse Leydig TM3 cells were treated with ddH$_2$O, 4 μg/mL ZnO NPs or 5 mM NAC, plus 4 μg/mL ZnO NPs for 24 h. Then, the autophagic vacuoles were observed by TEM as previously described [65]. The cells treated for 2 h by starvation media (140 mM NaCl, 1 mM CaCl$_2$, 1 mM MgCl$_2$, 5 mM glucose, and 20 mM HEPES at pH = 7.4 supplemented with 1 % BSA) were used as the positive control of autophagy.

4.12. Statistical Analysis

The data were represented as means ± SE. Statistical analyses were performed using a one-way ANOVA with Newman–Keuls multiple range test. $p < 0.05$ was considered statistically significant.

5. Conclusions

ZnO NPs could cause disruption and atrophy of seminiferous epithelium, and even damage to spermatogenesis in male mice. Apoptosis and autophagy were induced by ZnO NPs in the testis tissue with a decreased level of serum testosterone. In vitro studies demonstrated that ZnO NPs markedly inhibited the viability of mouse Leydig TM3 cells and induced apoptosis and autophagy. Oxidative stress was also induced after the cells were treated with ZnO NPs, while inhibition of oxidative stress could rescue the induction of apoptosis and autophagy, indicating that oxidative stress was involved in ZnO NPs-induced apoptosis and autophagy. However, suppression of autophagy further inhibited cell viability with increase of the apoptosis levels. Taken together, we have provided detailed evidence that oxidative stress is involved in ZnO NPs-induced apoptosis and autophagy of mouse Leydig TM3 cells, while autophagy contributes to counteract the reproductive toxicity of ZnO NPs in the testis.

Supplementary Materials: Supplementary materials can be found at http://www.mdpi.com/1422-0067/20/16/4042/s1.

Author Contributions: J.S., D.Y., R.C., and J.C. conceived and designed the study. J.S., D.Y., X.Z., Y.W., and S.T. carried out all the experiments. H.Y. and J.W. performed statistical analysis. J.S., H.Y., R.C., and J.C. drafted the paper and amended the paper. All authors read and approved the final manuscript.

Funding: We thank the financial support from the National Natural Science Foundation of China (No. 81660255, No. 81360098 and No. 21777036), the Young Scientist Training Project of Jiangxi Province, China (No. 20153BCB23032), the Research fund for postgraduate program of Nanchang University (No. cx2015179), and the Youth Plan of Beijing Academy of Science and Technology (YC201809).

Conflicts of Interest: The authors declare no conflict of interest.

References

1. Gambardella, C.; Ferrando, S.; Gatti, A.M.; Cataldi, E.; Ramoino, P.; Aluigi, M.G.; Faimali, M.; Diaspro, A.; Falugi, C. Review: Morphofunctional and biochemical markers of stress in sea urchin life stages exposed to engineered nanoparticles. *Environ. Toxicol.* **2016**, *31*, 1552–1562. [CrossRef] [PubMed]

2. Gambardella, C.; Aluigi, M.G.; Ferrando, S.; Gallus, L.; Ramoino, P.; Gatti, A.M.; Rottigni, M.; Falugi, C. Developmental abnormalities and changes in cholinesterase activity in sea urchin embryos and larvae from sperm exposed to engineered nanoparticles. *Aquat. Toxicol.* **2013**, *130*, 77–85. [CrossRef] [PubMed]

3. Piccinno, F.; Gottschalk, F.; Seeger, S.; Nowack, B. Industrial production quantities and uses of ten engineered nanomaterials in Europe and the world. *J. Nanopart. Res.* **2012**, *14*, 1–11. [CrossRef]

4. Mishra, P.K.; Mishra, H.; Ekielski, A.; Talegaonkar, S.; Vaidya, B. Zinc oxide nanoparticles: a promising nanomaterial for biomedical applications. *Drug. Discov. Today.* **2017**, *22*, 1825–1834. [CrossRef] [PubMed]

5. Oberdörster, G.; Oberdörster, E.; Oberdörster, J. Nanotoxicology: an emerging discipline evolving from studies of ultrafine particles. *Environ. Health. Perspect.* **2005**, *113*, 823–839. [CrossRef] [PubMed]

6. Singh, S. Zinc oxide nanoparticles impacts: cytotoxicity, genotoxicity, developmental toxicity, and neurotoxicity. *Toxicol. Mech. Methods* **2019**, *29*, 300–311. [CrossRef]

7. Yu, K.N.; Yoon, T.J.; Minai-Tehrani, A.; Kim, J.E.; Park, S.J.; Jeong, M.S.; Ha, S.W.; Lee, J.K.; Kim, J.S.; Cho, M.H. Zinc oxide nanoparticle induced autophagic cell death and mitochondrial damage via reactive oxygen species generation. *Toxicol. In Vitro* **2013**, *27*, 1187–1195. [CrossRef]

8. Ma, H.; Williams, P.L.; Diamond, S.A. Ecotoxicity of manufactured ZnO nanoparticles—A review. *Environ. Pollut.* **2013**, *172*, 76–85. [CrossRef]

9. Huo, L.; Chen, R.; Zhao, L.; Shi, X.; Bai, R.; Long, D.; Chen, F.; Zhao, Y.; Chang, Y.Z.; Chen, C. Silver nanoparticles activate endoplasmic reticulum stress signaling pathway in cell and mouse models: The role in toxicity evaluation. *Biomaterials* **2015**, *61*, 307–315. [CrossRef]

10. Chen, R.; Ling, D.; Zhao, L.; Wang, S.; Liu, Y.; Bai, R.; Baik, S.; Zhao, Y.; Chen, C.; Hyeon, T. Parallel comparative studies on mouse toxicity of oxide nanoparticle- and gadolinium-based T1 MRI contrast agents. *ACS. Nano.* **2015**, *9*, 12425–12435. [CrossRef]

11. Lin, L.; Xu, M.; Mu, H.; Wang, W.; Sun, J.; He, J.; Qiu, J.W.; Luan, T. Quantitative proteomic analysis to understand the mechanisms of zinc oxide nanoparticle toxicity to daphnia pulex (Crustacea: Daphniidae): comparing with bulk zinc oxide and zinc salt. *Environ Sci. Technol.* **2019**, *53*, 5436–5444. [CrossRef]

12. Chen, R.; Huo, L.; Shi, X.; Bai, R.; Zhang, Z.; Zhao, Y.; Chang, Y.; Chen, C. Endoplasmic reticulum stress induced by zinc oxide nanoparticles is an earlier biomarker for nanotoxicological evaluation. *ACS. Nano.* **2014**, *8*, 2562–2574. [CrossRef]

13. Yin, H.; Chen, R.; Casey, P.S.; Ke, P.C.; Davis, T.P.; Chen, C. Reducing the cytotoxicity of ZnO nanoparticles by a pre-formed protein corona in a supplemented cell culture medium. *RSC Adv.* **2015**, *5*, 73963–73973. [CrossRef]

14. Le, T.C.; Yin, H.; Chen, R.; Chen, Y.; Zhao, L.; Casey, P.S.; Chen, C.; Winkler, D.A. An experimental and computational approach to the development of ZnO nanoparticles that are safe by design. *Small* **2016**, *12*, 3568–3577. [CrossRef]

15. Huo, L.; Chen, R.; Shi, X.; Bai, R.; Wang, P.; Chang, Y.; Chen, C. High-content screening for assessing nanomaterial toxicity. *J. Nanosci. Nanotechnol.* **2015**, *15*, 1143–1149. [CrossRef]

16. Wen, X.; Wu, J.; Wang, F.; Liu, B.; Huang, C.; Wei, Y. Deconvoluting the role of reactive oxygen species and autophagy in human diseases. *Free. Radic. Biol. Med.* **2013**, *65*, 402–410. [CrossRef]

17. Jenkins, R.R.; Goldfarb, A. Introduction: oxidant stress, aging, and exercise. *Med. Sci. Sports Exerc.* **1993**, *25*, 210–212. [CrossRef]

18. Chen, R.; Qiao, J.; Bai, R.; Zhao, Y.; Chen, C. Intelligent testing strategy and analytical techniques for the safety assessment of nanomaterials. *Anal. Bioanal. Chem.* **2018**, *410*, 6051–6066. [CrossRef]

19. Chen, J.X.; Xu, L.L.; Mei, J.H.; Yu, X.B.; Kuang, H.B.; Liu, H.Y.; Wu, Y.J.; Wang, J.L. Involvement of neuropathy target esterase in tri-ortho-cresyl phosphate-induced testicular spermatogenesis failure and growth inhibition of spermatogonial stem cells in mice. *Toxicol. Lett.* **2012**, *211*, 54–61. [CrossRef]

20. Chen, J.X.; Xu, L.L.; Wang, X.C.; Qin, H.Y.; Wang, J.L. Involvement of c-Src/STAT3 signal in EGF-induced proliferation of rat spermatogonial stem cells. *Mol. Cell. Biochem.* **2011**, *358*, 67–73. [CrossRef]

21. Liu, M.L.; Wang, J.L.; Wei, J.; Xu, L.L.; Yu, M.; Liu, X.M.; Ruan, W.L.; Chen, J.X. Tri-ortho-cresyl phosphate induces autophagy of rat spermatogonial stem cells. *Reproduction* **2015**, *149*, 163–170. [CrossRef]

22. Talebi, A.R.; Khorsandi, L.; Moridian, M. The effect of zinc oxide nanoparticles on mouse spermatogenesis. *J. Assist. Reprod. Genet.* **2013**, *30*, 1203–1209. [CrossRef]

23. Zhang, Y.; Song, M.; Rui, X.; Pu, S.; Li, Y.; Li, C. Supplemental dietary phytosterin protects against 4-nitrophenol-induced oxidative stress and apoptosis in rat testes. *Toxicol. Rep.* **2015**, *2*, 664–676. [CrossRef]

24. Lone, Y.; Koiri, R.K.; Bhide, M. An overview of the toxic effect of potential human carcinogen Microcystin-LR on testis. *Toxicol. Rep.* **2015**, *2*, 289–296. [CrossRef]

25. Han, Z.; Yan, Q.; Ge, W.; Liu, Z.G.; Gurunathan, S.; De Felici, M.; Shen, W.; Zhang, X.F. Cytotoxic effects of ZnO nanoparticles on mouse testicular cells. *Int. J. Nanomed.* **2016**, *11*, 5187–5203. [CrossRef]

26. Bara, N.; Kaul, G. Enhanced steroidogenic and altered antioxidant response by ZnO nanoparticles in mouse testis Leydig cells. *Toxicol. Ind. Health.* **2018**, *34*, 571–588. [CrossRef]

27. Deepa, S.; Murugananthkumar, R.; Raj Gupta, Y.; Gowda, K.S.M.; Senthilkumaran, B. Effects of zinc oxide nanoparticles and zinc sulfate on the testis of common carp, Cyprinus carpio. *Nanotoxicology* **2019**, *13*, 240–257. [CrossRef]

28. Hu, Q.; Guo, F.; Zhao, F.; Fu, Z. Effects of titanium dioxide nanoparticles exposure on parkinsonism in zebrafish larvae and PC12. *Chemosphere* **2017**, *173*, 373–379. [CrossRef]

29. Kononenko, V.; Repar, N.; Marušič, N.; Drašler, B.; Romih, T.; Hočevar, S.; Drobne, D. Comparative in vitro genotoxicity study of ZnO nanoparticles, ZnO macroparticles and ZnCl2 to MDCK kidney cells: Size matters. *Toxicol. Vitro* **2017**, *40*, 256–263. [CrossRef]

30. Kim, J.; Kim, Y.C.; Fang, C.; Russell, R.C.; Kim, J.H.; Fan, W.; Liu, R.; Zhong, Q.; Guan, K.L. Differential regulation of distinct Vps34 complexes by AMPK in nutrient stress and autophagy. *Cell* **2013**, *152*, 290–303. [CrossRef]

31. Wong, P.M.; Puente, C.; Ganley, I.G.; Jiang, X. The ULK1 complex: Sensing nutrient signals for autophagy activation. *Autophagy* **2013**, *9*, 124–137. [CrossRef]

32. Levine, B.; Kroemer, G. Autophagy in the pathogenesis of disease. *Cell* **2008**, *132*, 27–42. [CrossRef]

33. Chen, J.X.; Sun, Y.J.; Wang, P.; Long, D.X.; Li, W.; Li, L.; Wu, Y.J. Induction of autophagy by TOCP in differentiated human neuroblastoma cells lead to degradation of cytoskeletal components and inhibition of neurite outgrowth. *Toxicology* **2013**, *310*, 92–97. [CrossRef]

34. Long, D.X.; Hu, D.; Wang, P.; Wu, Y.J. Induction of autophagy in human neuroblastoma SH-SY5Y cells by tri-ortho-cresyl phosphate. *Mol. Cell Biochem.* **2014**, *396*, 33–40. [CrossRef]

35. Chiarelli, R.; Martino, C.; Agnello, M.; Bosco, L.; Roccheri, M.C. Autophagy as a defense strategy against stress: focus on Paracentrotus lividus sea urchinembryos exposed to cadmium. *Cell Stress Chaperones* **2016**, *21*, 19–27. [CrossRef]

36. Xu, H.Y.; Wang, P.; Sun, Y.J.; Jiang, L.; Xu, M.Y.; Wu, Y.J. Autophagy in Tri-o-cresyl Phosphate-Induced Delayed Neurotoxicity. *J. Neuropathol. Exp. Neurol.* **2017**, *76*, 52–60. [CrossRef]

37. Xu, L.L.; Liu, M.L.; Wang, J.L.; Yu, M.; Chen, J.X. Saligenin cyclic-o-tolyl phosphate (SCOTP) induces autophagy of rat spermatogonial stem cells. *Reprod. Toxicol.* **2016**, *60*, 62–68. [CrossRef]

38. Lin, Y.F.; Chiu, I.J.; Cheng, F.Y.; Lee, Y.H.; Wang, Y.J.; Hsu, Y.H.; Chiu, H.W. The role of hypoxia-inducible factor-1alpha in zinc oxide nanoparticle-induced nephrotoxicity in vitro and in vivo. *Part. Fibre. Toxicol.* **2016**, *13*, 52. [CrossRef]

39. Johnson, B.M.; Fraietta, J.A.; Gracias, D.T.; Hope, J.L.; Stairiker, C.J.; Patel, P.R.; Mueller, Y.M.; McHugh, M.D.; Jablonowski, L.J.; Wheatley, M.A.; et al. Acute exposure to ZnO nanoparticles induces autophagic immune cell death. *Nanotoxicology* **2015**, *9*, 737–748. [CrossRef]

40. Jiang, L.; Li, Z.; Xie, Y.; Liu, L.; Cao, Y. Cyanidin chloride modestly protects Caco-2cells from ZnO nanoparticle exposure probably through the induction of autophagy. *Food Chem. Toxicol.* **2019**, *127*, 251–259. [CrossRef]

41. Giovanni, M.; Yue, J.; Zhang, L.; Xie, J.; Ong, C.N.; Leong, D.T. Pro-inflammatory responses of RAW264.7 macrophages when treated with ultralow concentrations of silver, titanium dioxide, and zinc oxide nanoparticles. *J. Hazard. Mater.* **2015**, *297*, 146–152. [CrossRef]

42. Wang, P.; Menzies, N.W.; Lombi, E.; McKenna, B.A.; Johannessen, B.; Glover, C.J.; Kappen, P.; Kopittke, P.M. Fate of ZnO nanoparticles in soils and cowpea (Vigna unguiculata). *Environ. Sci. Technol.* **2013**, *47*, 13822–13830. [CrossRef]

43. Chen, R.; Zhang, L.; Ge, C.; Tseng, M.T.; Bai, R.; Qu, Y.; Beer, C.; Autrup, H.; Chen, C. Subchronic toxicity and cardiovascular responses in spontaneously hypertensive rats after exposure to multiwalled carbon nanotubes by intratracheal instillation. *Chem. Res. Toxicol.* **2015**, *28*, 440–450. [CrossRef]

44. Zhao, F.; Meng, H.; Yan, L.; Wang, B.; Zhao, Y. Nanosurface chemistry and dose govern the bioaccumulation and toxicity of carbon nanotubes, metal nanomaterials and quantum dots in vivo. *Sci. Bull.* **2015**, *60*, 3–20. [CrossRef]

45. Wang, R.; Song, B.; Wu, J.; Zhang, Y.; Chen, A.; Shao, L. Potential adverse effects of nanoparticles on the reproductive system. *Int. J. Nanomed.* **2018**, *13*, 8487–8506. [CrossRef]

46. Elbialy, N.S.; Aboushoushah, S.F.; Alshammari, W.W. Long-term biodistribution and toxicity of curcumin capped iron oxide nanoparticles after single-dose administration in mice. *Life Sci.* **2019**, *230*, 76–83. [CrossRef]

47. Qian, L.; Cheng, X.; Ji, G.; Hui, L.; Mo, Y.; Tollerud, D.J.; Gu, A.; Zhang, Q. Sublethal effects of zinc oxide nanoparticles on male reproductive cells. *Toxicol. In Vitro* **2016**, *35*, 131.

48. Kim, Y.R.; Park, J.I.; Lee, E.J.; Park, S.H.; Seong, N.W.; Kim, J.H.; Kim, G.Y.; Meang, E.H.; Hong, J.S.; Kim, S.H.; et al. Toxicity of 100 nm zinc oxide nanoparticles: a report of 90-day repeated oral administration in Sprague Dawley rats. *Int. J. Nanomed.* **2014**, *9*, 109–126.

49. Ko, J.W.; Hong, E.T.; Lee, I.C.; Park, S.H.; Park, J.I.; Seong, N.W.; Hong, J.S.; Yun, H.I.; Kim, J.C. Evaluation of 2-week repeated oral dose toxicity of 100 nm zinc oxide nanoparticles in rats. *Lab. Anim. Res.* **2015**, *31*, 139–147. [CrossRef]

50. Hong, J.S.; Park, M.K.; Kim, M.S.; Lim, J.H.; Park, G.J.; Maeng, E.H.; Shin, J.H.; Kim, Y.R.; Kim, M.K.; Lee, J.K.; et al. Effect of zinc oxide nanoparticles on dams and embryo-fetal development in rats. *Int. J. Nanomed.* **2014**, *9*, 145–157.

51. Fei, G.; Ma, N.; Hong, Z.; Wang, Q.; Hao, Z.; Pu, W.; Hou, H.; Wen, H.; Li, L. Zinc oxide nanoparticles-induced epigenetic change and G2/M arrest are associated with apoptosis in human epidermal keratinocytes. *Int. J. Nanomed.* **2016**, *11*, 3859–3874.

52. Liang, S.; Sun, K.; Wang, Y.; Dong, S.; Wang, C.; Liu, L.; Wu, Y. Role of Cyt-C/caspases-9,3, Bax/Bcl-2 and the FAS death receptor pathway in apoptosis induced by zinc oxide nanoparticles in human aortic endothelial cells and the protective effect by alpha-lipoic acid. *Chem. Biol. Interact.* **2016**, *258*, 40–51. [CrossRef]

53. Xiao, L.; Liu, C.; Chen, X.; Yang, Z. Zinc oxide nanoparticles induce renal toxicity through reactive oxygen species. *Food Chem. Toxicol.* **2016**, *90*, 76–83. [CrossRef]

54. Chen, P.; Wang, H.; He, M.; Chen, B.; Yang, B.; Hu, B. Size-dependent cytotoxicity study of ZnO nanoparticles in HepG2 cells. *Ecotoxicol. Environ. Saf.* **2019**, *171*, 337–346. [CrossRef]

55. Kabeya, Y.; Mizushima, N.; Ueno, T.; Yamamoto, A.; Kirisako, T.; Noda, T.; Kominami, E.; Ohsumi, Y.; Yoshimori, T. LC3, a mammalian homologue of yeast Apg8p, is localized in autophagosome membranes after processing. *EMBO J.* **2000**, *19*, 5720–5728. [CrossRef]

56. Mizushima, N. Methods for monitoring autophagy. *Int. J. Biochem. Cell Biol.* **2004**, *36*, 2491–2502. [CrossRef]

57. Tremblay, J.J. Molecular regulation of steroidogenesis in endocrine Leydig cells. *Steroids.* **2015**, *103*, 3–10. [CrossRef]

58. Abarikwu, S.O.; Akiri, O.F.; Durojaiye, M.A.; Alabi, A.F. Combined effects of repeated administration of Bretmont Wipeout (Glyphosate) and Ultrazin (Atrazine) on testosterone, oxidative stress and sperm quality of Wistar rats. *Toxicol. Mech. Method* **2014**, *25*, 1–31. [CrossRef]

59. Asani, S.C.; Umrani, R.D.; Paknikar, K.M. Differential dose-dependent effects of zinc oxide nanoparticles on oxidative stress-mediated pancreatic beta-cell death. *Nanomedicine* **2017**, *12*, 745–759. [CrossRef]

60. Zhang, Q.; Zhang, Y.; Zhang, P.; Chao, Z.; Xia, F.; Jiang, C.; Zhang, X.; Jiang, Z.; Liu, H. Hexokinase II inhibitor, 3-BrPA induced autophagy by stimulating ROS formation in human breast cancer cells. *Genes. Cancer.* **2014**, *5*, 100–112.

61. Sharma, K.; Le, N.; Alotaibi, M.; Gewirtz, D.A. Cytotoxic autophagy in cancer therapy. *Int. J. Mol. Sci.* **2014**, *15*, 10034–10051. [CrossRef]

62. Denton, D.; Nicolson, S.; Kumar, S. Cell death by autophagy: facts and apparent artefacts. *Cell Death Differ.* **2012**, *19*, 87–95. [CrossRef]

63. Shintani, T.; Klionsky, D.J. Autophagy in health and disease: a double-edged sword. *Science* **2004**, *306*, 990–995. [CrossRef]

64. Sun, Y.; Shen, J.; Zeng, L.; Yang, D.; Shao, S.; Wang, J.; Wei, J.; Xiong, J.; Chen, J. Role of autophagy in di-2-ethylhexyl phthalate (DEHP)-induced apoptosis in mouse Leydig cells. *Environ Pollut.* **2018**, *243*, 563–572. [CrossRef]

65. Liu, X.; Xu, L.; Shen, J.; Wang, J.; Ruan, W.; Yu, M.; Chen, J. Involvement of oxidative stress in tri-ortho-cresyl phosphate-induced autophagy of mouse Leydig TM3 cells in vitro. *Reprod. Biol. Endocrin.* **2016**, *14*, 30. [CrossRef]

International Journal of
Molecular Sciences

Article

Fate Determination of ZnO in Commercial Foods and Human Intestinal Cells

Ye-Rin Jeon, Jin Yu and Soo-Jin Choi *

Division of Applied Food System, Major of Food Science & Technology, Seoul Women's University, Seoul 01797, Korea; yrjeon0715@swu.ac.kr (Y.-R.J.); ky5031@swu.ac.kr (J.Y.)
* Correspondence: sjchoi@swu.ac.kr; Tel.: +82-2-970-5634; Fax: +82-2-970-5977

Received: 13 December 2019; Accepted: 7 January 2020; Published: 9 January 2020

Abstract: (1) Background: Zinc oxide (ZnO) particles are widely used as zinc (Zn) fortifiers, because Zn is essential for various cellular functions. Nanotechnology developments may lead to production of nano-sized ZnO, although nanoparticles (NPs) are not intended to be used as food additives. Current regulations do not specify the size distribution of NPs. Moreover, ZnO is easily dissolved into Zn ions under acidic conditions. However, the fate of ZnO in commercial foods or during intestinal transit is still poorly understood. (2) Methods: We established surfactant-based cloud point extraction (CPE) for ZnO NP detection as intact particle forms using pristine ZnO-NP-spiked powdered or liquid foods. The fate determination and dissolution characterization of ZnO were carried out in commercial foods and human intestinal cells using in vitro intestinal transport and ex vivo small intestine absorption models. (3) Results: The results demonstrated that the CPE can effectively separate ZnO particles and Zn ions in food matrices and cells. The major fate of ZnO in powdered foods was in particle form, in contrast to its ionic fate in liquid beverages. The fate of ZnO was closely related to the extent of its dissolution in food or biomatrices. ZnO NPs were internalized into cells in both particle and ion form, but dissolved into ions with time, probably forming a Zn–ligand complex. ZnO was transported through intestinal barriers and absorbed in the small intestine primarily as Zn ions, but a small amount of ZnO was absorbed as particles. (4) Conclusion: The fate of ZnO is highly dependent on food matrix type, showing particle and ionic fates in powdered foods and liquid beverages, respectively. The major intracellular and intestinal absorption fates of ZnO NPs were Zn ions, but a small portion of ZnO particle fate was also observed after intestinal transit. These findings suggest that the toxicity of ZnO is mainly related to the Zn ion, but potential toxicity resulting from ZnO particles cannot be completely excluded.

Keywords: zinc oxide; fate; cloud point extraction; dissolution; commercial food; intestinal absorption

---

1. Introduction

Nanomaterials have been widely applied to diverse industries, such as electronics, medicine, pharmaceutics, cosmetics, and foods. In particular, food additive particles, including silicon dioxide (SiO_2), titanium dioxide (TiO_2), and zinc oxide (ZnO) are widely used as anticaking agents, coloring agents, and nutritional fortifiers, respectively [1–3]. Among them, ZnO can be added to milk, dairy products, cereals, and beverages. This is due to the fact that zinc (Zn) plays an essential role as a trace element in the immune system, cell function, enzyme activity, and signaling in human body [4–6]. ZnO is used as a food additive in the United States (US) and Europe, and classified as "Generally Recognized as Safe" by the US Food and Drug Administration [7]. The acceptable daily intake (ADI) for Zn in Zn-fortified functional foods is in the range of 2.55 to 12 mg in the Republic of Korea [8]. Nanotechnology developments have led to the production of nano-sized ZnO particles, although nanomaterials that range in size from 1 to 100 nm are not intended to be used as food additives. Indeed, the particle size distribution for ZnO as a food additive is unspecified [9].

The fate determination of ZnO is of importance for interpreting and understanding its potential toxicity, because the toxic effects resulting from ZnO nanoparticles (NPs), NP aggregates/agglomerates, micro-sized particles, or Zn ions are completely different [10,11]. A few reports have demonstrated the fates of ZnO in biological systems or environmental water samples [12–14]. ZnO NPs were reported to be primarily present as ionic forms in tissues after oral administration in rats [15–17]. Zn ion release from ZnO NPs in simulated gastric fluid increased to 14% has also been reported [18,19], suggesting their partial dissolution property. Studies have demonstrated that ZnO can be easily decomposed to Zn ions in acidic solutions or biological fluids [20–23]. Since most foods have a slightly acidic pH and Zn-fortified foods are taken up by the oral route, ZnO can be dissolved into Zn ions in food matrices to some extent, and does not present in its particle form in the body. However, there is no clear information about the fate of ZnO in commercial foods and human intestinal cells during intestinal transit.

Food additive ZnO is directly added to complex food matrices in solution or powdered form. Hence, the interactions between ZnO and food components should also be considered. Indeed, ZnO NPs have been demonstrated to interact with food components, such as proteins and saccharides, forming particle–matrix corona (nanocorona covered by matrices) [23–25]. These interactions were also found to affect cytotoxicity, intestinal transport efficiency, and oral absorption [24–27]. Thus, separating ZnO from complex food matrices in intact particle form is challenging. Currently, the most widely applied method for detecting ZnO is inductively coupled plasma–atomic emission spectroscopy (ICP-AES), ICP–mass spectroscopy, or atomic absorption spectroscopy. However, these methods require pre-digestion procedures with acids to digest organic matrices, and therefore are used for analyzing total Zn levels only and cannot be used to distinguish ZnO particles from Zn ions.

On the other hand, cloud point extraction (CPE) using Triton X-114 (TX-114) was first described and developed for the analysis of trace NPs, such as gold, silver, and copper as particle forms [28–30]. TX-114 is a surfactant, is cost-effective, and has the advantage of forming micelles at room temperature (20–25 °C), contributing to easy manipulation. To date, TX-114-based CPE studies have focused on the determination of silver NPs in water or cells [31,32]. ZnO separation by CPE in aqueous phase or environmental waters where little organic matrix is present has also been reported [33]. A CPE-based approach was used for the detection of aluminum (Al) or Zn in foodstuffs. However, these studies used ethylenediaminetetraacetic acid (EDTA) or 8-hydroxyquinoline as chelating agents for precipitation of Al or Zn ions in precipitated surfactant-rich phase by CPE [34,35]. Moreover, the CPE method was applied after pre-treatments such as filtering, nitric acid treatment, and dry-ashing were applied in order to digest organic matrices in foods [34,35]. No attempt has been made to detect ZnO in its particle form in complex systems, such as in foods or biomatrices.

In the present study, an analytical method for determining ZnO with intact particle size and shape in foods and human intestinal cells was established using surfactant-based CPE. The fates of ZnO in commercial processed foods, including liquid and powdered types, were determined based on the optimized method established with pristine ZnO-NP-spiked food matrices. Finally, fate determination was assessed in human intestinal cells, in vitro models of human intestinal barriers, and an ex vivo intestinal absorption model.

2. Results

2.1. Optimization and Characterization of CPE for ZnO NPs

TX-114-based CPE was first optimized utilizing pristine ZnO NPs. The optimization of ZnO dispersant was carried out with humic acid (HA) and glucose (Glc), and compared with NPs in distilled and deionized water (DDW) as a control at 37 and 45 °C, respectively. HA was chosen as a dispersing agent based on the research of Majedi et al. [33], and Glc is known to play a role in NP dispersion as well [25,36]. The physicochemical properties of ZnO NPs after dispersion (Step 1), during CPE (Step 2: surfactant-rich and aqueous phases without centrifugation), and after CPE (Step 3: surfactant-rich and aqueous phases separated by centrifugation) were characterized (Figure 1). When particle size, shape, and morphology were examined by field emission-scanning electron microscopy (FE-SEM), pristine ZnO NPs dispersed in DDW, HA, and Glc (Step 1) had an average primary particle size of ~60 nm with irregular shapes (Figure 2A). The precipitated, TX-114-rich phase of ZnO NPs after CPE (Step 3) had similar primary particle size distributions compared to pristine NPs in different dispersants, showing no significant differences in particle size between Step 1 and Step 3 (Figure 2B, $p > 0.05$). FE-SEM images during CPE process (Step 2) could not be clearly obtained due to the presence of high level of organic material TX-114.

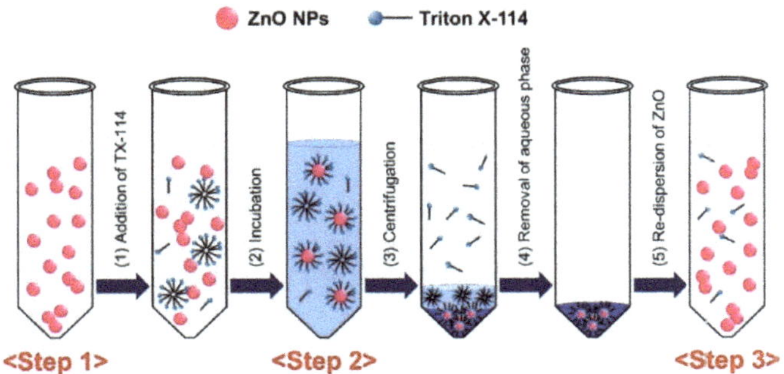

Figure 1. Schematic illustration of Triton X-114 (TX-114)-based cloud point extraction (CPE) procedure.

The hydrodynamic radii of ZnO NPs consequently increased when ZnO NPs were covered by TX-114 and after micelle formations (Step 2). Thus, the formation of ZnO NPs-TX-114 micelles was confirmed through dynamic light scattering (DLS) analysis. The hydrodynamic diameters of pristine ZnO NPs in DDW or HA were significantly smaller than those in Glc at Step 1 (Table 1). The hydrodynamic radii of ZnO NPs in different dispersants dramatically increased during the CPE process (Step 2) in all cases, but decreased to the same levels similar to that in pristine NPs after CPE (Step 3), only when ZnO NPs were dispersed in HA (Table 1). No statistical differences in temperatures (37 and 45 °C) were found ($p > 0.05$), except in ZnO NPs in HA during CPE (Step 2). On the other hand, the zeta potential values changed to slightly negative charges in all cases after CPE (Step 3, Table 1).

Table 1. Hydrodynamic radii and zeta potentials of pristine ZnO nanoparticles (NPs) in different CPE conditions.

Dispersant Type	Incubation Temperature (°C)	Z-Average Size (nm)			Zeta Potential (mV)		
		Step 1	Step 2	Step 3	Step 1	Step 2	Step 3
ZnO in DDW	37	261.27 ± 3.59 [A,a]	1716.33 ± 93.38 [A,c]	739.57 ± 113.49 [B,b]	25.9 ± 0.99 [A,a]	−0.98 ± 0.05 [B,b]	−1.62 ± 0.14 [A,b]
	45	261.27 ± 3.59 [A,a]	1717.00 ± 120.68 [A,c]	857.97 ± 56.61 [BC,b]			
ZnO in HA	37	260.43 ± 4.22 [A,a]	1653.67 ± 123.52 [A,b]	345.67 ± 17.05 [A,a]	−34.63 ± 1.33 [C,c]	−0.77 ± 0.02 [A,a]	−4.92 ± 0.44 [B,b]
	45	260.43 ± 4.22 [A,a]	2272.33 ± 279.40 [B,b]	338.90 ± 10.91 [A,a]			
ZnO in Glc	37	777.57 ± 19.03 [B,a]	1466.33 ± 22.72 [A,c]	897.13 ± 35.72 [B,C,b]	16.63 ± 0.42 [B,a]	−1.37 ± 0.10 [C,b]	−1.68 ± 0.17 [A,b]
	45	777.57 ± 19.03 [B,a]	1685.67 ± 147.97 [A,b]	1073.53 ± 171.95 [C,a]			

Different upper-case letters (A,B,C) indicate significant differences among different CPE conditions ($p < 0.05$). Different lower-case letters (a,b,c) indicate significant differences among different CPE steps ($p < 0.05$). Abbreviation: DDW, distilled and deionized water; HA, humic acid; Glc, glucose.

2.2. Recovery of Pristine ZnO NPs by CPE

The recovery (%) of ZnO NPs in different dispersants by CPE was checked by quantifying Zn amounts in both aqueous and TX-114-rich phases (supernatant and precipitate after CPE, respectively) in order to confirm the efficiency of the CPE procedure. As shown in Figure 2C, more than 82% of the

ZnO NPs were recovered from the precipitates as particles after CPE, regardless of CPE conditions. Only small portions (less than 2%) of ZnO NPs were detected in supernatants as ions after CPE. The highest total recovery (94.9%) of ZnO NPs in both supernatant and precipitate after CPE was found when ZnO NPs were dispersed in HA at 45 °C. It is worth noting that 93.4% of pristine ZnO NPs in HA were obtained in the precipitates as particles after CPE. Hence, this condition was further used for fate determination of ZnO in commercial foods and human intestinal cells.

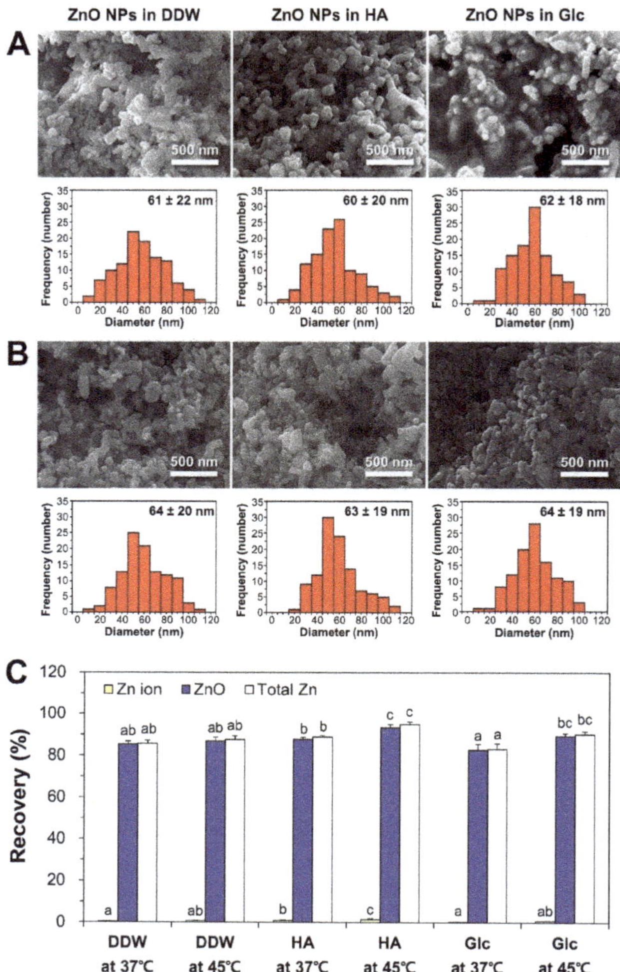

Figure 2. Field emission-scanning electron microscopy (FE-SEM) images and size distribution of pristine ZnO NPs in different dispersants (**A**) before CPE and (**B**) after CPE. (**C**) Recovery (%) of Zn ions, ZnO particles, and total Zn levels obtained from pristine ZnO NPs by CPE. Size distributions were obtained by randomly selecting 100 particles from FE-SEM images. Different lower-case letters (a,b,c) indicate significant differences among different CPE conditions ($p < 0.05$).

2.3. Dissolution Property of ZnO NPs in Food/Bio Matrices

The solubility of ZnO NPs in food matrices and cell culture medium was first checked prior to fate determination. Two different food types which were representative of powdered or liquid foods were used. Powdered foods (coffee mix and skim milk) and liquid beverages (milk and sports drink) were selected based on the potential utilization of ZnO in Zn-fortified foods. Figure 3A showed that the solubilities of ZnO NPs in different dispersants ranged from 0.8% to 1.7%. The pHs of ZnO dispersed in DDW, HA, and Glc were 8.6, 8.1, and 8.7, respectively.

On the other hand, the solubilities of ZnO in food matrices increased, reaching 39.4%, 30.0%, and 49.2% in coffee mix, skim milk, and milk, respectively. There was a dramatic increase in solubility observed in sports drink, which reached up to 90.9% (Figure 3B). The pHs of ZnO-spiked food matrices were 6.2, 6.9, 6.9, and 3.3 in coffee mix, skim milk, milk, and sports drink, respectively. The solubility of ZnO in cell culture minimum essential medium (MEM) (pH 7.0) was ~18% after 0.5 h and increased up to 24.8% after 6–24 h (Figure 3C).

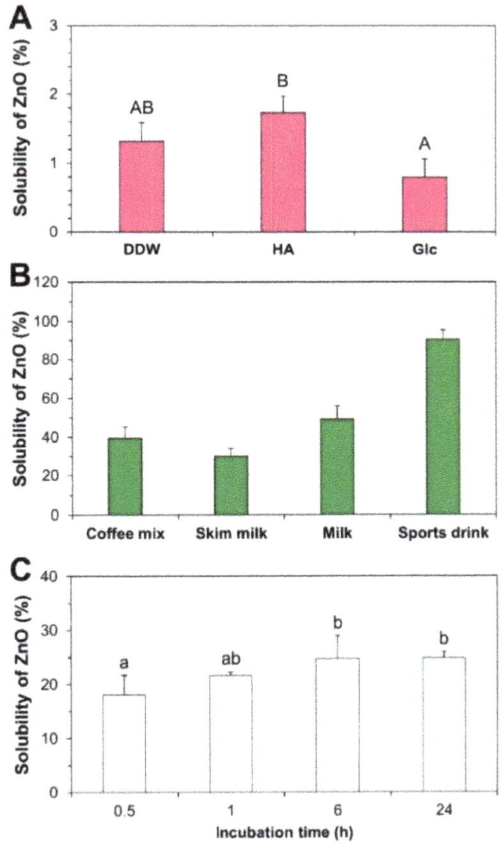

Figure 3. Dissolution properties of ZnO NPs in (**A**) dispersants, (**B**) food matrices, and (C) cell culture medium. Different upper-case letters (A,B) indicate significant differences among different dispersants ($p < 0.05$). Different lower-case letters (a,b) indicate significant differences among incubation times ($p < 0.05$).

2.4. Characterization and Fate of ZnO-NP-Spiked Foods

The reliability and accuracy of the CPE method for fate determination of ZnO in commercial foods were checked using ZnO-NP-spiked foods. DLS results demonstrated that all ZnO NPs recovered from the precipitates of ZnO-NP-spiked coffee mix, skim milk, and milk after CPE had statistically significant similarities in hydrodynamic radii compared to pristine ZnO NPs ($p > 0.05$, Figure S1). On the other hand, no particle forms were detected in the precipitates of ZnO-NP-spiked sports drink after CPE (Figure S1). SEM and energy dispersive X-ray spectroscopy (EDS) analysis revealed the presence of Zn elements recovered from the precipitated TX-114-rich phases of all ZnO-NP-spiked foods after CPE, except sports drink (Figure 4A,B).

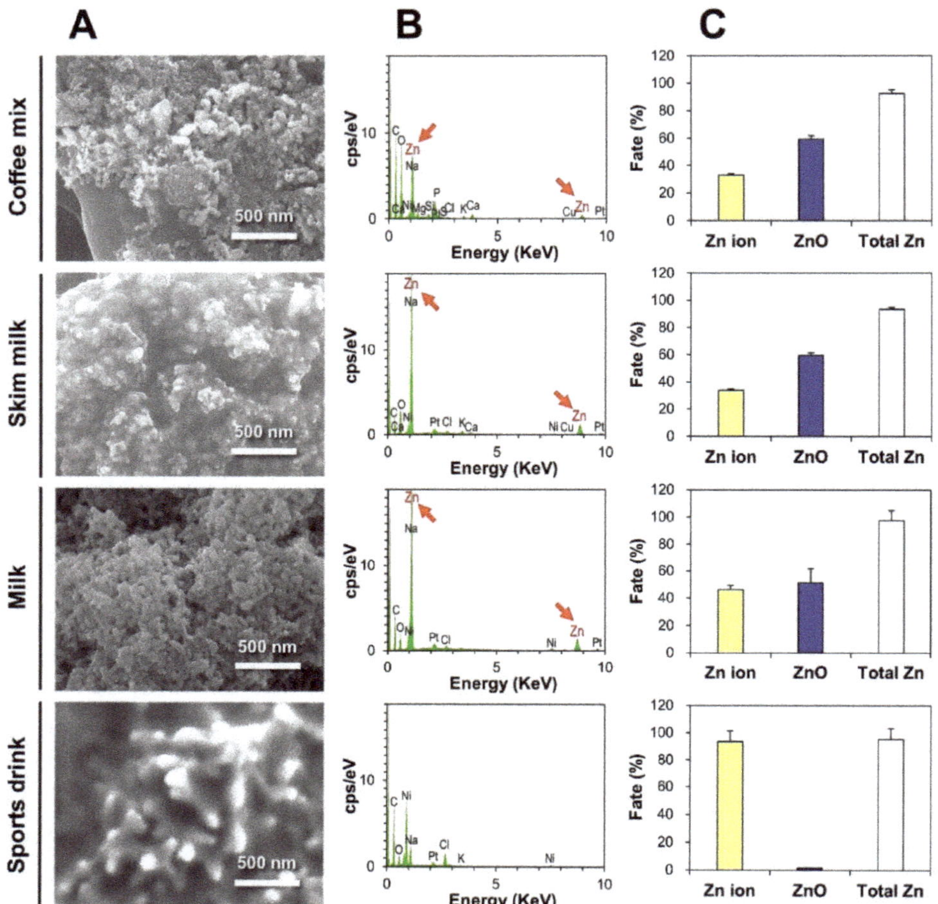

Figure 4. (**A**) FE-SEM images and (**B**) energy dispersive X-ray spectroscopy (EDS) analysis of the precipitates of ZnO-NP-spiked foods after CPE. (**C**) Fate (%) of ZnO NPs in ZnO-NP-spiked foods by CPE, followed by inductively coupled plasma–atomic emission spectroscopy (ICP-AES) quantification.

Quantitative ICP-AES analysis results on the supernatants and precipitates after CPE are presented in Figure 4C. About 59.5%, 59.5%, and 51.3% of ZnO were present as particle forms in coffee mix, skim milk, and milk, respectively, and 93.7% of ZnO NPs were dissolved and detected as Zn ions in sports drink. The total Zn recoveries (%) from both Zn ions and ZnO particles ranged from 92.6% to 97.5%.

2.5. Fate of ZnO in Commercial Foods

The presence of ZnO particles or Zn ions in Zn-fortified commercial foods was determined based on the established CPE method. Commercial foods that indicated ZnO as an additive as seen in product labeling were chosen, and another product containing Zn gluconate as a Zn fortifier was also used for comparative study. When the supernatants and precipitates after CPE were quantitatively analyzed by

ICP-AES, the fate of ZnO differed depending on food type (Figure 5). The highest concentration of ZnO was detected as particle forms (~77–99%) in powdered or dried foods (chocolate powder, powdered probiotics, and cereals). ZnO was determined to be mainly present as Zn ions (~92–98%) in liquid beverages (functional peptide beverage and fruit juice). It is worth noting that the Zn ionic fate of Zn gluconate after CPE was also found in another functional mineral beverage (Figure 5F). The pHs of peptide, fruit juice, and mineral beverages were 4.7, 3.4, and 2.7, respectively, whereas the pHs of powdered chocolate, probiotics, and cereals were 7.3, 7.5, and 6.5, respectively. The total Zn recoveries (%) from both ZnO particles and Zn ions ranged from 96.9% to 102.8%, which was calculated based on composition labeling in commercial foods.

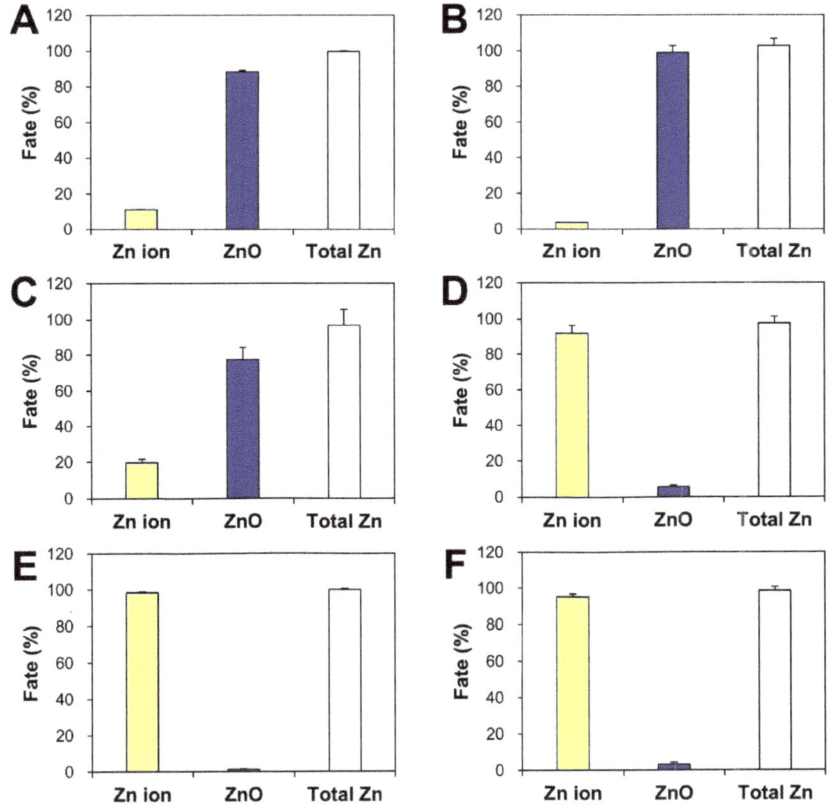

Figure 5. Fate (%) of ZnO in ZnO-added commercial foods: (**A**) chocolate powder, (**B**) powdered probiotics, (**C**) cereals, (**D**) functional peptide beverage, (**E**) fruit juice. (**F**) Fate of Zn gluconate in a functional mineral beverage by CPE.

2.6. Intracellular Fate of ZnO NPs

The fate of ZnO NPs was evaluated by CPE in human intestinal cells, and intracellular Zn levels in both supernatants and precipitates after CPE were quantified after cell lysis. Figure 6A shows that the uptake of ZnO NPs, as measured by total Zn levels, increased in concert with incubation time and reached a plateau at 1 h post-incubation. Intracellular ZnO NPs separated by CPE increased with time and reached a maximum level at 6 h, and then returned to normal control level after 24 h. On the other hand, intracellular Zn ion concentrations gradually increased and the highest Zn ion level was detected at 24 h, which was statistically similar to total Zn level at 24 h. When we compared ZnO NP/Zn ion ratios, most ZnO NPs were present as Zn ions (~71%) at 0.5 h, and both ZnO NPs and Zn ions were present at almost similar levels at 6 h (~53 vs. 47%). Zn ion forms (~87%) were primarily found at 24 h.

The intracellular fate of ZnO NPs was also checked with a Zn-selective fluorescent probe during incubation and examined by confocal microscopy. Figure 6B shows that the fluorescence intensity increased at 0.5–1 h and decreased thereafter. Magnified confocal images clearly demonstrated that the fluorescence intensity inside cells resulting from Zn ions was higher at 0.5–1 h than that at 6–24 h. Slightly increased fluorescence was also observed in control cells without ZnO NPs, attributed to basal intracellular Zn ion levels.

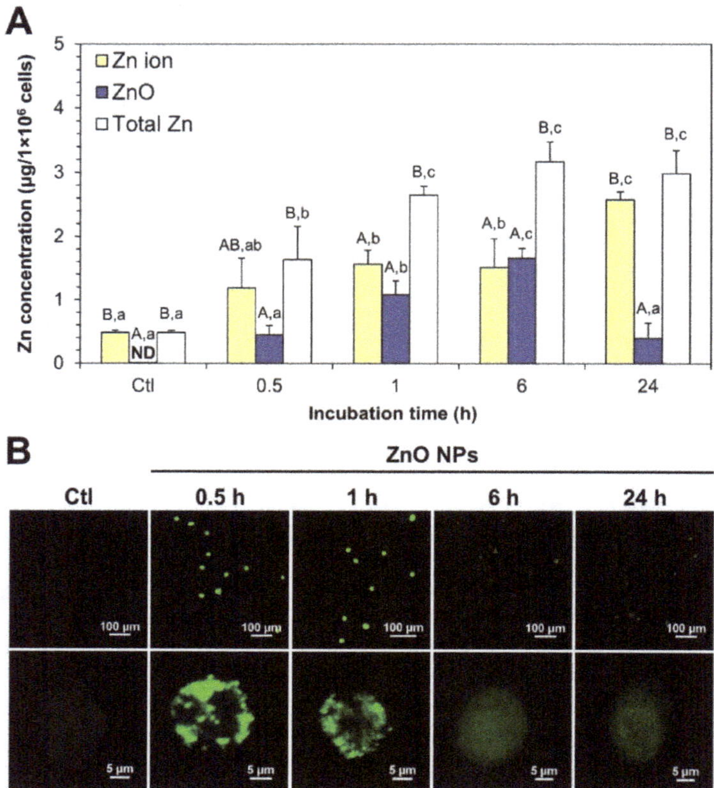

Figure 6. (**A**) Intracellular fate of ZnO NPs in human intestinal Caco-2 cells, as determined by CPE followed by ICP-AES analysis. (**B**) Confocal microscopic images of intracellular Zn ions, as determined by a Zn-selective fluorescent probe, TSQ in Caco-2 cells. Different upper-case letters (A,B) indicate significant differences among Zn ion, ZnO, and total Zn ($p < 0.05$). Different lower-case letters (a,b,c) indicate significant differences among different incubation times ($p < 0.05$). Abbreviation: ND, not detectable.

2.7. Intestinal Transport and Absorption Fate of ZnO NPs

The fate of ZnO NPs after intestinal transport was examined using in vitro Caco-2 monolayer and follicle-associated epithelium (FAE) models, representing the intestinal tight junction and microfold (M) cells in Peyer's patches, respectively. Figure 7A,B show that ZnO NPs were transported through both Caco-2 monolayer and M cells primarily in Zn ion form. No significant differences in transport amount were found between the Caco-2 monolayer and M cells ($p > 0.05$).

We also evaluated the intestinal absorption fate of ZnO NPs using ex vivo everted small intestine sacs, which reflect small intestinal absorption in vivo. These results demonstrated that ZnO NPs were primarily absorbed into the body across the mucosa of the small intestine as Zn ionic forms (Figure 7C). It is also worth noting that ZnO can be slightly but significantly transported by M cells and absorbed in the small intestine in particle form after 0.5–6 h (Figure 7B,C).

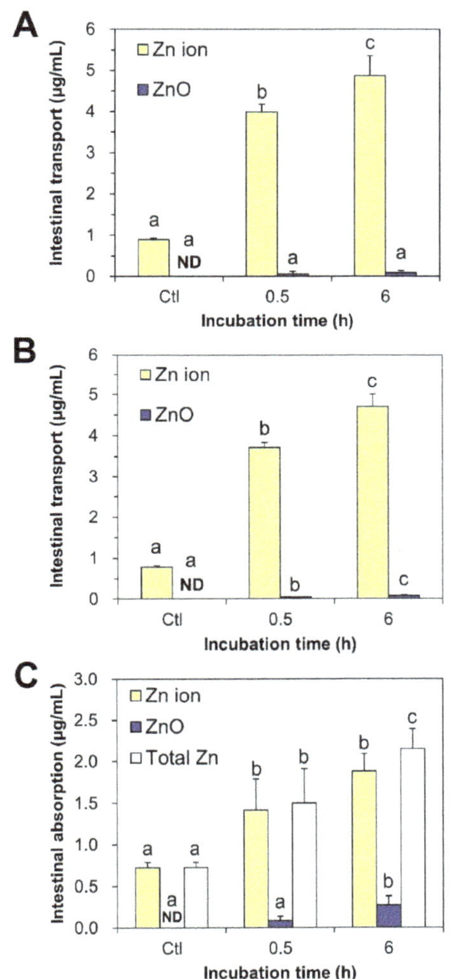

Figure 7. Intestinal transport fate of ZnO NPs obtained from in vitro models of (**A**) Caco-2 monolayer and (**B**) follicle-associated epithelium (FAE) by CPE. (**C**) Intestinal absorption fate of ZnO NPs using ex vivo everted small intestine sacs. Different lower-case letters (a,b,c) indicate significant differences among different incubation times ($p < 0.05$). Abbreviation: ND, not detectable.

3. Discussion

ZnO NPs have a wide range of applications in products intended for human consumption, such as foods and cosmetics, which raises increasing concerns about their potential toxicity. The fate determination of NPs is important, and can answer the fundamental question of whether NPs are present as intact particles or dissolved ionized forms. This type of research is also useful for understanding whether the toxicity of ZnO results from nano-sized particles or from Zn ions.

In particular, fate determination is crucial for partially soluble NPs in commercial products or biological environments, such as ZnO NPs and silver NPs [37,38]. However, separating NPs and ionized forms from complex food and biological matrices without affecting intact particle size and morphology is challenging.

In this study, TX-114-based CPE was first optimized using pristine ZnO NPs. The results obtained revealed that ZnO NPs can be effectively separated into ZnO particles and Zn ions: (1) ZnO NPs were recovered from the precipitated TX-114-rich phase after CPE as intact particles (Figure 2); (2) the hydrodynamic radii and primary particle sizes of pristine ZnO NPs obtained by CPE were statistically similar to those of pristine (Table 1, Figure 2), suggesting that particle size and size distribution were not affected during CPE procedure; (3) more than 82% of pristine ZnO NPs under all variable CPE conditions and ~93% of ZnO NPs in HA at 45 °C were recovered in the precipitates after CPE (Figure 2C), suggesting the efficacy of CPE for separating ZnO NPs as particle form; and (4) small amounts (less than 2%) of ZnO NPs were detected in supernatants as Zn ions (Figure 2C), indicating that only a minimum amount of Zn ions was released from ZnO NPs during CPE. This was also confirmed by low dissolution property (less than 2%) of ZnO NPs in different dispersants (Figure 3A). On the other hand, zeta potential values were affected and changed to slightly negative charges, approaching zero zeta potentials in all cases after CPE (Table 1). It is known that zeta potentials close to zero are optimal for the formation of NPs-TX-114 micelles [30,31]. The results obtained with pristine ZnO NPs suggest that TX-114-based CPE, especially in HA as a dispersant at 45 °C, can effectively separate ZnO particles and Zn ions without affecting primary particle size, size distribution, or morphology.

The dissolution property of ZnO NPs was highly affected by food matrices and cell culture medium (Figure 3B,C). The solubilities of ZnO NPs in coffee mix, skim milk, and milk ranged from ~30–49% (Figure 3B), although the pHs of ZnO-NP-spiked food matrices were close to neutral, except for sports drink (pH 3.3). It is known that ZnO can be dissolved into Zn ions under acidic conditions [39–41]. Thus, the high solubility (~91%) of ZnO in sports drink can be explained by the low pH of the sports drink. Increased solubilities of ZnO in other food matrices, such as coffee mix, skim milk, and milk, seem to be related to its interaction with food components. Meanwhile, the solubilities of ZnO NPs in MEM cell culture medium (pH 7.0) were ~18% to 25% over the incubation time (Figure 3C), which likely resulted from their interactions with various components found in the MEM [38,42]. The solubility of ZnO could, therefore, be highly affected by the presence of food or biomatrices.

The efficacy of CPE was also confirmed using ZnO-NP-spiked foods, showing the reliability and accuracy of the CPE for fate determination of ZnO NPs in foods: (1) DLS results recovered from the precipitates of ZnO-NP-spiked coffee mix, skim milk, and milk after CPE had similar hydrodynamic radii compared to pristine ZnO NPs (Supplementary Figure S1), indicating that ZnO NPs were well recovered as intact particle forms after CPE; (2) no particle forms were detected in the precipitates of ZnO-NP-spiked sports drink after CPE (Supplementary Figure S1), suggesting the complete dissolution of ZnO into Zn ions; (3) SEM-EDS analysis revealed the presence of Zn elements recovered from the precipitates of all ZnO-NP-spiked foods after CPE, except in sports drink (Figure 4A,B), which was in good agreement with DLS results; and (4) ~51–60% of ZnO were present as particles in coffee mix, skim milk, and milk, whereas almost all ZnO NPs were detected as Zn ions in sports drink after CPE, which is highly consistent with the dissolution properties of ZnO (Figure 3B). Indeed, no significant differences were found between solubilized Zn ions in food matrices (Figure 3B) and Zn ions recovered after CPE (Figure 4C, $p > 0.05$). Hence, the fate of ZnO NPs in food matrices can be determined as both intact particle and Zn ion forms by applying CPE, without affecting particle size and solubility. The total Zn recoveries (%) from both ZnO particles and Zn ions ranged from 92.6% to 97.5%, suggesting the accuracy of the analytical procedure.

The same trends were observed in commercial foods in which ZnO addition was indicated on product labeling. Most ZnO particles were present as Zn ions in liquid foods (functional peptide beverage and fruit juice), while the fate of ZnO was found to be mainly the particle forms in powdered foods (chocolate powder, powdered probiotics, and cereals, Figure 5). It is worth noting that another functional mineral beverage fortified with Zn gluconate was found to contain Zn ions after CPE (Figure 5F), supporting the reliability of the results. The pHs of liquid beverages were 2.7–4.7, which affected ZnO dissolution property due to the fact that ZnO dissolves more rapidly in acidic solutions. The slight presence of ZnO as particles in liquid beverages seems to be attributable to a Zn ion complex formed with other food components. The total Zn recoveries (%) from both ZnO and Zn ions ranged from 96.9% to 102.8%, and was calculated based on composition labeling in commercial foods and ICP-AES analysis, supporting the accuracy of the CPE procedure. Zn ion ratios in ZnO-NP-spiked powdered foods (Figure 4) were higher than those in commercial powdered or dried foods (Figure 5). The various processing steps used for commercial food products, such as formulation, mixing, and thermal or drying treatment, may increase the stability of ZnO in processed, powdered foods, contributing to its low dissolution in food matrices. Taken together, it is probable that ZnO as a food additive is primarily present as a particle in powdered or dried foods, but can be easily decomposed into Zn ions in liquid foods.

The intracellular fate of ZnO NPs, determined by CPE followed by ICP-AES analysis, revealed that ZnO NPs were taken up in particle forms, but slowly dissolved into Zn ions after a certain time inside cells (Figure 6A). A portion of dissolved Zn ions from ZnO in cell culture medium can be also rapidly taken up by cells, considering that the dissolution property of ZnO NPs in MEM is ~18–25% (Figure 3C). Vandebriel et al. also demonstrated that ZnO NPs can be taken up by cells by in both particle and ionic forms, which is consistent with our findings [43]. Paek et al. reported that Zn ions can be more rapidly and massively absorbed into the bloodstream after oral administration in rats [18], which may support the rapid cellular uptake of Zn ions compared to ZnO (Figure 6A). Taken together, both ZnO NPs and Zn ions can be internalized into cells, but the major fate of ZnO NPs is to become ionized inside cells over time. Gilbert et al. demonstrated that ZnO NPs were completely dissolved into Zn ions after cellular internalization [12]. Wang et al. also reported that ZnO NPs were internalized into cells by endocytosis and localized within acidic lysosomes, releasing Zn ions from internalized ZnO NPs [44–46]. ZnO NPs were reported to cause cytotoxicity associated with an increase in the Zn ions released inside cells [47,48]. However, contradictory results were obtained by confocal microscopy using a Zn–selective fluorescent probe, showing elevated Zn ion levels at 0.5–1 h and decreased Zn ions thereafter (Figure 6B). The discrepancy might be explained by complex formation between Zn ions and other molecular ligands inside cells, which was evidenced by Zn–S bond formation in tissues after oral administration of ZnO NPs in rats [15]. It is, therefore, strongly likely that ZnO NPs are dissolved into Zn ions and form Zn–molecular ligand complexes after internalization into cells, which is in good agreement with the results obtained by Gilbert et al. [12].

The intestinal transport and absorption fate of ZnO NPs, evaluated using in vitro models of human intestinal barriers and ex vivo everted small intestine sacs, was determined to be primarily Zn ion forms. This result suggests that ZnO NPs can be taken up by cells in both particle and ionized forms (Figure 6A), but most ZnO particles are dissolved into Zn ions during intestinal transit and absorption (Figure 7). Our previous report demonstrated that the ex vivo solubility of ZnO NPs in rat-extracted intestinal fluid was ~9% [23], supporting their high dissolution during intestinal transit. Thus, the major fate of absorbed ZnO NPs in the small intestine is likely to be the ionized forms. However, a small amount of ZnO can be also transported by M cells and absorbed as particle form (Figure 7B,C), suggesting different toxicokinetic behaviors of ZnO compared to those of Zn ions, as reported by previous research [18]. Intestinal transport of NPs by M cells was also demonstrated [23,49,50].

4. Materials and Methods

4.1. Materials

ZnO NPs (<100 nm), D-(+)-glucose, humic acid (sodium salt), TX-114, EDTA, formalin, ammonium chloride (NH_4Cl), and monosodium phosphate (NaH_2PO_4) were purchased from Sigma-Aldrich (St. Louis, MO, USA). Sodium hydroxide (NaOH), sodium chloride (NaCl), potassium chloride

(KCl), calcium chloride ($CaCl_2$), magnesium chloride ($MgCl_2$), sodium bicarbonate ($NaHCO_3$), nitric acid (HNO_3), and hydrogen peroxide (H_2O_2) were supplied by Samchun Pure Chemical Co., Ltd. (Pyeongtaek, Gyeonggi-do, Korea). Conical-bottom glass centrifuge tubes (15 mL) were obtained from Daeyoung Science (Seoul, Korea). MEM, Roswell Park Memorial Institute (RPMI) 1640 medium, Dulbecco's modified eagle's medium (DMEM), heat-inactivated fetal bovine serum (FBS), penicillin, streptomycin, Dulbecco's phosphate-buffered saline (DPBS), and phosphate-buffered saline (PBS) were purchased from Welgene Inc. (Gyeongsan, Gyeongsangbuk-do, Korea). N-(6-Methoxy-8-quinolyl)-p-toluenesulfonamide (TSQ) was obtained from Enzo Life Science Inc. (Farmingdale, NY, USA). Matrigel® was from Corning Inc. (Corning, NY, USA). Transwell® polycarbonate inserts were purchased from SPL Life Science Co., Ltd. (Pocheon, Gyeonggi-do, Korea).

Commercial foods used for the spiking experiment were as follows: coffee mix, skim milk, milk, and sports drink. Commercial Zn-fortified products analyzed were as follows: chocolate powder (ZnO added), powdered probiotics (ZnO added), cereals (ZnO added), functional peptide beverage (ZnO added), fruit juice (ZnO added), and functional mineral beverage (Zn gluconate added), all from international brands found in markets in Seoul, Republic of Korea, in 2019.

4.2. Characterization

Primary particle size, shape, and chemical characterization were determined by FE-SEM (JSM-7100F, JEOL, Tokyo, Japan), equipped with EDS (Aztec, Oxford Instruments, Abingdon, UK). Zeta potentials and hydrodynamic radii of particles (1 mg/mL) were measured with Zetasizer Nano System (Malvern Instruments, Worcestershire, UK) after stirring for 30 min, sonication for 15 min (Bransonic 5800, Branson Ultrasonics, Danbury, CT, USA), and dilution (0.1 mg/mL).

4.3. Optimization of CPE for Pristine ZnO NPs

ZnO NPs (100 µg/mL) were dispersed in 7 mL of DDW, HA (final concentration of 10 µg/mL), or Glc (final concentration of 1% (*w/v*)) solution by stirring for 30 min, followed by sonication for 15 min. CPE was processed as described by Majedi et al. [33]. The suspensions (7 mL) were transferred to conical-bottom glass centrifuge tubes and the pH was adjusted to 10 by adding an NaOH solution. Next, 5% (*w/v*) TX-114 (0.5 mL) and 0.2 M NaCl solution (0.75 mL) were added to the suspensions. After dilution to 10 mL with DDW, the mixtures were incubated for 30 min at 37 or 45 °C and centrifuged for 5 min at $2500 \times g$ at 25 °C. The precipitates and supernatants were analyzed by ICP-AES (JY2000 Ultrace, HORIBA Jobin Yvon, Longjumeau, France) after digestion with HNO_3 as described in Section 4.12.

4.4. Fate Determination of ZnO NPs in Food Matrices

ZnO NPs (10 mg) were spiked into 10 g of powdered foods, such as coffee mix and skim milk powder, and mixed well. Next, 0.1 g of the mixed powders were dispersed in 7 mL of HA solution, and the dispersions were stirred for 30 min followed by sonication for 15 min. In parallel, 10 mg of ZnO NPs were spiked with 100 mL of HA-added liquid foods, such as milk and sports drink. Spiked samples (1 mL) were then diluted to 7 mL by adding HA solution and stirring for 30 min, followed by sonication for 15 min. The Zn concentrations were chosen based on the ADI for Zn in Zn-fortified functional foods (2.55–12 mg) in the Republic of Korea [8]. The same procedure was applied as described in Section 4.3.

4.5. Dissolution Property of ZnO NPs in Food/Bio Matrices

ZnO NPs in DDW, different dispersants, or food matrices were prepared and stirred for 30 min. The same concentration of ZnO NPs for optimization of CPE and fate determination in food matrices were used. ZnO NPs (50 µg/mL) in cell culture medium MEM were incubated for 0.5, 1, 6, and 24 h with gentle shaking (180 rpm) at 37 °C. The ZnO suspensions were then centrifuged ($16,000 \times g$) for

15 min, and the supernatants were subjected to ICP-AES analysis after pre-digestion as described in Section 4.12.

4.6. Fate Determination of ZnO in Zn-Fortified Commercial Foods

Zn-fortified powdered foods (10 g) were homogenized in an agate mortar. Samples thus homogenized (0.1 g) were dispersed in 7 mL of HA solution, and the dispersions were stirred for 30 min and sonicated for 15 min. Commercial Zn-fortified beverages (1 mL) were diluted to 7 mL by adding HA solution, and the solutions were stirred for 30 min followed by sonication for 15 min. The same procedure was applied as described in Section 4.3.

4.7. Cell Culture

Human intestinal epithelial Caco-2 cells were purchased from the Korean Cell Line Bank (KCLB; Seoul, Korea). The cells were cultured in MEM containing 10% FBS, 100 units/mL of penicillin, and 100 µg/mL of streptomycin in a 5% CO_2 incubator at 37 °C.

4.8. Cellular Uptake and Intracellular Fate of ZnO NPs

The cells were plated at a density of 1×10^6 cells/well and incubated with ZnO NPs (50 µg/mL) for 0.5, 1, 6, and 24 h. After washing three times with DPBS, 5 mM EDTA in DPBS was used to treated the cells for 40 s in order to remove adsorbed NPs on the surface of cell membrane. After washing with PBS three times, the cells were harvested with a scraper, centrifuged, and re-suspended in 1 mL of DDW to determine the intracellular fate of ZnO NPs by CPE and cellular uptake quantification by ICP-AES. Cells in the absence of ZnO NPs were used as controls.

The suspended cells (1 mL) were transferred to conical-bottom glass centrifuge tubes and sonicated four times for 10 s on ice with a 150 W ultrasonic processor (Sonics & Materials Inc., Newtown, CT, USA). After dilution to 7 mL by adding HA solution, the same procedure was applied as described in Section 4.3.

4.9. Intracellular Fate of ZnO NPs by Confocal Microscopy

The cells were plated at a density of 2×10^4 cells on a glass coverslip, and ZnO NPs (50 µg/mL) were treated for 0.5, 1, 6, and 24 h. After washing with DPBS, the cells were fixed with 500 µL of freshly made 3.7% formalin (containing 1.5% methanol) in DPBS on ice for 20 min. Next, 50 mM NH_4Cl in DPBS was added and incubation was continued on ice for 30 min. After washing twice with DPBS, 50 µL of 30 µM fluorescent probe for Zn ions, TSQ, was added and incubated for 30 min in the dark at room temperature. Finally, the cells were rinsed three times with DPBS and visualized using a D-Eclipse C1 confocal microscope (Nikon Instech. Co., Kawasaki, Japan), equipped with Ar (488 nm) and HeNe (543 nm) lasers. Image acquisition and analysis were performed with EZ-C1 2.3 software (Nikon Instech. Co., Kawasaki, Japan). Each experiment was repeated twice on separate days.

4.10. Intestinal Transport Fate of ZnO NPs

Human Burkitt's lymphoma Raji B cells, supplied from the KCLB, were cultured in RPMI 1640 medium containing FBS (10%), non-essential amino acids (1%), L-glutamine (1%), penicillin (100 units/mL), and streptomycin (100 µg/mL) in a 5% CO_2 incubator at 37 °C. The FAE model, mimicking M cells, was established as described previously [23,51]. After coating Transwell® polycarbonate inserts with Matrigel® matrix prepared in serum-free DMEM for 2 h, supernatants were removed, and inserts were then washed with serum-free DMEM. Caco-2 cells (1×10^6 cells/well) were seeded on the apical sides and grown for 14 days. Lymphoma Raji B cells (1×10^6 cells/well) were added to the basolateral sides, and these co-cultures were maintained for 5 days until trans epithelial electrical resistance (TEER) values reached 150–200 Ω cm^2. The apical medium of the monolayers was then replaced by medium containing ZnO NPs (50 µg/mL), and incubation continued for 0.5 and 6 h.

Caco-2 monoculture was also used to evaluate the transported fate of ZnO NPs through intestinal epithelial tight junction barrier. Caco-2 cells (4.5×10^5 cells/well) were seeded on upper inserts and further cultured for 21 days (TEER $\geq 300 \ \Omega \ cm^2$). Apical medium of the monolayers was then replaced by medium containing ZnO NPs (50 µg/mL), and incubation continued for 0.5 and 6 h.

Basolateral solutions (1 mL) were collected in a conical-bottom glass centrifuge tube, and diluted to 7 mL with HA solution. The same procedure was applied as described in Section 4.3.

4.11. Intestinal Absorption Fate of ZnO NPs

Eight-week-old male Sprague Dawley (SD; 200–250 g) rats were purchased from Koatech Co. (Pyeongtaek, Gyeonggi-do, Korea). Animals were housed in plastic laboratory animal cages in a ventilated room, maintained at 20 ± 2 °C and $60\% \pm 10\%$ relative humidity with a 12 h light/dark cycle. Water and commercial complete laboratory food for rats were available ad libitum. Animals were environmentally acclimated for 5 days before treatment. All animal experiments were performed in compliance with the guideline issued by the Animal and Ethics Review Committee of Seoul Women's University (SWU IACUC-2019A-1), Republic of Korea.

Everted small intestine sacs were prepared as previously described [52]. Briefly, two male rats were fasted overnight (water available) and sacrificed by CO_2 euthanasia. The small intestines were collected, washed three times with Tyrode's solution (containing 0.8 g of NaCl, 0.02 g of KCl, 0.02 g of $CaCl_2$, 0.01 g of $MgCl_2$, 0.1 g of $NaHCO_3$, 0.005 g of NaH_2PO_4, and 0.1 g of glucose in 100 mL of distilled water), cut into sections (5 cm in length), and everted on a puncture needle (0.8 mm in diameter). After one end was clamped, the everted sacs were filled with 200 µL of Tyrode's solution and then tied using silk braided sutures. Each sac was placed in a six well plate containing 3 mL of ZnO NPs (50 µg/mL) for 0.5 and 6 h in a humidified 5% CO_2 atmosphere at 37 °C. The solution in the interior sac was collected and the same procedure was applied after dilution to 7 mL with HA solution, as described in Section 4.3.

4.12. ICP-AES Analysis

All samples were pre-digested with 10 mL of ultrapure HNO_3 at ~160 °C, and 1 mL of H_2O_2 solution was added and heated until the samples were colorless and until the solution was completely evaporated. The digested samples were diluted with 3 mL of DDW, and total Zn concentrations were determined by ICP-AES (JY2000 Ultrace, HORIBA Jobin Yvon).

4.13. Statistical Analysis

Results are expressed as means ± standard deviations. One-way analysis of variance with Tukey's test in SAS Ver.9.4 (SAS Institute Inc., Cary, NC, USA) was used to determine the significances of intergroup differences. Statistical significance was accepted for p values < 0.05.

5. Conclusions

Surfactant TX-114-based CPE was optimized and established for fate determination of ZnO in commercial foods and human intestinal cells. The solubility of ZnO was not affected by dispersants used for CPE, but was highly affected by food matrices or cell culture medium, showing dissolved fate to some extent. ZnO was found to be mainly present as particle forms in powdered or dried foods, whereas its major fate in liquid beverages was in Zn ionic form. On the other hand, ZnO NPs were internalized into cells as both particles and Zn ions, but slowly dissolved into Zn ions upon time, probably forming Zn–ligand complexes inside the cells. ZnO NPs were found to be transported through intestinal barriers and absorbed in the small intestine primarily as Zn ions. However, a portion of ZnO NPs could be absorbed into the body as particles. The toxicity of ZnO NPs is, therefore, likely to be mainly associated with Zn ion toxicity, but long-term potential toxicity resulting from particle forms cannot be completely excluded. These findings will be useful for understanding the potential toxicity of ZnO NPs and for their wide application to commercial foods at safe levels.

Supplementary Materials: Supplementary materials can be found at http://www.mdpi.com/1422-0067/21/2/433/s1. Figure S1: Hydrodynamic radii of pristine ZnO and ZnO obtained by CPE in ZnO-NP-spiked foods.

Author Contributions: S.-J.C. conceived and designed the study, and wrote the manuscript; Y.-R.J. performed all experiments; J.Y. contributed to the discussion. All authors have read and agreed to the published version of the manuscript.

Funding: This research was funded by the National Research Foundation of Korea (NRF) grant funded by the Korea government (MSIT) (2018R1A2B6001238), and by the Nano Material Technology Development Program (NRF-2018M3A7B6051668) of the National Research Foundation (NRF) funded by the Ministry of Science and ICT.

Conflicts of Interest: The authors declare no conflict of interest.

References

1. Peters, R.; Kramer, E.; Oomen, A.G.; Rivera, Z.E.; Oegema, G.; Tromp, P.C.; Fokkink, R.; Rietveld, A.; Marvin, H.J.; Weigel, S.; et al. Presence of nano-sized silica during in vitro digestion of foods containing silica as a food additive. *ACS Nano* **2012**, *6*, 2441–2451. [CrossRef] [PubMed]

2. Skocaj, M.; Filipic, M.; Petkovic, J.; Novak, S. Titanium dioxide in our everyday life; is it safe? *Radiol. Oncol.* **2011**, *45*, 227–247. [CrossRef] [PubMed]

3. Swain, P.S.; Rao, S.B.N.; Rajendran, D.; Dominic, G.; Selvaraju, S. Nano zinc, an alternative to conventional zinc as animal feed supplement: A review. *Anim. Nutr.* **2016**, *2*, 134–141. [CrossRef] [PubMed]

4. Bonaventura, P.; Benedetti, G.; Albarede, F.; Miossec, P. Zinc and its role in immunity and inflammation. *Autoimmun. Rev.* **2015**, *14*, 277–285. [CrossRef]

5. Hirano, T.; Murakami, M.; Fukada, T.; Nishida, K.; Yamasaki, S.; Suzuki, T. Roles of zinc and zinc signaling in immunity: Zinc as an intracellular signaling molecule. *Adv. Immunol.* **2008**, *97*, 149–176.

6. MacDonald, R.S. The role of zinc in growth and cell proliferation. *J. Nutr.* **2000**, *130*, 1500S–1508S. [CrossRef]

7. U.S. Food and Drug Administration (FDA). GRAS Notice. Available online: https://www.accessdata.fda.gov/scripts/cdrh/cfdocs/cfCFR/CFRSearch.cfm?fr=182.8991 (accessed on 1 April 2019).

8. Republic of Korea Ministry of Food and Drug Safety (MFDS). Criteria and Standard of Health Functional Foods. Available online: https://www.mfds.go.kr/brd/m_211/view.do?seq=14375 (accessed on 16 October 2019).

9. FDA. Food Additive Status List. Available online: https://www.fda.gov/food/food-additives-petitions/food-additive-status-list#ftnZ (accessed on 24 October 2019).

10. Kononenko, V.; Repar, N.; Marusic, N.; Drasler, B.; Romih, T.; Hocevar, S.; Drobne, D. Comparative in vitro genotoxicity study of ZnO nanoparticles, ZnO macroparticles and $ZnCl_2$ to MDCK kidney cells: Size matters. *Toxicol. In Vitro* **2017**, *40*, 256–263. [CrossRef]

11. Liu, J.; Kang, Y.; Yin, S.; Song, B.; Wei, L.; Chen, L.; Shao, L. Zinc oxide nanoparticles induce toxic responses in human neuroblastoma SHSY5Y cells in a size-dependent manner. *Int. J. Nanomed.* **2017**, *12*, 8085–8099. [CrossRef]

12. Gilbert, B.; Fakra, S.C.; Xia, T.; Pokhrel, S.; Madler, L.; Nel, A.E. The fate of ZnO nanoparticles administered to human bronchial epithelial cells. *ACS Nano* **2012**, *6*, 4921–4930. [CrossRef]

13. Majedi, S.M.; Lee, H.K.; Kelly, B.C. Role of water temperature in the fate and transport of zinc oxide nanoparticles in aquatic environment. *J. Phys. Conf. Ser.* **2013**, *429*, 012039. [CrossRef]

14. Sivry, Y.; Gelabert, A.; Cordier, L.; Ferrari, R.; Lazar, H.; Juillot, F.; Menguy, N.; Benedetti, M.F. Behavior and fate of industrial zinc oxide nanoparticles in a carbonate-rich river water. *Chemosphere* **2014**, *95*, 519–526. [CrossRef] [PubMed]

15. Baek, M.; Chung, H.E.; Yu, J.; Lee, J.A.; Kim, T.H.; Oh, J.M.; Lee, W.J.; Paek, S.M.; Lee, J.K.; Jeong, J.; et al. Pharmacokinetics, tissue distribution, and excretion of zinc oxide nanoparticles. *Int. J. Nanomed.* **2012**, *7*, 3081–3097.

16. Cho, W.S.; Kang, B.C.; Lee, J.K.; Jeong, J.; Che, J.H.; Seok, S.H. Comparative absorption, distribution, and excretion of titanium dioxide and zinc oxide nanoparticles after repeated oral administration. *Part. Fibre Toxicol.* **2013**, *10*, 9. [CrossRef] [PubMed]

17. Seok, S.H.; Cho, W.S.; Park, J.S.; Na, Y.; Jang, A.; Kim, H.; Cho, Y.; Kim, T.; You, J.R.; Ko, S.; et al. Rat pancreatitis produced by 13-week administration of zinc oxide nanoparticles: Biopersistence of nanoparticles and possible solutions. *J. Appl. Toxicol.* **2013**, *33*, 1089–1096. [CrossRef]

18. Paek, H.J.; Lee, Y.J.; Chung, H.E.; Yoo, N.H.; Lee, J.A.; Kim, M.K.; Lee, J.K.; Jeong, J.; Choi, S.J. Modulation of the pharmacokinetics of zinc oxide nanoparticles and their fates in vivo. *Nanoscale* **2013**, *5*, 11416–11427. [CrossRef]

19. Wang, B.; Feng, W.; Wang, M.; Wang, T.; Gu, Y.; Zhu, M.; Ouyang, H.; Shi, J.; Zhang, F.; Zhao, Y.; et al. Acute toxicological impact of nano- and submicro-scaled zinc oxide powder on healthy adult mice. *J. Nanopart. Res.* **2008**, *10*, 263–276. [CrossRef]

20. Avramescu, M.L.; Rasmussen, P.E.; Chenier, M.; Gardner, H.D. Influence of pH, particle size and crystal form on dissolution behaviour of engineered nanomaterials. *Environ. Sci. Pollut. Res.* **2017**, *24*, 1553–1564. [CrossRef]

21. Cho, W.S.; Duffin, R.; Howie, S.E.; Scotton, C.J.; Wallace, W.A.; Macnee, W.; Bradley, M.; Megson, I.L.; Donaldson, K. Progressive severe lung injury by zinc oxide nanoparticles; the role of Zn^{2+} dissolution inside lysosomes. *Part. Fibre Toxicol.* **2011**, *8*, 27. [CrossRef]

22. Liu, J.H.; Ma, X.; Xu, Y.; Tang, H.; Yang, S.T.; Yang, Y.F.; Kang, D.D.; Wang, H.; Liu, Y. Low toxicity and accumulation of zinc oxide nanoparticles in mice after 270-day consecutive dietary supplementation. *Toxicol. Res.* **2017**, *6*, 134–143. [CrossRef]

23. Yu, J.; Kim, H.J.; Go, M.R.; Bae, S.H.; Choi, S.J. ZnO interactions with biomatrices: Effect of particle size on ZnO-protein corona. *Nanomaterials* **2017**, *7*, 377. [CrossRef]

24. Bae, S.H.; Yu, J.; Lee, T.G.; Choi, S.J. Protein food matrix-ZnO nanoparticle interactions affect protein conformation, but may not be biological responses. *Int. J. Mol. Sci.* **2018**, *19*, 3926. [CrossRef] [PubMed]

25. Go, M.R.; Yu, J.; Bae, S.H.; Kim, H.J.; Choi, S.J. Effects of interactions between ZnO nanoparticles and saccharides on biological responses. *Int. J. Mol. Sci.* **2018**, *19*, 486.

26. Jo, M.R.; Chung, H.E.; Kim, H.J.; Bae, S.H.; Go, M.R.; Yu, J.; Choi, S.J. Effects of zinc oxide nanoparticle dispersants on cytotoxicity and cellular uptake. *Mol. Cell Toxicol.* **2016**, *12*, 281–288. [CrossRef]

27. Zukiene, R.; Snitka, V. Zinc oxide nanoparticle and bovine serum albumin interaction and nanoparticles influence on cytotoxicity in vitro. *Colloids Surf. B* **2015**, *135*, 316–323. [CrossRef] [PubMed]

28. Chao, J.B.; Liu, J.F.; Yu, S.J.; Feng, Y.D.; Tan, Z.Q.; Liu, R.; Yin, Y.G. Speciation analysis of silver nanoparticles and silver ions in antibacterial products and environmental waters via cloud point extraction-based separation. *Anal. Chem.* **2011**, *83*, 6875–6882. [CrossRef]

29. Hartmann, G.; Schuster, M. Species selective preconcentration and quantification of gold nanoparticles using cloud point extraction and electrothermal atomic absorption spectrometry. *Anal. Chim. Acta* **2013**, *761*, 27–33. [CrossRef]

30. Majedi, S.M.; Kelly, B.C.; Lee, H.K. Evaluation of a cloud point extraction approach for the preconcentration and quantification of trace CuO nanoparticles in environmental waters. *Anal. Chim. Acta* **2014**, *814*, 39–48. [CrossRef]

31. Liu, J.F.; Chao, J.B.; Liu, R.; Tan, Z.Q.; Yin, Y.G.; Wu, Y.; Jiang, G.B. Cloud point extraction as an advantageous preconcentration approach for analysis of trace silver nanoparticles in environmental waters. *Anal. Chem.* **2009**, *81*, 6496–6502. [CrossRef]

32. Yu, S.J.; Chao, J.B.; Sun, J.; Yin, Y.G.; Liu, J.F.; Jiang, G.B. Quantification of the uptake of silver nanoparticles and ions to HepG2 cells. *Environ. Sci. Technol.* **2013**, *47*, 3268–3274. [CrossRef]

33. Majedi, S.M.; Lee, H.K.; Kelly, B.C. Chemometric analytical approach for the cloud point extraction and inductively coupled plasma mass spectrometric determination of zinc oxide nanoparticles in water samples. *Anal. Chem.* **2012**, *84*, 6546–6552. [CrossRef]

34. Azizi, N.A.M.; Rahim, N.Y.; Raoov, M.; Asman, S. Optimisation and evaluation of zinc in food samples by cloud point extraction and spectrophotometric detection. *Sci. Res. J.* **2019**, *16*, 41–59. [CrossRef]

35. Tabrizi, A.B. Cloud point extraction and spectrofluorimetric determination of aluminium and zinc in foodstuffs and water samples. *Food Chem.* **2007**, *100*, 1698–1703. [CrossRef]

36. Park, K.; Lee, Y. The stability of citrate-capped silver nanoparticles in isotonic glucose solution for intravenous injection. *J. Toxicol. Environ. Health A* **2013**, *76*, 1236–1245. [CrossRef] [PubMed]

37. Hamilton, R.F.; Buckingham, S.; Holian, A. The effect of size on Ag nanosphere toxicity in macrophage cell models and lung epithelial cell lines is dependent on particle dissolution. *Int. J. Mol. Sci.* **2014**, *15*, 6815–6830. [CrossRef] [PubMed]

38. Reed, R.B.; Ladner, D.A.; Higgins, C.P.; Westerhoff, P.; Ranville, J.F. Solubility of nano-zinc oxide in environmentally and biologically important matrices. *Environ. Toxicol. Chem.* **2012**, *31*, 93–99. [CrossRef]

39. Adamcakova-Dodd, A.; Stebounova, L.V.; Kim, J.S.; Vorrink, S.U.; Ault, A.P.; O'Shaughnessy, P.T.; Grassian, V.H.; Thorne, P.S. Toxicity assessment of zinc oxide nanoparticles using sub-acute and sub-chronic murine inhalation models. *Part. Fibre Toxicol.* **2014**, *11*, 15. [CrossRef]

40. Bian, S.W.; Mudunkotuwa, I.A.; Rupasinghe, T.; Grassian, V.H. Aggregation and dissolution of 4 nm ZnO nanoparticles in aqueous environments: Influence of pH, ionic strength, size, and adsorption of humic acid. *Langmuir* **2011**, *27*, 6059–6068. [CrossRef]

41. Chaúque, E.F.C.; Zvimba, J.N.; Ngila, J.C.; Musee, N. Stability studies of commercial ZnO engineered nanoparticles in domestic wastewater. *Phys. Chem. Earth* **2014**, *67*, 140–144. [CrossRef]

42. Xia, T.; Kovochich, M.; Liong, M.; Madler, L.; Gilbert, B.; Shi, H.; Yeh, J.I.; Zink, J.I.; Nel, A.E. Comparison of the mechanism of toxicity of zinc oxide and cerium oxide nanoparticles based on dissolution and oxidative stress properties. *ACS Nano* **2008**, *2*, 2121–2134. [CrossRef]

43. Vandebriel, R.J.; De Jong, W.H. A review of mammalian toxicity of ZnO nanoparticles. *Nanotechnol. Sci. Appl.* **2012**, *5*, 61–71. [CrossRef]

44. Wang, B.; Zhang, Y.; Mao, Z.; Yu, D.; Gao, C. Toxicity of ZnO nanoparticles to macrophages due to cell uptake and intracellular release of zinc ions. *J. Nanosci. Nanotechnol.* **2014**, *14*, 5688–5696. [CrossRef] [PubMed]

45. Condello, M.; De Berardis, B.; Ammendolia, M.G.; Barone, F.; Condello, G.; Degan, P.; Meschini, S. ZnO nanoparticle tracking from uptake to genotoxic damage in human colon carcinoma cells. *Toxicol. In Vitro* **2016**, *35*, 169–179. [CrossRef] [PubMed]

46. Roy, R.; Parashar, V.; Chauhan, L.K.; Shanker, R.; Das, M.; Tripathi, A.; Dwivedi, P.D. Mechanism of uptake of ZnO nanoparticles and inflammatory responses in macrophages require PI3K mediated MAPKs signaling. *Toxicol. In Vitro* **2014**, *28*, 457–467. [CrossRef] [PubMed]

47. Shen, C.; James, S.A.; de Jonge, M.D.; Turney, T.W.; Wright, P.F.; Feltis, B.N. Relating cytotoxicity, zinc ions, and reactive oxygen in ZnO nanoparticle-exposed human immune cells. *Toxicol. Sci.* **2013**, *136*, 120–130. [CrossRef] [PubMed]

48. Shi, L.E.; Li, Z.H.; Zheng, W.; Zhao, Y.F.; Jin, Y.F.; Tang, Z.X. Synthesis, antibacterial activity, antibacterial mechanism and food applications of ZnO nanoparticles: A review. *Food Addit. Contam. A* **2014**, *31*, 173–186. [CrossRef] [PubMed]

49. Fievez, V.; Plapied, L.; Plaideau, C.; Legendre, D.; des Rieux, A.; Pourcelle, V.; Freichels, H.; Jerome, C.; Marchand, J.; Preat, V.; et al. In vitro identification of targeting ligands of human M cells by phage display. *Int. J. Pharm.* **2010**, *394*, 35–42. [CrossRef]

50. Gamboa, J.M.; Leong, K.W. In vitro and in vivo models for the study of oral delivery of nanoparticles. *Adv. Drug Deliv. Rev.* **2013**, *65*, 800–810. [CrossRef]

51. Des Rieux, A.; Fievez, V.; Theate, I.; Mast, J.; Preat, V.; Schneider, Y.J. An improved in vitro model of human intestinal follicle-associated epithelium to study nanoparticle transport by M cells. *Eur. J. Pharm. Sci.* **2007**, *30*, 380–391. [CrossRef]

52. Gu, T.; Yao, C.; Zhang, K.; Li, C.; Ding, L.; Huang, Y.; Wu, M.; Wang, Y. Toxic effects of zinc oxide nanoparticles combined with vitamin C and casein phosphopeptides on gastric epithelium cells and the intestinal absorption of mice. *RSC Adv.* **2018**, *8*, 26078–26088. [CrossRef]

International Journal of
Molecular Sciences

Article

Cytotoxicity of NiO and Ni(OH)$_2$ Nanoparticles Is Mediated by Oxidative Stress-Induced Cell Death and Suppression of Cell Proliferation

Melissa H. Cambre [1,†], Natalie J. Holl [1,†], Bolin Wang [1], Lucas Harper [1], Han-Jung Lee [2], Charles C. Chusuei [3], Fang Y.S. Hou [4], Ethan T. Williams [3], Jerry D. Argo [3], Raja R. Pandey [3] and Yue-Wern Huang [1,*]

[1]   Department of Biological Sciences, Missouri University of Science and Technology, Rolla, MO 65409-1120, USA; mhcxv8@mst.edu (M.H.C.); njhcm8@mst.edu (N.J.H.); bwnt5@mst.edu (B.W.); lh938@mst.edu (L.H.)
[2]   Department of Natural Resources and Environmental Studies, National Dong Hwa University, Hualien 97401, Taiwan; hjlee@gms.ndhu.edu.tw
[3]   Department of Chemistry, Middle Tennessee State University, Murfreesboro, TN 37132, USA; charles.chusuei@mtsu.edu (C.C.C.); etw2r@mtmail.mtsu.edu (E.T.W.); jda6c@mtmail.mtsu.edu (J.D.A.); rajaram.pandey@mtsu.edu (R.R.P.)
[4]   Department of Biomedical Sciences, University of Wisconsin-Milwaukee, Milwaukee, WI 53211, USA; fyshou@gmail.com
*   Correspondence: huangy@mst.edu; Tel.:+1-573-341-6589
†   M.H.C. and N.J.H. contributed equally.

Received: 12 February 2020; Accepted: 26 March 2020; Published: 28 March 2020

Abstract: The use of nanomaterial-based products continues to grow with advancing technology. Understanding the potential toxicity of nanoparticles (NPs) is important to ensure that products containing them do not impose harmful effects to human or environmental health. In this study, we evaluated the comparative cytotoxicity between nickel oxide (NiO) and nickel hydroxide (Ni(OH)$_2$) in human bronchoalveolar carcinoma (A549) and human hepatocellular carcinoma (HepG2) cell lines. Cellular viability studies revealed cell line-specific cytotoxicity in which nickel NPs were toxic to A549 cells but relatively nontoxic to HepG2 cells. Time-, concentration-, and particle-specific cytotoxicity was observed in A549 cells. NP-induced oxidative stress triggered dissipation of mitochondrial membrane potential and induction of caspase-3 enzyme activity. The subsequent apoptotic events led to reduction in cell number. In addition to cell death, suppression of cell proliferation played an essential role in regulating cell number. Collectively, the observed cell viability is a function of cell death and suppression of proliferation. Physical and chemical properties of NPs such as total surface area and metal dissolution are in agreement with the observed differential cytotoxicity. Understanding the properties of NPs is essential in informing the design of safer materials.

Keywords: nanoparticles; viability; cell proliferation; physicochemical properties; oxidative stress; caspase-3; mitochondrial membrane potential; cell cycle; apoptosis

---

1. Introduction

Nanomaterials have become increasingly popular in the production of a wide range of products and processes including, but not limited to, cosmetics [1], pharmaceuticals [2], medical research [3], semiconductor fabrication [4], food [5], and electronic manufacturing [6]. The global market for nanomaterial-based products is estimated to reach \$55 billion by 2022, with about a 20% compound annual growth rate [7]. Increased use of nanoparticles (NPs) heightens the potential for human exposure, especially from airborne particles or the consumption of products containing them. Workers in various industries are at higher risk of exposure to NPs used in manufacturing via inhalation [8]. While some

NPs are relatively harmless, others produce moderate to severe toxic effects. In vitro studies have demonstrated that NPs are cellularly internalized where they can cause injuries [9–13]. This is observed as increased reactive oxidative stress, mitochondrial dysfunction, severe DNA damage, cell cycle arrest, induction of apoptosis, and increased necrosis [14].

Toxicity depends on the physicochemical properties of NPs [15]. For instance, the crystal structure and morphology of NPs affect cytotoxicity. The amorphous form of TiO_2 generated more reactive oxygen species (ROS) than the anatase and rutile forms [16]. Rod-shaped CeO_2 produced toxic responses in RAW 264.7 cells while octahedron and cubic CeO_2 elicited little change [17]. Surface charge may also influence toxicity, with positively charged ZnO producing a higher degree of toxicity than negatively charged particles in A549 cells [18]. Three iron NPs (Fe_3O_4, oleic acid-coated Fe_3O_4, and carbon-coated Fe) with different positive charges also produced toxic responses in BEL-7402 cells, with higher positive charge correlating with worse toxic responses [19]. Dissolution rate, relative available binding sites on particle surfaces, and particle surface charge of various transition metal oxides correlated with toxicity in A549 cells [20]. It is important to note that the toxicity mechanisms of NPs are not always, but can be, cell line-dependent [9,21]. For instance, NiO NPs arrested BEAS-2B cells in the G_0/G_1 phase of cell cycle while arrest of A549 cells occurred in the G_2/M phase [21]. Furthermore, NiO NPs induced a higher rate of apoptosis in BEAS-2B cells than in A549 cells [21]. ZnO exposure also induced cell cycle alterations in A549 cells but not in BEAS-2B cells [9]. Collectively, morphology, surface charge, dissolution rate, and relative surface binding sites influence NP toxicity.

NiO NPs are used in coloring agents for enamels, nanowires, automotive rear-view mirrors, and more products [22]. $Ni(OH)_2$ NPs are used in rechargeable battery electrodes, nickel cadmium batteries, and nickel metal hydride batteries [23]. Nickel can be released to the environment via various anthropogenic processes. It exists in various oxidation states, though Ni(II), nickel in the +2 valence state, is its prevalent form [24]. The environmental levels of Ni-associated compounds and their effective toxic concentrations are influenced by nickel's oxidation state, agglomeration, and media (i.e., water, soil, foods, anaerobic, aerobic), as well as by interactions with other organic and inorganic matrixes. Although the environmental concentrations of nano-sized nickel are still unknown, their existence in the environment may impose risk to human health. The mechanism of toxicity is an essential element of and forms the base of risk assessment. Toxic responses upon exposure to NiO and $Ni(OH)_2$ NPs have been characterized to a limited extent in in vivo and in vitro settings. Treatment with NiO or $Ni(OH)_2$ NPs induced inflammation in the lungs of rodents [25,26]. NiO induced ROS and lipid peroxidation in A549 cells [27]. Oxidative stress, apoptosis, and reduced viability in the breast cancer cell line MCF-7 and the human airway epithelial cell line HEp-2 were also caused by NiO NP exposure [28]. Particulate and soluble nickel compound treatments (including NiO and $Ni(OH)_2$) led to a varying degree of toxicity in modified Chinese hamster ovary CHO-K1 (AS52) cells [29]. However, $Ni(OH)_2$ cytotoxicity has not been studied in human cell lines. Adverse responses to nickel NPs observed in animals and cells indicate human health could be threatened by exposure.

To date, there are no studies comparing the difference in cellular toxicity and toxicological mechanism upon exposure of HepG2 (a human hepatocellular carcinoma cell line) and A549 (a human bronchoalveolar carcinoma cell line) to NiO and $Ni(OH)_2$ NPs. Further, there have been no studies on the role of suppression of cell proliferation induced by these NPs on overall toxicity. Our preliminary data suggest that $Ni(OH)_2$ NPs decrease viability more significantly than NiO NPs in A549 cells. We thus hypothesized that (1) differential cytotoxicity of NiO and $Ni(OH)_2$ NPs is cell line-, particle-, time-, and dose-dependent, (2) cytotoxicity is mediated by oxidative stress and subsequent cellular events including modulation of mitochondrial membrane potential and caspase-3 enzyme activity, and (3) exposure to NiO and $Ni(OH)_2$ NPs alters cell cycle and suppresses cell proliferation. Our specific aims were to (1) demonstrate that cytotoxicity is cell line-, particle-, time-, and dose-dependent, (2) measure the differences in various biochemical responses upon NiO or $Ni(OH)_2$ exposure, and (3) investigate whether cell viability is a function of cell killing and inhibition of cell proliferation. To achieve our goals, we measured cell viability in HepG2 and A549 cells upon NiO or $Ni(OH)_2$ exposure. We then delineated the toxicological mechanism of action in the context of the physical and chemical properties of NPs and oxidative stress-mediated cellular injuries, including changes in mitochondrial membrane potential, caspase-3 activity, apoptosis, cell proliferation, and cell cycle, in A549 cells.

2. Results

2.1. Physiochemical Properties of NiO and Ni(OH)$_2$

The physiochemical properties of NiO and Ni(OH)$_2$ were analyzed to determine differences between the two NPs which may contribute to distinct cellular responses (Table 1). Transition electron microscope (TEM) analysis revealed the apparent morphology and crystalline structure of the NPs (Figure 1A–F). NiO was in the form of aggregated nanograins, while Ni(OH)$_2$ was a flakey aggregate. Selected area electron diffraction (SAED) patterns (Figure 1E,F) confirmed that NiO NPs have a cubic crystal symmetry with a Fm$\bar{3}$m space group, and Ni(OH)$_2$ NPs have a trigonal crystal symmetry with a P$\bar{3}$m1 space group, consistent with X-ray diffraction (XRD) data (Figure S1). Both NPs contained Ni and O peaks in energy disperse X-ray spectroscopy (EDX) analysis, as expected (Figure 1G,H). Ni(OH)$_2$ contained a small peak that correlated with Br. However, this contamination was estimated to be 1.01% by INCA software (ETAS, Stuttgart, Germany), excluding mass contributed by H, and was thus determined to be insignificant. Peaks corresponding to C and Cu are due to the TEM sample grid composition.

The two NPs were similar in approximate physical size (APS), with a size of 15.2 ± 4.9 nm for Ni(OH)$_2$ and 16.1 ± 4.8 nm for NiO. However, Ni(OH)$_2$ possessed a higher specific surface area (SSA) than NiO, with Ni(OH)$_2$ having an area of 103.2 m^2/g and NiO having 73.5 m^2/g. In our experiments, cytosolic pH conditions were considered as 7.4 pH and lysosomal conditions were considered as 4.5 pH. Ni(OH)$_2$ had a larger number of relative binding sites than NiO based on X-ray photoelectron spectroscopy (XPS) analysis, with a physisorbed to chemisorbed O ratio of 1.956 at pH 7.4 and 2.018 at pH 4.5 compared to NiO with 1.245 at pH 7.4 and 0.385 at pH 4.5 (Table 2, Figure S2). Point of zero charge (PZC) analysis revealed both NiO and Ni(OH)$_2$ had positive surface charge at cytosolic and lysosomal pH, with NiO at 8.7 pH and Ni(OH)$_2$ at 7.9 pH, respectively (Figure 2, Table S1). Ni(OH)$_2$ was more positively charged than NiO, with zeta potentials of 35.8 ± 0.7 mV and 29.1 ± 0.8 mV, respectively. Ni(OH)$_2$ was also more soluble compared to NiO. The NP had a higher dissolution rate at cytosolic and lysosomal pH compared to NiO (Figure 3). The dissolution of Ni(OH)$_2$ after 48 h at pH 7.4 was 2.3% and at pH 4.5 it was 26.2%. After 48 h, NiO only had a dissolution of 0.18% at pH 7.4 and 1.5% at pH 4.5. Thus, Ni(OH)$_2$ had an almost 13-fold increase in dissolution at pH 7.4 compared to NiO and over a 17-fold increase at pH 4.5.

Cell culture medium was a well-buffered solution. However, to understand whether NPs can influence pH in cell culture medium, we added NPs to cell-containing medium (with supplements). The starting pH was 7.7. When incubated with cells for 24 or 48 h, with or without 100 μg/mL of NPs, the medium stayed in the range of 7.2 to 7.7 pH, which was consistent with our previous study; the slight change is negligible in consideration of cytotoxicity [20].

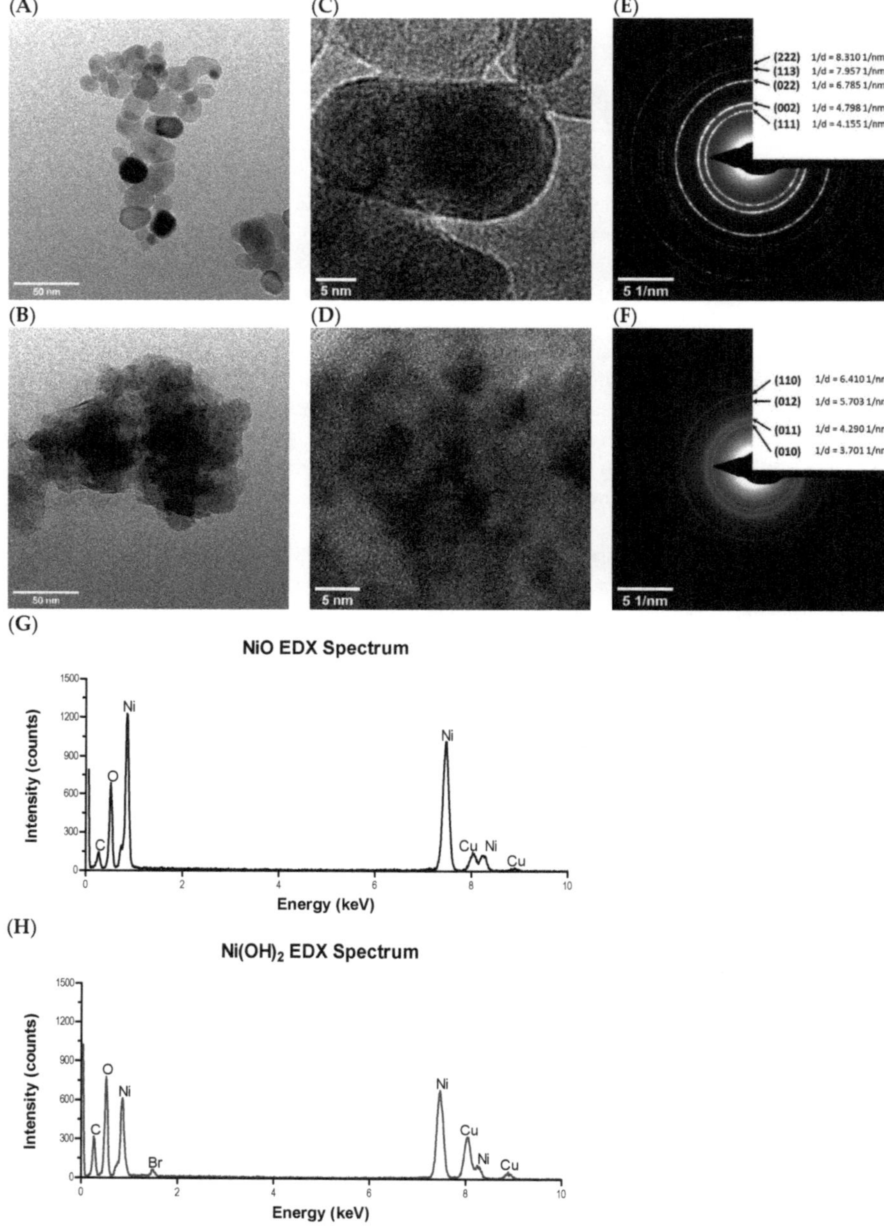

Figure 1. Morphology of (**A**) NiO and (**B**) Ni(OH)$_2$ nanoparticles (NPs) from TEM bright field images. The lattice fringes of (**C**) NiO and (**D**) Ni(OH)$_2$ polycrystals are visible from high resolution transition electron microscopy (HRTEM) images. Selected area electron diffraction (SAED) patterns of (**E**) NiO and (**F**) Ni(OH)$_2$ are shown with (hkl) and 1/distance (1/nm) values for prominent rings indicated. The energy disperse X-ray spectroscopy (EDX) spectra of (**G**) NiO and (**H**) Ni(OH)$_2$ show that both particles are composed of Ni and O, with C and Cu peaks being attributed to the TEM grid material.

Table 1. Physical characteristics of NiO and Ni(OH)$_2$ NPs. Data are expressed as mean ± SD.

Characteristic	NiO	Ni(OH)$_2$
APS * (nm)	16.1 ± 4.8	15.2 ± 4.9
SSA ** (m^2/g)	73.5	103.2
Apparent morphology	Aggregated nanograins	Flakey aggregate
PZC *** (pH)	8.7	7.9
Zeta potential (mV)	29.1 ± 0.8	35.8 ± 0.7

* APS, approximate physical size, ** SSA denotes specific surface area. *** PZC, point of zero charge is defined as the pH value at which the surface is electrostatically neutral.

Table 2. Relative number of binding sites from integrated X-ray photoelectron spectroscopy (XPS) O 1s peak areas.

NP Condition	Metal Oxide or Chemisorbed OH	Physisorbed O	Physisorbed to Chemisorbed O Ratio
NiO pH 7.4	912	1135	1.245
NiO pH 4.5	4874	1875	0.385
Ni(OH)$_2$ pH 7.4	1330	2601	1.956
Ni(OH)$_2$ pH 4.5	2020	4076	2.018

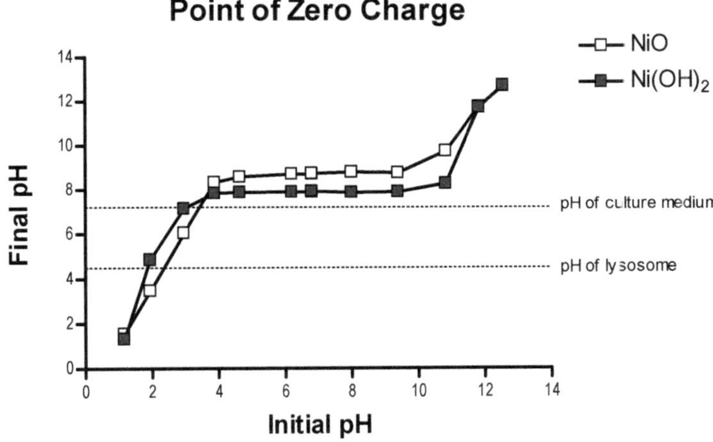

Figure 2. Point of zero charge (PZC) analysis of NiO and Ni(OH)$_2$. Results indicate that the PZCs of NiO and Ni(OH)$_2$ are 8.7 pH and 7.9 pH.

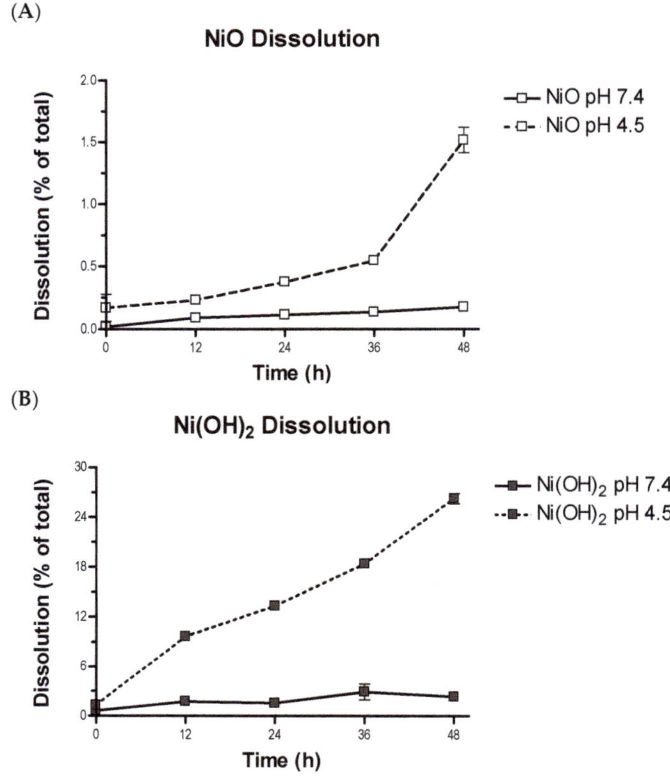

Figure 3. Dissolution of (**A**) NiO and (**B**) Ni(OH)$_2$ at 7.4 and 4.5 pH after 12, 24, 36, and 48 h. Ni in solution was analyzed using inductively coupled plasma-optical emission spectrometry (ICP-OES) and compared to the initial mass of Ni in constant composition experiments. Values are expressed as the mean ± SD from three measurements.

2.2. Cell Viability

A549 and HepG2 cells were treated with NiO and Ni(OH)$_2$ to elucidate the cell-line specific and NP-dependent characteristics of viability. A549 cells experienced an increased loss of viability with time and concentration after exposure to both NiO and Ni(OH)$_2$ (Figure 4A). Exposure for 48 h resulted in a steeper decrease in viability with increasing concentration. Each concentration tested of the NPs (0, 10, 25, 50, 75, 100 µg/mL) resulted in significantly decreased viability in A549 cells ($N = 3$, $p < 0.05$). NiO exposure of 100 µg/mL resulted in a decrease in viability of 42.2% and 73.0% after 24 and 48 h, respectively. Exposure to Ni(OH)$_2$ at 100 µg/mL resulted in a 60.8% decrease in viability after 24 h and an 88.9% decrease after 48 h. The HepG2 cell line only experienced a notable decrease in viability after 48-h Ni(OH)$_2$ exposure at 75 and 100 µg/mL ($N = 3$, $p < 0.05$), with a 27.9% drop in viability at 100 µg/mL (Figure 4B, Table S2). Ni(OH)$_2$ resulted in a drop of 3.1% at 100 µg/mL after 24 h. NiO caused a drop of 0.8% and 6.3% after 24 and 48 h at 100 µg/mL in HepG2, respectively. A549 cells were more susceptible to the toxicity of NiO and Ni(OH)$_2$ than HepG2. Ni(OH)$_2$ was more toxic in both cell lines. Overall, NiO and Ni(OH)$_2$ affected cell viability in a concentration-, time-, particle-, and cell line-dependent manner. Due to the significant differences in toxicity upon NiO or Ni(OH)$_2$ exposure, A549 cells were subject to subsequent mechanistic studies of cytotoxicity.

(**A**)

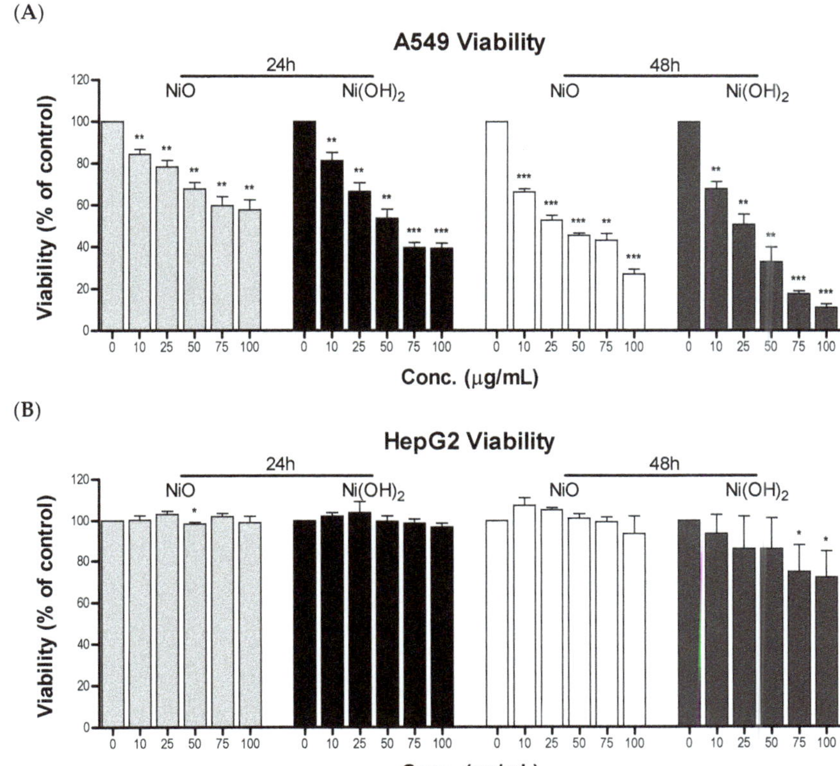

Figure 4. Viability of (**A**) A549 cells and (**B**) HepG2 cells upon exposure to various concentrations of NiO or Ni(OH)$_2$ for 24 and 48 h. Untreated cells were normalized to 100% viable and treated cells were the percentage of viable cells compared to the control, * $p < 0.05$, ** $p < 0.01$, *** $p < 0.001$ vs. control using a one-tailed, unpaired *t*-test. Values are expressed as the mean ± SD from three independent experiments each with three trials.

2.3. Oxidative Stress

2.3.1. Elevation of Oxidative Stress (OS)

Oxidative stress was measured after NiO or Ni(OH)$_2$ exposure for 24 and 48 h in A549 cells to determine its role in the decrease of cell viability (Figure 5). Five concentrations of each NP were tested, being 0, 10, 25, 50, 75, and 100 µg/mL. Longer exposure times resulted in a steeper increase in OS with increasing concentration of NPs. All concentrations produced OS significantly higher than the OS observed in untreated cells ($N = 4$, $p < 0.05$). A549 cells exposed to 100 µg/mL of NiO had a 2.5- and a 12.7-fold increase of OS after 24 and 48 h, respectively. Ni(OH)$_2$ at 100 µg/mL caused a 4.9-fold increase in OS after 24 h and a 27.8-fold increase after 48 h. Ni(OH)$_2$ induced higher levels of OS at both 24 and 48 h.

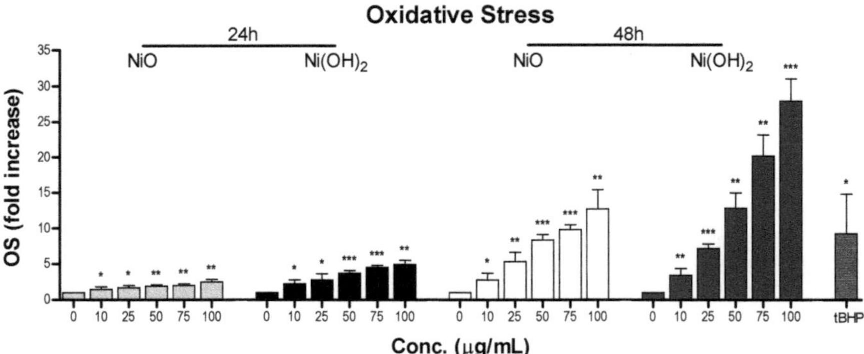

Figure 5. Reactive oxygen species (ROS) produced in A549 cells upon exposure to various concentrations of NiO or Ni(OH)$_2$ for 24 and 48 h. Untreated cells were considered 1-fold of activity and treated cells were the relative fold increase in ROS. Tert-butyl hydroperoxide (tBHP) served as a positive control, * $p < 0.05$, ** $p < 0.01$, *** $p < 0.001$ vs. control using a one-tailed, unpaired t-test. Values are expressed as the mean ± SD from four independent experiments.

2.3.2. Perturbation of Mitochondrial Membrane Potential (MMP)

The dissipation of mitochondrial membrane potential was observed to determine its role in loss of viability in A549 cells upon exposure to NiO or Ni(OH)$_2$ at 10 and 100 µg/mL. In the untreated control cells, an abundance of red color is indicative of healthy mitochondria (Figure 6). Cells treated with Ni(OH)$_2$ or NiO experience OS and have a noticeable decrease in dysfunctional mitochondria, seen as a decrease in red color. Exposure to Ni(OH)$_2$ appears to decrease the abundance of healthy mitochondria more than exposure to NiO. This is likely a result of a higher OS production upon exposure to Ni(OH)$_2$, inducing a greater dissipation in MMP.

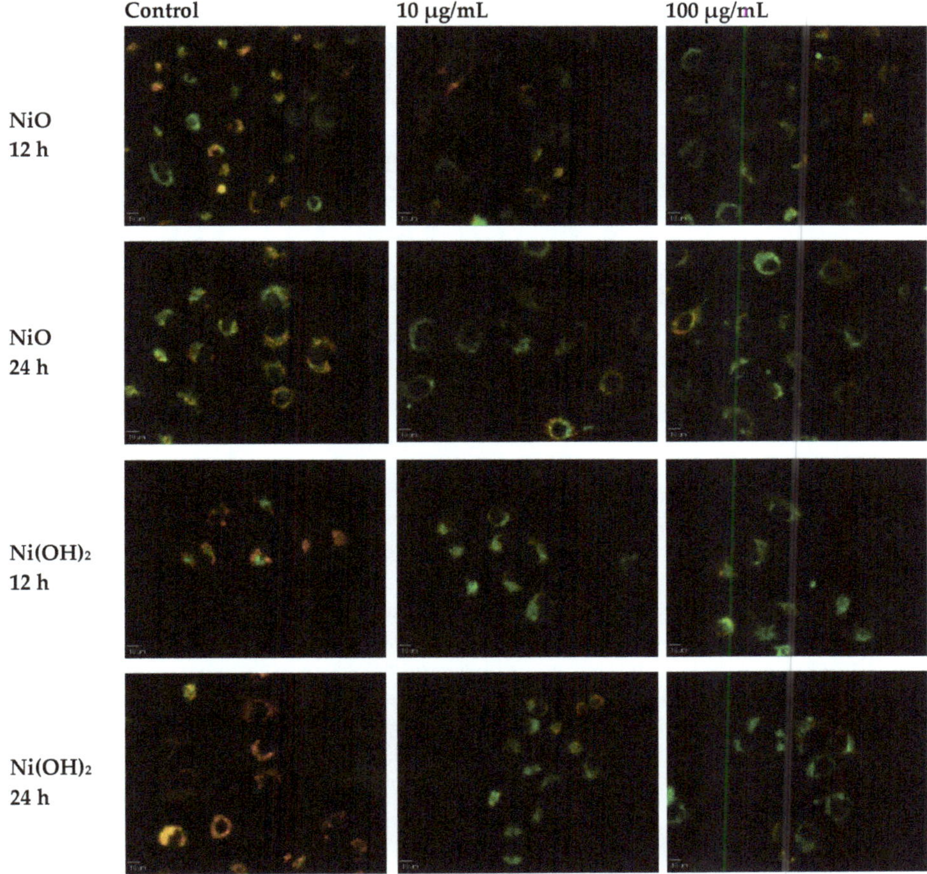

Figure 6. Fluorescence microscopy images of mitochondria membrane potential (MMP) after exposure to NiO or Ni(OH)$_2$ for 12 or 24 h. Scale bars are 10 μm.

2.4. Apoptosis

2.4.1. Elevation of Caspase-3 Enzymatic Activity

Caspase-3 enzymatic activity was measured to determine if programmed cell death was activated in A549 cells upon exposure to NiO or Ni(OH)$_2$ (Figure 7). The same concentrations of NPs, 0, 10, 25, 50, 75, and 100 μg/mL, were used. Exposure to NiO significantly increased caspase-3 activity in all groups except for 10 μg/mL at 24 and 48 h ($N = 3$, $p < 0.05$). NiO exposure of 100 μg/mL caused a 1.4- and a 1.9-fold increase in caspase-3 activity at 24 and 48 h, respectively. Caspase-3 activity was significantly increased at all tested concentrations of Ni(OH)$_2$ ($N = 3$, $p < 0.05$). This increase in activity was higher than NiO, with 100 μg/mL of Ni(OH)$_2$ producing 1.7- and 2.2-fold increases for 24 and 48 h, respectively. Increased caspase-3 enzymatic activity for Ni(OH)$_2$ compared to NiO is consistent with the increased loss of viability.

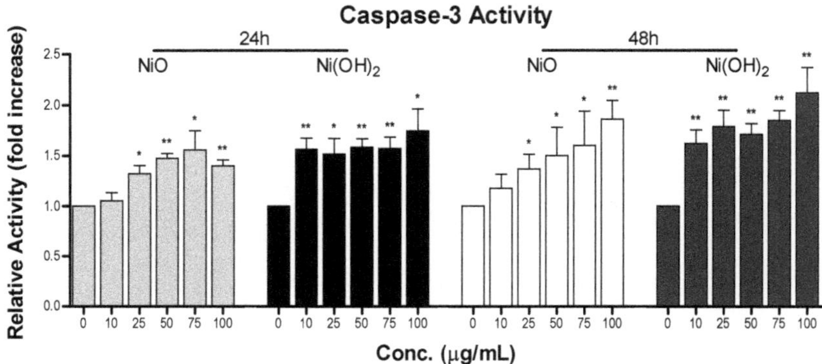

Figure 7. Measurement of caspase-3 activity after A549 cell exposure to various concentrations of NiO or Ni(OH)$_2$ for 24 and 48 h. Untreated cells were considered 1-fold of activity and treated cells are the relative fold increase in caspase-3 activity, * $p < 0.05$, ** $p < 0.01$ vs. control using a one-tailed, unpaired *t*-test. Values are expressed as the mean ± SD from three independent experiments.

2.4.2. Induction of Apoptosis

The levels of induced apoptosis were quantified using flow cytometry to elucidate programmed cell death in A549 cells when exposed to the two NPs (Figure 8 and Figure S3A,B). For our purpose, the total apoptotic percentage of each population was the summation of the subpopulations of cells undergoing early apoptosis and late apoptosis. Measurement of apoptosis in cells using a flow cytometer reflects the number of cells currently undergoing apoptosis and not total viability. NiO exposure for 24 h resulted in a statistically significant decrease in apoptosis for 25, 50, and 75 µg/mL; however, this decrease was considered negligible as the numbers were relatively low and there was no significant change in apoptosis at 100 µg/mL compared to the control. Apoptosis was increased significantly at 75 and 100 µg/mL for 48 h NiO, 24 h Ni(OH)$_2$, and 48 h Ni(OH)$_2$ exposure ($N = 3$, $p < 0.01$). Exposure to 100 µg/mL of NiO resulted in a 6.3% increase in apoptosis after 48 h. There was a 3.8% and a 69.9% increase in apoptosis after exposure to 100 µg/mL of Ni(OH)$_2$ at 24 and 48 h, respectively.

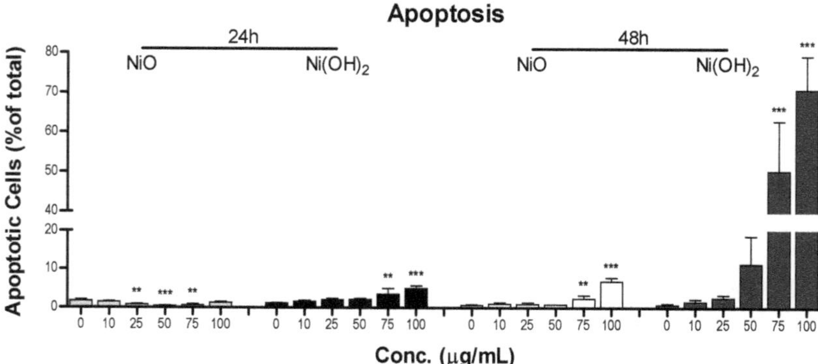

Figure 8. Flow cytometer analysis of total apoptosis in A549 cells after exposure to various concentrations of NiO or Ni(OH)$_2$ for 24 and 48 h, ** $p < 0.01$, *** $p < 0.001$ compared to each respective control using a one-way ANOVA with a Dunnett comparison. Values are expressed as the mean ± SD from three independent experiments.

2.5. Cell Cycle and Proliferation

2.5.1. Suppression of Cell Proliferation

Proliferation was analyzed to determine proliferation's role in A549 cell viability (Figure 9). Exposure to NiO or Ni(OH)$_2$ significantly reduced the rate of proliferation at all tested concentrations at both time points ($N = 4$, $p < 0.05$). A steady decrease in proliferation was seen as concentration increased and the decrease was steeper for longer exposure. Consistent with other results, Ni(OH)$_2$ produced stronger suppression of proliferation. NiO exposure of 100 µg/mL caused a decrease in cell proliferation of 53.9% and 78.4% after 24 and 48 h, respectively. Exposure to 100 µg/mL of Ni(OH)$_2$ resulted in a decrease of 72.9% and 95.7% for 24- and 48-h cell proliferation, respectively.

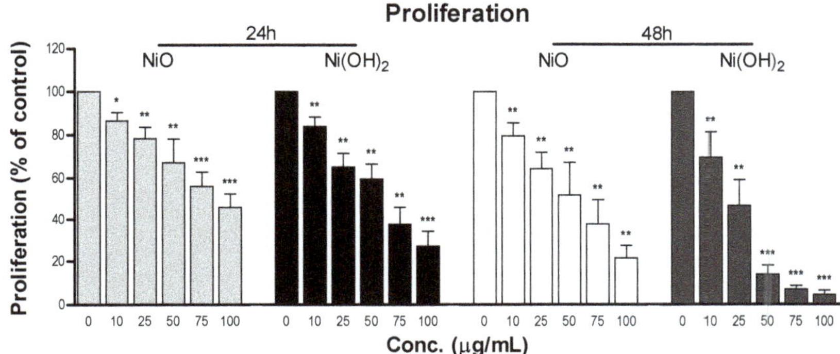

Figure 9. Inhibition of proliferation of A549 cells upon exposure to various concentrations of NiO or Ni(OH)$_2$ for 24 and 48 h. Unexposed cells were normalized to 100% proliferative and exposed cells were the percentage of proliferating cells compared to the control, * $p < 0.05$, ** $p < 0.01$, *** $p < 0.0001$ vs. control using a one-tailed, unpaired *t*-test. Values are expressed as the mean ± SD from four independent experiments.

2.5.2. Alteration of Cell Cycle

The alteration of cell cycle was measured in A549 cells to determine whether cells became arrested in various phases of the cell cycle upon exposure to NiO or Ni(OH)$_2$ (Figure 10). The changes induced by NiO were not significant at any concentration or at either time point, except for the decrease in G$_0$/G$_1$ observed at 48-h 100 µg/mL exposure (Figure 10A,B). NiO exposure of 100 µg/mL after 24 h resulted in a 3.4% decrease in G$_0$/G$_1$, a 3.5% increase in S, and a 1.2% increase in G$_2$/M. Exposure to 100 µg/mL of NiO for 48 h resulted in a decrease of 5.8% in G$_0$/G$_1$, an increase of 3.4% in S, and an increase of 2.4% in G$_2$/M ($N = 3$, $p < 0.05$ for G$_2$/M). Exposure to Ni(OH)$_2$ led to cell cycle dysregulation, with decreasing cells in G$_0$/G$_1$, increasing cells in S, and increasing cells in G$_2$/M with increasing Ni(OH)$_2$ concentration (Figure 10C,D and Figure S3C,D). Ni(OH)$_2$ produced a more dramatic shift in cell cycle, with significant changes in G$_0$/G$_1$ at 100 µg/mL and in G$_2$/M at 75 and 100 µg/mL after 24 h ($N = 3$, $p < 0.05$). G$_0$/G$_1$ was decreased by 23.1%, S was increased by 10.2%, and G$_2$/M was increased by 12.9% for 24-h Ni(OH)$_2$ 100 µg/mL exposure. After 48 h, 75 and 100 µg/mL of Ni(OH)$_2$ exposure induced significant changes in G$_0$/G$_1$, S, and G$_2$/M, with G$_2$/M also experiencing a significant change at 50 µg/mL ($N = 3$, $p < 0.05$). Exposure of 100 µg/mL of Ni(OH)$_2$ after 48 h produced a G$_0$/G$_1$ decrease of 34.1%, an S increase of 15.5%, and a G$_2$/M increase of 18.6%. The increase in the proportion of cells in S and G$_2$/M indicates cells are arresting in these phases. Collectively, arrest is much more prevalent and prominent when cells are treated with Ni(OH)$_2$.

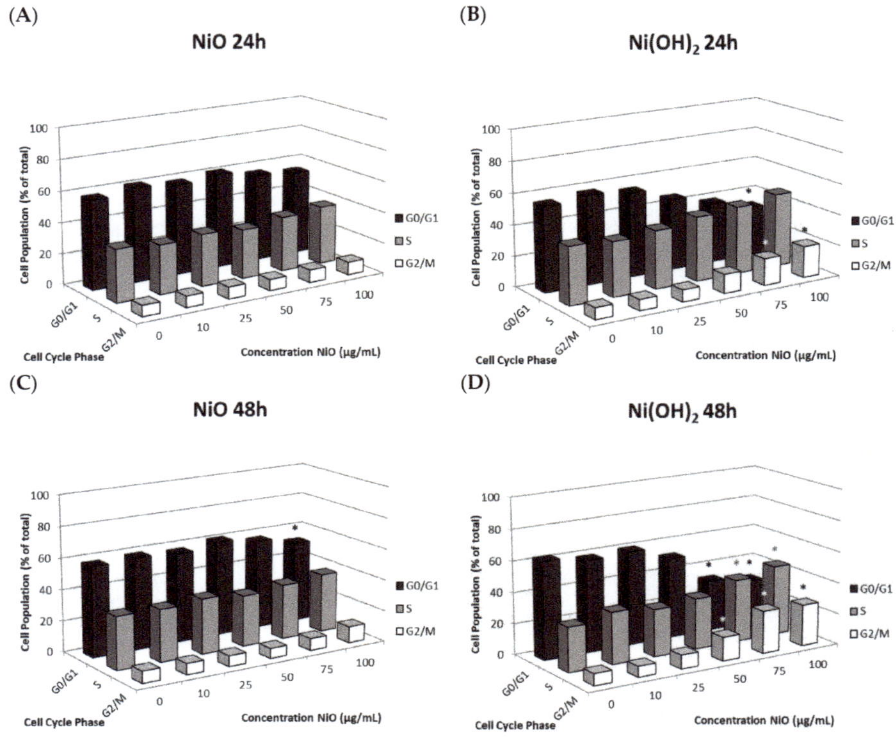

Figure 10. Flow cytometer analysis of cell cycle phase distribution of A549 cells. Analysis was measured after exposure to various concentrations of nanoparticle at (**A**) 24 h NiO, (**B**) 24 h Ni(OH)$_2$, (**C**) 48h NiO, and (**D**) 48h Ni(OH)$_2$, * $p < 0.05$ compared to each respective control using a one-way ANOVA with a Dunnett comparison. Values are expressed as the mean ± SD from three independent experiments.

3. Discussion

In this study, we investigated and compared the cytotoxicity of two nickel NPs. Several cellular responses were explored as components of this cytotoxicity. We hypothesized that (1) the differential cytotoxicity of NiO and Ni(OH)$_2$ NPs is cell line-, particle-, time-, and dose-dependent, (2) cytotoxicity is mediated by oxidative stress and subsequent cellular events including modulation of mitochondrial membrane potential and caspase-3 enzyme activity, and (3) exposure to NiO and Ni(OH)$_2$ NPs alters cell cycle and suppresses cell proliferation.

Cell viability is cell line-dependent. A549 cells (a lung cell line) are much more sensitive to NPs than HepG2 cells (a liver cell line). As A549 cells are epithelial cells in a respiratory organ, it is reasonable that they would be more sensitive to particle exposure than hepatic cells, which are suited to interact with toxic compounds. Other studies are in agreement with this notion. For instance, A549 cells experienced greater induction of OS, lactate dehydrogenase leakage, reduction in glutathione levels, dissipation of MMP, elevation of apoptotic gene expression, and decline in cellular viability compared to HepG2 cells upon exposure to CuFe$_2$O$_4$ and ZnFe$_2$O$_4$ NPs [30,31]. Upon exposure to a variety of sizes and concentrations of silica NPs, HepG2 cells were less susceptible than A549 cells and exhibited a lower degree of toxic responses, including decreased ROS induction, lower decline in glutathione (GSH), and less reduction of cell viability [32]. A549 cells also experienced a greater reduction in MMP and reduction of viability than HepG2 cells upon treatment with silver NPs [33]. One possible explanation regarding the discrepancy between the in vitro toxic responses of the two cell types may be due to the fact that the liver is primarily responsible for removing toxic compounds from the body and has a higher capacity for detoxification (i.e., phase I & II enzymes) than the lung. On a different note, previously we conducted a study on comparative cytotoxicity between two lung cell lines using seven transition metal oxide nanoparticles [20]. BEAS-2B is an immortalized, but not cancerous, human bronchial epithelial cell line, whereas A549 is a human bronchoalveolar carcinoma-derived cell line. Both cell types showed similar trends of toxicity. The issue of organ-specific and cell type-specific cytotoxicity is still unsettled and deserves further attention.

We found that NiO- and $Ni(OH)_2$-induced cytotoxicity is concentration-, time-, and particle-specific in A549 cells. A549 cells experienced concentration-dependent viability at all tested dosages. $Ni(OH)_2$ consistently produced more severe outcomes compared to NiO and increased treatment time led to increased cytotoxic effects for both particles. Previously, the toxicity of $Ni(OH)_2$ had not been examined in human cells. However, our viability results have a similar trend to a study conducted in AS52 cells, where viability was found to be concentration-dependent and the median lethal concentration (LC_{50}) of $Ni(OH)_2$ was found to be six times greater than that of NiO [29].

OS was elevated upon exposure to NiO and $Ni(OH)_2$ and had a strong correlation with cell viability at both 24 and 48 h (Figure 11). This indicates that the generation of free radicals and oxidants is a hallmark of NP toxicity and stress from these oxidative species triggers consequential molecular events leading to cell death. OS-mediated dissipation of MMP due to the presence of both NPs was supported by the apparent reduction in influx of cationic JC-1 into mitochondria. Reduction in the number of healthy mitochondria, or their general functionality, in a cell plays a substantial role in perturbing the homeostasis of bioenergetics and multiple signaling pathways pertaining to cell survival. One such signaling alteration is the increase of caspase-3 enzymatic activity and subsequent apoptosis. Our data demonstrate that both NiO and $Ni(OH)_2$ elevate caspase-3 enzymatic activity and apoptosis in a time- and concentration-dependent manner. NP-induced cell death is complex. In the present study, $Ni(OH)_2$ NPs imposed a much more significant elevation of apoptosis than NiO NPs, particularly towards 48-h exposure. Besides apoptosis, necrosis is also involved in NP-induced cell death. Capasso et al. found a concentration-dependent increase of necrosis mediated by NiO NPs [21]. Another study found that CuO, ZnO, and Mn_2O_3 induced apoptosis in A549 cells but apoptotic cell populations increased in various increments, with the apoptotic rate staying relatively the same between two concentrations before drastically increasing in subsequent concentrations [20]. Additionally, poly vinyl pyrrolidone-coated Ag and Ag^+ NPs induced both apoptosis and necrosis in time- and particle-dependent manners in THP-1 monocyte cells [34]. Collectively, the roles of apoptosis and necrosis seem to be dynamic in the context of acute response and prolonged exposure.

Although cell death induced by NPs has been demonstrated by a wealth of literature, suppression of cell proliferation is relatively under-studied. We hypothesized that the degree of cell viability imposed by exposure to NPs would be a function of cell death and cell proliferation. Our tritiated thymidine incorporation assays provided proliferation data which possessed a very strong linear correlation with cell viability for NiO and $Ni(OH)_2$ over a period of 24 and 48 h (Figure 12). These correlations indicate that suppression of proliferation is a key factor in determining the reduction of cell viability. Modulation of cell proliferation has multiple causations, alteration of cell cycle being one. Our results showed that $Ni(OH)_2$ arrests cells in the S and G_2/M phases while NiO did not influence the cycle significantly. This also indicated that more factors play into proliferation rate. Although studies have found NP-mediated, phase-specific alterations of cell cycle, the effects are not all the same. For instance, exposure to TiO_2 caused HaCat cells to arrest in the S phase while ZnO and CuO exposure caused the same cell type to arrest in G_2/M [35–37]. NPs may be composed of the same elements but have different properties based on their structure, which can also influence phase-specific arrest. Spherical TiO_2 NPs arrested A549 cells in the G_0/G_1 phase while needle-like TiO_2 NPs arrested A549 cells in G_2/M [38,39]. The mechanism of how cell cycle deregulation eventually influences cell proliferation remains to be elucidated.

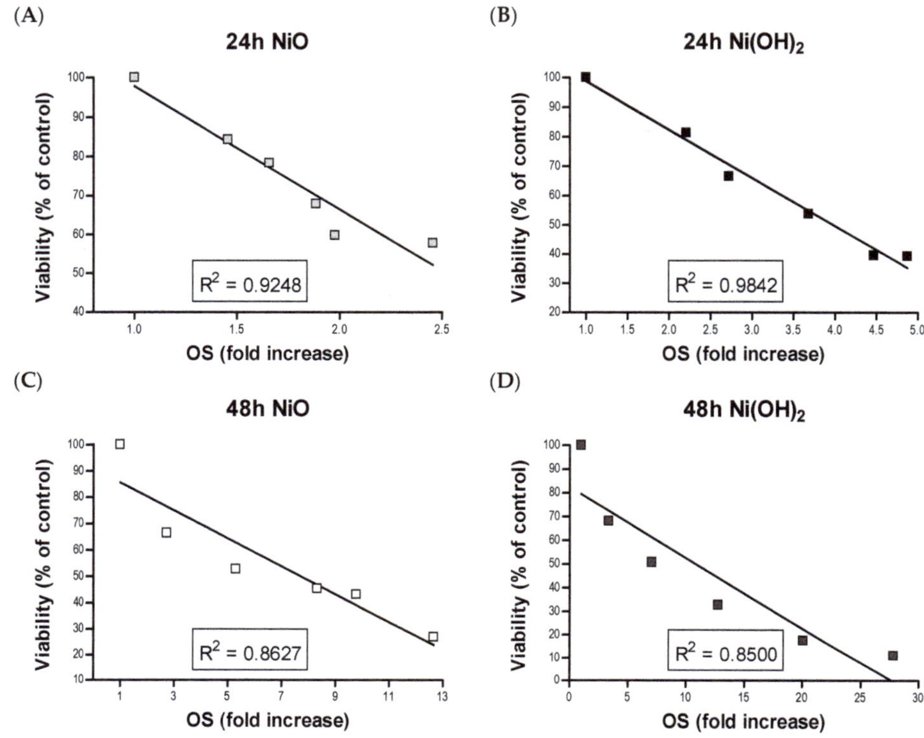

Figure 11. Linear correlation between viability and oxidative stress (OS) rankings for (**A**) 24 h NiO, (**B**) 24 h Ni(OH)$_2$, (**C**) 48 h NiO, and (**D**) 48 h Ni(OH)$_2$.

Among the measured physical and chemical properties, specific surface area, metal dissolution, and surface charge are different between NiO and Ni(OH)$_2$ NPs. Ni(OH)$_2$ NPs have a higher specific surface area, indicative of more available binding sites (Table 2) to interact with biomolecules such as protein, lipids, and nucleic acids. Consequentially, more interactions can lead to a higher degree of observed cellular injuries. One remaining issue is identification of chemical mechanism(s) (e.g., oxidation-reduction potential between NPs and biomolecules) that may damage biomolecules. In our previous study [20], ions dissolved from transition metal oxides correlated with the differential toxicity of seven NPs. Therefore, we suspect that dissolution and the effects of ions may play a role in the observed differential toxicity between NiO and Ni(OH)$_2$. Compared with NiO, a higher degree of metal dissolution from Ni(OH)$_2$ might lead to Ni^{2+}-mediated toxicity. It is likely that once these particles are internalized and reside in the acidic lysosomal environment, the generation and intracellular action of Ni^{2+} would be more exacerbated for Ni(OH)$_2$ than NiO as Ni(OH)$_2$ has been shown to have a much higher dissolution at lysosomal pH as compared to NiO (Figure 3). Both NPs possess positive surface charge, which allows for electrostatic interactions with negative molecules, such as glycosaminoglycans, leading to endocytosis [40]. The slight surface charge difference between these two NPs is not likely a key factor of cellular availability.

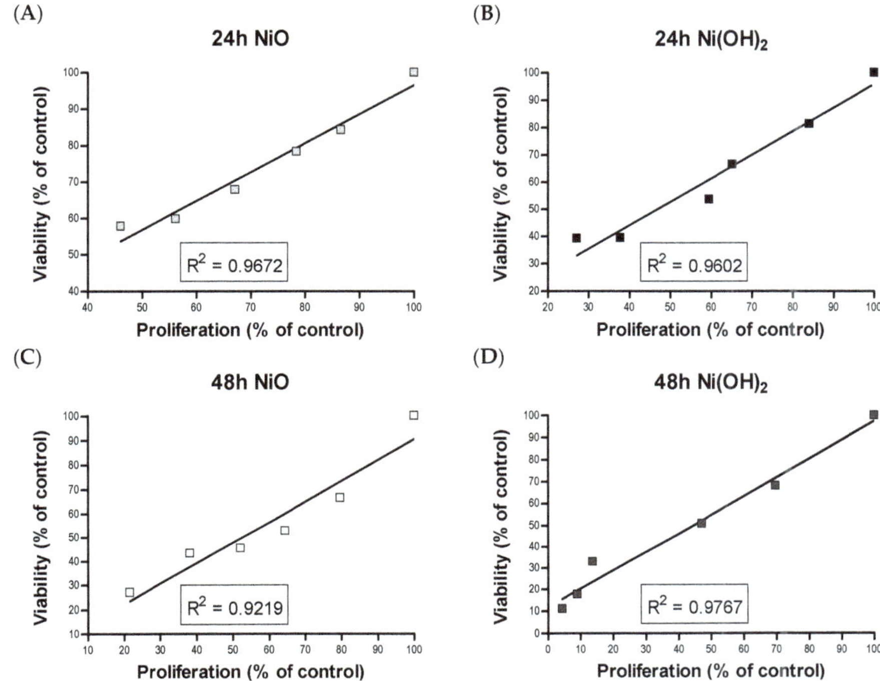

Figure 12. Linear correlation between viability and proliferation rankings for (**A**) 24 h NiO, (**B**) 24 h Ni(OH)$_2$, (**C**) 48 h NiO, and (**D**) 48 h Ni(OH)$_2$.

4. Materials and Methods

4.1. Material Sources

NiO NPs were purchased from Nanostructured and Amorphous Materials (Los Alamos, Houston, TX, USA) and Ni(OH)$_2$ NPs were purchased from US Research Nanomaterials (Houston, TX, USA). A549 cells and HepG2 cells were acquired from the American Tissue Culture Collection (Manassas, VA, USA). 2′,7′-dichlorodihydrofluorescein diacetate (H$_2$DCFDA) and propidium iodide (PI) were obtained from Fisher Scientific (St. Peters, MO, USA). The JC-1 Mitochondrial Membrane Potential Detection Kit and sulforhodamine B were purchased from Biotium (Freemont, CA, USA). Ac-DEVD-pNA was obtained from Anaspec (Fremont, CA, USA). Annexin V-FITC and 7-aminoactinomycin D (7-AAD) were acquired from BD Biosciences (Franklin Lakes, NJ, USA). Tritiated thymidine was purchased from Perkin-Elmer (Waltham, MA, USA). Other chemicals used for experiments were of the highest purity that could be obtained.

4.2. Storage and Characterization of Nanoparticles

NPs were stored in an amber desiccator under a pure nitrogen atmosphere to protect them from moisture, oxidation, and UV damage. The instrumentation and protocols used to characterize NPs followed our previous publication [20]. SSA and APS of NPs in non-aqueous conditions were measured by Brunauer–Emmett–Teller (BET) and TEM, respectively. Morphology and crystal structure, surface charge, metal dissolution, and relative available surface binding sites of NPs in aqueous conditions were measured by high resolution transition electron microscopy (HRTEM), PZC, inductively coupled plasma-optical emission spectrometry (ICP-OES), and XPS.

4.3. TEM and HRTEM

NiO and Ni(OH)$_2$ NPs were suspended in ethanol and dropped onto copper grids with amorphous carbon film. Grids were allowed to dry overnight before insertion into a Tecnai F20 TEM (Thermo Fisher Scientific, Hillsboro, OR, USA) equipped with an energy-dispersive detector (EDS) detector. TEM and HRTEM images were captured, as well as the SAED patterns for both samples.

4.4. Quantification of Available Binding Sites

XPS integrated peak areas of the O 1s binding energy core level were used to quantify the relative number of available binding sites for material in the extracellular matrix to interact with on the NiO and Ni(OH)$_2$ nanoparticle surfaces. Quantification was achieved by noting the physisorbed versus chemisorbed oxygen on the NP surfaces as described in our previous study [20]. In the case of NiO NPs, the deconvoluted O 1s peak denoting the metal oxide represents the underlying substrate while those peak areas of those oxidation states not from the metal oxide denote physisorbed O. The number of available binding sites were quantified by dividing the XPS peak area of the physisorbed O by that of chemisorbed O.

The O 1s core levels of the NiO NPs were observed at 529.0 eV and 531.0 eV, denoting the metal oxide [41–43] and physisorbed O on the NP surfaces, respectively. For the Ni(OH)$_2$ NP surfaces, the binding energy centers of the O 1s orbitals appear at 531.2 and 532.9 eV, denoting chemisorbed OH groups and physisorbed OH/H$_2$O [41,44] on the NP surfaces, respectively. While binding energies for adsorbed OH/H$_2$O are well established, they are not for chemisorbed OH on Ni(OH)$_2$; this information was obtained via XPS scans on the dried Ni(OH)$_2$ NPs performed in our laboratory as a reference standard. The binding energy peak center for this oxidation state was found to be 531.2 eV, having a full width at half maximum (FWHM) of 3.4 km/sec, the parameters of which were used in peak area deconvolution of the O 1s envelopes.

4.5. Cell Culture and Nanoparticle Treatment

A549 cells were maintained in Ham's F-12 modified medium supplemented with 10% HyClone FetalClone serum (GE Healthcare Life Sciences, Marlborough, MA, USA) and 1% penicillin/streptomycin. HepG2 cells were maintained in Eagle's minimum essential medium supplemented with 10% FetalClone serum and 1% penicillin/streptomycin. Both cell lines were grown in 10-cm tissue culture dishes at 37 °C in a 5% CO$_2$ humidified incubator. Upon reaching a confluence of ca. 70–80%, cells were trypsinized, and appropriate numbers of cells were seeded into tissue culture dishes or plates for various experiments. NPs were suspended in cell culture media to create a working concentration of 1 mg NP per 1 mL media. The working suspension of NPs was sealed with parafilm and sonicated for 3 min to break up aggregates. The suspension was vortexed to achieve a homogenous mixture before being added to cells and was diluted in cellular media to achieve a series of desired concentrations. Each treatment group was completed in at least triplicate and appropriate controls were included, with the most common controls being untreated cells.

4.6. Cell Viability

Cell viability was measured using the sulforhodamine B (SRB) assay. A549 cells and HepG2 cells were seeded into 24-well tissue culture plates and allowed to grow for 24 h before compound exposure. For A549 cells, 45,000 and 22,000 cells were seeded per well for 24- and 48-h exposure, respectively. For HepG2 cells, 120,000 cells were seeded per well for both 24 and 48 h. Cells were treated with nickel NPs (0, 10, 25, 50, 75, or 100 µg/mL) for 24 or 48 h. Upon termination of experiments, cell medium was discarded from the cells. The cells were fixed with cold 10% trichloroacetic acid (TCA) for 1 h at 4 °C. The fixed cells were then washed three times with distilled water and then allowed to dry completely. Cells were incubated with 0.5 mL of SRB staining solution (0.2% SRB in 1% acetic acid) for 30 min at room temperature. The cells were then washed three times with 1 mL of 1% acetic acid for 20 min on a rocker to eliminate excess dye. A Q-tip was used to remove excess solution stuck to the sides of the wells. Acetic acid removal was followed by addition of 400 µL of cold 10 mM Tris-HCl solution to each well for 20 min. Aliquots of 250 µL each were transferred onto a 96-well plate and absorbance was measured at 510 nm using a microplate reader (FLUOstar Omega, BMG Labtechnologies, Cary, NC, USA). Cell viability of treatment groups was calculated based on the percent absorbance relative to the control group with appropriate blanks subtracted.

4.7. Oxidative Stress (OS)

Reactive oxidative species were measured with H_2DCFDA. Upon entry of cells, H_2DCFDA was deacetylated by esterases to a non-fluorescent compound. When H_2DCFDA is oxidized by reactive oxidative species, it is converted to the highly fluorescent compound 2′,7′-dichlorofluorescein (DCF) and can be detected by fluorescence spectroscopy. A549 cells were exposed to a series of concentrations of NPs (0, 10, 25, 50, 75, or 100 µg/mL) for 24 or 48 h. Cells were seeded into 96-well plates at 1500 or 750 cells per well for 24- and 48-h treatments, respectively, and grown for 24 h before exposure. As a positive control, cells were incubated with 400 µM tert-butyl hydroperoxide (tBHP) at 37 °C for 1 h before termination of the experiment. Upon termination, the media was removed from the cells followed by a wash with phosphate-buffered saline (PBS). Eighty µL of 0.87 mM H_2DCFDA in ethanol was added to each well and the plate was incubated for 1 h. Cells were then washed with PBS three times followed by addition of 100 µL of PBS. Fluorescence was measured at 510 nm using a microplate reader. The florescence intensity of cells in experimental plates was divided by the fluorescence intensity of the control group to determine the percent increase in ROS with appropriate blank intensities considered.

4.8. Mitochondrial Membrane Potential

Mitochondrial membrane potential was determined with fluorescence microscopy using the JC-1 MMP Detection Kit (Cayman Chemical Company, Ann Arbor, MI, USA). JC-1 (5,5′,6,6′-Tetrachloro-1,1′,3,3′-tetraethylbenzimidazolylcarbocyanine iodide, CAS#: 3520-43-2) monomers fluoresce green when in the cytosol of cells. Accumulation of JC-1 in mitochondria allows the compound to form aggregates that exhibit red fluorescence in a concentration-dependent manner. Normal mitochondrial membrane potential, in healthy cells, allows JC-1 to influx and form aggregates in the mitochondria. In unhealthy cells, mitochondrial membrane potential is decreased, JC-1 concentration cannot increase high enough to form aggregates, and thus the compound remains green [45]. The comparison of red and lack of red fluorescence allows the loss of mitochondrial membrane potential to be observed.

A549 cells were grown for 24 h in 35-mm glass-bottom culture dishes at 15,000 cells per dish. Cells were exposed to several concentrations of nickel NPs (0, 10, or 100 µg/mL) for 12 or 24 h. Following NP treatment, the plates were incubated with 100 µL of JC-1 staining solution per mL of medium at 37 °C for 15 min. Each plate was then washed with 1 mL of PBS followed by addition of 1 mL of PBS before fluorescence detection under an epifluorescence microscope (Olympus Corporation, Tokyo, Japan). Red fluorescence was observed with a Texas Red filter (excitation/emission: 590/610 nm) while green fluorescence was with a FITC filter (excitation/emission: 490/520 nm).

4.9. Caspase-3 Activity

Caspase-3 enzymatic activity was measured using Ac-DEVD-pNA as a substrate. A549 cells were grown for 24 h in 24-well plates at seeding densities of 45,000 or 22,000 cells per well for 24- or 48-h exposure, respectively. Cells were treated with a series of concentrations of NPs (0, 10, 25, 50, 75, or 100 µg/mL) for 24 or 48 h. After incubation with NPs, cells were washed once with 0.5 mL of PBS. Two hundred µL of cold lysis buffer (50 mM Tris-HCl, 1.5 mL of 5 M NaCl, 0.25 g of sodium deoxycholate, 1 mM ethylenediaminetetraacetic acid (EDTA), 0.5 mL of Triton-100, and 50 mL of distilled water) was added to each well, cells were scrapped off the well bottoms, resuspended in the lysis buffer, and incubated at 4 °C for 10 min. Samples were centrifuged at 15,000× *g* for 20 min at 4 °C. Total protein in each sample was measured using a Pierce BCA protein assay (Thermo Scientific, Rockford, IL, USA). Cell lysate and reaction buffer (20% glycerol, 0.5 mM EDTA, 5 mM dithiothreitol, and 100 mM 4-(2-hydroxyethyl)-1-piperazineethanesulfonic acid (HEPES), pH 7.5) were combined in a 96-well plate to have 20 μg of cellular protein and a total volume of 198 μL in each well. Following, 2 μL of 0.5 mg/mL Ac-DEVD-pNA substrate was added to each well. Samples were incubated at 37 °C for 6 h. Absorbance of enzyme-catalyzed release of p-nitroanilide was measured at 405 nm with a microplate reader.

4.10. Apoptosis

Apoptosis was measured with flow cytometry using annexin V-FITC and 7-AAD. A549 cells were seeded into 6-cm tissue culture dishes at densities of 250,000 and 120,000 for 24- and 48-h exposure, respectively, and allowed to grow for 24 h before treatment. Cells were treated with a series of NP concentrations (0, 10, 25, 50, 75, or 100 μg/mL) for 24 or 48 h. Upon termination of the NP exposure period, cells were washed with PBS and harvested with trypsinization. NP treatment medium, PBS washes, and trypsinized cells were all collected in the same centrifuge tube for each concentration of NP in order to avoid loss of floating cells undergoing apoptosis. The samples were centrifuged and the supernatant was discarded. The pellet was washed with 1 mL of ice-cold PBS, centrifuged again, and the supernatant was removed. The cells were resuspended in 100 μL of 1x annexin V binding buffer, 2 μL of annexin V-FITC, and 2 μL of 7-AAD. The cells were then incubated for 20 min in the dark. The stained cell solutions were transferred to the wells of a 96-well plate for flow cytometry analysis on a CytoFLEX flow cytometer (Beckman-Coulter, Brea, CA, USA). Cells in different stages of apoptosis were quantified using FCS Express 6 (DeNovo software, Pasadena, CA, USA). Early and late apoptotic cells were added to represent the total percentage of apoptotic cells.

4.11. Cell Proliferation

Proliferation was determined with a tritiated thymidine ([5′-^{3}H]-thymidine) incorporation assay. A549 cells were seeded in 24-well plates with 45,000 and 22,000 cells per well for 24- and 48-h exposure, respectively. Cells were grown for 24 h before being dosed with nickel compounds. Cells were exposed to a series of concentrations of NPs (0, 10, 25, 50, 75, or 100 μg/mL) and treated with [5′-^{3}H]-thymidine (Perkin-Elmer) simultaneously for 24 or 48 h. A working solution of [5′-^{3}H]-thymidine was prepared with 20 μL of [5′-^{3}H]-thymidine (1 μCi/μL) in 500 μL of PBS. Each well of the 24-well plate was treated with 20 μL of the [5′-^{3}H]-thymidine working solution. Upon termination of each experiment, cells were washed twice with ice-cold PBS. The cells were then fixed in 0.5 mL of ice-cold 10% TCA for 5 min on ice. TCA fixation was repeated once. Cells were brought to room temperature and lysed using 0.5 mL of room temperature 1 N NaOH for 5 min. The solution was neutralized by adding an equal amount of 1 N HCl. The lysed cell solution was thoroughly mixed by pipetting up and down and then transferred to liquid scintillation counting vials with 4 mL of Econo-Safe scintillation counting fluid (Research Products International, Mt. Prospect, IL, USA). Sample vials were then subjected to scintillation counting using a Beckman liquid scintillation counter LS6500 (Beckman-Coulter). The total count of radioactivity was divided by the radioactivity from the 0 μg/mL control cells to determine the percentage of proliferating cells compared to cells not exposed to nickel compounds. All radioactive waste was disposed of following Missouri S&T's Department of Environmental Health and Safety procedures.

4.12. Cell Cycle

Alteration of cell cycle due to nickel NP exposure was measured with flow cytometry and PI staining. A549 cells were grown in 6-cm tissue culture dishes for 24 h before treatment. The seeding density of cells per dish was 250,000 cells for 24-h treatment and 120,000 cells for 48-h treatment. Cells were exposed to a series of concentrations of NPs (0, 10, 25, 50, 75, or 100 μg/mL) for 24 or 48 h.

Following incubation with NP treatments, cells were washed with PBS, harvested using trypsinization, and centrifuged. The cell pellet was then resuspended in 1 mL of PBS and 3 mL of ice-cold absolute methanol was added drop-wise to the solution while vortexing. The cells were stored at 4 °C for at least 24 h to fix. After fixation, the cells were centrifuged and washed once with PBS. The cells were then suspended in PI staining solution (50 µg/mL PI, 0.1% RNase A, and 0.05% Triton X-100 in PBS) for 20 min in the dark. One mL of PBS was added to each sample before centrifuging, the supernatant was removed, and cells were resuspended in 250 µL of PBS. The stained samples were then pipetted into a 96-well plate and analyzed with a flow cytometer. FCS Express 6 was used to determine the distribution of cells in different cell cycle phases. The number of cells in each phase of the cell cycle (G_0/G_1, S, and G_2/M) was totaled and the percentage in each phase was calculated.

4.13. Statistical Analysis

Each experiment was repeated at least three times independently with each treatment group having at least triplicate samples. Data are presented as mean ± standard deviation. Statistical analysis was performed in Minitab 19. One-tailed unpaired t-tests were used to compare experimental groups to the control group in normalized data sets, with μ > control or μ < control depending on the experimental hypothesis. Analysis of variance (ANOVA) with Dunnett comparison was used to determine significant differences against the control group. Significance was set at $p < 0.05$. P-values less than 0.05, 0.01, and 0.001 are noted in figure legends. Linear regression to analyze correlations between data was completed using GraphPad Prism 4. All figures were produced using GraphPad Prism 4 except for the cell cycle distribution graphs, which were produced by Microsoft Excel 2016.

5. Conclusions

Figure 13 depicts the interwoven pathways of toxic events for NPs. Toxicity exerted by NiO and $Ni(OH)_2$ NPs is cell line-, concentration-, time-, and particle-dependent in the range of 10 to 100 µg/mL. $Ni(OH)_2$ is more cytotoxic than NiO. NP-induced oxidative stress triggered subsequent dissipation of mitochondrial membrane potential and induction of caspase-3 enzyme activity. The subsequent apoptotic events led to the reduction of cell number. In addition to cell death, cell cycle deregulation and suppression of cell proliferation also regulated cell number. Thus, the observed cell viability is a function of cell death and suppression of proliferation. Differences in physical and chemical properties of the NPs, such as metal dissolution and total surface area, are in agreement with the observed differential toxicity of these two NPs.

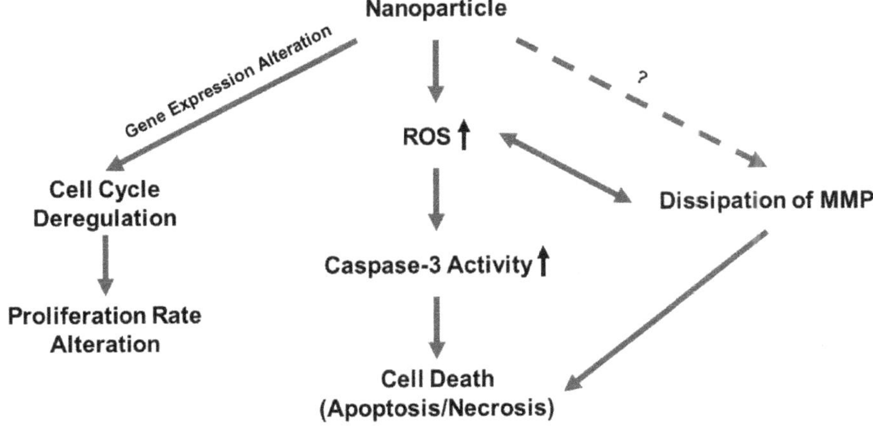

Figure 13. Cell viability is a function of cell death and suppression of proliferation.

Supplementary Materials: Supplementary materials can be found at http://www.mdpi.com/1422-0067/21/7/2355/s1, Figure S1: XRD spectra of NiO and Ni(OH)2. Figure S2: XPS core levels of O 1s orbitals of NiO and Ni(OH)2. Figure S3: Exemplary flow cytometer data in FCS Express 6. Table S1: Point of Zero Charge. Table S2: HepG2 Viability.

Author Contributions: Conceptualization, Y.-W.H. and H.-J.L.; methodology and investigation, M.H.C., N.J.H., B.W., L.H., E.T.W., J.D.A., R.R.P., F.Y.S.H., C.C.C., and Y.-W.H.; validation, N.J.H., C.C.C., F.Y.S.H., and Y.-W.H.; formal analysis, M.H.C., N.J.H., F.Y.S.H., and Y.-W.H.; data curation, M.H.C., N.J.H., B.W., C.C.C., and F.Y.S.H.; writing—original draft preparation, M.L.C., N.J.H., and Y.-W.H.; writing—review and editing, N.J.H., Y.-W.H., F.Y.S.H.; C.C.C., and H.-J.L.; supervision, C.C.C., F.Y.S.H., and Y.-W.H.; project administration, Y.-W.H.; funding acquisition, Y.-W.H. All authors have read and agreed to the published version of the manuscript.

Funding: This research was supported by Missouri S&T cDNA Resource Center.

Acknowledgments: We thank Teresa Carter for assistance in carrying out the dissolution experiments. We thank technical supports from Missouri S&T Material Research Center and Missouri S&T Center for Energy and Environment. We thank Wei-Ting Chen for assistance of HRTEM, EDX, and XRD analyses.

Conflicts of Interest: The authors declare no conflict of interest. The funders had no role in the design of the study; in the collection, analyses, or interpretation of data; in the writing of the manuscript, or in the decision to publish the results.

Abbreviations

NP	Nanoparticle
APS	Approximate physical size
SSA	Specific surface area
PZC	Point of zero charge
HRTEM	High-resolution transmission electron microscopy
EDX	Energy disperse X-ray spectroscopy
XRD	X-ray diffraction
SAED	Selected area electron diffraction
XPS	X-ray photoelectron spectroscopy
BET	Brunauer–Emmett–Teller
ICP-OES	Inductively coupled plasma-optical emission spectrometry
OS	Oxidative stress
MMP	Mitochondrial membrane potential
SRB	Sulforhodamine B
7-AAD	7-aminoactinomycin D
DCF	2′,7′-dichlorofluorescein
PI	Propidium iodide

References

1. Nohynek, G.J.; Dufour, E.K.; Roberts, M.S. Nanotechnology, cosmetics and the skin: Is there a health risk? *Skin Pharmacol. Physiol.* **2008**, *21*, 136–149. [CrossRef]

2. De Jong, W.H.; Borm, P.J. Drug delivery and nanoparticles:applications and hazards. *Int. J. Nanomed.* **2008**, *3*, 133–149. [CrossRef]

3. O'Neal, D.P.; Hirsch, L.R.; Halas, N.J.; Payne, J.D.; West, J.L. Photo-thermal tumor ablation in mice using near infrared-absorbing nanoparticles. *Cancer Lett.* **2004**, *209*, 171–176. [CrossRef]

4. Kowshik, M.; Deshmukh, N.; Vogel, W.; Urban, J.; Kulkarni, S.K.; Paknikar, K.M. Microbial synthesis of semiconductor CdS nanoparticles, their characterization, and their use in the fabrication of an ideal diode. *Biotechnol. Bioeng.* **2002**, *78*, 583–588. [CrossRef]

5. Weir, A.; Westerhoff, P.; Fabricius, L.; Hristovski, K.; von Goetz, N. Titanium dioxide nanoparticles in food and personal care products. *Environ. Sci. Technol.* **2012**, *46*, 2242–2250. [CrossRef]

6. Li, Y.; Wu, Y.; Ong, B.S. Facile synthesis of silver nanoparticles useful for fabrication of high-conductivity elements for printed electronics. *J. Am. Chem. Soc.* **2005**, *127*, 3266–3267. [CrossRef]

7. Inshakova, E.; Inshakov, O. World market for nanomaterials: Structure and trends. In *MATEC Web of Conferences*; EDP Sciences: Julius, France, 2017; Volume 129, p. 02013. [CrossRef]

8. Van Broekhuizen, P.; van Broekhuizen, F.; Cornelissen, R.; Reijnders, L. Workplace exposure to nanoparticles and the application of provisional nanoreference values in times of uncertain risks. *J. Nanopart Res.* **2012**, *14*, 770. [CrossRef]

9. Lai, X.; Wei, Y.; Zhao, H.; Chen, S.; Bu, X.; Lu, F.; Qu, D.; Yao, L.; Zheng, J.; Zhang, J. The effect of Fe2O3 and ZnO nanoparticles on cytotoxicity and glucose metabolism in lung epithelial cells. *J. Appl. Toxicol.* **2015**, *35*, 651–664. [CrossRef] [PubMed]

10. Di Bucchianico, S.; Gliga, A.R.; Akerlund, E.; Skoglund, S.; Wallinder, I.O.; Fadeel, B.; Karlsson, H.L. Calcium-dependent cyto- and genotoxicity of nickel metal and nickel oxide nanoparticles in human lung cells. *Part Fibre Toxicol.* **2018**, *15*, 32. [CrossRef] [PubMed]

11. Vamanu, C.I.; Cimpan, M.R.; Hol, P.J.; Sornes, S.; Lie, S.A.; Gjerdet, N.R. Induction of cell death by TiO2 nanoparticles: Studies on a human monoblastoid cell line. *Toxicol. Vitro* **2008**, *22*, 1689–1696. [CrossRef]

12. Han, J.W.; Gurunathan, S.; Jeong, J.K.; Choi, Y.J.; Kwon, D.N.; Park, J.K.; Kim, J.H. Oxidative stress mediated cytotoxicity of biologically synthesized silver nanoparticles in human lung epithelial adenocarcinoma cell line. *Nanoscale Res. Lett.* **2014**, *9*, 459. [CrossRef] [PubMed]

13. Limbach, L.K.; Li, Y.; Grass, R.N.; Brunner, T.J.; Hintermann, M.A.; Muller, M.; Gunther, D.; Stark, W.J. Oxide nanoparticle uptake in human lung fibroblasts: Effects of particle size, agglomeration, and diffusion at low concentrations. *Environ. Sci. Technol.* **2005**, *39*, 9370–9376. [CrossRef] [PubMed]

14. Martindale, J.L.; Holbrook, N.J. Cellular response to oxidative stress: Signaling for suicide and survival. *J. Cell. Physiol.* **2002**, *192*, 1–15. [CrossRef] [PubMed]

15. Huang, Y.W.; Cambre, M.; Lee, H.J. The Toxicity of Nanoparticles Depends on Multiple Molecular and Physicochemical Mechanisms. *Int. J. Mol. Sci.* **2017**, *18*, 2702. [CrossRef]

16. Jiang, J.; Oberdorster, G.; Elder, A.; Gelein, R.; Mercer, P.; Biswas, P. Does nanoparticle activity depend upon size and crystal phase? *Nanotoxicology* **2008**, *2*, 33–42. [CrossRef]

17. Forest, V.; Leclerc, L.; Hochepied, J.F.; Trouve, A.; Sarry, G.; Pourchez, J. Impact of cerium oxide nanoparticles shape on their In Vitro cellular toxicity. *Toxicol. Vitro* **2017**, *38*, 136–141. [CrossRef]

18. Baek, M.; Kim, M.K.; Cho, H.J.; Lee, J.A.; Yu, J.; Chung, H.E.; Choi, S.J. Factors influencing the cytotoxicity of zinc oxide nanoparticles: Particle size and surface charge. *J. Phys. Conf. Ser* **2011**, *304*, 012044. [CrossRef]

19. Kai, W.; Xiaojun, X.; Ximing, P.; Zhenqing, H.; Qiqing, Z. Cytotoxic effects and the mechanism of three types of magnetic nanoparticles on human hepatoma BEL-7402 cells. *Nanoscale Res. Lett.* **2011**, *6*, 480. [CrossRef]

20. Chusuei, C.C.; Wu, C.H.; Mallavarapu, S.; Hou, F.Y.; Hsu, C.M.; Winiarz, J.G.; Aronstam, R.S.; Huang, Y.W. Cytotoxicity in the age of nano: The role of fourth period transition metal oxide nanoparticle physicochemical properties. *Chem. Biol. Interact.* **2013**, *206*, 319–326. [CrossRef]

21. Capasso, L.; Camatini, M.; Gualtieri, M. Nickel oxide nanoparticles induce inflammation and genotoxic effect in lung epithelial cells. *Toxicol. Lett.* **2014**, *226*, 28–34. [CrossRef]

22. AZoNano. AZoNano. Nickel Oxide (NiO) Nanoparticles—Properties, Applications. Available online: https://www.azonano.com/article.aspx?ArticleID=3378 (accessed on 21 October 2017).

23. Uğurlu, B. Nickel Hydroxide Nanoparticles and Usage Areas of Nickel Hydroxide Nanopowders. Available online: https://nanografi.com/blog/nickel-hydroxide-nanoparticles-and-usage-areas-of-nickel-hydroxide-nanopowders/ (accessed on 28 July 2018).

24. Cempel, M.; Nikel, G. Nickel: A Review of its sources and environmental toxicology. *Pol. J. Environ. Stud.* **2006**, *15*, 375–382.

25. Kadoya, C.; Ogami, A.; Morimoto, Y.; Myojo, T.; Oyabu, T.; Nishi, K.; Yamamoto, M.; Todoroki, M.; Tanaka, I. Analysis of bronchoalveolar lavage fluid adhering to lung surfactant. Experiment on intratracheal instillation of nickel oxide with different diameters. *Ind. Health* **2012**, *50*, 31–36. [CrossRef] [PubMed]

26. Kang, G.S.; Gillespie, P.A.; Chen, L.C. Inhalation exposure to nickel hydroxide nanoparticles induces systemic acute phase response in mice. *Toxicol. Res.* **2011**, *27*, 19–23. [CrossRef] [PubMed]

27. Horie, M.; Fukui, H.; Nishio, K.; Endoh, S.; Kato, H.; Fujita, K.; Miyauchi, A.; Nakamura, A.; Shichiri, M.; Ishida, N.; et al. Evaluation of acute oxidative stress induced by NiO nanoparticles In Vivo and In Vitro. *J. Occup. Health* **2011**, *53*, 64–74. [CrossRef]

28. Siddiqui, M.A.; Ahamed, M.; Ahmad, J.; Majeed Khan, M.A.; Musarrat, J.; Al-Khedhairy, A.A.; Alrokayan, S.A. Nickel oxide nanoparticles induce cytotoxicity, oxidative stress and apoptosis in cultured human cells that is abrogated by the dietary antioxidant curcumin. *Food Chem. Toxicol.* **2012**, *50*, 641–647. [CrossRef]

29. Fletcher, G.G.; Rossetto, F.E.; Turnbull, J.D.; Nieboer, E. Toxicity, uptake, and mutagenicity of particulate and soluble nickel compounds. *Environ. Health Perspect.* **1994**, *102* (Suppl. 3), 69–79. [CrossRef]

30. Alhadlaq, H.A.; Akhtar, M.J.; Ahamed, M. Zinc ferrite nanoparticle-induced cytotoxicity and oxidative stress in different human cells. *Cell Biosci.* **2015**, *5*, 55. [CrossRef]

31. Ahmad, J.; Alhadlaq, H.A.; Alshamsan, A.; Siddiqui, M.A.; Saquib, Q.; Khan, S.T.; Wahab, R.; Al-Khedhairy, A.A.; Musarrat, J.; Akhtar, M.J.; et al. Differential cytotoxicity of copper ferrite nanoparticles in different human cells. *J. Appl. Toxicol.* **2016**, *36*, 1284–1293. [CrossRef]

32. Kim, I.Y.; Joachim, E.; Choi, H.; Kim, K. Toxicity of silica nanoparticles depends on size, dose, and cell type. *Nanomedicine* **2015**, *11*, 1407–1416. [CrossRef] [PubMed]

33. Xin, L.; Wang, J.; Fan, G.; Che, B.; Wu, Y.; Guo, S.; Tong, J. Oxidative stress and mitochondrial injury-mediated cytotoxicity induced by silver nanoparticles in human A549 and HepG2 cells. *Environ. Toxicol.* **2016**, *31*, 1691–1699. [CrossRef]

34. Foldbjerg, R.; Olesen, P.; Hougaard, M.; Dang, D.A.; Hoffmann, H.J.; Autrup, H. PVP-coated silver nanoparticles and silver ions induce reactive oxygen species, apoptosis and necrosis in THP-1 monocytes. *Toxicol. Lett.* **2009**, *190*, 156–162. [CrossRef] [PubMed]

35. Gao, X.; Wang, Y.; Peng, S.; Yue, B.; Fan, C.; Chen, W.; Li, X. Comparative toxicities of bismuth oxybromide and titanium dioxide exposure on human skin keratinocyte cells. *Chemosphere* **2015**, *135*, 83–93. [CrossRef] [PubMed]

36. Gao, F.; Ma, N.; Zhou, H.; Wang, Q.; Zhang, H.; Wang, P.; Hou, H.; Wen, H.; Li, L. Zinc oxide nanoparticles-induced epigenetic change and G2/M arrest are associated with apoptosis in human epidermal keratinocytes. *Int. J. Nanomed.* **2016**, *11*, 3859–3874. [CrossRef]

37. Luo, C.; Li, Y.; Yang, L.; Zheng, Y.; Long, J.; Jia, J.; Xiao, S.; Liu, J. Activation of Erk and p53 regulates copper oxide nanoparticle-induced cytotoxicity in keratinocytes and fibroblasts. *Int. J. Nanomed.* **2014**, *9*, 4763–4772. [CrossRef]

38. Moschini, E.; Gualtieri, M.; Gallinotti, D.; Pezzolato, E.; Fascio, U.; Camatini, M.; Mantecca, P. Metal oxide nanoparticles induce cytotoxic effects on human lung epithelial cells A549. *Chem. Eng. Trans.* **2010**, *22*, 29–35. [CrossRef]

39. Wang, Y.; Cui, H.; Zhou, J.; Li, F.; Wang, J.; Chen, M.; Liu, Q. Cytotoxicity, DNA damage, and apoptosis induced by titanium dioxide nanoparticles in human non-small cell lung cancer A549 cells. *Environ Sci Pollut Res Int* **2015**, *22*, 5519–5530. [CrossRef]

40. Gump, J.M.; June, R.K.; Dowdy, S.F. Revised role of glycosaminoglycans in TAT protein transduction domain-mediated cellular transduction. *J. Biol. Chem.* **2010**, *285*, 1500–1507. [CrossRef]

41. Khawaja, E.E.; Salim, M.A.; Khan, M.A.; Al-del, F.F.; Khattak, G.D.; Hussain, Z. XPS, auger, electrical and optical studies of vanadium phosphate glasses doped with nickel oxide. *J. Non Cryst. Solids* **1989**, *110*, 33–43. [CrossRef]

42. Venezia, A.M.; Bertoncello, R.; Deganello, G. X-ray photoelectron investigation of pumice-supported nickel catalysts. *Surf. Interface Anal.* **1995**, *23*, 239–247. [CrossRef]

43. Lian, K.; Thorpe, S.J.; Kirk, D.W. Electrochemical and surface characterization of electrocatalytically active amorphous Ni CO alloys. *Electrochim. Acta* **1992**, *37*, 2029–2041. [CrossRef]

44. Wagner, C.D.; Zatka, D.A.; Raymond, R.H. Use of the oxygen KLL Auger lines in identification of surface chemical states by electron spectroscopy for chemical analysis. *Anal. Chem.* **1980**, *52*, 1445–1451. [CrossRef]

45. Sivandzade, F.; Bhalerao, A.; Cucullo, L. Analysis of the mitochondrial membrane potential using the cationic JC-1 Dye as a sensitive fluorescent probe. *Bio. Protoc.* **2019**, *9*, e3128. [CrossRef] [PubMed]

International Journal of
Molecular Sciences

Article

Differential Cytotoxicity Induced by Transition Metal Oxide Nanoparticles is a Function of Cell Killing and Suppression of Cell Proliferation

Larry M. Tolliver [1], Natalie J. Holl [1], Fang Yao Stephen Hou [2], Han-Jung Lee [3], Melissa H. Cambre [1] and Yue-Wern Huang [1,*]

[1]   Department of Biological Sciences, Missouri University of Science and Technology, Rolla, MO 65409, USA; larry.tolliver@thermofisher.com (L.M.T.); njhcm8@mst.edu (N.J.H.); mhcxv8@mst.edu (M.H.C.)
[2]   Department of Biomedical Sciences, University of Wisconsin-Milwaukee, Milwaukee, WI 53211, USA; fyshou@gmail.com
[3]   Department of Natural Resources and Environmental Studies, National Dong Hwa University, Hualien 97401, Taiwan; hjlee@gms.ndhu.edu.tw
*   Correspondence: huangy@mst.edu; Tel.: 1-573-341-6589

Received: 12 February 2020; Accepted: 1 March 2020; Published: 3 March 2020

Abstract: The application of nanoparticles (NPs) in industry is on the rise, along with the potential for human exposure. While the toxicity of microscale equivalents has been studied, nanoscale materials exhibit different properties and bodily uptake, which limits the prediction ability of microscale models. Here, we examine the cytotoxicity of seven transition metal oxide NPs in the fourth period of the periodic table of the chemical elements. We hypothesized that NP-mediated cytotoxicity is a function of cell killing and suppression of cell proliferation. To test our hypothesis, transition metal oxide NPs were tested in a human lung cancer cell model (A549). Cells were exposed to a series of concentrations of TiO_2, Cr_2O_3, Mn_2O_3, Fe_2O_3, NiO, CuO, or ZnO for either 24 or 48 h. All NPs aside from Cr_2O_3 and Fe_2O_3 showed a time- and dose-dependent decrease in viability. All NPs significantly inhibited cellular proliferation. The trend of cytotoxicity was in parallel with that of proliferative inhibition. Toxicity was ranked according to severity of cellular responses, revealing a strong correlation between viability, proliferation, and apoptosis. Cell cycle alteration was observed in the most toxic NPs, which may have contributed to promoting apoptosis and suppressing cell division rate. Collectively, our data support the hypothesis that cell killing and cell proliferative inhibition are essential independent variables in NP-mediated cytotoxicity.

Keywords: nanoparticle; cell proliferation; transition metal oxide; cell cycle; apoptosis

---

1. Introduction

Nanotoxicology is the study of nanomaterial toxicity. Nanomaterials are defined as any particulate or agglomerate that has at least one dimension in the size range from 1 to 100 nm [1]. Nanomaterials are being used with an increasing frequency in a variety of industries. Their use is common in semiconductors [2], electronics [3], pharmaceuticals [4], cosmetics [5], consumables [6], and drug delivery platforms being studied for cancer therapy [7]. It is estimated that by the year 2020 the global market for nanomaterial-based applications will reach approximately $3 trillion, and there will be six million workers in the nanotechnology sector worldwide [8,9]. With the use of nanomaterials increasing in frequency and application, exposure amongst the general public and occupational workers has become a concern. While toxicological data exists for some of the microscale equivalents of nanoparticles (NPs), the information cannot predict nanotoxicity, as nanoscale materials have different physical and chemical properties than their microscale equivalents [10]. Particles of smaller size, specifically NPs, can be inhaled deeper into the lungs than larger particles and may enter the circulatory system, resulting in inflammation, systemic distribution, cardiovascular disease, and potential neurological effects [11]. To date, there have not been any epidemiological studies or clinical evidence of NPs causing adverse health effects in humans [12]. However, studies have shown toxicity of select NPs in animal models or in vitro studies [13]. The National Institute for Occupational Safety and Health has designated workplace and occupation exposure limit recommendations for some particles based on their size. For example, TiO_2 exposure is suggested to be limited to 2.4 mg/m^3 for particles less than 2.5 μm and 0.3 mg/m^3 for particles less than 0.1 μm during work days lasting up to 10 h during a 40-h work week [12]. Certain regulations on consumer exposure through food have also been established [6,14].

Our previous studies have demonstrated relationships between NPs, production of reactive oxygen species (ROS), and perturbation of intracellular Ca^{2+} concentrations ($[Ca^{2+}]_{in}$) [15–17]. NPs increase $[Ca^{2+}]_{in}$. The moderation of this increase is attributed to the influx of extracellular calcium, membrane integrity disruption, and perturbation of store-operated calcium entry. The increases in intracellular ROS levels may also have multiple sources. There exist synergistic relationships between $[Ca^{2+}]_{in}$ and oxidative stress (OS) as the increases in both can be reduced by antioxidants. Finally, while $[Ca^{2+}]_{in}$ and OS affect the activity of each other, they both induce cell death by distinct pathways [17]. By systematically studying seven oxides of transition metals in the fourth-period of the periodic table of elements (Ti, Cr, Mn, Fe, Ni, Cu, Zn), we delineated that cytotoxicity is a function of particle surface charge, relative number of particle surface binding sites, and metal ion dissolution rate [18].

NP-mediated toxicity is a rather complicated process and more factors, other than the ones we have determined, are at work. There have been numerous reports on titanium oxide [19], nanogold [20], carbon nanotubes [21], silica oxide [22], aluminum oxide [15], and cerium oxide [23] relating to cell killing. The final cell number that researchers observed was more than just the effect of cell death. Indeed, the outcome of reduced cell number could be the consequence of cell proliferation rate alteration. Herein, we hypothesize that cell viability is a function of cell killing and suppression of cell proliferation. To test our hypothesis, we conducted time- and dose-dependent cytotoxicity studies using human bronchoalveolar carcinoma (A549) cells as a model. Particle properties such as shape, size, and specific surface area were characterized by transmission electron microscopy (TEM) and the Brunauer–Emmett–Teller (BET) method. Apoptosis measurement was performed with flow cytometry and verified with cellular imaging. Tritiated thymidine incorporation was used to determine the rate of cell proliferation. Toxicity was ranked according to severity of cellular responses and a correlation analysis between viability, proliferation, and apoptosis was conducted to visualize the strength of relation between the toxic responses induced by each NP. Alteration of cell cycle was assessed to ascertain whether it contributed to changes in cell proliferation in the two most toxic NPs.

2. Results

2.1. Characterization of Nanoparticles

Physical properties of the seven transition metal oxide NPs were previously characterized (Table S1) [18]. The approximate physical sizes (APS) of the NPs ranged from 16 ± 5 nm (NiO) to 82 ± 31 nm (Mn$_2$O$_3$). The APS measured in the report were similar to those in the data sheet provided by the manufacture. TEM revealed needle-like (TiO$_2$), nearly spherical (Cr$_2$O$_3$, NiO, CuO, ZnO), or spherical (Mn$_2$O$_3$, Fe$_2$O$_3$) morphology for each of the NPs. The lowest specific surface area (SSA) of the NPs was 8.7 m^2/g (Mn$_2$O$_3$) and SSA ranged to a high of 179 m^2/g (TiO$_2$). It was noted that while TiO$_2$, Fe$_2$O$_3$, and CuO had similar APS, they possessed distinctly different SSA. This could be due to differences in surface porosity, morphology, or particle aggregation.

2.2. Reduction of Cytotoxicity

Concentration- and time-dependent cytotoxicity of the seven transition metal oxides is summarized in Figure 1. These NPs can be arranged in three cytotoxicity "tiers" with Fe_2O_3, Cr_2O_3 and TiO_2 being nontoxic to mildly toxic, NiO and Mn_2O_3 being moderately toxic, and ZnO and CuO being highly toxic. The two highly toxic particles were so devastating that the concentration ranges for these particles had to be lowered to 0–20 µg/mL from 0–100 µg/mL that was used for all other particles. Specific viability percentages can be found in Supplementary Table S2 (24-h) and Supplementary Table S3 (48-h). In the low toxicity group, Fe_2O_3 and Cr_2O_3 did not produce notable changes in viability ($n = 3$, p's > 0.05). The 24- and 48-h viabilities of 100 µg/mL of TiO_2 were similar, being $76.2 \pm 5.1\%$ at 24 h and $72.8 \pm 4.7\%$ at 48 h. A significant decrease in viability upon TiO_2 exposure was observed at both time points and all concentrations except for 24-h 10 µg/mL ($n = 3$, p's < 0.05). Both moderately toxic particles (NiO and Mn_2O_3) demonstrated greater viability decline in the 48-h group compared to the 24-h group and had significant decrease in viability at all times and concentrations ($n = 3$, p's < 0.05). The 100 µg/mL of NiO dose resulted in $36.8 \pm 4.6\%$ at 24 h and $9.7 \pm 1.8\%$ at 48 h, with the 100 µg/mL of Mn_2O_3 producing similar changes of $40.4 \pm 5.5\%$ at 24 h and $15.7 \pm 5.1\%$ at 48 h. In the high toxicity group, greater toxicity was also observed with ZnO at 48 h. ZnO produced significant changes at 16 and 20 µg/mL after 24 h and all concentrations except 4 µg/mL after 48 h ($n = 3$, p's < 0.05). The responses varied between concentrations for ZnO with the values at 8 µg/mL being $94.0 \pm 3.4\%$ for 24 h but $48.4 \pm 3.4\%$ for 48 h. By doubling the concentration, the viability at 16 µg/mL dropped to $48.6 \pm 2.1\%$ for 24 h and $1.6 \pm 0.1\%$ for 48 h. At 20 µg/mL, almost no living cells were detected. CuO viabilities were similar at both 24 and 48 h across all concentrations but were all significantly different from the controls ($n = 3$, p's < 0.05). CuO exposure of 20 µg/mL resulted in a viability of $3.5 \pm 1.0\%$ at 24 h and $6.6 \pm 2.9\%$ at 48 h. The toxicity trend is similar to that of our previous findings [18]. This indicates that 1) the NPs have been stable in the storage condition specified in the Materials and Methods and 2) variabilities between multiple experimenters are negligible.

(A)

24h Viability

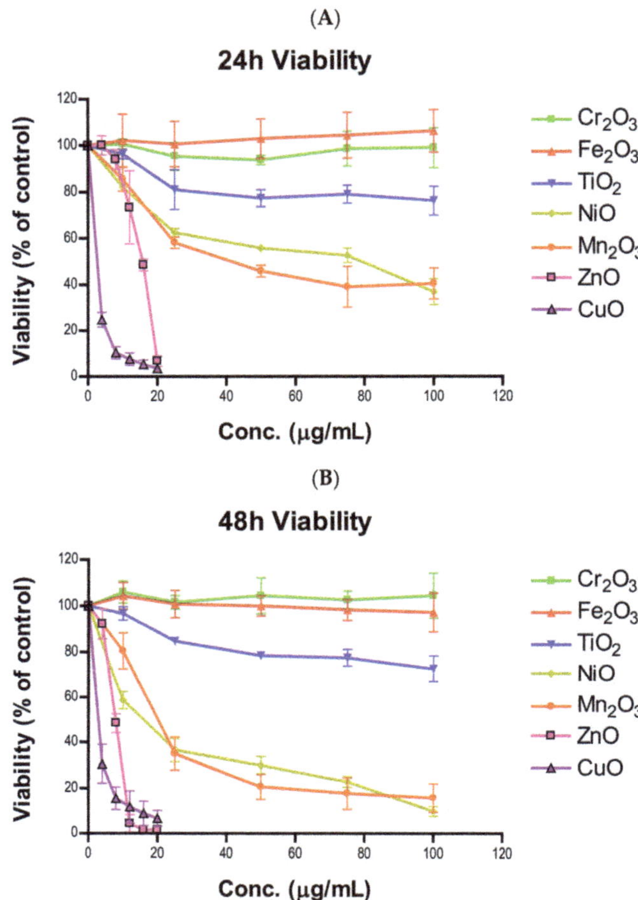

(B)

48h Viability

Figure 1. Viability of A549 cells after (**A**) 24- or (**B**) 48-h exposure to various concentrations of one of seven nanoparticles.

2.3. Induction of Apoptosis

We combined early apoptotic cells with late apoptotic cells to determine the total proportion of cells undergoing apoptosis at each time point. The total apoptotic percentage in the treatment groups was compared to the corresponding control groups. The total apoptotic percentages for all seven NPs at 24 and 48 h are summarized in Figure 2 (Supplementary Tables S4 and S5). For the low to mildly toxic tier, both time points and all concentrations of Cr_2O_3 and Fe_2O_3 showed no difference in apoptosis when compared to the control group ($n = 3$, p's > 0.05). TiO_2 at 24 h exhibited elevated apoptosis at 100 µg/mL (12.5 ± 2.9%) and at all concentrations at 48 h ($n = 3$, p's > 0.05). For the moderate tier, NiO had significant apoptosis at 100 µg/mL (13.8 ± 2.5%) at 24 h and Mn_2O_3 induced significant apoptosis at 50 µg/mL (17.3 ± 4.6%) as well as at 100 µg/mL (21.6 ± 5.4%) ($n = 3$, p's < 0.05). NiO at 48 h exhibited significant apoptosis induction at all concentrations tested while Mn_2O_3 only displayed significant increase in apoptosis at 100 µg/mL (23.8 ± 7.2%) compared to the control ($n = 3$, p's < 0.05). In the high toxicity tier, ZnO had significant induction at only 20 µg/mL at both 24-h (77.5 ± 7.8%) and 48-h (68.4 ± 6.3%) time points ($n = 3$, p's < 0.05). All tested doses of CuO produced significant apoptosis in both time groups, with 100 µg/mL causing apoptosis in extremely high numbers at 24 (88.2 ± 5.3%) and 48 h (86.6 ± 4.6%) ($n = 3$, p's< 0.05).

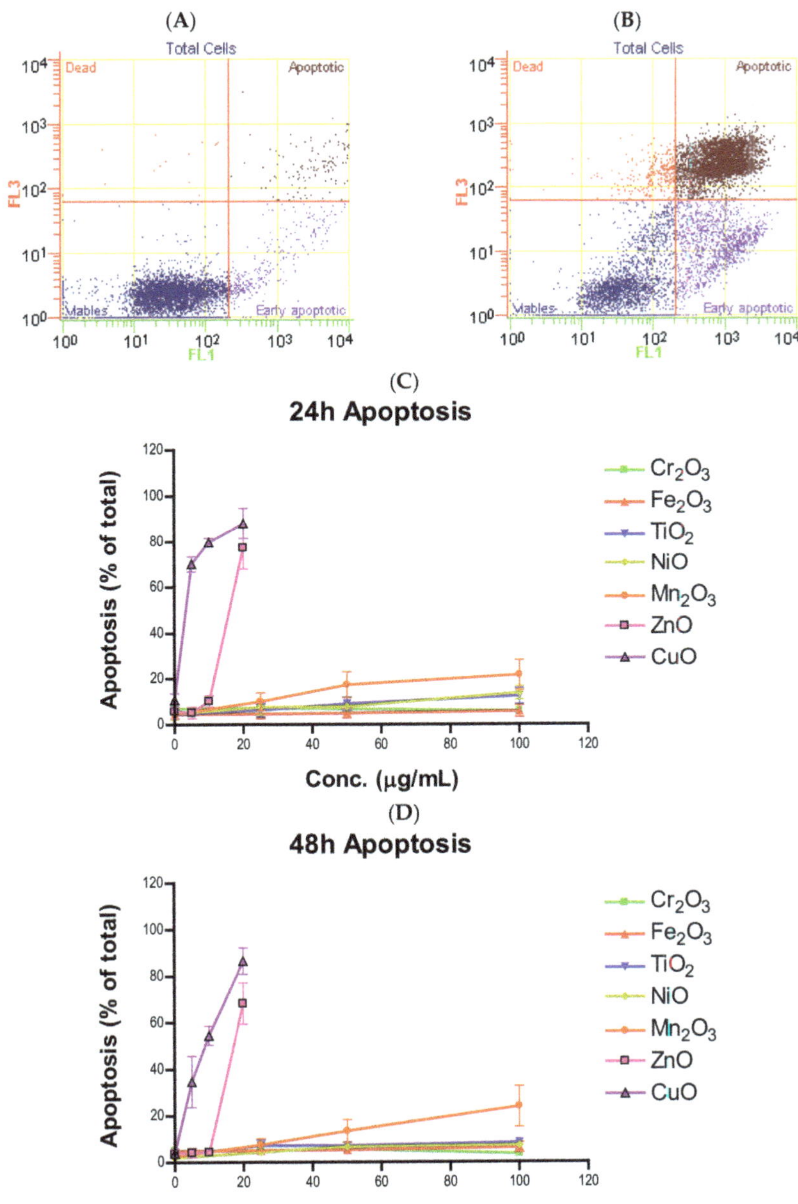

Figure 2. Flow cytometry gating of cells treated with (**A**) 0 and (**B**) 20 μg/mL of ZnO after 24 h. Total apoptosis of A549 cells after (**C**) 24- or (**D**) 48-h exposure to various concentrations of one of seven nanoparticles.

2.4. Alteration of Cell Morphology

Alterations in cell morphology due to NP exposure were observed both by epifluorescence microscopy (Figure 3A) and scanning electron microscopy (SEM) (Figure 3B). Apoptotic morphologies such as membrane blebbing, nuclear fragmentation, and apoptotic bodies were clearly observed.

Membrane blebbing was particularly noticeable under the SEM close-up image. The severity of apoptosis observed was similar to the trend set by viability results, with Cr_2O_3 and Fe_2O_3 treatment resulting in none to minimal toxicity and small numbers of apoptotic cells, Mn_2O_3 and NiO having moderate numbers, and ZnO and CuO having high numbers, with almost no healthy cells remaining.

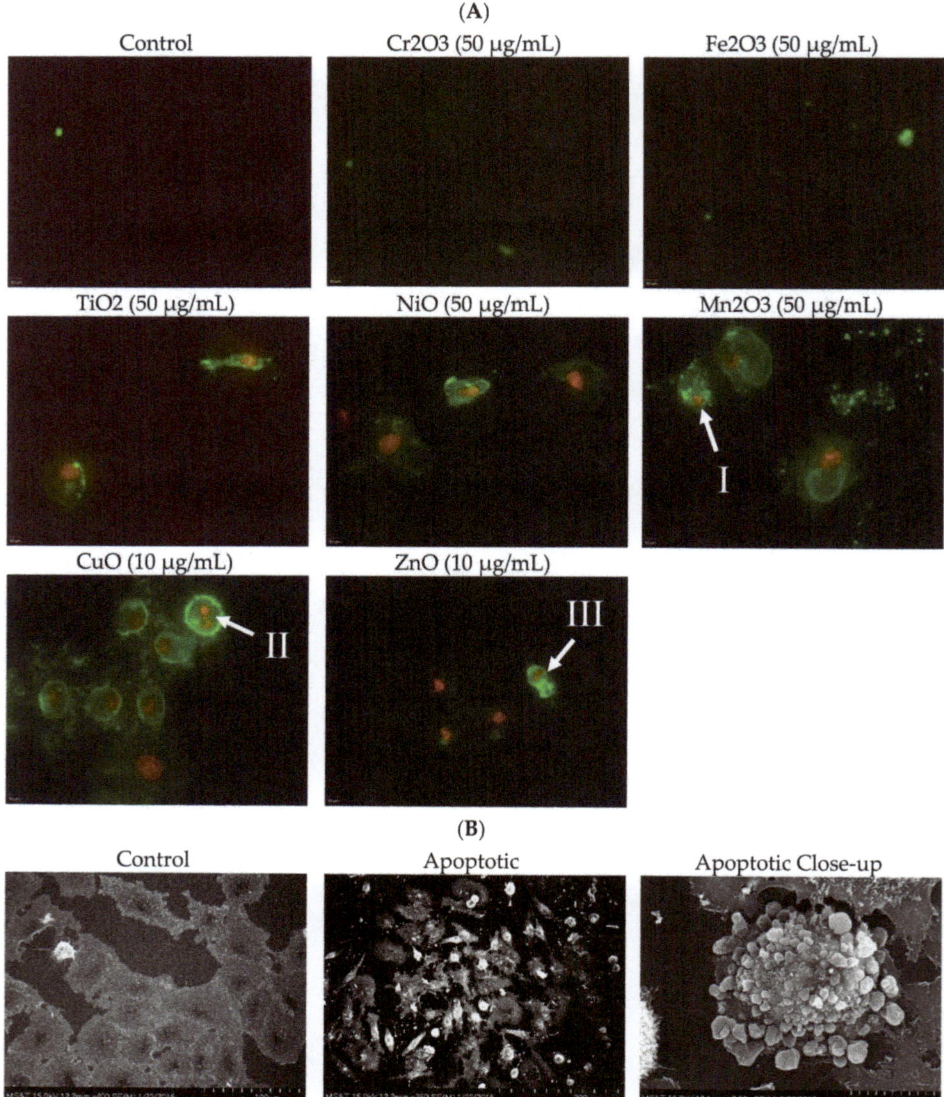

Figure 3. (**A**) Fluorescence apoptotic stains with Annexin V-FITC and 7-aminoactinomycin D (7-AAD) after cells were exposed to nanoparticles at 50 or 10 µg/mL. Green color alone indicates cells undergoing early apoptosis. Red and green in combination indicate cells undergoing late apoptosis. Examples of (I) blebbing, (II) nuclear fragmentation, and (III) apoptotic bodies are marked. Scale bar is 10 µm. (**B**) SEM images of A549 cells after exposure to 50 µg/mL MnO. Membrane blebbing is quite noticeable in the close-up image.

2.5. Suppression of Cell Proliferation

Tritiated thymidine incorporation is indicative of cell proliferation rate. The percentage of cells proliferating was calculated by taking the average of all radioactive counts in a dosage group and dividing it by the average of the radioactive counts of the untreated group. Figure 4 shows time- and concentration-dependent suppression of cell proliferation. Significant inhibition of proliferation was observed at high doses in all seven NPs for both time points (Supplementary Tables S6 and S7). Cr_2O_3 and Fe_2O_3 treated cells were the least affected with 24-h 100 µg/mL doses having cells actively proliferating at 74.5 ± 10.4% and 79.6 ± 0.2%, respectively. These values were similar to the 48-h values, with 73.9 ± 5.6% for Cr_2O_3 and 75.3 ± 0.3% for Fe_2O_3. Although, the decreases for Cr_2O_3 and Fe_2O_3 were significant at 50, 75, and 100 µg/mL at both time points, as well as at 10 and 25 µg/mL of Fe_2O_3 after 24 h ($n = 4$, p's < 0.05). TiO_2 differed from the rest of the low toxicity group by having greater decreased proliferative percentages at 100 µg/mL (69.6 ± 8.9% for 24 h and 61.6 ± 2.9% for 48 h), with significant decreases observed at all concentrations and times ($n = 4$, p's < 0.05). The moderately toxic group also produced a significant decrease at all times and concentrations ($n = 4$, p's < 0.05). While NiO had high time-dependent proliferative reduction at 100 µg/mL (43.8 ± 4.6% for 24 h and 21.6 ± 4.9% for 48 h), Mn_2O_3 inhibition of proliferation at 24 (15.2 ± 5.5%) and 48 h (5.8% ± 2.2%) was much less prominent. ZnO had complete inhibition of proliferation at both 24 (1.7 ± 0.3%) and 48 h (1.0 ± 0.7%) at 20 µg/mL, which should be considered background level radiation. ZnO doses of 12, 16, and 20 µg/mL exhibited significantly decreased proliferation at both time points ($n = 4$, p's < 0.05). Cells exposed to 20 µg/mL of CuO experienced a similar degree of inhibition at both 24 (1.7 ± 0.6%) and 48 h (0.6 ± 0.3%), though all times and concentrations were significantly decreased ($n = 4$, p's < 0.05).

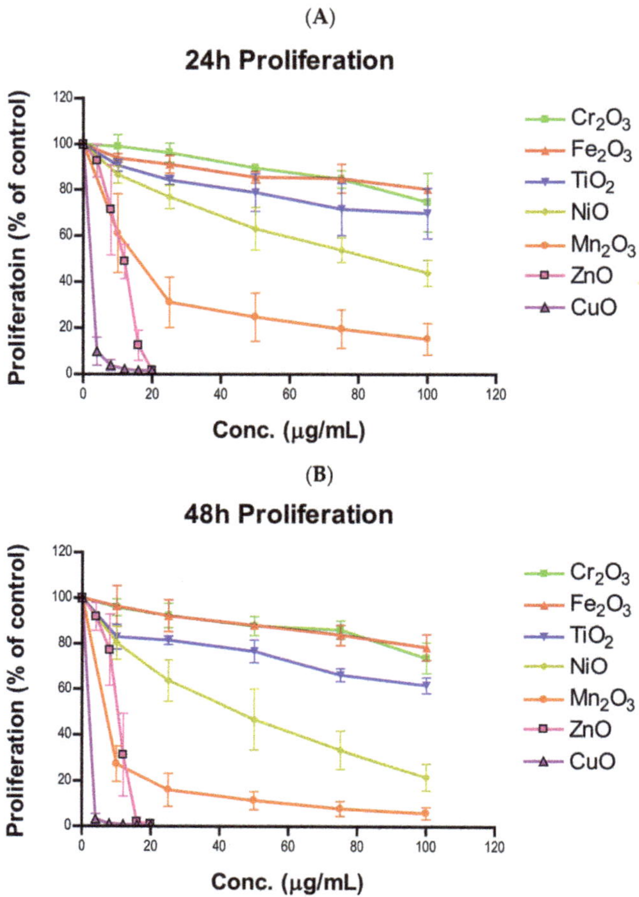

Figure 4. Proliferation of A549 cells after (**A**) 24-h or (**B**) 48-h exposure to various concentrations of one of seven nanoparticles.

2.6. Alteration of Cell Cycle

We selected ZnO and CuO to study the alteration of the cell cycle (Figure 5, Supplementary Figure S1, and Table S8). An increase in the S phase and a decrease in the G_2/M phase at 20 µg/mL of ZnO were observed in both 24 and 48 h ($n = 3$, p's < 0.05). Compared to 24 h (+9.3%), the increase in the S phase was more prominent at 48 h (+17.1%), which might have contributed to the more significant decrease in G_0/G_1 (−11.0%, $n = 3$, p's < 0.05). Similar patterns occurred upon exposure to CuO ($n = 3$, p's < 0.05). However, CuO exhibited significant differences from the control at lower concentrations, with almost no cells observed in G_2/M at any concentration. At 24 h, S phase increased significantly at 20 µg/mL (+11.6%) and G_2/M decreased at all concentrations of CuO, with 20 µg/mL having a change of −7.7% ($n = 3$, p's < 0.05). All phases exhibited significant differences compared to the controls after 48-h CuO exposure, with 20 µg/mL producing changes in G_0/G_1, S, and G_2/M of −15.2%, +21.7%, and −6.6%, respectively ($n = 3$, p's < 0.05).

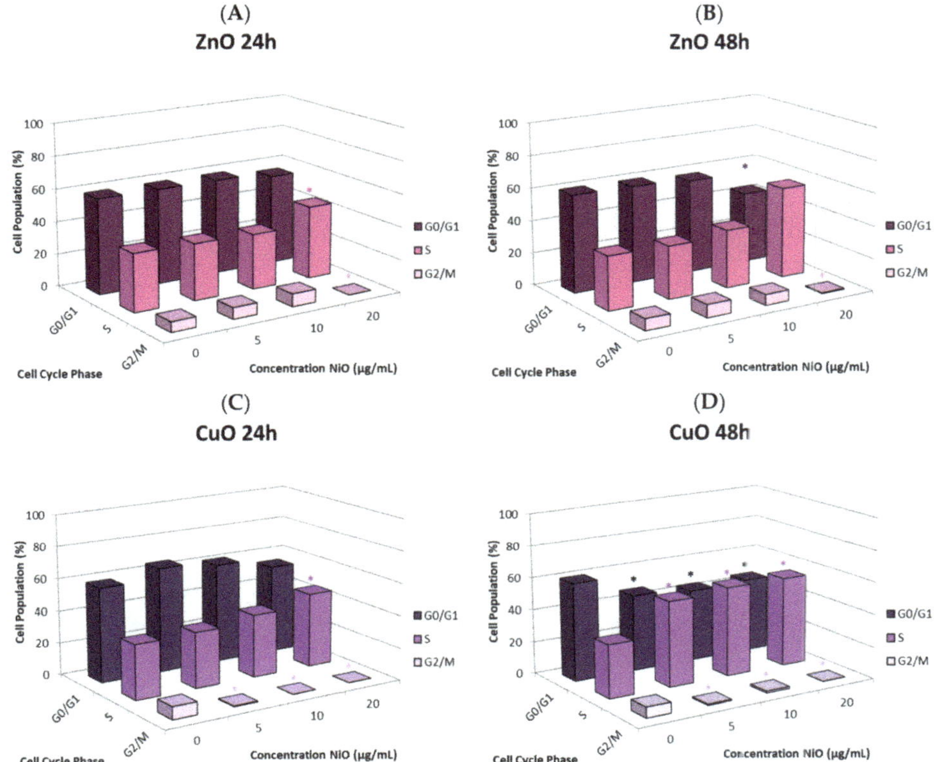

Figure 5. Alteration of cell cycle of A549 cells after exposure to various concentrations of ZnO NPs for (**A**) 24 h and (**B**) 48 h or CuO NPs for (**C**) 24 h or (**D**) 48 h. Values significantly different from the control (p's < 0.05) are indicated with *.

2.7. Correlation

The results of linear regression analyses between 24- and 48-h viability, apoptosis, and proliferation rankings are illustrated in Figure 6.

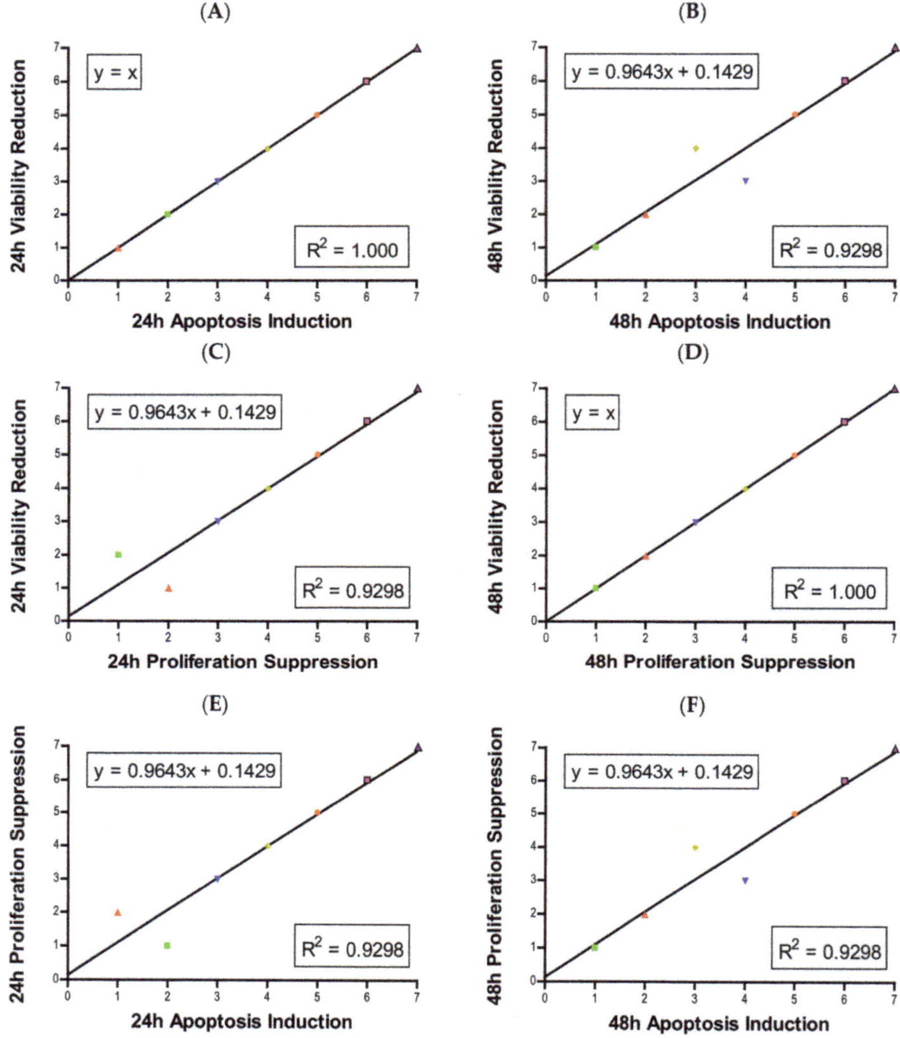

Figure 6. Linear regression analysis of (**A**) 24 h viability vs. proliferation (**B**) 48 h viability vs. proliferation (**C**) 24 h viability vs. apoptosis (**D**) 48 h viability vs. apoptosis (**E**) 24 h proliferation vs. apoptosis (**F**) 48 h proliferation vs. apoptosis.

3. Discussion

In this study, we investigated the cytotoxicity of seven fourth-period transition metal oxide NPs and explored several cellular responses as components of cytotoxicity. We hypothesized that cytotoxicity is a function of cell killing and suppression of cell proliferation and, as such, we assayed for cell viability, apoptosis, cellular proliferation, and cell cycle progression. Data from cell viability indicated the response to NPs is concentration- and time-dependent. Importantly, data revealed that NPs exert a higher degree of toxicity towards 48-h and longer exposure. To our knowledge, the temporal stability of NPs in storage has not been addressed. We took the opportunity to compare our current 24-h viability data to the data from our previous study, which was conducted by a different experimenter [18]. Concentration-dependent patterns from both studies resembled each other, with some slight variations. This is indicative of particle stability in storage conditions. However, we did notice that TiO_2 seemed to be more toxic in these sets of experiments (approximately 96% in previous vs. 76% in current). Due to this discrepancy, multiple researchers were asked to perform the same experiment to ensure our results were reliable, which yielded almost the same outcome. We are unsure of what caused the difference, but particle-specific aging is speculated.

Induction of apoptosis is a hallmark of NP toxicity [24]. For our purposes, early and late apoptotic events were combined into a total apoptotic value. Concentration- and time-dependent apoptotic effects were observed in TiO_2, NiO, Mn_2O_3, CuO, and ZnO, with the last two showing a much more severe degree of programmed cell death. Epifluorescent microscopy and SEM images confirmed apoptotic morphologies, such as membrane blebbing and nuclear fragmentation, after exposure to NPs. Cell viability reduction correlated with apoptotic events, indicating a close pathway-dependent relationship (Figure 6A,B).

Plenty of nanotoxicological studies have focused on NP-imposed cell killing [25–27]. We believe that cell killing may not be the only reason that results in cell number reduction. The suppression of cell proliferation can have the same effect as cell killing in driving down cell numbers. Whether transition metal oxides have influences over cell proliferative inhibition has not been systematically investigated. In this study, all seven oxides of transition metal NPs in the fourth period of the periodic table showed time- and concentration-dependent proliferative inhibition in A549 cells. In general, proliferative inhibition followed the same tier trend as cell viability. NiO and Mn_2O_3 exhibited more prominent time-dependent inhibition than TiO_2, Cr_2O_3, and Fe_2O_3. Strikingly, the degrees of time-dependent suppression of proliferation by CuO and ZnO were much steeper at 12 and 16 µg/mL as time progressed towards 48 h. A significant correlation between cell viability reduction and suppression of cell proliferation suggests a proliferative inhibition is an independent variable influencing cell number, in addition to cell killing (Figure 6C,D). Previously we developed a model to estimate the number of cells in the second generation [28]:

Cell # in Generation 2 = 2 (Proliferating cells) + Non-proliferating cells - Dead cells (via killing) (1)

This model assumes the doubling time of a cell line is 24 h and the rate of doubling time is not altered by the NPs. Future studies should investigate the differential contribution of these two components to the change in cell number.

The cause of the observed cell proliferation suppression may be multiple. Alteration of cell cycle induced by the NPs could have a major influence. We selected CuO and ZnO to further investigate this issue. Time- and concentration-dependent effects on cell cycle in response to NP exposure were observed. In general, there was an increase in the proportion of cells in S phase and a decrease in the G_0/G_1 and G_2/M phases. Cells arresting in certain phases of cell cycle either attempt to fix damage or accumulate too much damage and undergo apoptosis. The increase in S phase indicates that cells exposed to CuO and ZnO are being stalled in S phase; the NPs may directly damage DNA or be influencing DNA replication machinery. In this study, significant correlation between suppression of proliferation and apoptosis may suggest cell cycle alteration-mediated cell death plays a role in proliferative inhibition (Figure 6E,F). Other studies also have observed cell cycle alteration induced by NPs and the phenomenon is quite dynamic. Arrest can occur in any phase, or in multiple phases of the cell cycle. Phase-specific arrest depends on cell line, particle, and particle concentration of [28–33]. For instance, in one study [34] exposure to NiO NPs resulted in a significant decrease in G_0/G_1 and an increase in G_2/M in A549 cells. In contrast, exposure to NiO NPs led to a significant increase in G_0/G_1 and a decrease in G_2/M in BEAS-2B cells [34]. Exposure to NiO NPs caused BEAS-2B cells arrest in the G_2/M phase, while ZnO and Fe_2O_3 did not affect the cell cycle [34,35]. It is important to be aware of the fact that NPs composed of the same elements may have quite different physical and chemical properties, such as surface charge, surface area, dissolution of ions, morphology, and crystalline structure, in different studies. Consequentially, interpretations of the dynamic outcome in cell cycle alteration becomes a complicated issue if NPs are not well characterized.

4. Materials and Methods

4.1. Material Sources

A549 cells were obtained from the American Type Culture Collection (ATCC CCL-185, Manassas, VA, USA). The seven transition metal oxide NPs (TiO_2, Cr_2O_3, Mn_2O_3, Fe_2O_3, NiO, CuO, and ZnO) were purchased from Nanostructured and Amorphous Materials (Houston, TX, USA). Sulforhodamine B (SRB) dye was procured from Biotium (Freemont, CA, USA). Annexin V-FITC and 7-AAD were obtained from BD Biosciences (Franklin Lakes, NJ, USA). Tritiated thymidine came from Perkin-Elmer (Waltham, MA, USA) and propidium iodide (PI) was purchased from Fisher Scientific (Pittsburgh, PA, USA).

4.2. Storage and Characterization of Nanoparticles

NPs were stored in an amber desiccator under a pure nitrogen atmosphere to protect them from moisture, oxidation, and UV damage. Before being used, the NPs were further dried in an oven. Characterization of the NPs followed our previous publication [18]. The shape and APS of NPs were determined by TEM. The BET method was used to measure the SSA of the NPs.

4.3. Cell Culture

A549 are a common in vitro cell line model for nanotoxicity testing. Cells were maintained in 10 cm tissue culture dishes at 37 °C in a 5% CO_2 humidified incubator. The growth media was Ham's F-12 modified medium (Corning Inc., Corning, NY, USA) supplemented with 10% HyClone FetalClone serum (GE Healthcare Life Sciences, Marlborough, MA, USA) and 1% of a combination of penicillin/streptomycin antibiotics (MP Biomedicals, Irvine, CA, USA). Cells were allowed to grow to a confluence of approximately 70–80% before being passaged or seeded for experiments. Cells were only grown for approximately 20 passages before a new vial of cells was brought up from liquid nitrogen storage.

4.4. Nanoparticle Treatment

NPs were prepared as a one mg per mL working solution by weighing out particles on an analytical balance and suspending them in a corresponding amount of cell culture medium. The suspensions were sonicated in sealed polyethylene vials for three minutes to break up aggregates and ensure an even mixture of NPs. NP suspensions were immediately used for experiments following preparation and were diluted to the desired concentrations in experimental dishes. For mildly and moderately toxic NPs (TiO_2, Cr_2O_3, Mn_2O_3, Fe_2O_3, and NiO), 0, 10, 25, 50, 75, and 100 µg/mL were used while cells were only exposed to 0, 4, 8, 12, 16 and 20 µg/mL of highly toxic particles (CuO and ZnO). Apoptosis and cell cycle experiments were limited to 4 doses of NPs, being 0, 25, 50, and 100 µg/mL for mildly and moderately toxic and 0, 5, 10, and 20 µg/mL for highly toxic particles. Untreated cells were used as a negative control in all experiments.

4.5. Cell Viability Assay

A549 cells were seeded into 24-well plates at a density of 45,000 cells per well for 24-h exposure and 22,000 cells per well for 48-h exposure. Cells were treated with 0, 10, 25, 50, 75 and 100 µg/mL or 0, 4, 8, 12, 16 and 20 µg/mL for mildly and moderately or highly toxic NPs, respectively. At the end of cell exposure (24 or 48 h) to NP suspensions, the medium was discarded and the SRB assay was used to determine cell viability relative to the control group, with untreated cells being considered 100% viable [36]. Briefly, the cells were fixed with cold 10% trichloroacetic acid (TCA) for 1 h at 4 °C. The TCA solution was then discarded, and the fixed cells were washed three times with distilled water, followed by complete drying. SRB solution (0.2% in 1% acetic acid) was added to stain the cells for 30 min at room temperature. The solution containing stain was pipetted off and excess dye was eliminated from the cells by rinsing three times with 1% acetic acid. Sample wells were allowed to dry before dissociating the dye in cold 10-mM Tris buffer (pH 10.5). Stained sample solutions were transferred onto a 96-well plate in aliquots of 100 µL and absorbance was read at 510 nm using a microplate reader (FLUORstar Omega, BMG Labtechnologies, Cary, NC, USA).

4.6. Apoptosis Analysis

The quantification of apoptosis due to NP exposure was measured using the fluorescent dyes Annexin V-FITC and 7-AAD on a Cell Lab Quanta SC MPL flow cytometer (Beckman-Coulter, Brea, CA, USA). Annexin V-FITC binds to and labels phosphatidylserine (PS), a cell membrane phospholipid that flips to the extracellular surface during apoptosis. 7-AAD is a DNA stain that is only able to penetrate the permeable membranes of dying cells. A549 cells were allowed to grow for 24 h in 6 cm tissue culture dishes and then treated with varying concentrations of transition metal oxide NPs for 24 and 48 h. The seeding densities were 250,000 cells per dish and 120,000 cells per dish for 24- and 48-h treatments, respectively. The range of NP concentrations tested included 0, 25, 50, and 100 µg/mL for TiO_2, Cr_2O_3, Fe_2O_3, NiO, and Mn_2O_3 and 0, 5, 10, and 20 µg/mL for ZnO and CuO. At the end of the exposure period, the media was removed, and the dishes were washed with phosphate buffered saline (PBS). The cells were harvested using 0.25% trypsin-EDTA (Gibco, Life Technologies, Carlsbad, CA, USA) and transferred to a centrifuge tube. The tubes were centrifuged, the supernatant was discarded, and 1 mL of ice-cold PBS was used to wash the pellet. The tubes were then centrifuged again, and PBS wash was removed. Each sample was resuspended in 5 µL of Annexin V-FITC, 5 µL of 7-AAD, and 100 µL of 1x Annexin V binding buffer (BD Biosciences, Franklin Lakes, NJ, USA). These tubes were incubated in the dark for 15 min. After incubation, another 400 µL of Annexin V binding buffer was added to each sample and 250 µL of this cell suspension was transferred to a 96-well plate (Corning Inc., Corning, NY, USA) for analysis via flow cytometry. Operating conditions for the flow cytometer were the stock apoptosis protocol included with the software. The data was exported to Microsoft Excel 2016 using the Quanta SC Analysis software and calculations of averages and standard deviations were performed in Excel. The total fraction of apoptotic cells was determined by summing the populations in early and late apoptosis.

4.7. Scanning Electron Microscopy Imaging

Microscopic examination of apoptotic and control cells was performed using a Hitachi S-4700 SEM (Hitachi, Tokyo, Japan). Cells were grown for 24 h at an initial density of 15,000 cells per dish in 35 mm glass bottom tissue culture dishes (MatTek, Ashland, MA, USA) before treatment with 50 µg/mL of mildly or moderately toxic NPs or 10 µg/mL of highly toxic NPs for 24 h. At the end of the exposure period, the cell culture media was removed. The dish was washed with PBS and the cells were fixed in a solution of 3% glutaraldehyde in PBS overnight. After fixation, the cells were dehydrated with an increasing series of ethanol concentrations from 50% to 100% for 15 min at a time, with the final concentration (100%) repeated once. The cells were then dried using a mixture of ethanol and hexamethyldisilazane (HMDS, Acros Organics, Fisher Scientific, Pittsburgh, PA, USA) for 15 min at a time using a ratio of 1:2 HMDS to ethanol, then 2:1, and finally ending with pure HMDS. Like the ethanol, the final HMDS step was repeated once. The dishes were then mounted on pin stubs with carbon dot adhesives or carbon paint. Then the dishes were coated with a mixture of gold and palladium for 1 min in a Hummer VI sputter coater (Anatech, Sparks, NV, USA) to provide contrast and prevent charging of the sample once in the SEM. The plastic sides of the culture dish were then pried off, leaving only the cell-containing glass cover slip on the pin stub. A piece of copper tape was run from the pin stub to the Au/Pd coated slip to provide a path to ground for the electron beam. The mounted sample was then placed into the scanning electron microscope for imaging. Operating conditions for the SEM followed the protocol of the Missouri S&T Advanced Materials Characterization Laboratory.

4.8. Epifluorescence Microscopy

Qualitative imaging of apoptosis induced by exposure to NPs was observed using the fluorescent dyes Annexin V-FITC and 7-AAD on an Olympus IX51 inverted epifluorescence microscope (Olympus Corporation, Tokyo, Japan). Cells were seeded on 35 mm glass bottom microscopy tissue culture dishes at 15,000 cells per dish and allowed to grow to approximately 70% confluence, around 24 h. Cells were then treated with 50 µg/mL of mildly or moderately toxic NPs or 10 µg/mL of highly toxic NPs for 24 h. Following incubation with NPs, the cell culture medium was removed, and the dishes were washed with PBS. Annexin V binding buffer mixed with 5 µL of Annexin V-FITC and 5 µL of 7-AAD was added and the dishes were incubated in the dark for 15 min. After incubation, the plates were washed with Annexin V binding buffer twice and enough buffer to cover the bottom of the dish was left on the cells to prevent drying. The plates were placed into an opaque container to keep them out of the light before imaging. The cells were imaged using the Olympus microscope, using green and red filters for Annexin V-FITC and 7-AAD, respectively.

4.9. Tritiated Thymidine Incorporation Assay

The tritiated thymidine ($[5'\text{-}^3\text{H}]$-thymidine) incorporation assay has been widely used to study cell proliferation [37] and was used to determine the proliferation of NP-treated cells. A549 cells were plated into 24-well tissue culture plates with seeding densities of 45,000 and 22,000 cells per well for 24- and 48-h treatment periods, respectively. Cells were exposed to 0, 10, 25, 50, 75, and 100 µg/mL or 0, 4, 8, 12, 16, and 20 µg/mL for mildly and moderately or highly toxic transition metal oxide NPs, respectively. At the same time as the cells were dosed with NPs, they were also treated with 20 µL tritiated thymidine. Thymidine working solution was prepared in 500 µL of PBS with 20 µL of $[5'\text{-}^3\text{H}]$-thymidine (1 µCi/µL). After 24 or 48 h of exposure, the cell culture medium was removed, and the wells were washed with ice-cold PBS twice. Following the PBS wash, the cells were quickly fixed in ice-cold 10% TCA for 5 min on ice. The TCA fixation was repeated once. After fixation, the cells were lysed using 0.5 mL of room-temperature 1 N NaOH. The same volume of 1 N HCl was used to neutralize the cell solution. The lysed cell solution was thoroughly mixed by pipetting up and down and transferred to liquid scintillation counting vials with 4 mL of Econo-Safe scintillation counting fluid (Research Products International, Mt Prospect, IL, USA). Sample vials were capped, labeled and racked for analysis in a Beckman liquid scintillation counter LS6500 (Beckman-Coulter, Brea, CA, USA). Untreated cells were considered to have 100% possible proliferative potential and treatment groups were presented as relative to the control. All radioactive waste was disposed of following Missouri S&T's Department of Environmental Health and Safety procedures.

4.10. Cell Cycle Analysis

The alteration of the cell cycle due to NP exposure was measured using PI staining and subsequent analysis via flow cytometry. A549 cells were seeded into 6 cm tissue culture dishes with initial densities of 250,000 cells per dish for 24-h treatment and 120,000 cells per dish for 48-h treatment. Cells were exposed to varying concentrations of transition metal oxide NPs (0, 25, 50, and 100 µg/mL for mildly and moderately toxic and 0, 5, 10, and 20 µg/mL for highly toxic NPs) for 24- or 48-h periods. After incubation with NPs, the cells were washed, harvested using trypsin, and centrifuged down into a pellet. The cell pellet was then resuspended in 1 mL of PBS and 3 mL of ice-cold absolute methanol was added dropwise to samples being stirred with a vortex. The cells suspensions were left at least overnight in 75% methanol to ensure complete fixation. After fixation, the cells were washed once with PBS and centrifuged. The cells were then suspended in PI staining solution (50 µg/mL PI, 0.1% RNase A, and 0.05% Triton X-100 in PBS) and incubated in the dark for 20 min. RNase A was included to destroy any RNA present, as PI can also bind to RNA, and to ensure PI staining was limited to DNA. Triton X-100 ensured cell membranes were permeable to PI. After incubation, 1 mL of PBS was added to each sample. Samples were centrifuged, supernatant was removed, and pellet was resuspended in 250 mL of PBS. The stained samples were transferred into a 96-well plate and analyzed with a CytoFLEX flow cytometer (Beckman-Coulter, Brea, CA, USA). Alterations of cell cycle due to NP exposure were determined through PI fluorescent intensity, as different phases of the cell cycle have differing amounts of DNA content. The percentage distribution of cells in each phase of the cell cycle (G_0/G_1, S, and G_2/M) was determined using FCS Express 6 (DeNovo software, Pasadena, CA, USA).

4.11. Statistical Analysis

Each experiment was repeated at least three times independently with treatment groups having multiple samples. Data are presented as mean ± standard deviation. Statistical analysis was completed in Minitab 19. One-tailed unpaired t-tests were used to compare experimental groups to the control group in normalized data sets, with $\mu >$ control or $\mu <$ control depending on the experimental hypothesis. Analysis of variance (ANOVA) with Dunnett comparison was used to determine values statistically significant from control groups. Significance was set at $p < 0.05$. The majority of figures were produced using GraphPad Prism 4 except for cell cycle distribution graphs, which were produced by Microsoft Excel 2016. For correlation analysis, each NP in each assay was assigned a rank from 1 to 7, with 1 having the least effect and 7 having the most severe effect. Once the particles were ranked for every assay, ranks were compared against one another using linear regression analysis in GraphPad Prism 4. The R^2-value was calculated and displayed on each plot. A high R^2 value indicates strong correlation while a low R^2 value is indicative of little-to-no correlation.

5. Conclusions

Our data support the hypothesis that NP-imposed cytotoxicity is a function of cell killing and cell proliferative inhibition. The adverse effects are time and concentration dependent. Compared to our study in 2013, cytotoxicity pattern is comparable indicating NPs have been stable in our storage conditions over the years. Apoptosis and signature morphological changes were confirmed via flow cytometry, fluorescent microscopy, and SEM observation. Cell arrest in the cell cycle leading to apoptosis may play a role in suppressing cell division rate.

Supplementary Materials: Supplementary materials can be found at http://www.mdpi.com/1422-0067/21/5/1731/s1, or by contacting the corresponding author.

Author Contributions: Conceptualization, Y.-W.H., H.-J.L.; methodology, L.M.T., Y.-W.H., M.H.C., N.J.H., F.Y.S.H.; formal analysis, L.M.T., Y.-W.H., M.H.C., N.J.H., F.Y.S.H.; investigation, L.M.T., M.H.C., N.J.H.; data curation, L.M.T., M.H.C., N.J.H.; writing—original draft preparation, L.M.T., N.J.H., Y.-W.H.; writing—review and editing, N.J.H., Y.-W.H., H.-J.L., F.Y.S.H.; supervision, Y.-W.H., F.Y.S.H.; project administration, Y.-W.H.; funding acquisition, Y.-W.H. All authors have read and agreed to the published version of the manuscript.

Funding: This research was funded by the Missouri S&T cDNA Resource Center.

Abbreviations

NP	Nanoparticles
OS	Oxidative Stress
TEM	Transmission electron microscopy
SEM	Scanning electron microscopy
BET	Brunauer-Emmett-Teller
APS	Approximate physical size
7-AAD	7-aminoactinomycin D

References

1. Peer, D.; Karp, J.M.; Hong, S.; Farokhzad, O.C.; Margalit, R.; Langer, R. Nanocarriers as an emerging platform for cancer therapy. *Nat. Nanotechnol.* **2007**, *2*, 751–760. [CrossRef] [PubMed]
2. Kowshik, M.; Deshmukh, N.; Vogel, W.; Urban, J.; Kulkarni, S.K.; Paknikar, K.M. Microbial synthesis of semiconductor CdS nanoparticles, their characterization, and their use in the fabrication of an ideal diode. *Biotechnol. Bioeng.* **2002**, *78*, 583–588. [CrossRef]
3. Li, Y.; Wu, Y.; Ong, B.S. Facile synthesis of silver nanoparticles useful for fabrication of high-conductivity elements for printed electronics. *J. Am. Chem. Soc.* **2005**, *127*, 3266–3267. [CrossRef] [PubMed]
4. De Jong, W.H.; Borm, P.J. Drug delivery and nanoparticles: Applications and hazards. *Int. J. Nanomed.* **2008**, *3*, 133–149. [CrossRef] [PubMed]
5. Nohynek, G.J.; Dufour, E.K.; Roberts, M.S. Nanotechnology, cosmetics and the skin: Is there a health risk? *Skin Pharmacol. Physiol.* **2008**, *21*, 136–149. [CrossRef] [PubMed]
6. Weir, A.; Westerhoff, P.; Fabricius, L.; Hristovski, K.; von Goetz, N. Titanium dioxide nanoparticles in food and personal care products. *Environ. Sci. Technol.* **2012**, *46*, 2242–2250. [CrossRef]
7. Farokhzad, O.C.; Langer, R. Impact of nanotechnology on drug delivery. *ACS Nano.* **2009**, *3*, 16–20. [CrossRef]
8. Roco, M.C.; Mirkin, C.A.; Hersam, M.C. *Nanotechnology Research Directions for Societal Needs in 2020: Retrospective and Outlook*; Springer: New York, NY, USA, 2011. [CrossRef]
9. Iavicoli, I.; Leso, V.; Beezhold, D.H.; Shvedova, A.A. Nanotechnology in agriculture: Opportunities, toxicological implications, and occupational risks. *Toxicol. Appl. Pharmacol.* **2017**, *329*, 96–111. [CrossRef]
10. Carlson, C.; Hussain, S.M.; Schrand, A.M.; Braydich-Stolle, L.K.; Hess, K.L.; Jones, R.L.; Schlager, J.J. Unique cellular interaction of silver nanoparticles: Size-dependent generation of reactive oxygen species. *J. Phys. Chem. B* **2008**, *112*, 13608–13619. [CrossRef]
11. Terzano, C.; Di Stefano, F.; Conti, V.; Graziani, E.; Petroianni, A. Air pollution ultrafine particles: Toxicity beyond the lung. *Eur. Rev. Med. Pharmacol. Sci.* **2010**, *14*, 809–821. [PubMed]
12. NIOSH. Current Intelligence Bulletin 63: Occupational Exposure to Titanium Dioxide. 2011. Available online: https://statnano.com/standard/niosh/1145 (accessed on 19 December 2019).
13. Graham, U.M.; Jacobs, G.; Yokel, R.A.; Davis, B.H.; Dozier, A.K.; Birch, M.E.; Tseng, M.T.; Oberdorster, G.; Elder, A.; DeLouise, L. From Dose to Response: In Vivo Nanoparticle Processing and Potential Toxicity. *Adv. Exp. Med. Biol.* **2017**, *947*, 71–100. [CrossRef] [PubMed]
14. Jain, A.; Ranjan, S.; Dasgupta, N.; Ramalingam, C. Nanomaterials in food and agriculture: An overview on their safety concerns and regulatory issues. *Crit. Rev. Food Sci. Nutr.* **2018**, *58*, 297–317. [CrossRef] [PubMed]
15. Lin, W.; Stayton, I.; Huang, Y.W.; Zhou, X.D.; Ma, Y. Cytotoxicity and cell membrane depolarization induced by aluminum oxide nanoparticles in human lung epithelial cells A549. *Toxicol. Environ. Chem.* **2008**, *90*, 983–996. [CrossRef]
16. Lin, W.; Xu, Y.; Huang, C.C.; Ma, Y.; Shannon, K.B.; Chen, D.R.; Huang, Y.W. Toxicity of nano- and micro-sized ZnO particles in human lung epithelial cells. *J. Nanoparticle Res.* **2009**, *11*, 25–39. [CrossRef]
17. Huang, C.C.; Aronstam, R.S.; Chen, D.R.; Huang, Y.W. Oxidative stress, calcium homeostasis, and altered gene expression in human lung epithelial cells exposed to ZnO nanoparticles. *Toxicol. In Vitro* **2010**, *24*, 45–55. [CrossRef] [PubMed]
18. Chusuei, C.C.; Wu, C.H.; Mallavarapu, S.; Hou, F.Y.; Hsu, C.M.; Winiarz, J.G.; Aronstam, R.S.; Huang, Y.W. Cytotoxicity in the age of nano: The role of fourth period transition metal oxide nanoparticle physicochemical properties. *Chem. Biol. Interact.* **2013**, *206*, 319–326. [CrossRef]
19. Biola-Clier, M.; Gaillard, J.C.; Rabilloud, T.; Armengaud, J.; Carriere, M. Titanium Dioxide Nanoparticles Alter the Cellular Phosphoproteome in A549 Cells. *Nanomaterials* **2020**, *10*, 185. [CrossRef]
20. Botha, T.L.; Brand, S.J.; Ikenaka, Y.; Nakayama, S.M.M.; Ishizuka, M.; Wepener, V. How toxic is a non-toxic nanomaterial: Behaviour as an indicator of effect in Danio rerio exposed to nanogold. *Aquat. Toxicol.* **2019**, *215*, 105287. [CrossRef]
21. Duke, K.S.; Bonner, J.C. Mechanisms of carbon nanotube-induced pulmonary fibrosis: A physicochemical characteristic perspective. *Wiley Interdiscip. Rev. Nanomed. Nanobiotechnol.* **2018**, *10*, e1498. [CrossRef]
22. Lin, W.; Huang, Y.W.; Zhou, X.D.; Ma, Y. In vitro toxicity of silica nanoparticles in human lung cancer cells. *Toxicol. Appl. Pharmacol.* **2006**, *217*, 252–259. [CrossRef]

23. Lin, W.; Huang, Y.W.; Zhou, X.D.; Ma, Y. Toxicity of cerium oxide nanoparticles in human lung cancer cells. *Int. J. Toxicol.* **2006**, *25*, 451–457. [CrossRef] [PubMed]

24. Fumelli, C.; Marconi, A.; Salvioli, S.; Straface, E.; Malorni, W.; Offidani, A.M.; Pellicciari, R.; Schettini, G.; Giannetti, A.; Monti, D.; et al. Carboxyfullerenes protect human keratinocytes from ultraviolet-B-induced apoptosis. *J. Invest. Dermatol.* **2000**, *115*, 835–841. [CrossRef] [PubMed]

25. Lockman, P.R.; Koziara, J.M.; Mumper, R.J.; Allen, D.D. Nanoparticle surface charges alter blood-brain barrier integrity and permeability. *J. Drug Target* **2004**, *12*, 635–641. [CrossRef] [PubMed]

26. Shiohara, A.; Hoshino, A.; Hanaki, K.; Suzuki, K.; Yamamoto, K. On the cyto-toxicity caused by quantum dots. *Microbiol. Immunol.* **2004**, *48*, 669–675. [CrossRef] [PubMed]

27. Renwick, L.C.; Brown, D.; Clouter, A.; Donaldson, K. Increased inflammation and altered macrophage chemotactic responses caused by two ultrafine particle types. *Occup. Environ. Med.* **2004**, *61*, 442–447. [CrossRef]

28. Huang, Y.W.; Cambre, M.; Lee, H.J. The Toxicity of Nanoparticles Depends on Multiple Molecular and Physicochemical Mechanisms. *Int. J. Mol. Sci.* **2017**, *18*, 2702. [CrossRef]

29. Gao, X.; Wang, Y.; Peng, S.; Yue, B.; Fan, C.; Chen, W.; Li, X. Comparative toxicities of bismuth oxybromide and titanium dioxide exposure on human skin keratinocyte cells. *Chemosphere* **2015**, *135*, 83–93. [CrossRef]

30. Hanagata, N.; Zhuang, F.; Connolly, S.; Li, J.; Ogawa, N.; Xu, M. Molecular responses of human lung epithelial cells to the toxicity of copper oxide nanoparticles inferred from whole genome expression analysis. *ACS Nano.* **2011**, *5*, 9326–9338. [CrossRef]

31. Luo, C.; Li, Y.; Yang, L.; Zheng, Y.; Long, J.; Jia, J.; Xiao, S.; Liu, J. Activation of Erk and p53 regulates copper oxide nanoparticle-induced cytotoxicity in keratinocytes and fibroblasts. *Int. J. Nanomed.* **2014**, *9*, 4763–4772. [CrossRef]

32. Patel, P.; Kansara, K.; Senapati, V.A.; Shanker, R.; Dhawan, A.; Kumar, A. Cell cycle dependent cellular uptake of zinc oxide nanoparticles in human epidermal cells. *Mutagenesis* **2016**, *31*, 481–490. [CrossRef]

33. Valdiglesias, V.; Costa, C.; Kilic, G.; Costa, S.; Pasaro, E.; Laffon, B.; Teixeira, J.P. Neuronal cytotoxicity and genotoxicity induced by zinc oxide nanoparticles. *Environ. Int.* **2013**, *55*, 92–100. [CrossRef]

34. Capasso, L.; Camatini, M.; Gualtieri, M. Nickel oxide nanoparticles induce inflammation and genotoxic effect in lung epithelial cells. *Toxicol. Lett.* **2014**, *226*, 28–34. [CrossRef]

35. Lai, X.; Wei, Y.; Zhao, H.; Chen, S.; Bu, X.; Lu, F.; Qu, D.; Yao, L.; Zheng, J.; Zhang, J. The effect of Fe_2O_3 and ZnO nanoparticles on cytotoxicity and glucose metabolism in lung epithelial cells. *J. Appl. Toxicol.* **2015**, *35*, 651–664. [CrossRef]

36. Monteiro-Riviere, N.A.; Inman, A.O.; Zhang, L.W. Limitations and relative utility of screening assays to assess engineered nanoparticle toxicity in a human cell line. *Toxicol. Appl. Pharmacol.* **2009**, *234*, 222–235. [CrossRef]

37. Taylor, J.H.; Woods, P.S.; Hughes, W.L. The Organization and Duplication of Chromosomes as Revealed by Autoradiographic Studies Using Tritium-Labeled Thymidinee. *Proc. Natl. Acad. Sci. USA* **1957**, *43*, 122–128. [CrossRef]

International Journal of
Molecular Sciences

Article

The Vitamin A and D Exposure of Cells Affects the Intracellular Uptake of Aluminum Nanomaterials and Its Agglomeration Behavior: A Chemo-Analytic Investigation

Fabian L. Kriegel *, Benjamin-Christoph Krause, Philipp Reichardt, Ajay Vikram Singh, Jutta Tentschert, Peter Laux, Harald Jungnickel and Andreas Luch

German Federal Institute for Risk Assessment, Department of Chemical & Product Safety, Max-Dohrn-Straße 8-10, 10589 Berlin, Germany; Benjamin-Christoph.Krause@bfr.bund.de (B.-C.K.); Philipp.Reichardt@bfr.bund.de (P.R.); Ajay-Vikram.Singh@bfr.bund.de (A.V.S.); Jutta.Tentschert@bfr.bund.de (J.T.); peter.laux@bfr.bund.de (P.L.); Harald.Jungnickel@bfr.bund.de (H.J.); Andreas.Luch@bfr.bund.de (A.L.)
* Correspondence: Fabian.kriegel@bfr.bund.de

Received: 17 December 2019; Accepted: 12 February 2020; Published: 14 February 2020

Abstract: Aluminum (Al) is extensively used for the production of different consumer products, agents, as well as pharmaceuticals. Studies that demonstrate neurotoxicity and a possible link to Alzheimer's disease trigger concern about potential health risks due to high Al intake. Al in cosmetic products raises the question whether a possible interaction between Al and retinol (vitamin A) and cholecalciferol (vitamin D3) metabolism might exist. Understanding the uptake mechanisms of ionic or elemental Al and Al nanomaterials (Al NMs) in combination with bioactive substances are important for the assessment of possible health risk associated. Therefore, we studied the uptake and distribution of Al oxide (Al_2O_3) and metallic Al^0 NMs in the human keratinocyte cell line HaCaT. Possible alterations of the metabolic pattern upon application of the two Al species together with vitamin A or D3 were investigated. Time-of-flight secondary ion mass spectrometry (ToF-SIMS) imaging and inductively coupled plasma mass spectrometry (ICP-MS) were applied to quantify the cellular uptake of Al NMs.

Keywords: nanoparticle uptake; ICP-MS; ToF-SIMS; aluminum; vitamin; metabolomics

---

1. Introduction

Aluminum is one of the most abundant metals that is used in a wide range of industrial manufacturing processes. It is also present in numerous consumer products such as cosmetics or food contact materials. Studies on Al toxicity revealed a potential risk for neuronal toxicity in humans following chronic Al exposure [1,2] and a possible relation of enhanced Al intake to the development of Alzheimer's disease [3].

The uptake of Al may occur via different routes of exposure. Al NMs (nanomaterials) may cross the barriers of the body because of their small size and thus can significantly increase the overall Al burden [4]. Humans are most likely exposed to Al through cosmetic products due to skin contacts or via food additives [3]. To lower the intake of Al, the first legal actions were taken by the EU regulation No. 380/2012 amending Annex II to Regulation (EC) No. 1333/2008, which became applicable at 1st of August 2014 [5]. Further to this, the use of Al-containing food additives is restricted by the recommendation of the European Food Safety Authority (EFSA) to lower the tolerable weekly intake (TWI) of Al to ≤1 mg/kg body weight. However, it was suggested that this TWI might be significantly exceeded, especially in children [6].

To avoid such a high intake, the conditions simulating use and tolerable quantities for food additives containing Al were adapted in the EFSA regulation. The body's uptake of Al is influenced by several processes. For example, vitamin D3 not only enhances the uptake of essential inorganic elements but also of non-essential and toxic elements such as lead or Al [7]. Vitamin D3 also interferes with retinol metabolism and its uptake in human epidermal keratinocytes [8]. The studies related to the interaction between vitamin D3/A and Al may comprise, among other biological endpoints, Al-mediated toxicity, its distribution pattern at the cellular level, as well as its capability to modulate metabolic patterns of skin cells. The complexation behavior of Al is highly sophisticated and certainly adds to the challenges and elucidation of uptake mechanisms. Furthermore, the ubiquitous occurrence of Al resulting in a high background level hampers the application of sensitive analytical methods. The relatively low density of the light metal, its low mass and the complexity of biological matrices complicate the use of widespread analytical techniques like Raman spectroscopy or transmission electron microscopy (TEM). ToF-SIMS analysis [9] has been previously applied for the detection of cerium dioxide particle clusters in rat lung tissue [10] and for the characterization of Al particles in artificial saliva [11]. In this study, we tried to overcome the mentioned analytical limitations by using ToF-SIMS to explore uptake and distribution patterns of Al and Al_2O_3 NMs in the human keratinocyte cell line HaCaT. Furthermore, metabolic profile changes of the cell membrane constituents were investigated. Aluminum chloride ($AlCl_3 \cdot 6H_2O$) was used as soluble ionic control. All three Al species were tested with regard to uptake, cellular distribution patterns, and possible cell membrane alterations upon uptake. In addition to single applications, combinations with vitamin A and D3 were tested as well. ICP-MS measurements were utilized to determine the uptake efficacy of each Al species in the different scenarios.

2. Results and Discussion

2.1. Characterization of Al and Al_2O_3 NMs

Both, Al^0 and Al_2O_3 NMs have been extensivly characterized by our group [12] (see Table 1). The core particle diameter was determined by means of TEM to be between 2–50 nm for the rather spherical Al^0 NMs. Single particle (SP) ICP-MS showed a primary particle size range of 50–80 nm for both Al NMs. The results of the small angle x-ray scattering (SAXS) measurments confirmed the findings of TEM and SP-ICP-MS.

Table 1. Characterization data for Al^0 and Al_2O_3 nanomaterials (NM). Modified from [12].

Methods	Al^0 NM	Al_2O_3 NM
TEM	Primary particle size and shape: 2–50 nm, nearly spherical	Primary particle size and shape: 10 × 20–50 nm, grain-like shape
EELS-TEM	Core-shell structure, thin (2–5 nm) oxide layer	Fully oxidized particle
XRD	Aluminum surface; partially oxidized	Fully oxidized surface
SAXS	Particle size: >20 nm	Primary particle size: 14.2 nm Aggregates' size: >20 nm
SP-ICP-MS	Primary particle size: 54–80 nm	Primary particle size: 50–80 nm
ICP-MS	Ion release: 0.2–0.5% (in 0.05% BSA)	Ion release: 0.2–0.4% (in 0.05% BSA)

The more rod shaped Al_2O_3 NMs display a width of 10 nm and a length between 20–50 nm as determined by TEM. Dynamic light scattering (DLS) determined the hydrodynamic diameter in dispersion of 250 nm for Al^0 NMs, while Al_2O_3 NMs showed a smaller diameter of 180 nm. Zeta Potential measurements of Al^0 and Al_2O_3 NM showed comparable results with −17.2 mV and −17.3 mV in DMEM. X-ray diffraction (XRD) as well as TEM in electron energy loss spectroscopy mode (EELS-TEM) demonstrated the difference of the Al^0 and Al_2O_3 NMs composition and especially their surface. While Al^0 NMs had a core-shell structure with Al core and a 2–5 nm oxygen shell (see Figure 1), the Al_2O_3 NMs were homogeneously oxidized.

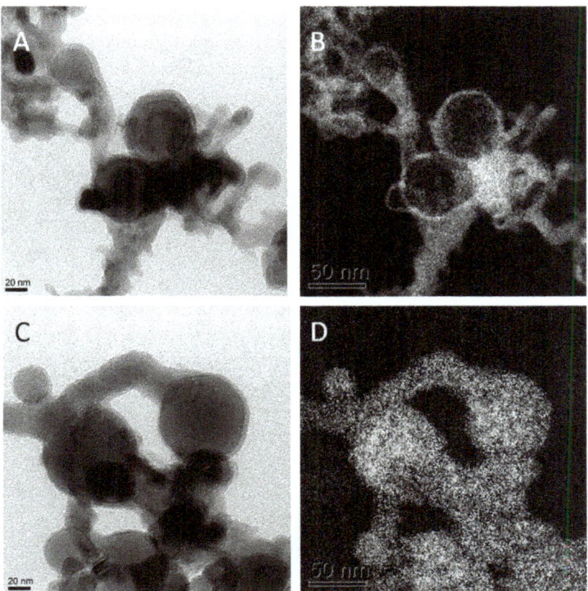

Figure 1. Transmission electron microscopy (TEM) results: (**A**) TEM pictures of Al^0 NMs; (**B**) oxygen mapping of left TEM picture; (**C**) TEM picture of Al^0 NMs; (**D**) aluminum mapping of image in (**C**).

2.2. Cellular Uptake

To investigate the possible uptake of Al NMs and Al_2O_3 NMs, we exposed HaCaT cells to 100 µg/mL Al NMs for 24 h in the presence or absence of 1 µmol/L retinol, 5.12 µmol/L of vitamin D3 (high vitamin D3), or a lower vitamin D3 concentration of 80 nmol/L (low vitamin D3). Subsequently we analyzed the exposed cells using ICP-MS to quantify the NM uptake. Quantitative results are shown in Figure 2. Untreated cells had Al levels comparable to ICP-MS blank samples.

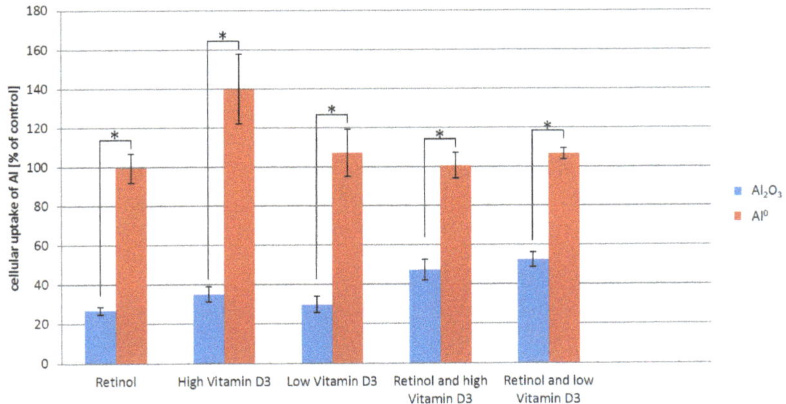

Figure 2. Inductively coupled plasma mass spectrometry (ICP-MS) measurements of Al content of HaCaT cells exposed to either Al_2O_3 or Al^0 NMs as well as retinol and/vitamin D3. The cellular uptake is normalized to the Al uptake of cells exposed to Al^0 NMs or Al_2O_3 NMs only. * $p < 0.05$.

For the interpretation of the results, it has to be considered that the data shown are normalized to the Al^0 uptake by HaCaT cells exposed to Al^0 only (without co-exposure to vitamins). The ICP-MS results show that retinol and vitamin D3 appear to have no detectable effect on the uptake of the Al^0 NM. In contrast, treatment of HaCaT cells with retinol or vitamin D3 significantly lowered the uptake of the Al_2O_3 NM (Figure 2).

The different uptake behavior of the two kinds of Al NMs after treatment of the HaCaT cells with vitamins is likely due to differences in their physicochemical properties. It is known that the complexation behavior and thus the agglomeration rate of the used Al NMs differ strongly [12]. Al_2O_3 is characterized as a rather insoluble and rod shaped oxidized NM, whereas Al^0 NMs are partially soluble and quasi-spherical. Differences in the particle physicochemical properties (e.g., surface area, solubility, etc.) may lead to differences in the uptake mechanism preferred by the cells.

These findings presented in this study are in good accordance with the results of an uptake study of polystyrene NMs on Caco-2 cells [13]. Furthermore, the composition of the NMs might also influence their uptake and distribution [14]. The core-shell structure of the Al^0 NM (see Table 1) contains approximately 85% Al. In contrast, Al_2O_3 is fully oxidized and shows a homogenous distribution of Al as well as oxygen on its surface. Our group was able to demonstrate that the protein corona that is formed during the contact of NMs with cell culture media is less complex for Al^0 NM compared to the protein corona of Al_2O_3 NM [15]. The surface properties of NMs facilitate interactions with the surrounding environment, which also affects the bioavailability and the interactions of the NM with the cell.

For the assessment of changes of the metabolite patterns of the HaCaT cell membranes and the overall particle distribution, ToF-SIMS analytics was employed. The acquired ToF-SIMS mass spectra for HaCaT cells showed a strong Al peak indicating the presence of Al NMs (Figure 3). Cellular uptake can also be observed in HaCaT cell cultures co-incubated with retinol, vitamin D3, and its combinations. The Al NM uptake could also be observed after treatment with Al_2O_3 as well as in cultures cultivated with retinol plus vitamin D3 low or high for both NM species.

3D reconstruction of ToF-SIMS images from single HaCaT cells reveals the intracellular presence of Al^0 and Al_2O_3 NMs at all exposure scenarios: NM alone, NM in combination with retinol and/or vitamin D3.

HaCaT cells treated with Al NMs only store the particles in large agglomerates (Figure 4a,e), whereas cells treated with Al NMs in combination with retinol or vitamin D3 show a different uptake and distribution behavior (compare Figure 4b,f vs. Figure 4c,g). Treatment with retinol leads to the accumulation of Al close to the cell membrane with large agglomerates (Figure 4b,f). This process might be due to the increased collagen synthesis and reduced matrix metalloproteinase expression which is known to occur because of vitamin A treatment [16]. The resulting collagen increase facilitates NM collagen interactions [17] which might hinder the NM allocation. Upon co-application of vitamin D3 the ToF-SIMS analyses revealed a different cellular deposition pattern of Al^0 NMs when compared to that after co-exposure to retinol. Co-exposure with vitamin D3 led to an even distribution of particles throughout the entire cell and to the formation of much smaller aggregates (Figure 4c,d,g,h). In addition of being responsible for an enhanced uptake of metals (Figure 2), vitamin D3 also seems to affect particulate distribution patterns within the cell. The ToF-SIMS results are in accordance with the ICP-MS findings, where high vitamin D3 is responsible for an enhanced intracellular uptake (Figure 2).

Int. J. Mol. Sci. **2020**, *21*, 1278

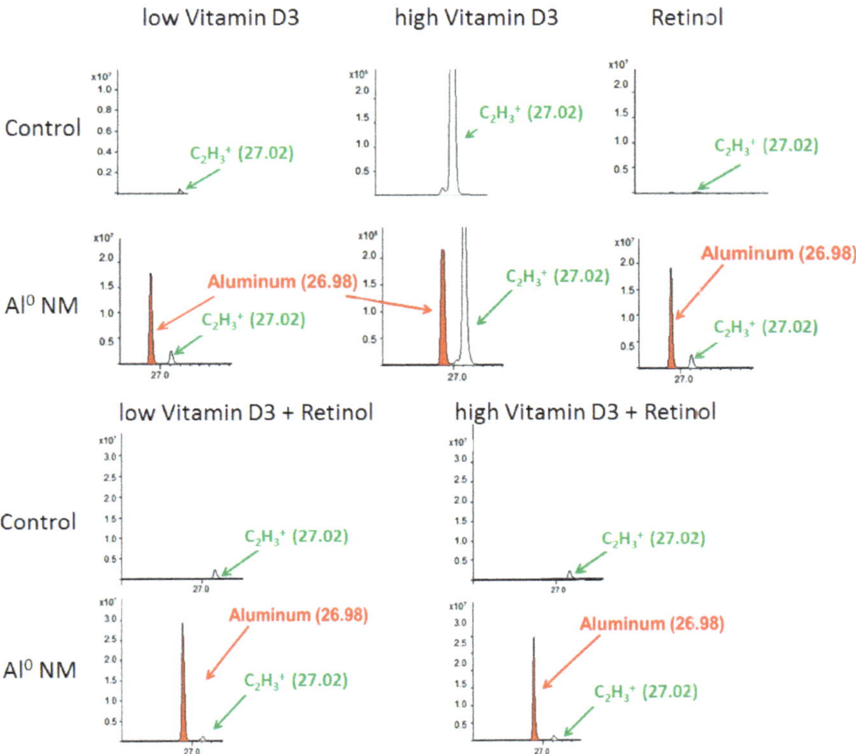

Figure 3. Time-of-flight secondary ion mass spectrometry (ToF-SIMS) mass spectrum (positive mode), showing the Al peak in red color (at *m/e* 26.98 u) and a peak in green color (at 27.02 u = $C_2H_3^+$), resulting from organic matter in HaCaT cells. The upper line shows the spectra for control HaCaT cells, the lower line for HaCaT cells which were exposed to Al NMs (about 20 nm) for 24 h in addition to high or low vitamin D3, and retinol or their combinations. The x-axis shows the mass to charge ratio (*m/z*); y-axis the ion intensities.

The ToF-SIMS results for the exposure of HaCaT cells with Al_2O_3 NMs show that either treatment with retinol or vitamin D3 leads to a strong decrease in the NM uptake and to smaller agglomerate sizes when compared to cells treated with Al_2O_3 NMs only (Figure 2). Comparison of Al^0 and Al_2O_3 NM treatments of HaCaT cells reveals a generally smaller size of NM agglomerates in the latter ones (Figure 5). Based on this it can be assumed that the number of particle agglomerates per cell is much higher for Al_2O_3 than for Al^0 NMs. Therefore Al_2O_3 NM aggregation is being largely compromised resulting in a more even intracellular distribution of Al_2O_3 NMs when compared with Al^0 NM (compare Figure 4b,f vs. Figure 5b,f).

When HaCaT cells were exposed to retinol in combination with either low or high levels of vitamin D3 and Al, the localization of the particles was restricted to the cell membrane region (Figure 6a,c), which corroborates above findings in Figures 4 and 5. ICP-MS results show, however, a similar mass balance for all exposure experiments with Al NMs (Figure 2), indicating again a much larger number of smaller agglomerates in the case of exposure to vitamin D3.

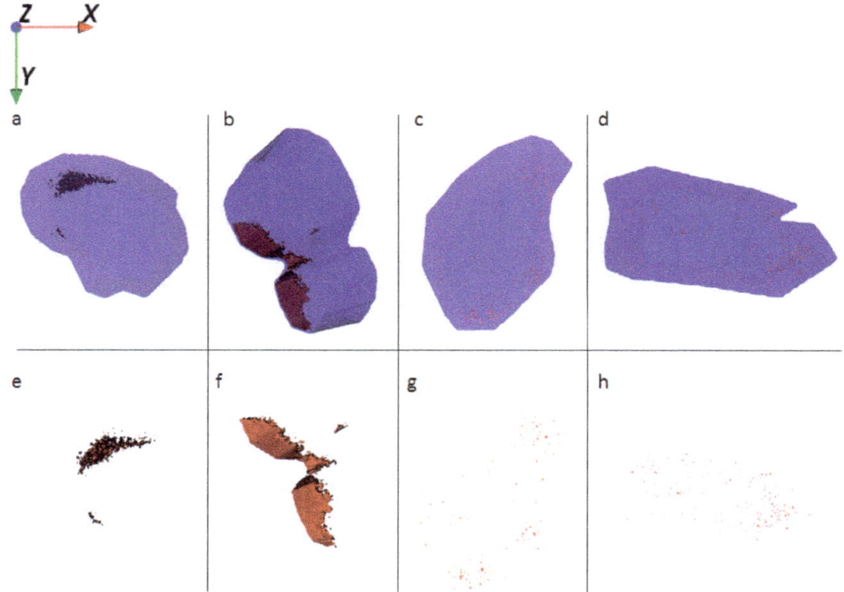

Figure 4. Ion reconstruction of a 3D depth profile (depth layer numbers: 50–250) of one single HaCaT cell, which was exposed to Al^0 NMs for 24 h. The images show the top-down view of the outline of a cell of a depth profile. The translucent blue outline was reconstructed based on the $C_3H_8N^+$ signal that originates from intracellular amino acids. (**a**) Control cells treated with Al^0 NM only (red color); and corresponding intracellular localization of Al^0 NM agglomerates (**e**). (**b**) Cells treated with retinol and Al^0 NM (red color); and corresponding intracellular localization of Al NM agglomerates (**f**). (**c**) Cells treated with low vitamin D3 and Al^0 NM (red color); and corresponding intracellular localization of Al^0 NM agglomerates (**g**). (**d**) Cells treated with high vitamin D3 and Al^0 NM (red color); and corresponding intracellular localization of Al^0 NM agglomerates (**h**).

Uptake of Al_2O_3 NMs alone and in combination with retinol, vitamin D3, or both vitamins revealed wide distribution and agglomeration of Al nanomaterials throughout the cytoplasm. In contrast to all co-exposure experiments, formation of larger NM clusters was observed after exposure to Al_2O_3 NMs alone (Figure 5a,e). ICP-MS data show a significant increase of Al_2O_3 NMs being present in HaCaT cells following their straight application in contrast to the co-exposure experiments (Figure 2). In addition, ToF-SIMS results show a significant reduction in the sizes of NM agglomerates after the co-exposure experiments (Figure 5 vs. Figure 6).

2.3. Metabolic Changes after Nanomaterial Uptake

In addition to NM uptake and distribution, we assessed the alterations of the cell membrane constituents of HaCaT cells caused by Al and Al_2O_3 NM exposures. Treatments with ionic $AlCl_3*6H_2O$ and unexposed HaCaT cells were used for comparison (Figure 7).

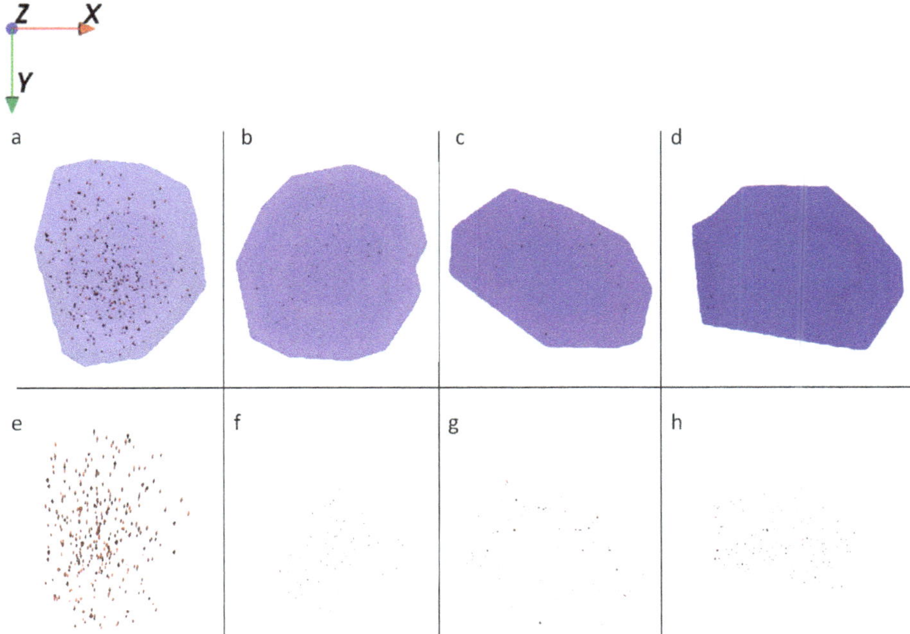

Figure 5. Ion reconstruction of a 3D depth profile (depth layer numbers: 50–250) of one single HaCaT cell exposed to Al_2O_3 NMs for 24 h. The images show the top-down view of the outline of a cell of a depth profile. The translucent blue outline was reconstructed based on the $C_3H_8N^+$ signal that originates from intracellular amino acids. (**a**) Control cells exposed to Al_2O_3 NMs only (red color); and corresponding intracellular localization of Al_2O_3 NM agglomerates (**e**). (**b**) Cells treated with retinol and Al_2O_3 NMs (red); and corresponding intracellular localization of Al_2O_3 NM agglomerates (**f**). (**c**) Cells treated with low vitamin D3 and Al_2O_3 NMs (red); and corresponding intracellular localization of Al_2O_3 NM agglomerates (**g**). (**d**) Cells treated with high vitamin D3 and Al_2O_3 NMs (red); and corresponding intracellular localization of Al_2O_3 NM agglomerates (**h**).

The results obtained for cells exposed to retinol, low vitamin D3, and high vitamin D3 reveal significant differences in the composition of the respective cell membranes (Figure 7). Significant differences in the cell membrane composition could also be observed in HaCaT cells co-exposed to retinol and low vitamin D3 or high vitamin D3. In order to determine the differences of lipid membrane constituents we further investigated the significant changed metabolites upon co-epxosure of Al^0 or Al_2O_3 NMs with retinol or high or low vitamin D3 (see Table 2 for Al^0 and Table 3 for Al_2O_3). For further details please see also the Supplementary Material (Figures S1–S6).

The further investigation of changes in the membrane composition was carried out via ToF-SIMS analysis of the major altered cell membrane lipids. Diacylglycerols (DAGs) and phosphatidic acids (PAs) were found to be increased following treatment with either Al^0 or Al_2O_3 NMs in combination with vitamins similarly (see Tables 2 and 3). DAGs were significantly increased after administration of retinol or vitamin D3 together with both Al NMs. DAGs are the product of hydrolysis of the phospholipid phosphatidylinositol 4,5-bisphosphate (PIP2) that serve as activators of the protein kinase C (PKC) pathway [18]. Alterations in the PKC signaling in HaCaT cells lead to strong morphological changes including shape of this cell type [19]. Different PKC isoforms have also different effects on the proliferation and differentiation behavior of HaCaT cells [20]. The PAs and its precursor substances lyso-phosphatidic acid were significantly increased over all treatment regimes. The PAs can be degraded to DAGs and serve as precursors of other membrane lipids [21]. Therefore, we conclude a comparable metabolic alteration of DAGs and PAs in response to Al NM uptake. Phosphatidylethanolamines (PEs) are another important group of membrane constituents. The PEs serve as substrates for phosphatidylcholine (PC) biosynthesis and are crucial for cell surface signaling [22]. Investigations on the ratio of PC to PE in mice showed decreased membrane integrity and the development of a leaky membrane when the ratio decreases [23]. In D3-treatments an increase of the level of PEs was detected because of a shift of the PCs toward PEs thereby causing a leaky membrane of the cells affected. High vitamin D3 addition revealed significantly increased levels of lyso-PC, which is a degradation product of PC and therefore indicates reduced PC levels. This process lowers the PC to PE ratio even more drastically and an enhanced uptake of Al may occur. These findings fit well with the ICP-MS results (see Figure 2).

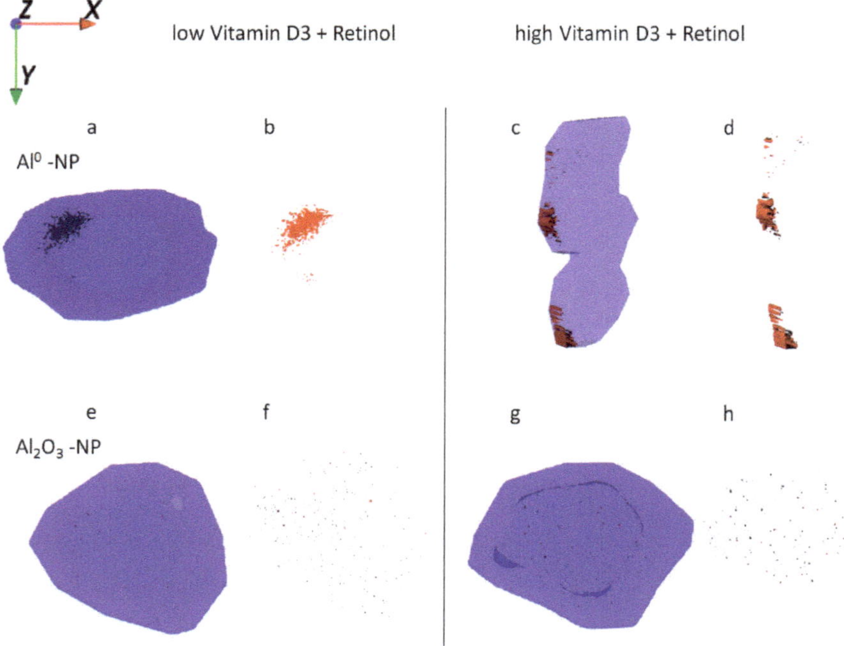

Figure 6. Ion reconstruction of a 3D depth profile (depth layer numbers: 50–250) of one single HaCaT cell exposed to Al NMs (upper panel **a–d**) or Al_2O_3 NMs (lower panel **e–h**). The 3D depth profile of the cell is depicted as translucent blue. The images show the top-down view into the outline of a cell of a depth profile. In addition to NM treatment low vitamin D3 (**a,b,e,f**) and high vitamin D3 (**c,d,g,h**) were administered together with retinol. NM agglomerates are shown in detail next to their respective cell.

The significantly increased metabolites after treatment with Al_2O_3 and retinol or vitamin D3 are more diverse when compared to Al^0-treated cells (see Table 3).

In addition to the above mentioned findings, another metabolite, dihydroceramides (DCs), was found to be enhanced after the treatment of HaCaT cells with Al_2O_3 and vitamin D3 (see Table 3). The increased levels of this compound in HaCaT cells exposed to Al_2O_3 NM and vitamin D3 provides another explanation for the decreased nanoparticle uptake in cells exposed to Al_2O_3 NM as DCs enhance the rigidity of the plasma membrane [24]. This could lead to changes in the active transport, vesicle formation, diffusion, and activation of the cell-signaling pathway, all of which are representing processes that depend on plasma membrane dynamics [25].

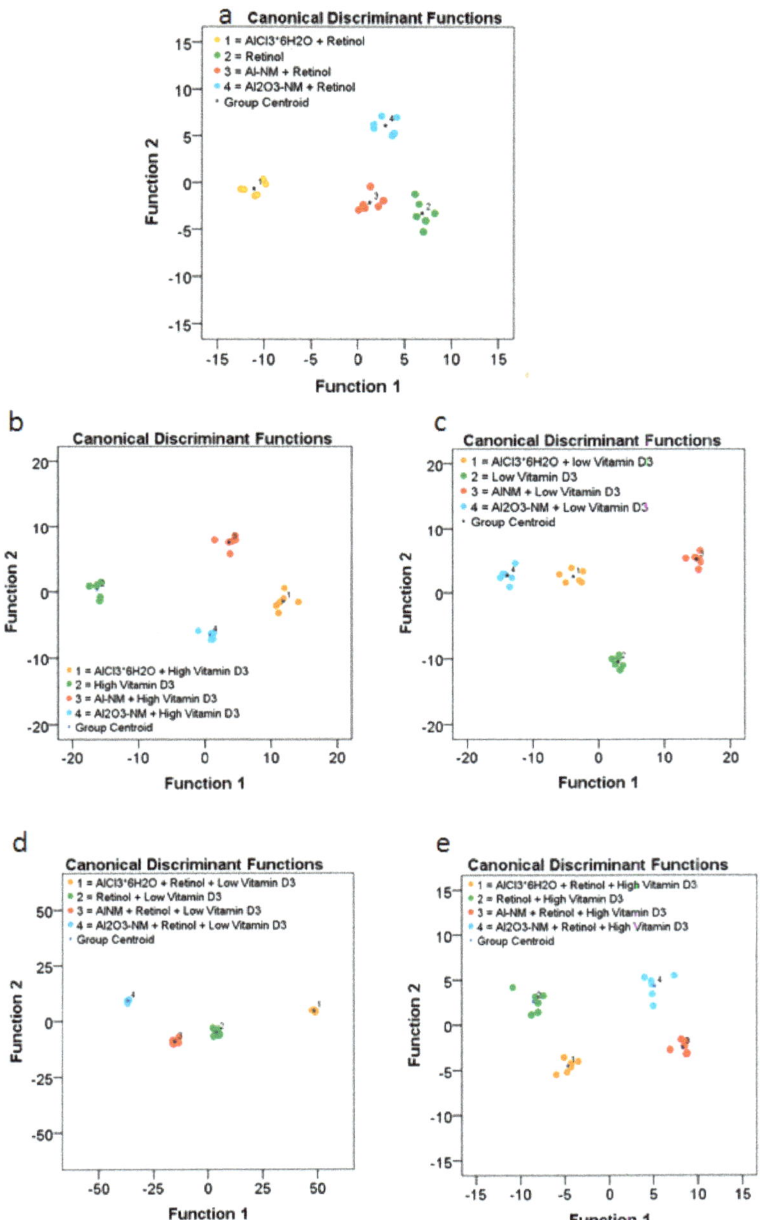

Figure 7. ToF-SIMS analysis of changes in the composition of the cell membranes of HaCaT cells after treatment with Al or Al_2O_3 NMs and ionic $AlCl_3 \cdot 6H_2O$ without or in combination with vitamins ((**a**) retinol; (**b**) high vitamin D3; (**c**) low vitamin D3; (**d**) retinol plus low vitamin D3; (**e**) retinol plus high vitamin D3). The diagram shows the values of the discriminant scores obtained from Fisher's discriminant analysis of 24 single HaCaT cells for each experiment. The performance of the discriminant model was verified by applying the cross-validation procedure based on the "leave-one-out" cross-validation formalism (100%).

Table 2. Significantly increased cell membrane constituents upon treatment of HaCaT cells with Al^0 NMs divided by the additional treatment of either retinol low or high vitamin D3.

Retinol	Low Vitamin D3	High Vitamin D3
diacylglycerols	phosphatidylethanolamines	lyso-phosphatidylcholines
lyso-phosphatidic acids	-/-	-/-

Table 3. Significantly increased cell membrane constituents upon treatment of HaCaT cells with Al_2O_3 NMs divided by the additional treatment of either retinol low or high vitamin D3.

Retinol	Low Vitamin D3	High Vitamin D3
diacylglycerols	diacylglycerols	Diacylglycerols
lyso-phosphatidic acids	lyso-phosphatidylcholines	phosphatidic acids
-/-	dihydroceramides	diacylglycerol phosphates

The changes in the membrane lipid composition described above serve as an explanation for the decreased uptake of Al_2O_3 NMs after vitamin A/D3 treatment. The increased levels of DCs after treatment with vitamin D3 could also derive from a protective mechanism of the cell to secure the physiological integrity of mitochondria. It has been previously shown that DCs block the permeabilization of the mitochondrial outer membrane [26].

The comparison of the changes in the constituents of the cell membrane that were introduced upon treatment with either Al^0 or Al_2O_3 NMs clearly shows the necessity to distinguish between the different types of aluminum NMs.

3. Materials and Methods

3.1. Cell Culture and NM Exposure

Al NMs (18 nm, 99.9%) and Al_2O_3 NMs (20 nm, 99+%) were purchased from IoLiTec Ionic Liquids Technologies GmbH, Heilbronn, Germany. The following chemicals were purchased from Sigma-Aldrich (Sigma-Aldrich Corp., St. Louis, MO, USA): $AlCl_3 \cdot 6H_2O$ (hexahydrate, ≥97%), retinol (≥97.5%), cholecalciferol (certified reference material), and calcipotriol (European Pharmacopoeia Reference Standard). The chemicals were diluted to the respective concentrations (high vitamin D3 high concentration: 5.12 µmol/L; low vitamin D3 concentration: 80 nmol/L; retinol: 1 µmol/L) in DMSO (≥99.7%), obtained from Sigma-Aldrich (Sigma-Aldrich Corp., St. Louis, MO, USA).

The human immortalized keratinocyte cell line HaCaT was cultured in Dulbecco's modified Eagle's medium (DMEM), 10% fetal bovine serum, and 1% antibiotics (10,000 µg/mL streptomycin and 10,000 units/mL penicillin) at 37 °C with 5% CO_2. Cells were passaged at 70–80% confluence two times a week.

For ToF-SIMS measurements 0.05×10^6 cells were seeded on 1 cm^2 silica wafers and left in the incubator for 24 h. Afterwards cells were treated with the respective NM with or without vitamin derivates for 24 h. The wafers were then washed using 150 mM ammonium bicarbonate solution. Samples were fast frozen and lyophilized prior to ToF-SIMS measurements.

NM dispersions were prepared following the NanoGenoTox dispersion protocol: "Final protocol for producing suitable manufactured NMs exposure media" (October, 2011). In brief, a 2.56-mg/mL stock dispersion of each NM was prepared by pre-wetting the powder with 0.5% (vol/vol) ethanol (96%) followed by addition of Millipore water containing 0.05% BSA. Dispersion occurred for 5 min and 9 s at an amplitude of 10% with a probe sonifier, either 200 W Bandelin Sonopuls HD 2200, (BANDELIN Electronic GmbH & Co. KG, Berlin, Germany) (at BfR); or 400 W Branson Sonifier S-450 CE Digital, (Branson Ultrasonics, St. Louis, Missouri, USA) (IMPB). The sample was cooled in an ice-water bath during sonication [27].

3.2. ICP-MS Analysis

Measurements were performed with a quadrupole ICP mass spectrometer (iCAP Q, Thermo Fisher Scientific GmbH, Dreieich, Germany) equipped with a PrepFast system (ESI Elemental Service & Instruments GmbH, Mainz, Germany), PFA ST Nebulizer, a quartz cyclonic spray chamber, and a 2.5 mm quartz injector (all from Thermo Fisher Scientific, Waltham, MA, USA) using the following isotopes: ^{27}Al and, as an internal standard, ^{103}Rh. Calibrations were performed using ionic standards of Al in a 3.5% HNO_3 solution ranging from 2 to 500 µg/L. The internal standard was added using the ICP-MS PrepFast system. The gas flows for the cooling gas and the auxiliary gas were set to 14 L/min and 0.65 L/min, respectively. The sample flow rate was 0.4 mL/min. All isotopes were analyzed using the collision cell technique at 5 mL/min collision gas flow (93% He and 7% H_2). Analysis of NM uptake was studied with five replicates per dose. Results were presented as mean values ± standard error of the mean (SEM).

3.3. ToF-SIMS Analysis

A dedicated cryogenic sample preparation technique with a high cooling rate was used for sample analysis [28,29]. Liquid propane was cooled using liquid nitrogen, thus preventing evaporation of propane at the contact surface of the immersed specimen. Specimens were in contact with liquid propane for 10 s and were afterwards kept in a frozen state using dry ice. The condenser of the Christ Beta 2-8 lyophilizer (Martin Christ GmbH, Osterode am Harz, Germany) was cooled to −80 °C. The frozen samples were placed on the frozen heating plate, which was inserted in the freeze-drying chamber and heated to −20 °C. Afterwards, vacuum was applied to achieve a pressure of 1.65 mbar. The main drying process started, in which the water is sublimated within 2 h by vacuum and heated to 23 °C. The temperature of 23 °C is maintained for 30 min. The instrument was ventilated and the freeze-dried samples were stored at −80 °C prior to the ToF-SIMS analysis. A ToF-SIMS instrument (ION-TOF V; Ion-TOF GmbH, Münster, Germany) was used for mass spectrometry analyses with a pulsed 30 keV Bi^{3+} liquid metal ion gun (LMIG, direct current (dc), 16 nA). Measurement of cell samples was performed at room temperature. Each spectrum was acquired by scanning the ion beam over a sample area of 400 × 400 µm. Positive secondary ions were collected in the mass range up to *m/z* 1200 using 10^6 Bi^{3+} pulses. Instrument and analysis conditions were used as described elsewhere for the ToF-SIMS analysis of cell membrane lipids [30].

All depth profiles were performed in dual beam mode on a TOF.SIMS V instrument (ION-TOF GmbH, Münster, Germany) of the reflectron-type, equipped with a 30 keV Bi_3^+ LMIG as primary ion source, a 20 keV argon gas cluster ion source both mounted at 45° with respect to the sample surface and an electron flood gun. Bi^{3+} was selected as primary ion by appropriate mass filter settings. Primary and sputter ion currents were directly determined at 200 µs cycle time (i.e., a repetition rate of 5.0 kHz) using a Faraday cup located on a grounded sample holder. Scanning area for analysis was $200 × 200$ $µm^2$ with 512 × 512 pixels. The sputter area for each measurement was 1000 µm × 1000 µm. Surface charging was compensated by flooding with low energy electrons.

ToF-SIMS depth profiles were acquired in positive ion mode. The mass scale was internally calibrated using a number of well-defined and easily assignable secondary ions ($C_2H_5^+$, $C_3H_7^+$, and $C_4H_9^+$) keeping the error of calibration for all spectra below 5 ppm. The data were evaluated using the Surface Lab software (ION-TOF GmbH, Münster, Germany).

Statistical analyses of the ToF-SIMS data were performed as described in detail elsewhere [30–34]. In brief, the acquired data were binned to 1 u. Data processing was carried out with the statistical package SPSS + (version 21) using the mass range between 200 and 1200 mass units to detect significant differences between treated and untreated cells. Ions lower than mass 200 were excluded from the study to avoid contamination of the ions from salts, system contaminants, and other medium components. Each acquired spectrum was then normalized, setting the peak sum to 100%. A principal component analysis (PCA) was performed using all ions. To show that data sets could be separated with a supervised model from each other a Fisher's discriminant analysis was performed. The performance of the discriminant model was verified by applying the cross-validation procedure based on the "leave-one-out" cross-validation formalism.

4. Conclusions

The ToF-SIMS measurements have shown that aggregation and incorporation of Al NMs in HaCaT cells are influenced by treatment with vitamins. Experiments with retinol led to the formation of large Al aggregates, which are intercalated in the membrane regions of the cells, while vitamin D3 treatments resulted in the formation of small agglomerates within the entire cell. These findings suggest that, depending on the vitamin treatment, different pathways are used for the uptake of Al NMs.

Furthermore, the results show a decreased uptake rate of Al_2O_3 NMs after treatment with vitamins (retinol and vitamin D3) in comparison with the control as well as with the exposure to Al^0 NMs in combination with both vitamins. A likely explanation for this behavior is the change in the lipid membrane composition, which might lead to an enhanced rigidity of the membrane. Treatment with either retinol or vitamin D3 leads to a drastic decrease in the uptake of Al_2O_3 NMs.

In contrast to the protective effect observed for Al_2O_3 NMs, ToF-SIMS measurements revealed a changed lipid metabolite profile for cells in response to co-exposure to vitamin D3 and Al^0 NMs. The metabolic changes led to a shift in the PC-to-PE ratio, which contributes to an increased uptake of Al NMs due to a leaky cell membrane.

We conclude a protective effect of the vitamins A and D3 for cells, which are in contact with nanoparticulate oxides such as Al_2O_3. On the other hand, the presence of these substances may slightly promote the uptake of metallic NMs. Our findings also reflect the high importance of a thorough physicochemical particle characterization, as parameters like agglomeration, solubility, and biokinetics may affect the uptake of NMs with the same elemental constituent.

Supplementary Materials: The following are available online at http://www.mdpi.com/1422-0067/21/4/1278/s1.

Author Contributions: F.L.K. and B.-C.K. conceived and designed the experiments. The project and research was supervised by A.V.S., P.L., and A.L. F.L.K. conducted the cell culture experiments and prepared together with P.R. the ToF-SIMS samples. B.-C.K. carried out the ICP-MS measurements and performed the statistical evaluation of the results. ToF-SIMS measurements and subsequent data analysis was performed by H.J., J.T., and P.R. The manuscript was written by F.L.K. and B.E.K. with input from all authors. All authors have read and approved the final manuscript.

Funding: The BfR provided intramural support SFP1322-642 for F.L.K., P.R., H.J. and A.L.

Acknowledgments: The authors thank Yves U. Hachenberger for fruitful discussions and his support.

Conflicts of Interest: The authors declare no conflict of interest.

Abbreviations

Al	aluminum
DAG	diacylglycerol
DLS	dynamic light scattering
EELS	electron energy loss spectroscopy
EFSA	European Food Safety Authority
HaCaT	human keratinocyte cell line
ICP-MS	inductively coupled plasma mass spectrometry
NMs	nanomaterials
PA	phosphatidic acids
PC	phosphatidylcholine
PCA	principal component analysis
PE	phosphatidylethanolamine
PIP2	phosphatidylinositol 4,5-bisphosphate

PKC   protein kinase C
SAXS   small angle X-ray scattering
SEM   standard error of the mean
SP   single particle
TEM   transmission electron microscopy
ToF-SIMS   time-of-flight secondary ion mass spectrometry
TWI   tolerable weekly intake
XRD   X-ray diffraction

References

1. Lu, X.; Liang, R.; Jia, Z.; Wang, H.; Pan, B.; Zhang, Q.; Niu, Q. Cognitive disorders and tau-protein expression among retired aluminum smelting workers. *J. Occup. Env. Med.* **2014**, *56*, 155–160. [CrossRef] [PubMed]
2. Sheykhansari, S.; Kozielski, K.; Bill, J.; Sitti, M.; Gemmati, D.; Zamboni, P.; Singh, A.V. Redox metals homeostasis in multiple sclerosis and amyotrophic lateral sclerosis: a review. *Cell Death Dis.* **2018**, *9*, 348. [CrossRef] [PubMed]
3. Kandimalla, R.; Vallamkondu, J.; Corgiat, E.B.; Gill, K.D. Understanding Aspects of Aluminum Exposure in Alzheimer's Disease Development. *Brain Pathol.* **2016**, *26*, 139–154. [CrossRef] [PubMed]
4. Oberdorster, G.; Oberdorster, E.; Oberdorster, J. Nanotoxicology: An emerging discipline evolving from studies of ultrafine particles. *Environ. Health Perspect.* **2005**, *113*, 823–839. [CrossRef] [PubMed]
5. Scotter, M.J. 3-Overview of EU regulations and safety assessment for food colours. In *Colour Additives for Foods and Beverages*; Woodhead Publishing: Oxford, UK, 2015; pp. 61–74. [CrossRef]
6. Tietz, T.; Lenzner, A.; Kolbaum, A.E.; Zellmer, S.; Riebeling, C.; Gürtler, R.; Jung, C.; Kappenstein, O.; Tentschert, J.; Giulbudagian, M.; et al. Aggregated aluminium exposure: risk assessment for the general population. *Arch. Toxicol.* **2019**. [CrossRef] [PubMed]
7. Schwalfenberg, G.K.; Genuis, S.J. Vitamin D, Essential Minerals, and Toxic Elements: Exploring Interactions between Nutrients and Toxicants in Clinical Medicine. *Scientific World J.* **2015**, *2015*, 318595. [CrossRef]
8. Torma, H.; Rollman, O.; Binderup, L.; Michaelsson, G. Vitamin D analogs affect the uptake and metabolism of retinol by human epidermal keratinocytes in culture. *J. Investig. Dermatol. Symp. Proc.* **1996**, *1*, 49–53.
9. Jungnickel, H.; Laux, P.; Luch, A. Time-of-Flight Secondary Ion Mass Spectrometry (ToF-SIMS): A New Tool for the Analysis of Toxicological Effects on Single Cell Level. *Toxics* **2016**, *4*, 5. [CrossRef]
10. Laux, P.; Riebeling, C.; Booth, A.M.; Brain, J.D.; Brunner, J.; Cerrillo, C.; Creutzenberg, O.; Estrela-Lopis, I.; Gebel, T.; Johanson, G.; et al. Biokinetics of nanomaterials: The role of biopersistence. *NanoImpact* **2017**, *6*, 69–80. [CrossRef]
11. Sieg, H.; Kästner, C.; Krause, B.; Meyer, T.; Burel, A.; Böhmert, L.; Lichtenstein, D.; Jungnickel, H.; Tentschert, J.; Laux, P.; et al. Impact of an Artificial Digestion Procedure on Aluminum-Containing Nanomaterials. *Langmuir* **2017**, *33*, 10726–10735. [CrossRef]
12. Krause, B.; Meyer, T.; Sieg, H.; Kästner, C.; Reichardt, P.; Tentschert, J.; Jungnickel, H.; Estrela-Lopis, I.; Burel, A.; Chevance, S.; et al. Characterization of aluminum, aluminum oxide and titanium dioxide nanomaterials using a combination of methods for particle surface and size analysis. *RSC Adv.* **2018**, *8*, 14377–14388. [CrossRef]
13. Banerjee, A.; Qi, J.; Gogoi, R.; Wong, J.; Mitragotri, S. Role of nanoparticle size, shape and surface chemistry in oral drug delivery. *J. Control. Release* **2016**, *238*, 176–185. [CrossRef] [PubMed]
14. Fadeel, B.; Fornara, A.; Toprak, M.S.; Bhattacharya, K. Keeping it real: The importance of material characterization in nanotoxicology. *Biochem. Biophys. Res. Commun.* **2015**, *468*, 498–503. [CrossRef] [PubMed]
15. Sieg, H.; Braeuning, C.; Kunz, B.M.; Daher, H.; Kastner, C.; Krause, B.C.; Meyer, T.; Jalili, P.; Hogeveen, K.; Bohmert, L.; et al. Uptake and molecular impact of aluminum-containing nanomaterials on human intestinal caco-2 cells. *Nanotoxicology* **2018**, *12*, 992–1013. [CrossRef] [PubMed]
16. Varani, J.; Warner, R.L.; Gharaee-Kermani, M.; Phan, S.H.; Kang, S.; Chung, J.H.; Wang, Z.Q.; Datta, S.C.; Fisher, G.J.; Voorhees, J.J. Vitamin A antagonizes decreased cell growth and elevated collagen-degrading matrix metalloproteinases and stimulates collagen accumulation in naturally aged human skin. *J. Investig. Dermatol.* **2000**, *114*, 480–486. [CrossRef] [PubMed]

17. Kandamachira, A.; Selvam, S.; Marimuthu, N.; Janardhanan Kalarical, S.; Fathima Nishter, N. Collagen-nanoparticle Interactions: Type I Collagen Stabilization Using Functionalized Nanoparticles. *Soft Mater.* **2015**, *13*, 59–65. [CrossRef]

18. Chauhan, A.; Chauhan, V.P.S.; Brockerhoff, H. Activation of Protein-Kinase-C by Phosphatidylinositol 4,5-Bisphosphate - Possible Involvement in Na+/H+ Antiport down-Regulation and Cell-Proliferation. *Biochem. Biophys. Res. Commun.* **1991**, *175*, 852–857. [CrossRef]

19. Hegemann, L.; Wevers, A.; Bonnekoh, B.; Mahrle, G. Changes of epidermal cell morphology and keratin expression induced by inhibitors of protein kinase C. *J. Dermatol. Sci.* **1992**, *3*, 103–110. [CrossRef]

20. Papp, H.; Czifra, G.; Lazar, J.; Gonczi, M.; Csernoch, L.; Kovacs, L.; Biro, T. Protein kinase C isozymes regulate proliferation and high cell density-mediated differentiation in HaCaT keratinocytes. *Exp. Dermatol.* **2003**, *12*, 811–824. [CrossRef]

21. Athenstaedt, K.; Daum, G. Phosphatidic acid, a key intermediate in lipid metabolism. *Eur. J. Biochem.* **1999**, *266*, 1–16. [CrossRef]

22. Vance, J.E.; Tasseva, G. Formation and function of phosphatidylserine and phosphatidylethanolamine in mammalian cells. *Bba-Mol. Cell Biol. L* **2013**, *1831*, 543–554. [CrossRef] [PubMed]

23. Li, Z.; Agellon, L.B.; Allen, T.M.; Umeda, M.; Jewell, L.; Mason, A.; Vance, D.E. The ratio of phosphatidylcholine to phosphatidylethanolamine influences membrane integrity and steatohepatitis. *Cell Metab.* **2006**, *3*, 321–331. [CrossRef] [PubMed]

24. Vieira, C.R.; Munoz-Olaya, J.M.; Sot, J.; Jimenez-Baranda, S.; Izquierdo-Useros, N.; Abad, J.L.; Apellaniz, B.; Delgado, R.; Martinez-Picado, J.; Alonso, A.; et al. Dihydrosphingomyelin Impairs HIV-1 Infection by Rigidifying Liquid-Ordered Membrane Domains. *Chem. Biol.* **2010**, *17*, 766–775. [CrossRef] [PubMed]

25. Rodriguez-Cuenca, S.; Barbarroja, N.; Vidal-Puig, A. Dihydroceramide desaturase 1, the gatekeeper of ceramide induced lipotoxicity. *Bba-Mol. Cell Biol. L* **2015**, *1851*, 40–50. [CrossRef] [PubMed]

26. Stiban, J.; Fistere, D.; Colombini, M. Dihydroceramide hinders ceramide channel formation: Implications on apoptosis. *Apoptosis* **2006**, *11*, 773–780. [CrossRef]

27. Tavares, A.M.; Louro, H.; Antunes, S.; Quarre, S.; Simar, S.; De Temmerman, P.J.; Verleysen, E.; Mast, J.; Jensen, K.A.; Norppa, H.; et al. Genotoxicity evaluation of nanosized titanium dioxide, synthetic amorphous silica and multi-walled carbon nanotubes in human lymphocytes. *Toxicol. Vitr.* **2014**, *28*, 60–69. [CrossRef]

28. Vickerman, J.C. Molecular Surface Mass Spectrometry by SIMS. In *Surface Analysis – The Principal Techniques*; John Wiley & Sons, Ltd.: Chichester, UK, 2009; pp. 113–205. [CrossRef]

29. Singh, A.V.; Jungnickel, H.; Leibrock, L.; Tentschert, J.; Reichardt, P.; Katz, A.; Laux, P.; Luch, A. ToF-SIMS 3D imaging unveils important insights on the cellular microenvironment during biomineralization of gold nanostructures. *Sci Rep.-Uk* **2020**, *10*, 261. [CrossRef]

30. Tentschert, J.; Draude, F.; Jungnickel, H.; Haase, A.; Mantion, A.; Galla, S.; Thunemann, A.F.; Taubert, A.; Luch, A.; Arlinghaus, H.F. TOF-SIMS analysis of cell membrane changes in functional impaired human macrophages upon nanosilver treatment. *Surf. Interface Anal.* **2013**, *45*, 483–485. [CrossRef]

31. Booth, A.; Storseth, T.; Altin, D.; Fornara, A.; Ahniyaz, A.; Jungnickel, H.; Laux, P.; Luch, A.; Sorensen, L. Freshwater dispersion stability of PAA-stabilised cerium oxide nanoparticles and toxicity towards Pseudokirchneriella subcapitata. *Sci. Total Environ.* **2015**, *505*, 596–605. [CrossRef]

32. Haase, A.; Arlinghaus, H.F.; Tentschert, J.; Jungnickel, H.; Graf, P.; Mantion, A.; Draude, F.; Galla, S.; Plendl, J.; Goetz, M.E.; et al. Application of Laser Postionization Secondary Neutral Mass Spectrometry/Time-of-Flight Secondary Ion Mass Spectrometry in Nanotoxicology: Visualization of Nanosilver in Human Macrophages and Cellular Responses. *ACS Nano* **2011**, *5*, 3059–3068. [CrossRef]

33. Jungnickel, H.; Jones, E.A.; Lockyer, N.P.; Oliver, S.G.; Stephens, G.M.; Vickerman, J.C. Application of TOF-SIMS with chemometrics to discriminate between four different yeast strains from the species Candida glabrata and Saccharomyces cerevisiae. *Anal. Chem.* **2005**, *77*, 1740–1745. [CrossRef] [PubMed]

34. Thompson, C.E.; Jungnickel, H.; Lockyer, N.P.; Stephens, G.M.; Vickerman, J.C. ToF-SIMS studies as a tool to discriminate between spores and vegetative cells of bacteria. *Appl. Surf. Sci.* **2004**, *231–232*, 420–423. [CrossRef]

International Journal of
Molecular Sciences

Article

Manifestation of Systemic Toxicity in Rats after a Short-Time Inhalation of Lead Oxide Nanoparticles

Marina P. Sutunkova [1], Svetlana N. Solovyeva [1], Ivan N. Chernyshov [1], Svetlana V. Klinova [1], Vladimir B. Gurvich [1], Vladimir Ya. Shur [2], Ekaterina V. Shishkina [2], Ilya V. Zubarev [2], Larisa I. Privalova [1] and Boris A. Katsnelson [1,*]

[1] The Medical Research Center for Prophylaxis and Health Protection in Industrial Workers, 30 Popov Str., 620014 Ekaterinburg, Russia; marinasutunkova@yandex.ru (M.P.S.); solovyevasn@ymrc.ru (S.N.S.); chernishov@ymrc.ru (I.N.C.); klinovasv@ymrc.ru (S.V.K.); gurvich@ymrc.ru (V.B.G.); privalovali@yahoo.com (L.I.P.)

[2] The Institute of Natural Sciences, the Ural Federal University, 620000 Ekaterinburg, Russia; vladimir.shur@urfu.ru (V.Y.S.); ekaterina.shishkina@urfu.ru (E.V.S.); i.v.zubarev@urfu.ru (I.V.Z.)

* Correspondence: bkaznelson@etel.ru Tel.: +7-343-253-04-21; Fax: +7-343-371-77-40

Received: 6 November 2019; Accepted: 19 January 2020; Published: 21 January 2020

Abstract: Outbred female rats were exposed to inhalation of lead oxide nanoparticle aerosol produced right then and there at a concentration of 1.30 ± 0.10 mg/m^3 during 5 days for 4 h a day in a nose-only setup. A control group of rats were sham-exposed in parallel under similar conditions. Even this short-time exposure of a relatively low level was associated with nanoparticles retention demonstrable by transmission electron microscopy in the lungs and the olfactory brain. Some impairments were found in the organism's status in the exposed group, some of which might be considered lead-specific toxicological outcomes (in particular, increase in reticulocytes proportion, in δ-aminolevulinic acid (δ-ALA) urine excretion, and the arterial hypertension's development).

Keywords: nanoparticles; lead oxide; inhalation exposure; toxicity

1. Introduction

Lead oxide nanoparticles (PbO-NP) are engineered for some essential technical applications (e.g., magnetic data storage and magnetic resonance imaging). However, from the standpoint of human health risk assessment, it is much more important that PbO particles pollute workplace and ambient air in aerosol form in such long-established and major industries as copper and lead smelters and refineries. These polluting aerosols, resulting from the evaporation of molten metal in pyrometallurgy and its condensation in the air due to cooling and oxidation, contain a substantial proportion of particles falling within the nano-range [1].

Thus, PbO-NPs pose a real inhalation exposure threat, which makes it essential to assess the latter in toxicological experiments. However, we are aware of only one study of this kind published in 2017 by a group of Czech researchers [2].

The study involved an experiment on ICR white mice continuously exposed in a whole-body inhalation setup for 6 weeks to PbO-NP aerosol generated in situ with a mean particle diameter of 25.9 nm (in the size range of 8–230 nm) with an average concentration of 121.7 μ/m^3. The researchers measured the lead contents of organs and tissues with the visualization of retained nanoparticles and described pathological changes in them, concluding that "subchronic exposure to lead oxide nanoparticles has profound negative effects at both cellular and tissue levels". Particularly noteworthy are changes in the hippocampus associated with the penetration of nanoparticles into the brain from the nose as a primary site for their deposition. Such penetration is characteristic of inhaled NPs of any chemical composition, as has been discovered in a number of studies [3–5], including ours [6–8].

However, it is well known that (in contrast to changes in the brain) in a whole-body inhalation chamber, animals' fur is inevitably soiled with particles, which are then licked off by them; thus, not all toxic effects can be attributed precisely and only to inhalation exposure. It is also regrettable that this study did not consider any of the functional or biochemical indicators of organ and systemic toxicity.

There is no doubt that this toxicity is an inherent feature of PbO-NP at all biological levels—from cellular to systemic. It has been shown, in particular, under subchronic intoxication caused in vivo by repeated intraperitoneal injections of PbO nanosuspension to rats [9] and in vitro by its addition to the incubation medium for human fibroblasts [10]. It would be difficult to think of any a priori causes why the inhalation of PbO-NP should not cause lead poisoning. However, this question is so important for occupational medicine and health risk assessment in the above-mentioned industrial conditions that it must be resolved through experimental studies rather than speculatively.

As the first step in this direction, we are presenting the results of an inhalation experiment in which rats were exposed to PbO-NP aerosol at a concentration around 10 times higher than in the above-mentioned experiment [2]. However, in contrast to the latter, it lasted only five days with daily exposures, which were relatively short because of the restrained positioning of the animals in the "nose only" inhalation setup (see Figure 5). We are convinced, though, that this disadvantage is offset by a very important advantage of such setups—they exclude the possibility of nanoparticles penetrating into the organism by routes other than inhalation and the related uncertainty of the causal relationships between just inhalation exposure at a given level and toxic effects.

Overall, the inhalation period in our experiment was 50 times shorter compared with the experiment [2], and the total exposure to nanoparticles may be estimated as about five times lower, even though the concentration of nanoparticles was 10 times higher.

The purpose of this paper is to demonstrate that signs of intoxication were detectable even under an exposure as low and short as in our experiment.

2. Results and Discussion

Previously, we demonstrated that the deposition of nanoparticles in the lungs as a result of intratracheal administration [11–14] or inhalation exposure [6–8] caused a response in the lower airways that was essentially similar to but a lot more intensive than the one provoked by similar exposures to low-soluble highly cytotoxic micrometer particles, such as, for instance, quartz dust [15,16]. This response manifests itself as an increase in the cellularity of the bronchoalveolar lavage fluid (BALF), mainly at the expense of neutrophil leukocytes. The increase in proteins, in some enzymes (particularly of lysosomal or partly lysosomal origin) and in a number of other biochemical components of the BALF extracellular fraction also testifies for inhaled PbO-NP's pulmonary toxicity.

As follows from the results presented in Tables 1 and 2, we observed the same in our experiment. Although the differences from the corresponding indices of the control group's BALF appear to be rarely statistically significant, the mutual correspondence of the resulting changes allows us to consider them to be an actual effect of inhalation exposure to nanoparticles.

Table 1. Cell counts in the bronchoalveolar lavage fluid (BALF) in rats after repeated inhalation of lead oxide nanoparticles (x ± s.e.). PbO-NP: lead oxide nanoparticles.

Group and Number of Animals	Number of Cells $\times 10^6$		
	Total	Neutrophil Leukocytes	Alveolar Macrophages
Control (sham-exposed), 7 rats	3.29 ± 0.67	0.35 ± 0.13	2.94 ± 0.64
Exposed to PbO-NP Aerosol, 7 rats	5.29 ± 1.33	0.62 ± 0.12	4.66 ± 1.24

Table 2. Some biochemical indices in bronchoalveolar lavage fluid (BALF) supernatant in rats after repeated inhalation of lead oxide nanoparticles (x ± s.e.).

Indices	Group and Number of Animals	
	Control (Sham-Exposure), 7 Rats	Exposed to PbO-NP Aerosol, 7 Rats
Total protein, mg/L	81.68 ± 6.19	133.38 ± 31.52
Alanine aminoransferase (ALT) U/L	5.52 ± 0.80	7.64 ± 0.58
Aspartate aminotransferase (AST), U/L	1.96 ± 0.16	1.42 ± 0.21
De-Ritis coefficient (AST/ALT ratio)	2.77 ± 0.31	5.92 ± 1.00 *
Alkaline phosphatase, U/L	13.62 ± 2.85	45.16 ± 5.17 *
Gamma-glutamyl transpeptidase, U/L	0.46 ± 0.16	7.26 ± 1.07 *
Lactate dehydrogenase, U/L	29.50 ± 5.72	38.20 ± 5.64
Urea, mmol/L	0.22 ± 0.04	0.24 ± 0.07
Glucose, mmol/L	0.00 ± 0.00	0.02 ± 0.02

*—values statistically significantly different from the corresponding values of the control group ($p \leq 0.05$ by Student's *t*-test).

As was stated in the Introduction, neither the total exposure to lead nor the period for related adverse changes to develop in the organism might be estimated as considerable, which reduced the probability of development of lead poisoning. Indeed, many of the indices listed in Table 3 indicate that it was of low to moderate intensity.

Table 3. Body and inner organs mass and some functional indices in rats after repeated inhalation of lead oxide nanoparticles (x ± s.e.).

Indices	Group and Number of Animals	
	Control (Sham-Exposure), 7 Rats	Exposed to PbO-NP Aerosol, 7 Rats
Body and organs mass indices		
Initial body mass, g	251.43 ± 2.19	251.79 ± 1.54
Final body mass, g	247.50 ± 2.71	250.71 ± 2.82
Body mass gain, %	−1.56 ± 0.70	−0.43 ± 0.92
Heart mass, g	0.77 ± 0.03	0.83 ± 0.03
Lung mass, g	1.67 ± 0.09	1.89 ± 0.10
Liver mass, g	9.03 ± 0.46	9.09 ± 0.42
Kidney mass, g	1.67 ± 0.05	1.75 ± 0.06
Spleen mass, g	0.54 ± 0.02	0.56 ± 0.02
Brain mass, g	1.96 ± 0.04	1.90 ± 0.04
Heart mass, g/100 g of body mass	0.31 ± 0.01	0.33 ± 0.01
Lung mass, g/100 g of body mass	0.68 ± 0.04	0.75 ± 0.03
Liver mass, g/100 g of body mass	3.65 ± 0.16	3.61 ± 0.12
Kidney mass, g/100 g of body mass	0.68 ± 0.02	0.70 ± 0.02
Spleen mass, g/100 g of body mass	0.22 ± 0.01	0.22 ± 0.01
Brain mass, g/100 g of body mass	0.79 ± 0.01	0.76 ± 0.03

Table 3. *Cont.*

Indices	Group and Number of Animals	
	Control (Sham-Exposure), 7 Rats	Exposed to PbO-NP Aerosol, 7 Rats
Neurobehavioral tests		
Number of head-dips into holes during 3 min	13.86 ± 1.46	15.07 ± 1.51
Number of squares crossed during 3 min	31.64 ± 2.17	32.21 ± 2.27
Total number of movements on the "open field" during 3 min	51.21 ± 3.90	56.36 ± 4.54
Temporal summation of sub-threshold impulses, sec	10.24 ± 0.29	13.11 ± 0.20 *
Hematological indices		
Hemoglobin, g/L	140.33 ± 2.16	142.57 ± 2.13
Hematocrit, %	21.05 ± 0.27	21.54 ± 0.41
Erythrocytes, 10^{12} cells/L	6.83 ± 0.17	6.86 ± 0.23
Mean corpuscular volume, μm^3	61.77 ± 0.76	62.99 ± 1.34
Mean corpuscular hemoglobin, 10^{-12} g	20.60 ± 0.24	20.86 ± 0.48
Mean corpuscular hemoglobin concentration, g/L	333.33 ± 2.50	331.14 ± 1.68
Red cell distribution width, %	12.50 ± 0.32	13.40 ± 0.38
Reticulocytes, ‰	26.44 ± 4.57	71.80 ± 7.87 *
Number of micronuclei in the polychromatophilic erythrocytes of bone marrow, ‰	0.25 ± 0.25	0.40 ± 0.24
Thrombocytes, 10^6 cells/mL	705.00 ± 63.71	813.71 ± 42.92
Thrombocrit, %	0.20 ± 0.01	0.22 ± 0.01
Mean platelet volume, μm^3	5.27 ± 0.13	5.33 ± 0.10
Leukocytes, 10^6/mL	6.40 ± 0.34	6.94 ± 0.52
Banded neutrophils, 10^6/mL	0.07 ± 0.01	0.08 ± 0.01
Segmented neutrophils, 10^6/mL	1.48 ± 0.12	1.68 ± 0.19
Basophils, 10^6/mL	0.00 ± 0.00	0.00 ± 0.00
Eosinophils, 10^6/mL	0.37 ± 0.12	0.28 ± 0.07
Lymphocytes, 10^6/mL	4.12 ± 0.27	4.45 ± 0.28
Monocytes, 10^6/mL	0.37 ± 0.03	
Serum biomarkers		
Total protein content of blood serum, g/L	81.33 ± 1.21	81.50 ± 1.05
Albumin content of blood serum, g/L	49.56 ± 0.75	50.94 ± 0.95
Globulins of blood serum, g/L	31.77 ± 1.05	30.56 ± 0.62
A/G index	1.57 ± 0.05	1.67 ± 0.05
Alanine aminoransferase (ALT) activity in blood serum, mM/h×L	54.07 ± 2.62	56.91 ± 2.55
Aspartate aminotransferase (AST) activity in blood serum, mM/h×L	226.70 ± 10.22	265.60 ± 14.03 *

Table 3. *Cont.*

Indices	Group and Number of Animals	
	Control (Sham-Exposure), 7 Rats	Exposed to PbO-NP Aerosol, 7 Rats
de Ritis coefficient (AST/ALT ratio)	4.22 ± 0.20	4.72 ± 0.31
Total bilirubin in blood serum, µmol/L	1.46 ± 0.14	1.21 ± 0.11
Glucose in blood serum, mmol/L	0.39 ± 0.22	0.54 ± 0.16
Thiol groups (SH-groups), mmol/L	0.37 ± 0.02	0.37 ± 0.02
Gamma-glutamyl transpeptidase (GGTP) in blood serum, U/L	4.83 ± 0.25	3.97 ± 0.19 *
Urea in blood serum, mmol/L	7.27 ± 0.49	6.91 ± 0.45
Uric acid in blood serum, µmol/L	92.57 ± 5.97	91.57 ± 6.52
Alkaline phosphatase in blood serum, nmol/(s×L)	198.80 ± 23.43	190.00 ± 20.32
Catalase in blood serum, µmol/L	0.67 ± 0.01	0.65 ± 0.01
Reduced glutathione in the blood hemolysate, µmol/L	35.33 ± 3.11	33.41 ± 3.47
Ceruloplasmin in blood serum, mg/%	123.03 ± 7.22	127.58 ± 6.06
Malonyl dialdehyde (MDA) in blood serum, µmol/L	3.71 ± 0.14	3.98 ± 0.18
Lactate dehydrogenase (LDH), U/L	2590.00 ± 177.21	3171.43 ± 129.31 *
Amilase in blood serum, U/L	3443.71 ± 218.91	4148.83 ± 362.11
Total Ca in blood serum, mmol/L	2.75 ± 0.02	2.75 ± 0.03
Endothelin-1, pg/mL	26.11 ± 2.72	25.62 ± 1.15
Myoglobin, ng/mL	67.35 ± 9.02	35.51 ± 16.39
Troponin, ng/mL	0.059 ± 0.052	0.105 ± 0.103
Natriuretic peptide, pg/mL	1.09 ± 0.08	1.24 ± 0.15
Vascular endothelial growth factor (VEGF), 10^6 U/mL	3.64 ± 0.69	5.51 ± 2.29
Urinary biomarkers		
Diuresis, mL	40.36 ± 2.29	38.17 ± 2.77
Urine specific gravity, g/mL	1.012 ± 0.001	1.010 ± 0.000 *
Urine pH	6.91 ± 0.06	6.63 ± 0.11 *
Protein in urine, g/L	104.43 ± 5.31	123.63 ± 12.36
Total coproporphyrin in urine, µmol	71.28 ± 27.58	120.86 ± 49.59
δ-aminolevulinic acid (ALA) in urine, µg/mL	5.65 ± 0.86	13.97 ± 1.25 *
Urea in urine, mmol/L	115.19 ± 6.63	132.53 ± 7.37
Uric acid in urine, µmol/L	1.55 ± 0.88	9.17 ± 4.93
Creatinine in blood serum, µmol/L	58.60 ± 1.84	56.87 ± 2.98
Creatinine in urine, µmol/L	0.71 ± 0.03	0.93 ± 0.10 *
Endogenous creatinine clearance	0.50 ± 0.04	0.59 ± 0.04

*—values statistically significantly different from the corresponding values of the control group ($p \leq 0.05$ by Student's *t*-test).

Note, first of all, that body mass reduction, which was observed in the sham-exposed group as well, was not enhanced by lead exposure. This reduction was likely to be an immediate consequence of the subacute stress caused by immobilization at the beginning of the experiment when the rats were not yet sufficiently well adapted to it. Incidentally, we did not observe this effect during months of exposure in the same setup in three different experiments involving other NP species [6–8].

The masses of the internal organs (both absolute and related to body mass) did not differ statistically significantly from the control values. However, still, these indices were noticeably, even if not significant statistically, increased for lungs, the first target organ for inhaled nanoparticles; for kidneys, the main organ eliminating lead from the blood, which is well-known to suffer damage under lead intoxication (e.g., [17–19]); and for heart, whose electrocardiogram (ECG) signs of impairment will be described below.

The shift in the balance between the processes of excitation and inhibition in the central nervous system (CNS) toward inhibition is evidenced by a statistically significant increase in the index of temporal summation of sub-threshold impulses. Although none of the three behavior indices that we recorded was statistically significantly different from the control values, still, the unidirectional increase in these indices (number of head-dips into holes, number of crossed squares, total number of movements on the "open field") suggests enhanced exploratory activity.

The impairment of the porphyrin metabolism, which is specific to lead toxicodynamics, manifested itself in a more than twofold and statistically significant increase in the concentration of δ-aminolevulinic acid (δ-ALA) in urine, which is one of the early signs of lead poisoning. Typical of it is also an increased concentration of coproporphyrin in urine, which in this study was noticeable, even if statistically not significant. As is known, such impairment is a prerequisite to hem synthesis suppression; however, in this experiment, it did not reach a clearly detectable level, since neither the whole blood nor erythrocyte hemoglobin content was decreased at all. Lead anemia did not develop contrary to expectations judging also by no reduction in the erythrocyte count; however, already there was an appreciable and statistically significant compensatory enhancement of erythropoiesis which manifested itself as a considerable increase in the proportion of reticulocytes, another one of the most sensitive effects of lead intoxication.

It should be noted also that the number of micronuclei in the polychromatophilic erythrocytes of the bone marrow was more than doubled. Although this shift was not statistically significant, it is noteworthy since we practically always discover a systemic genotoxic effect under exposure to various metal oxide nanoparticles judging by increased DNA fragmentation coefficient, which is another informative genotoxicity index. For instance, it was significantly increased under subchronic intoxication caused by repeated intraperitoneal injections of PbO-NPs [9].

Turning back to the hematological features of lead intoxication, only some of the white blood indices revealed a noticeable, although statistically insignificant increase. These shifts are also noteworthy because they are fairly typical of experimental lead intoxications and, in some studies, were even more substantial than in the present one and statistically significant at that (e.g., [20–24]).

Proceeding now to the blood serum, note that the total protein and protein fractions in it were not different from the control value. At the same time, some biochemical indices point to damage to hepatocytes, which, as is well known, may show itself not only as enzyme biosynthesis inhibition (which manifested itself as a statistically significant decrease in the serum activity of γ-glutamyl transpeptidase and an insignificant decrease in the activity of alkaline phosphatase) but also as enhanced enzyme release (which led to an increased serum activity of aminotransferases, lactate dehydrogenase, and amilase). As can be seen in Table 3, some of these shifts are statistically significant.

The small increase in the kidney mass, although statistically insignificant, may assumingly be associated with nephrotoxicity, which is lead's other highly characteristic feature [17–19]. We can interpret in the same way both the increased protein content of the urine and the statistically significant, albeit small decrease in urine's specific density (despite a somewhat lower volume of diuresis) and significantly increased creatinine content (while the creatinine content of the blood serum was not changed). The δ-ALA content of the urine was increased statistically significantly and much higher than in the blood serum. Note that such combination of signs of damage to both tubular epithelium and Malpighian glomeruli (confirmed morphologically) had been observed under subchronic intraperitoneal exposure to PbO-NP [9].

The indices of calcium, myoglobin, troponin, natriuretic peptide, endothelin-1, and vascular endothelial growth factor (VEGF) content of the blood serum were studied as markers of a possible vasocardiotoxic effect of lead intoxication discussed below, but none of them revealed any substantial shift.

The data in Table 4 provide evidence that the lead exposure caused interrelated, although statistically insignificant, hemodynamic changes in the rat tail measured post exposure. All the three blood pressure levels (systolic, diastolic, and mean) were elevated compared with the control values. The natural assumption that this arterial hypertension was caused by increased systemic resistance to blood flow is in agreement with the decelerated blood flow and decreased blood volume. It is important to note that we had also discovered that all these shifts under subchronic intoxication were caused by intraperitoneal injections of lead acetate [24] but, to the best of our knowledge, nobody had previously estimated hemodynamic parameters under PbO-NP exposure of any kind.

Table 4. Some hemodynamic indices of rats after repeated inhalation of lead oxide nanoparticles (x ± s.e.).

Groups and Number of Animal	Control (Sham-Exposure), 7 Rats	Exposed to PbO-NP Aerosol, 7 Rats
Systolic blood pressure, mmHg	128.32 ± 3.38	135.26 ± 6.09
Diastolic blood pressure, mmHg	91.20 ± 2.63	98.31 ± 4.86
Mean blood pressure, mmHg	103.19 ± 2.86	110.25 ± 5.22
Heart rate, bpm	335.77 ± 7.15	338.85 ± 7.80
Blood flow in the tail, μL/min	32.70 ± 3.07	29.27 ± 1.84
Tail blood volume, μL	122.90 ± 12.12	108.60 ± 7.97

It should be noted that many of the epidemiological studies have provided evidence of a cause–response relationship between human exposure to lead and the prevalence of arterial hypertension [25–29]. Animal experiments have also been performed, seeking mainly to identify the possible mechanisms of lead-induced hypertension [30–34]. However, the author of a relatively recent overview [35] concluded that "in an occupational setting, the effect of lead exposure on blood pressure remains controversial". In this context, it is to be recalled that the air inhaled by metallurgy workers is contaminated with lead in aerosol form, containing a considerable fraction of nanoparticles (see the Introduction). Therefore, the evidence obtained by us for the first time ever that even a very moderate intoxication caused by exposure to PbO nanoaerosol provokes shifts of a hypertensive type cannot be ignored.

Table 5 presents ECG analysis results obtained in two standard leads, and it is worthwhile to compare them with the data of our ECG study under subchronic lead acetate intoxication [24]. Whereas in the latter we observed elongation of the majority of the interwave intervals pointing to the slowing of the heart rate, in the current experiment, changes in the ECG intervals in the absence of bradycardia were of different types, with the only shift that was statistically significant being of opposite sign (QRS shortening). The statistically significant increase of the P and T amplitudes discovered in both ECG leads in this experiment was not observed in the previous experiment with lead acetate. At the same time, both studies had a common feature, a lowered isoelectric ECG line in the second lead, which can point to some impairment of the myocardium or, at least, metabolic disturbances in it. Such disturbances were indeed revealed in the hearts of rats suffering subchronic lead intoxication [24].

Table 5. Electrocardiogram (ECG) readings in rats after repeated inhalation of lead oxide nanoparticles (x ± s.e.).

Indices	Control (Sham-Exposure), 7 Rats,		Exposed to PbO-NP Aerosol, 7 Rats	
	ECG in the 1st Lead	ECG in the 2nd Lead	ECG in the 1st Lead	ECG in the 2nd Lead
Intervals, msec				
RR	161.25 ± 3.75	166.22 ± 3.54	167.27 ± 3.82	162.31 ± 3.71
PQ	45.33 ± 1.56	45.23 ± 1.01	45.20 ± 1.07	46.70 ± 1.10
QRS	33.31 ± 3.02	22.83 ± 0.88	25.58 ± 0.80 *	21.75 ± 0.62
QT	61.12 ± 1.70	70.88 ± 1.70	65.48 ± 2.38	72.56 ± 1.70
QT corrected using Basett's formula.	152.49 ± 2.86	174.46 ± 3.72	161.29 ± 6.82	180.63 ± 4.30
QT corrected using Friderica's formula	112.41 ± 2.41	129.18 ± 2.79	119.36 ± 4.77	133.26 ± 3.08
P duration	15.55 ± 0.29	15.89 ± 0.76	15.98 ± 0.41	17.19 ± 1.33
Amplitudes, mV				
Isoelectric line	−0.02 ± 0.00	−0.06 ± 0.00	−0.03 ± 0.00	−0.08 ± 0.00 *
P	0.03 ± 0.00	0.09 ± 0.01	0.05 ± 0.01 *	0.11 ± 0.01 *
Q	−0.0005 ± 0.0006	−0.0008 ± 0.0006	−0.0003 ± 0.0004	−0.0035 ± 0.0016
R	0.27 ± 0.06	0.44 ± 0.03	0.28 ± 0.05	0.50 ± 0.06
S	−0.05 ± 0.02	−0.05 ± 0.03	−0.03 ± 0.01	−0.09 ± 0.03
QRS	0.22 ± 0.06	0.39 ± 0.05	0.25 ± 0.05	0.41 ± 0.06
T	0.03 ± 0.01	0.14 ± 0.02	0.08 ± 0.01 *	0.21 ± 0.02 *

*—values statistically significantly different from the corresponding values of the control group ($p \leq 0.05$ by Student's *t*-test).

Transmission electron microscopy of the lung tissue revealed nanoparticles in the cytoplasm of type 1 and type 2 alveolocytes (Figures 1 and 2).

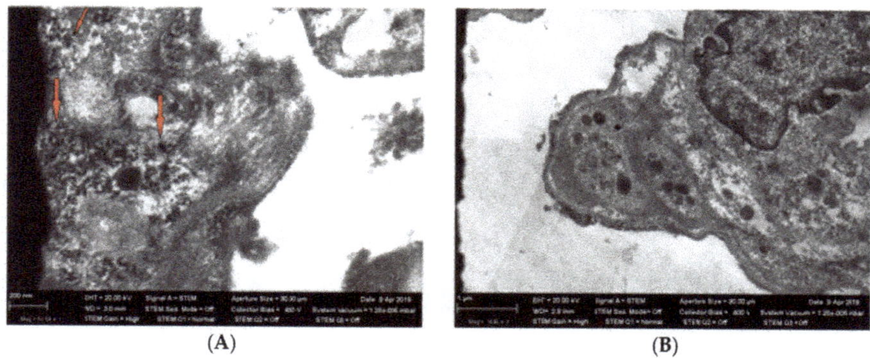

(**A**) (**B**)

Figure 1. Panel (**A**)—nanoparticles (shown by arrows) in alveolocytes of type 1 (TEM, magnification ×51,640) in rat lungs from the exposed group. Panel (**B**)—alveolocytes of type 1 (TEM, magnification ×16,480) in rat lungs from the control (sham-exposed) group.

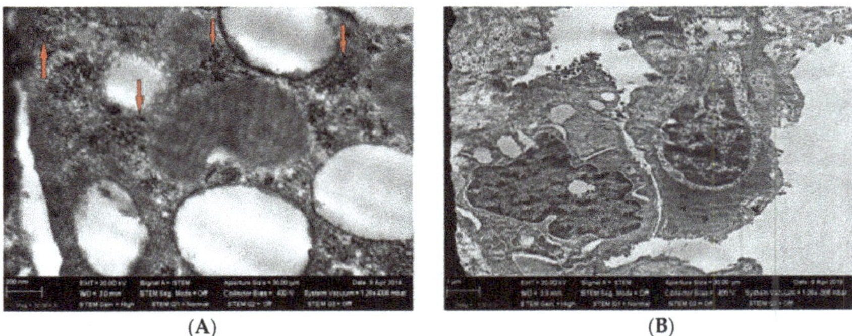

(A) **(B)**

Figure 2. Panel (**A**)—nanoparticles (shown by red arrows) in alveolocytes of type 2 (TEM, magnification ×50,580) in rat lungs from the exposed group. Panel (**B**)—alveolocytes of type 2 (TEM, magnification ×9000) in rat lungs from the control (sham-exposed) group.

Electron microscopy of the olfactory region of the brain showed cytoplasmic vacuolization of neurons, numerous nanoparticles in the neurons' bodies (Figure 3), and pronounced demyelinization of axon membranes (Figure 4).

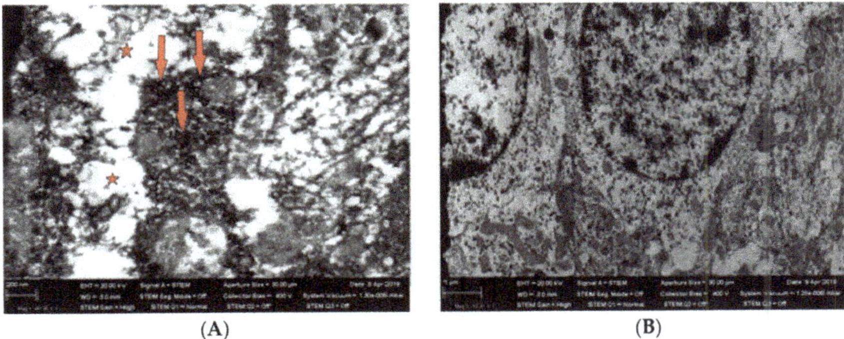

(A) **(B)**

Figure 3. Panel (**A**)—a neuron body with numerous nanoparticles (shown by red arrows) and vacuolized cytoplasm (shown by a red asterisk); TEM, magnification ×41,950; Panel (**B**)—non-damaged neuron bodies in the olfactory region of a rat brain from the control (sham-exposed) group; TEM, magnification ×10,410.

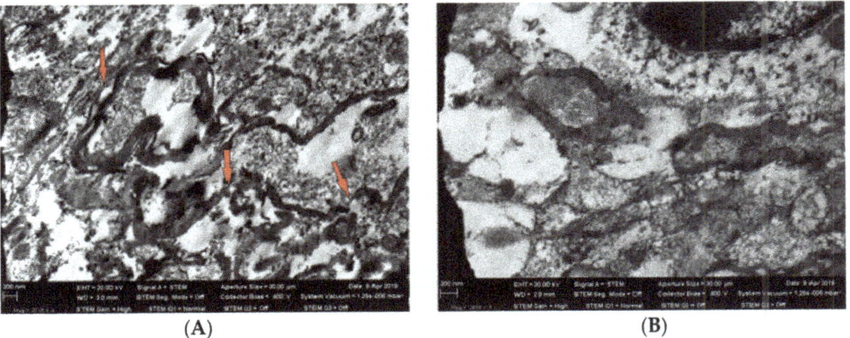

(A) **(B)**

Figure 4. Panel (**A**)—an axon with signs of membrane demyelinization (shown by red arrows) in the olfactory region of a rat brain from the exposed group; TEM, magnification ×20,050. Panel (**B**)—non-damaged axon myelin sheath in the olfactory region of a rat brain from the control (sham-exposed) group; TEM, magnification ×24,680.

The mechanism underlying the penetration of inhaled nanoparticles into the brain neurons was mentioned in the Introduction. We should just emphasize that such penetration is also associated with some ultra-structural changes in the brain tissue. It can be assumed that the somewhat controversial shifts in the CNS functional indices that we discussed above are connected not only and even not so much with systemic lead intoxication as with a direct effect of cytotoxic nanoparticles on the brain structures. In this connection, we should recall the key role of the sense of smell in animal behavior control.

3. Materials and Methods

Airborne Pb-NPs were obtained by sparking from 99.9999% pure lead rods with a diameter of 5.6 mm (supplied by "Giredmet Ltd."—Moscow, Russia) using a Palas DNP-3000 generator set at "Medium energy" regime with the current strength of 5 A and nitrogen flow of 8 L/min. Then, this flow was being mixed with air (6 L/min) for cooling and for oxidizing Pb into PbO-NPs, which were fed into a nose-only exposure tower (CH Technologies, Westwood, NJ, USA) with rats placed into individual restrainers (Figure 5).

Figure 5. The nose-only inhalation setup (photographed without the door wings of the draught cupboard) and a similar tower for the sham exposure of the control group.

A setup of the same design obtained from the same supplier was used for a sham exposure of control rats. Particles collected on filters and inspected under a scanning electron microscope (SEM) had a spherical shape and either were singlets or formed small aggregates (Figure 6). The latter, if compact, were measured as one particle. Even so, the particle size distribution (Figure 7) proved fairly clean-cut and restricted to the nanometric range with a mean (± s.d.) diameter of 36 ± 4 nm. The chemical composition of particles sampled on the filters was confirmed by Raman spectroscopy to be PbO (the obtained spectrum had two characteristic peaks at wavelengths 82 cm^{-1} and 147 cm^{-1} which, according to [36] correspond to PbO).

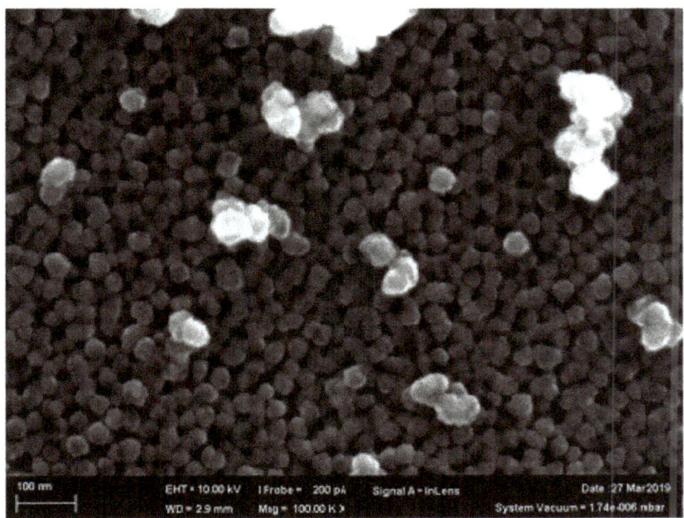

Figure 6. Nanoparticles retained by the Whatman Anodisc membrane filter (mesh diameter 20 nm) of the inhalation setup. SEM, magnification ×100,000.

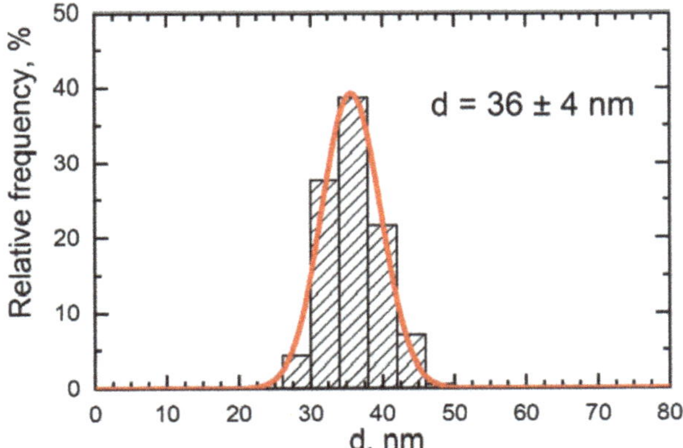

Figure 7. Particle or particle aggregate size normal distribution function (red curve) obtained by statistical processing of 360 measured SEM images of particles accumulated on a polycarbonate filter from the exposed rats' breathing zone.

Our experiment was carried out on outbred white female rats from our own breeding colony with the initial body weight of 252 ± 1.5 g in the NP-exposed 14 rats and 251 ± 2.2 g in the sham-exposed 14 rats. All these rats were housed in conventional conditions, breathed unfiltered air, and were fed standard balanced food. The experiments were planned and implemented in accordance with the "International guiding principles for biomedical research involving animals" developed by the Council for International Organizations of Medical Sciences (1985) and were approved by the Ethics Committee of the Ekaterinburg Medical Research Center for Prophylaxis and Health Protection in Industrial Workers.

After a preliminary training, the rats were exposed or sham-exposed for 4 h a day, 5 times during one working week. Along with each single exposure, a sample of airborne nanoparticles was collected on an acetyl cellulose fine fiber filter attached to the inhalation setup instead of a rat's nose while monitoring the volume velocity of air drawn through the filter. Each daily filter was being sampled during 4 hr in parallel with exposure of rats. The mass of Pb retained on it was determined with an atomic absorption spectrometer, ContrAA 700 (Analytic Jena A, Jena, Germany), and translated into the mass of PbO and then into its air concentration as mg/m^3, which proved to be equal to 1.30 ± 0.10 mg/m^3 (The OSHA Permissible Exposure Limit for Pb as well as the respective Russian national standard is 50 $\mu g/m^3$ averaged over an 8-h period which corresponds to 54 μg PbO/m^3. For the 4 h exposure, the equivalent level would be 108 $\mu g/m^3$. We chose a one order of magnitude higher level to presumably ensure some toxicity outcomes, a short total exposure period notwithstanding).

The mass deposition of inhaled particles in the lower airways can be estimated only very roughly because this estimation is based on physiological parameters that are rather uncertain:

(1) The rat's minute respiratory ventilation as assessed experimentally by different authors varies between 78 mL [37] and 210 mL [38].

The so-called multi-path particle dosimetry (MPPD) model for rats used in [39] assumes the breathing frequency to be equal to 102 min^{-1} and tidal volume to be 2.1 mL. It gives minute ventilation, 214 mL, which almost exactly corresponds to the above given highest experimental value and thus seems to be an extreme estimate. In our previous inhalation studies with iron oxide [6], silica [40], and nickel oxide [41] nanoparticles, we calculated particle deposition based on the minute ventilation value of 100 mL, which is close to the median of the above experimental range but of course somewhat arbitrary.

(2) The pulmonary deposition fraction of inhaled NPs may be estimated to be equal to 0.52, which is close to the lowest of the differently substantiated assessments given by [42] but much higher as compared with 0.124, which value is assumed to be the alveolar region deposition fraction for 20 nm particles according to the same MPPD model.

Thus, the pulmonary PbO-NP deposition per each single exposure may be tentatively estimated as mean NP concentration in the inhaled air, mg/m^3 × minute respiratory ventilation, mL × exposure time, min × mL to m^3 ratio × mcg to mg ratio × deposited fraction = $1.30 \times 100 \times 240 \times 10^{-6} \times 10^3 \times 0.52 = 16.2$ mcg.

After the end of the exposure week, the following procedures were performed for all rats:

(1) weighing of the body;
(2) estimation of the CNS ability to evoke temporal summation of sub-threshold impulses (a variant of the withdrawal reflex and its facilitation by repeated electrical stimulations in an intact, conscious rat) [9,19,43–45];
(3) recording of the number of head-dips into the holes of a hole-board (which is a simple but informative index of exploratory activity frequently used for studying the behavioral effects of toxicants and drugs) (e.g., [9,19,44–47]);
(4) collection of daily urine (from rats put into metabolic cages) for analysis of protein, urine specific gravity, pH, urea, uric acid, total coproporphyrin, and δ-aminolevulinic acid (δ-ALA);
(5) bronchoalveolar lavage to obtain fluid (BALF) for cytological and biochemical characterization.

Bronchoalveolar lavage was carried out 24 h after the last inhalation exposure. A cannula connected to a Luer's syringe containing 10 mL of normal saline was inserted into the surgically prepared trachea of a rat under hexenal anaesthesia. The fluid entered the lungs slowly under the gravity of the piston, with the animal and syringe positioned vertically. Then, the rat and the syringe were turned 180°, and the fluid flowed back into the syringe. The extracted BALF was poured into siliconized refrigerated tubes. An aliquot sample of the BALF was drawn into a white blood cells (WBC) count pipette together with 3% acetic acid and methylene blue. Cell count was performed in a standard hemocytometer (the so-called Goryayev's Chamber). For cytological examination, the BALF was centrifuged for 4 min at 1000 rpm; then, the fluid was decanted, and the sediment was used for preparing smears on two microscope slides. After air drying, the smears were fixed with methyl alcohol and stained with azure eosin. The smears were microscoped with immersion at a magnification of

×1000. The differential count for determining the percentage of alveolar macrophages (AM), neutrophil leucocytes (NL), and other cells counted together was conducted up to a total number of 100 cells. Allowing for the number of cells in the BALF, these percentages were recalculated in terms of absolute AM and NL counts.

Seven out of 14 not rats not subjected to that lavage in both exposed and sham-exposed groups were killed by cervical dislocation under ether anesthesia. The liver, spleen, kidneys, heart, lungs, and brain were weighed. Blood was collected from the tail vein under ether anesthesia at the end of the experiment. The biochemical indices determined in the blood included total serum protein, albumin, globulin, alanine and aspartate transaminases (ALT, AST), bilirubin, catalase, glutathione, ceruloplasmin, malondialdehyde (MDA), lactate dehydrogenase (LDH), alkaline phosphatase, amilase, glucose, thiol groups (SH-groups), gamma-glutamyl transpeptidase (GGTP), urea, acid, calcium, myoglobin, troponin, natriuretic peptide, endothelin-1, and vascular endothelial growth factor (VEGF).

For determining the hemoglobin content, hematocrit, thrombocrit, mean erythrocyte volume, and for counting a red blood cells (RBC), WBC, and thrombocytes, we used a MYTHIC-18 auto-hematology analyzer (C2 Diagnostic, Montperllier, France). Reticulocyte percentage was counted on smears under optical microscopy after supravital staining with brilliant cresyl blue. The number of micronucleus in the polychromatophilic erythrocytes of the bone marrow was counted in smears under optical microscopy after fixing in methanol and staining by Pappenheim's stain.

All the clinical laboratory blood and urine tests with the exception of the specially considered ones were performed using the well-known techniques described in many manuals (e.g., [48]). Before performing euthanasia, we measured the heart rate, arterial pressure, blood flow rate, and blood volume in the rat tails using the noninvasive blood pressure system CODA-HT8 (Kent Scientific, Torrington, CT, USA) and recorded electrocardiograms (ECG) in the first and second leads with programmed analysis using the ecgTUNNEL system (emka TECHNOLOGIES, Paris, France).

In addition, the pulmonary and brain accumulation of NPs and the ultrastructure of respective tissues were visualized by means of transmission electron microscopy (TEM). To this end, pieces of an organ were fixed in 2% paraformaldehyde and 2.5% glutaraldehyde in a cacodylate buffer with 5% sucrose at pH 7.3, post-fixed in 1% osmium tetroxide, contrasted with uranyl acetate en bloc, and embedded in epoxy resin (Spurr). This sample preparation procedure was carried out in a microwave tissue processor, HISTOS REM (Milestone, Milan, Italy). Semi-thin (900 nm thick) sections of epoxy blocks were stained in toluidine blue with the addition of 1% borax and examined under the optical microscope for choosing a site for TEM. The 60 nm ultrathin sections of this site obtained with the help an ultramicrotome (Power Tome, RMC, Tucson, AZ, USA) were contrasted with uranyl acetate and lead citrate. Grid-mounted sections were investigated in an electron microscope, AURIGA (Carl Zeiss; MT, Oberkochen, Germany) in the STEM mode in the range of magnifications 1200–200,006.

4. Conclusions

The experimental results and their discussion with reference to the literature data enable us to argue that inhalation exposure to lead oxide nanoparticles associated with their retention in the organism (olfactory brain included) demonstrated with TEM, led (even though the exposure level was relatively low and was of short duration) to disturbances in the organism, some of which are specific to lead intoxication (in particular, increase in reticulocytes proportion, in δ-ALA urine excretion, and signs of arterial hypertension).

Author Contributions: Formal analysis, I.N.C.; Investigation, S.N.S., S.V.K., V.Y.S. and E.V.S.; Project administration, M.P.S.; Resources, V.B.G.; Visualization, I.V.Z.; Writing – original draft, B.A.K.; Writing – review & editing, L.I.P. All authors have read and agreed to the published version of the manuscript.

Funding: This work was funded by the budget of the Ekaterinburg Medical Research Center for Prophylaxis and Health Protection in Industrial Workers.

Conflicts of Interest: The authors declare no conflict of interest.

References

1. Katsnelson, B.A.; Privalova, L.I.; Sutunkova, M.P.; Minigalieva, I.A.; Gurvich, V.B.; Shur, V.Y.; Shishkina, E.V.; Makeyev, O.H.; Valamina, I.E.; Varaksin, A.N.; et al. Experimental research into metallic and metal oxide nanoparticle toxicity in vivo. In *Bioactivity of Engineered Nanoparticles*; Yan, B., Zhou, H., Gardea-Torresdey, J., Eds.; Springer Nature Switzerland AG: Basel, Switzerland, 2017; Chapter 11, pp. 259–319.
2. Dumková, J.; Smutná, T.; Vrlíková, L.; Le Coustumer, P.; Večeřa, Z.; Dočekal, B.; Mikuška, P.; Čapka, L.; Fictum, P.; Hampl, A.; et al. Sub-chronic inhalation of lead oxide nanoparticles revealed their broad distribution and tissue-specific subcellular localization in target organs. *Part. Fibre Toxicol.* **2017**, *14*, 55. [CrossRef]
3. Oberdörster, G.; Sharp, Z.; Atudore, V.; Elder, A.; Gelein, R.; Kreylin, W. Translocation of inhaled ultrafine particle to the brain. *Inhal. Toxicol.* **2004**, *16*, 437–445. [CrossRef]
4. Elder, A.; Gelein, R.; Silva, V.; Feikert, T.; Opanashuk, L.; Carter, J.; Potter, R.; Maynard, A.; Ito, Y.; Finkelstein, J.; et al. Translocation of inhaled ultrafine manganese oxide particles to the central nervous system. *Environ. Health Perspect.* **2006**, *114*, 1172–1178. [CrossRef]
5. Kao, Y.-Y.; Cheng, T.-J.; Yang, D.-M.; Liu, P.-S. Demonstration of an olfactory bulb–brain translocation pathway for ZnO nanoparticles in rodent ells in vitro and in vivo. *J. Mol. Neurosci.* **2012**, *48*, 464–471. [CrossRef]
6. Sutunkova, M.P.; Katsnelson, B.A.; Privalova, L.I.; Gurvich, V.B.; Konysheva, L.K.; Shur, V.Y.; Shishkina, E.V.; Minigalieva, I.A.; Solovjeva, S.N.; Grebenkina, S.V.; et al. On the contribution of the phagocytosis and the solubilization to the iron oxide nanoparticles retention in and elimination from lungs under long-term inhalation exposure. *Toxicology* **2016**, *363–364*, 19–28. [CrossRef]
7. Sutunkova, M.P.; Solovyeva, S.N.; Katsnelson, B.A.; Gurvich, V.B.; Privalova, L.I.; Minigalieva, I.A.; Slyshkina, T.V.; Valamina, I.E.; Shur, V.Y.; Zubarev, I.V.; et al. A paradoxical rat organism's response to a long-term inhalation of silica-containing submicron (predominantly, nanoscale) particles of an actual industrial aerosol at realistic exposure levels. *Toxicology* **2017**, *384*, 59–68. [CrossRef]
8. Sutunkova, M.; Solovyeva, S.; Minigalieva, I.; Gurvich, V.; Valamina, I.; Makeyev, O.; Shur, V.; Shishkina, E.; Zubarev, I.; Saatkhudinova, R.; et al. Toxic Effects of Low-Level Long-Term Inhalation Exposures of Rats to Nickel Oxide Nanoparticles. *Int. J. Mol. Sci.* **2019**, *20*, 1778. [CrossRef]
9. Minigalieva, I.A.; Katsnelson, B.A.; Panov, V.G.; Privalova, L.I.; Varaksin, A.N.; Gurvich, V.B.; Sutunkova, M.P.; Shur, V.Y.; Shishkina, E.V.; Valamina, I.E.; et al. In vivo toxicity of copper oxide, lead oxide and zinc oxide nanoparticles acting in different combinations and its attenuation with a complex of innocuous bio-protectors. *Toxicology* **2017**, *380*, 72–93. [CrossRef]
10. Bushueva, T.; Minigalieva, I.; Panov, V.; Kuznetsova, A.; Naumova, A.; Shur, V.; Shishkina, E.; Gurvich, V.; Privalova, L.; Katsnelson, B. More data on in vitro assessment of comparative and combined toxicity of metal oxide nanoparticles. *Food Chem. Toxicol.* **2019**, *133*, 110753. [CrossRef]
11. Katsnelson, B.A.; Privalova, L.I.; Degtyareva, T.D.; Sutunkova, M.P.; Minigalieva, I.A.; Kireyeva, E.P.; Khodos, M.Y.; Kozitsina, A.N.; Shur, V.Y.; Nikolaeva, E.V.; et al. Experimental estimates of the toxicity of Iron oxide Fe_3O_4 (Magnetite) nanoparticles. *Cent. Eur. J. Occup. Environ. Med.* **2010**, *16*, 47–63.
12. Katsnelson, B.A.; Privalova, L.I.; Sutunkova, M.P.; Tulakina, L.G.; Pichugova, S.V.; Beykin, J.B.; Khodos, M.J. The "in vivo" interaction between iron oxide Fe_3O_4 nanoparticles and alveolar macrophages. *J. Bull. Exp. Biol. Med.* **2012**, *152*, 627–631. [CrossRef]
13. Katsnelson, B.A.; Privalova, L.I.; Gurvich, V.B.; Makeyev, O.H.; Shur, V.Y.; Beikin, Y.B.; Sutunkova, M.P.; Kireyeva, E.P.; Minigalieva, I.A.; Loginova, N.V.; et al. Comparative in vivo assessment of some adverse bioeffects of equidimensional gold and silver nanoparticles and the attenuation of nanosilver's effects with a complex of innocuous bioprotectors. *Int. J. Mol. Sci.* **2013**, *14*, 2449–2483. [CrossRef]
14. Sutunkova, M.P.; Privalova, L.I.; Minigalieva, I.A.; Gurvich, V.B.; Panov, V.G.; Katsnelson, B.A. The most important inferences from the Ekaterinburg nanotoxicology team's animal experiments assessing adverse health effects of metallic and metal oxide nanoparticles. *Toxicol. Rep.* **2018**, *5*, 363–376. [CrossRef]
15. Morosova, K.I.; Aronova, G.V.; Katsnelson, B.A.; Velichkovsski, B.T.; Genkin, A.M.; Elnichnyh, L.N.; Privalova, L.I. On the defensive action of glutamate against the cytotoxicity and fibrogenicity of quartz dust. *Br. J. Ind. Med.* **1982**, *39*, 244–252. [CrossRef]

16. Katsnelson, B.A.; Konysheva, L.K.; Privalova, L.I.; Sharapova, N.Y. Quartz dust retention in rat lungs under chronic exposure simulated by a multicompartmental model: Further evidence of the key role of the cytotoxicity of quartz particles. *Inhal. Toxicol.* **1997**, *9*, 703–715.

17. IPCS. *Principles Amd Methods for the Assessment of Nephrotoxicity Associated with Exposure to Chemicals*; International Programme on Chemical Safety. Environmental Health Criteria 119; World Health Organization: Geneva, Switzerland, 1991.

18. IPCS. *Inorganic Lead*; International Programme on Chemical Safety. Environmental Health Criteria 165; World Health Organization: Geneva, Switzerland, 1995.

19. Varaksin, A.N.; Katsnelson, B.A.; Panov, V.G.; Privalova, L.I.; Kireyeva, E.P.; Valamina, I.E.; Beresneva, O.Y. Some considerations concerning the theory of combined toxicity: A case study of subchronic experimental intoxication with cadmium and lead. *Food Chem. Toxicol.* **2014**, *64*, 144–156. [CrossRef]

20. Mugahi, M.N.; Heidari, Z.; Mahmoudzadeh-Sagheb, H.; Barbarestani, M. Effects of chronic lead acetate intoxication on blood indices of male adult rat. *DARU* **2003**, *11*, 147–151.

21. Al Momen, A. Thrombocytosis secondary to chronic lead poisoning. *Platelets* **2010**, *21*, 297–299. [CrossRef]

22. Farkhondeh, T.; Boskabady, M.H.; Kohi, M.K.; Sadeghi-Hashjin, G.; Moin, M. Lead exposure affects inflammatory mediators, total and differential white blood cells in sensitized guinea pigs during and after sensitization. *Drug Chem. Toxicol.* **2014**, *37*, 329–335. [CrossRef]

23. Protsenko, Y.L.; Katsnelson, B.A.; Klinova, S.V.; Lookin, O.N.; Balakin, A.A.; Nikitina, L.V.; Gerzen, O.P.; Minigalieva, I.A.; Privalova, L.I.; Gurvich, V.B.; et al. Effects of subchronic lead intoxication of rats on the myocardium contractility. *Food Chem. Toxicol.* **2018**, *120*, 378–389. [CrossRef]

24. Klinova, S.V.; Minigalieva, I.A.; Privalova, L.I.; Valamina, I.E.; Makeyev, O.H.; Shuman, E.A.; Korotkov, A.A.; Panov, V.G.; Sutunkova, M.P.; Riabova, J.V.; et al. Further verification of some postulates of the combined toxicity theory: New animal experimental data on separate and joint adverse effects of lead and cadmium. *Food Chem. Toxicol.* **2019**, in press. [CrossRef]

25. Glenn, B.S.; Stewart, W.F.; Schwartz, B.S.; Bressler, J. Relation of alleles of the sodium-potassium adenosine triphosphatase alpha 2 gene with blood pressure and lead exposure. *Am. J. Epidemiol.* **2001**, *153*, 537–545. [CrossRef]

26. Glenn, B.S.; Stewart, W.F.; Links, J.M.; Todd, A.C.; Schwartz, B.S. The longitudinal association of lead with blood pressure. *Epidemiology* **2003**, *14*, 30–36. [CrossRef]

27. Glenn, B.S.; Bandeen-Roche, K.; Lee, B.K.; Weaver, V.M.; Todd, A.C.; Schwartz, B.S. Changes in systolic blood pressure associated with lead in blood and bone. *Epidemiology* **2006**, *17*, 538–544. [CrossRef]

28. Navas-Acien, A.; Guallar, E.; Silbergeld, E.K.; Rothenberg, S.J. Lead exposure and cardiovascular disease–a systematic review. *Environ. Health Perspect.* **2007**, *115*, 472–482. [CrossRef]

29. Fiorim, J.; Ribeiro, R.F.; Silveira, E.A.; Padilha, A.S.; Vescovi, M.V.; de Jesus, H.C.; Stefanon, I.; Salaices, M.; Vassallo, D.V. Low-level lead exposure increases systolic arterial pressure and endothelium-derived vasodilator factors in rat aortas. *PLoS ONE* **2011**, *6*, e17117. [CrossRef]

30. Carmignani, M.; Boscolo, P.; Poma, A.; Volpe, A.R. Kininergic system and arterial hypertension following chronic exposure to inorganic lead. *Immunopharmacology* **1999**, *44*, 105–110. [CrossRef]

31. Carmignani, M.; Volpe, A.R.; Boscolo, P.; Qiao, N.; Di Gioacchino, M.; Grilli, A.; Felaco, M. Catecholamine and nitric oxide systems as targets of chronic lead exposure in inducing selective functional impairment. *Life Sci.* **2000**, *68*, 401–415. [CrossRef]

32. Simões, M.R.; Ribeiro Júnior, R.F.; Vescovi, M.V.; de Jesus, H.C.; Padilha, A.S.; Stefanon, I.; Vassallo, D.V.; Salaices, M.; Fioresi, M. Acute lead exposure increases arterial pressure: Role of the renin-angiotensin system. *PLoS ONE* **2011**, *6*, e18730. [CrossRef]

33. Vaziri, N.D.; Norris, K. Lipid disorders and their relevance to outcomes in chronic kidney disease. *Blood Purif.* **2011**, *31*, 189–196. [CrossRef]

34. Silveira, E.A.; Siman, F.D.; de Oliveira, F.T.; Vescovi, M.V.; Furieri, L.B.; Lizardo, J.H.; Stefanon, I.; Padilha, A.S.; Vassallo, D.V. Low-dose chronic lead exposure increases systolic arterial pressure and vascular reactivity of rat aortas. *Free Radic. Biol. Med.* **2014**, *67*, 366–376. [CrossRef] [PubMed]

35. Gidlow, D.A. Lead toxicity. *Occup. Med.* **2015**, *65*, 348–356. [CrossRef] [PubMed]

36. Burgio, L.; Clark, R.J.H. Library of FT-Raman spectra of pigments, minerals, pigment media and varnishes, and supplement to existing library of Raman spectra of pigments with visible excitation. *Spectrochim. Acta A* **2001**, *57*, 1491–1521. [CrossRef]

37. Ramahandran, G. *Assessing Nanoparticle Risk to Human Health*; Elsevier: Amsterdam, The Netherlands, 2016; 280p.

38. Maulderly, J.L.; McCunney, R.G. *Particle Overload in the Rat Lung and Lung Cancer*; Implications for Human Risk Assessment; Taylor & Francis: Philadelphia, PA, USA, 1997; 298p.

39. Anderson, D.S.; Patchin, E.S.; Silva, R.M.; Uyeminami, D.L.; Sharmah, A.; Guo, T.; Das, G.K.; Brown, J.M.; Shannahan, J.; Gordon, T.; et al. Influence of particle size on persistence and clearance of aerosolized silver nanoparticles in the rat lung. *Toxicol. Sci.* **2015**, *144*, 366–381. [CrossRef]

40. Solovyeva, S.N.; Sutunkova, M.P.; Katsnelson, B.A.; Gurvich, V.B.; Privalova, L.I.; Minigalieva, I.A.; Slyshkina, T.V.; Valamina, I.E.; Makeyev, O.H.; Shur, V.Y.; et al. Interplay of the pulmonary phagocytosis response to, and the in vivo solubilization of amorphous silica nanoparticles deposited in lungs of rats under long-term inhalation exposures as determinants of their modest fibrogenicity and low systemic toxicity (experimental and mathematical modeling). Abstract: International Conference on Occupational Safety and Health, 4th Edition, London, U.K. *J. Nurs. Health Stud.* **2018**, *3*, 2574–2825.

41. Katsnelson, B.A.; Minigalieva, I.A.; Gurvich, V.B.; Privalova, L.I. Consequent stages of developing a multi-compartmental mechanistic model for chronically inhaled nanoparticles pulmonary retention. *Toxicol. Rep.* **2019**, *6*, 279–287. [CrossRef]

42. Kolanjiyil, A.V. Deposited Nanomaterial Mass Transfer from Lung Airways to Systemic Regions. Master's Thesis, North Carolina State University, Raleigh, NC, USA, 2013.

43. Rylova, M.L. *Methods of Investigating Long-Term Effects of Noxious Environmental Agents in Animal Experiments*; Meditsina: Leningrad, Russia, 1964; 228p. (In Russian)

44. Panov, V.G.; Katsnelson, B.A.; Varaksin, A.N.; Privalova, L.I.; Kireyeva, E.P.; Sutunkova, M.P.; Valamina, I.E.; Beresneva, O.Y. Further development of mathematical description for combined (a case study of lead–fluoride combination). *Toxicol. Rep.* **2015**, *2*, 297–307. [CrossRef]

45. Minigalieva, I.A.; Katsnelson, B.A.; Privalova, L.I.; Sutunkova, M.P.; Gurvich, V.B.; Shur, V.Y.; Shishkina, E.V.; Valamina, I.E.; Makeyev, O.H.; Panov, V.G.; et al. Attenuation of combined nickel (II) oxide and manganese (II, III) oxide nanoparticles' adverse effects with a complex of bioprotectors. *Int. J. Mol. Sci.* **2015**, *16*, 22555–22583. [CrossRef]

46. Adeyemi, O.O.; Yemitan, O.K.; Taiwo, A.E. Neurosedative and muscle-relaxant activities of ethyl acetate extract of Baphianitida nitida AFZEL. *Ethnopharmacology* **2006**, *106*, 312–316. [CrossRef]

47. Fernandez, S.P.; Wasowski, C.; Loscalzo, L.M.; Granger, R.E.; Johnston, G.A.; Paladini, A.C.; Marder, M. Central nervous system depressant action of flavonoid glycosides. *Eur. J. Pharmacol.* **2006**, *539*, 168–176. [CrossRef]

48. Willard, M.D.; Tvedten, H. *Small Animal Clinical Diagnosis by Laboratory Methods*, 5th ed.; Saunders Company: Philadelphia, PA, USA, 2012; 911p.

International Journal of
Molecular Sciences

Article

The Effect of the Chorion on Size-Dependent Acute Toxicity and Underlying Mechanisms of Amine-Modified Silver Nanoparticles in Zebrafish Embryos

Zi-Yu Chen [1,2,†], Nian-Jhen Li [1], Fong-Yu Cheng [3], Jian-Feng Hsueh [1], Chiao-Ching Huang [4], Fu-I Lu [5,6], Tzu-Fun Fu [7], Shian-Jang Yan [2,†], Yu-Hsuan Lee [4,8,*] and Ying-Jan Wang [1,9,*]

[1] Department of Environmental and Occupational Health, College of Medicine, National Cheng Kung University, Tainan 70101, Taiwan; q781001@gmail.com (Z.-Y.C.); lavender6328@hotmail.com (N.-J.L.); seed123465@gmail.com (J.-F.H.)

[2] Department of Physiology, College of Medicine, National Cheng Kung University, Tainan 70101, Taiwan; drosouv@gmail.com

[3] Department of Chemistry, Chinese Culture University, Taipei 11114, Taiwan; zfy3@ulive.pccu.edu.tw

[4] Department of Food Safety/Hygiene and Risk Management, College of Medicine, National Cheng Kung University, Tainan 70101, Taiwan; winds031215@gmail.com

[5] Department of Biotechnology and Bioindustry Sciences, College of Bioscience and Biotechnology, National Cheng Kung University, Tainan 70101, Taiwan; fuilu@mail.ncku.edu.tw

[6] The Integrative Evolutionary Galliforms Genomics Research (iEGG) and Animal Biotechnology Center, National Chung Hsing University, Taichung 40227, Taiwan

[7] Department of Medical Laboratory Science and Biotechnology, Medical College, National Cheng Kung University, Tainan 70101, Taiwan; tffu@mail.ncku.edu.tw

[8] Department of Cosmeceutics, China Medical University, Taichung 40402, Taiwan

[9] Department of Medical Research, China Medical University Hospital, China Medical University, Taichung 40402, Taiwan

*   Correspondence: bmm175@hotmail.com (Y.-H.L.); yjwang@mail.ncku.edu.tw (Y.-J.W.); Tel.: +886-4-2205-3366 (ext. 6509) (Y.-H.L.); +886-6-235-3535 (ext. 5804) (Y.-J.W.); Fax: 886-6-275-2484 (Y.-J.W.)

†   These authors contributed equally to this work.

Received: 28 November 2019; Accepted: 16 April 2020; Published: 20 April 2020

Abstract: As the worldwide application of nanomaterials in commercial products increases every year, various nanoparticles from industry might present possible risks to aquatic systems and human health. Presently, there are many unknowns about the toxic effects of nanomaterials, especially because the unique physicochemical properties of nanomaterials affect functional and toxic reactions. In our research, we sought to identify the targets and mechanisms for the deleterious effects of two different sizes (~10 and ~50 nm) of amine-modified silver nanoparticles (AgNPs) in a zebrafish embryo model. Fluorescently labeled AgNPs were taken up into embryos via the chorion. The larger-sized AgNPs (LAS) were distributed throughout developing zebrafish tissues to a greater extent than small-sized AgNPs (SAS), which led to an enlarged chorion pore size. Time-course survivorship revealed dose- and particle size-responsive effects, and consequently triggered abnormal phenotypes. LAS exposure led to lysosomal activity changes and higher number of apoptotic cells distributed among the developmental organs of the zebrafish embryo. Overall, AgNPs of ~50 nm in diameter exhibited different behavior from the ~10-nm-diameter AgNPs. The specific toxic effects caused by these differences in nanoscale particle size may result from the different mechanisms, which remain to be further investigated in a follow-up study.

Keywords: zebrafish embryos; acute toxicity; silver nanoparticles; chorion pore size; lysosomal activity; apoptosis

---

1. Introduction

The rapid development of nanotechnology has stimulated the use of nanomaterials in various fields, including medical imaging, new drug delivery technologies and a variety of industrial products [1]. According to the survey of "The Community Research and Development Information Service, CORDIS" [2], the global market for nanomaterials is estimated at approximately 11 million tons/year with a market value of EUR 20 billion and growing. Among them, AgNPs are the most widely used nanoparticles (NPs) in consumer applications and account for approximately 24% of the commercial products [3]. Due to their superior antimicrobial activity, AgNPs are used in medical products such as catheters, implants and other materials to prevent infection [4,5]. In addition, AgNPs are frequently used in clothing, the food industry, paints, household products and other fields [6–8]. Although AgNPs are widely used, exposure to humans and their release into the environment and accumulation in aquatic ecosystems has become a concern [9,10]. Nevertheless, the comprehensive biological and toxicological effects of AgNPs in humans and the environment has still not been studied in detail compared to their commercialization in various consumer and medical products [11]. Thus, it is important to analyze the effects of AgNPs on ecological systems as well as human health.

The toxicity of AgNPs has been investigated using a variety of model systems from mammalian cell lines to aquatic organisms and rodent species [12–14]. Despite the enormous amount of published research, the understanding of the mechanism of action is still hindered by the variability of the NP characteristics used in the different studies. These differences include the size, coating, surface interaction, contact time, bioaccumulation and transformation under realistic environmental conditions [15,16]. These factors may change the toxicity profile and the interpretation of the results [15]. Specifically, the toxicity of AgNPs in most living organisms is size- and coating-dependent. Particle size determines and influences the way that AgNPs enter cells and their toxic effects on organisms [17,18]. It has also been reported that AgNPs coated with sodium citrate were more toxic than those coated with polyvinylpyrrolidone [19,20]. Regarding the size-dependent toxicity, Gliga et al. reported that small AgNPs have faster rates of Ag dissolution and higher cytotoxic potential than large AgNPs in a human lung epithelial cell model [21]. Nonetheless, Kim et al. showed greater cytotoxicity of 100-nm-sized AgNPs compared to smaller-sized particles (10 and 50 nm) in MC3T3-E1 and PC12 cells [22]. Conflicting observations have also been reported by using zebrafish embryos as the testing model. Kim and Tanguay indicated that the smaller 20 nm AgNPs were more toxic than the larger 110 nm AgNPs, regardless of the presence of the embryo chorion and test media [23]. In contrast, when compared with smaller AgNPs (13.1 ± 2.5 nm), larger AgNPs (97 ± 13 nm) incite more striking size-, stage-, and dose-dependent toxic effects of AgNPs upon embryonic development [24–26]. Thus, a debate persists regarding the size-dependent toxic effects and underlying mechanisms triggered by AgNPs.

The zebrafish (Danio rerio) embryo (ZFE) test is an appealing in vivo model to assess the hazards of both conventional chemicals and NPs in (eco)toxicology [27–31]. The ZFE, as a vertebrate model species, combines the advantages of rapid development and optical transparency, allowing for easy observations of phenotypic responses in the internal organs, including the brain, jaw, eye, heart, yolk sac, trunk, tail, and so on [23]. In addition, it has been considered an alternative testing model to animals, allowing for the observation of lethal, acute, and sublethal toxicological endpoints [32,33]. In addition, massive amounts of ZFEs can be generated rapidly at a very low cost, permitting them to serve as a high-throughput in vivo assay for the study of developmental processes upon exposure to nanomaterials [24,25]. To estimate the toxic potency of AgNPs, the uptake of metals by ZFEs and the underlying kinetic processes may play a pivotal role [24]. It is expected that the distribution of metal-based nanoparticles follows a temporal and spatial pattern [32,34], whereas little is known about the association of NPs and ZFEs with regard to the quantification of the internalized amounts and the visualization of particle distribution around the embryo. Basically, the chorion is considered to be a barrier to the entry of NPs into zebrafish embryos [35,36]. Although the pore size of chorion canals (approximately 0.6–0.7 μm) is larger than the size of the NPs, the effect of the chorion on NP transport and subsequent biological toxicity may become complicated when NPs agglomerate or interact with chorion surface proteins.

Single types of in vitro cell culture assays are widely used to study the cytotoxic effects of nanomaterials, which can overlook vital and specific cell–cell interactions [37]. We previously demonstrated that AgNPs can be taken into fibroblast cells through endocytosis. The internalized AgNPs eventually accumulated in lysosomes or autophagosomes. The mechanisms of AgNP-induced autophagy dysregulation could be mediated by activation of oxidative stress and endoplasmic reticulum (ER) stress signaling pathways [37–39]. When compared with cell culture models, the ZFE test can simultaneously study the effects of NPs on a wide variety of cells and detect all related pathways, including oxidative and ER stress, involvement of apoptotic pathways and disruption of autophagy regulatory signaling in developing embryos. Most impressively, toxicological outcomes obtained from zebrafish and/or their embryos may be extrapolated to human biology because of their 70% of DNA similarity to humans [40,41].

The objective of the present study was to explore the toxic effects and mechanisms of AgNPs at the organismal and cellular levels by using the ZFE model system. To perform exposure studies, we selected AgNPs of two different sizes (10 and 50 nm) with amine surface modifications prepared using the same synthetic pathway. The specific focuses of our study were designed to assess (1) the effect of the chorion on AgNP transportation, (2) size-dependent mortality and lethal dose, (3) morphological defects of the embryos through the early developing stage, and (4) oxidative stress, apoptosis and autophagy-related lysosomal activity through in vivo monitoring of specific biomarkers in embryonic tissues. Our results showed that larger-sized (50 nm) AgNPs can transport and distribute into embryos more efficiently than smaller (10 nm) AgNPs, leading to severe developmental toxicity by inducing ROS-mediated stress responses. These findings are essential for a better understanding of AgNPs in aquatic ecosystems and provide important underlying mechanisms for ecological risk assessments of AgNPs and other nanoparticles.

2. Results

2.1. AgNP Properties

Two different sizes of amine-modified AgNPs were used to investigate the toxic effects on zebrafish embryos. The physicochemical characteristics of the small and large AgNPs were fully characterized with respedt to their size, morphology, surface charge and composition, as summarized in Figure 1. Transmission electron microscopy (TEM) images indicated that the AgNPs appeared mostly spherical with a mean pristine particle diameter of 13.04 ± 1.3 nm (SAS) or 52.3 ± 6.3 nm (LAS) (Figure 1A,B). Dynamic light scattering (DLS) analysis of SAS (average 30.0 nm, Figure 1C) and LAS (average 55.6 nm, Figure 1D) showed narrow peaks, indicating that the SAS and LAS both possessed a homogeneous dispersion. The scanning TEM and electron dispersive X-ray (EDX) analyses revealed that except for TEM-coated grid materials (carbon and copper), the SAS and LAS were composed of silver, which indicates the high purity of the nanomaterials (Figure 1E,F). According to DLS, the hydrodynamic sizes of the SAS and LAS were larger than in the dry state, which might be attributable to the aggregation of some nanoparticles in solution, and the zeta potentials for both types of particles were positive in water. We also measured the dispersity value; in general, the smaller the value of dispersity, the more stable the nanosuspensions. In our results, the dispersities of SAS and LAS were 0.263 and 0.185, indicating that the particles were stable following dispersion. In addition, the synthesized SAS and LAS were analyzed using UV-vis absorbance spectrophotometry to confirm their properties. As shown in Figure 1G, the maximum absorbance of the SAS was lower (399 nm) compared to the LAS at 430 nm.

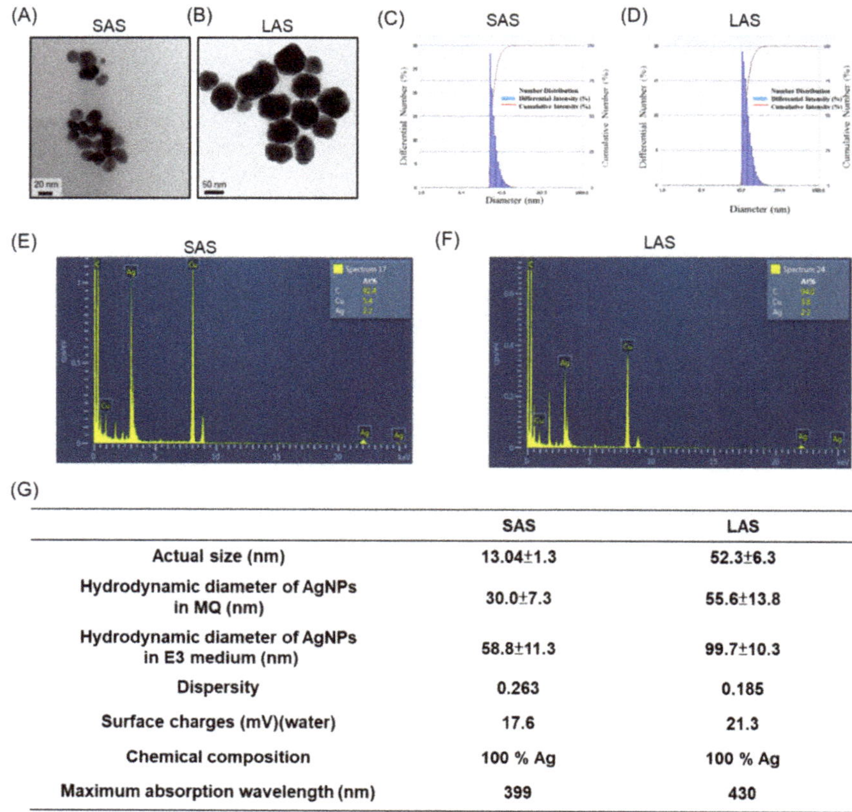

	SAS	LAS
Actual size (nm)	13.04±1.3	52.3±6.3
Hydrodynamic diameter of AgNPs in MQ (nm)	30.0±7.3	55.6±13.8
Hydrodynamic diameter of AgNPs in E3 medium (nm)	58.8±11.3	99.7±10.3
Dispersity	0.263	0.185
Surface charges (mV)(water)	17.6	21.3
Chemical composition	100 % Ag	100 % Ag
Maximum absorption wavelength (nm)	399	430

Figure 1. The physical-chemical properties of the SAS and LAS. (**A,B**) Transmission electron microscopy (TEM) images of SAS and LAS. (**C,D**) The diameter and number distribution of SAS/LAS were measured by dynamic light scattering. (**E,F**) The composition of SAS and LAS were analyzed via energy-dispersive X-ray spectroscopy. (**G**) Measurement of physical-chemical properties, including primary size, hydrodynamic diameter, dispersity, surface charges, chemical composition and maximum absorption wavelength.

2.2. Distribution of the AgNPs in Developing Zebrafish Embryos

To investigate the distribution of AgNPs in zebrafish embryos, we used rhodamine B isothiocyanate (R6G)-conjugated SAS and LAS to provide dynamic imaging of zebrafish embryos. Figure 2A indicates that R6G-SAS accumulated in the outer layer of the chorion of the zebrafish embryos at 24 hpf and 48 hpf. Conversely, the LAS-exposed group tended to penetrate into the zebrafish embryo at 48 hpf (Figure 2B). To demonstrate whether AgNPs can penetrate into the zebrafish embryo, zebrafish embryos were exposed to SAS and LAS at different concentrations at 24, 72, and 96 hpf. Atomic absorption spectroscopy (AAS) was employed to quantify SAS and LAS (Figure 2C–E) in the zebrafish. The embryos exposed to SAS and LAS significantly increased AgNPs accumulative levels. Interestingly, accumulation of LAS was significantly higher than SAS. At 72 hpf, the accumulative AgNPs dramatically dropped to 0.018 (SAS 1 µg/mL), 0.03 (SAS 1 µg/mL), 0.15 (SAS10 µg/mL) and 1.69 (LAS 10 µg/mL) due to dechorionation upon embryos hatching. The zebrafish embryos exposed to SAS and LAS showed a concentration-dependent increase, but there was more silver ion content in the zebrafish body after exposure to LAS.

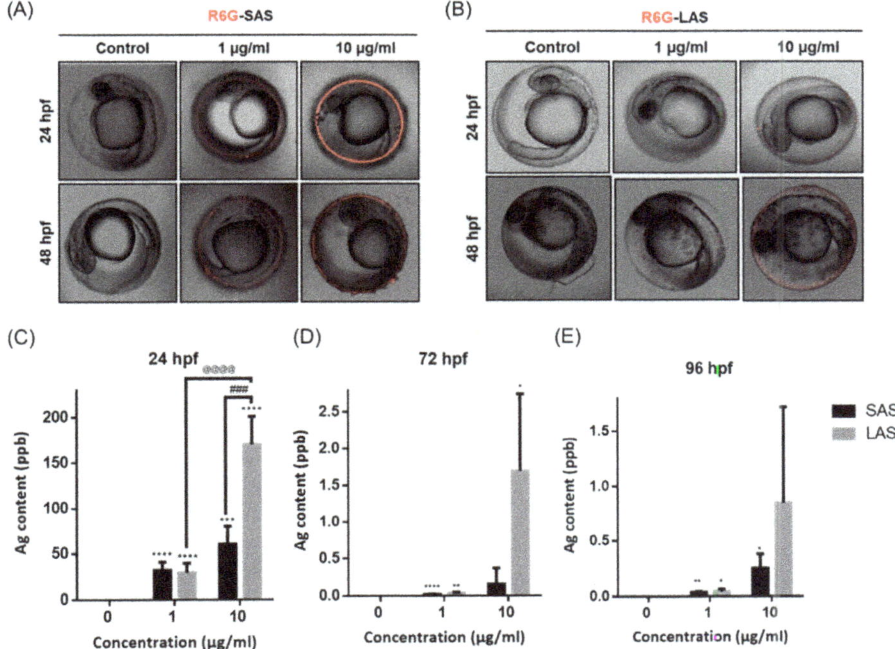

Figure 2. The metal accumulation of SAS and LAS. Zebrafish embryo were exposed to 1, 10 µg/mL R6G conjugated (**A**) SAS (**B**) LAS at 24–48 hpf. Red fluorescence signals represent deposition of AgNPs. The retention of silver in embryos at (**C**) 24, (**D**) 72, (**E**) 96 hpf. The level of silver ion was detected by atomic absorption spectrophotometer analyses. The experiment was performed thrice using 30 embryos each. The values are presented as the mean ± SEM. Values significantly different from the control are indicated by asterisks (one-way ANOVA, followed by a *t*-test: * $p < 0.05$, ** $p < 0.01$, *** $p < 0.001$ and **** $p < 0.0001$). The significantly difference between SAS and LAS are indicated by pound sign (one-way ANOVA, followed by a *t*-test ### $p < 0.001$). The significantly difference between LAS 1 µg/mL and LAS 10 µg/mL groups are indicated by the at sign (one-way ANOVA, followed by a *t*-test @@@@ $p < 0.0001$).

2.3. Evaluation of Zebrafish Chorion Pore Size after Exposure to Different Sizes of AgNPs

The chorion protects the embryo and is the first barrier to come into contact with the SAS and LAS upon exposure. Using scanning electron microscopy (SEM), we observed that the inner layer of the chorion appeared to have evenly sized pores and a smoother surface in the control group. Both the SAS and LAS groups showed agglomerates around the inner membrane. By amplifying the LAS group, we could more clearly observe the granular agglomerates around the inner membrane pores (Figure 3A). In addition, we used backscattered electrons (BSE) to analyze the differences in the sample surface composition and to observe the accumulation of LAS in the outer or inner layers of the zebrafish embryo. As shown in Figure 3B,C, there was significant LAS accumulation in the inner or outer layer of the chorion. Interestingly, only exposure of LAS revealed an enlarged pore size in the chorion inner layer (Figure 3D). In the results of the pore size measurement, SAS exposure did not increase the pore size of the chorion compared to the control group, and its pore size was 0.78 µm. Instead, LAS exposure enlarged the pore size of the chorion (0.94 µm). The pores of the chorion are necessary for oxygen and nutrient transportation from the outer aquatic environment to the embryo and for the elimination of waste. However, AgNPs may interfere with the material exchange between the inner and outer layers of the chorion, resulting in toxic effects to zebrafish embryos.

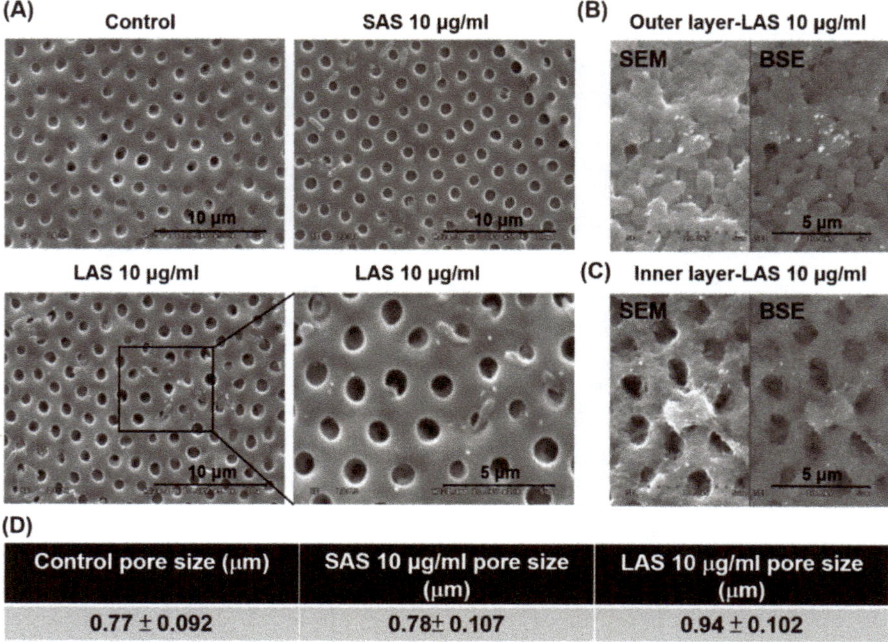

Control pore size (µm)	SAS 10 µg/ml pore size (µm)	LAS 10 µg/ml pore size (µm)
0.77 ± 0.092	0.78 ± 0.107	0.94 ± 0.102

Figure 3. The chorion inner/outer membrane of SAS/LAS-treated zebrafish embryos. Zebrafish embryos were observed via SEM after exposure to (**A**) 10 µg/mL SAS and LAS at 48 hpf. The backscattered electrons of the (**B**) outer layer and (**C**) inner layer membrane. The white dots indicate the accumulation of LAS. (**D**) The diameter of the chorion inner membrane pores was approximately 0.77 µm for the control, 0.78 µm for SAS and 0.94 µm for LAS.

2.4. Identifying the Effects of Different Sizes of AgNPs on Embryo Mortality and Morphological Abnormalities

According to previous studies, the particle size and surface area correlated with NP-induced toxic effects and interactions with living organisms. Therefore, we determined the effects of particle size on mortality and developmental toxicity. The mortality of the AgNP-treated zebrafish embryos is demonstrated in Figure 4A,B. The embryos were exposed to 0, 0.05, 0.1, 0.5, 0.75, 1, 10, and 100 µg/mL SAS or LAS and the cumulative mortality was recorded at 24, 48, 72, 96, and 120 hpf. SAS and LAS dose-dependently decreased the survival rate. The lethal concentration 50 (LC_{50}) of the SAS and LAS showed similar results (Figure 4C). Compared with the SAS groups, the mortality of LAS was higher than that of SAS. To investigate developmental toxicity, we observed body length, head-trunk angle and malformation. The body length and head-trunk angle are important indices of developmental toxicity. We exposed the embryos with AgNPs both in deionized water (Figure 5A,B) and E3 medium (Figure 5C,D), a medium usually applied to developmental study, and the results demonstrated that LAS suspended in deionized water significantly decreased the body length (Figure 5A) and head-trunk angle (Figure 5B), whereas SAS and LAS suspended in E3 medium elicit mild or none toxic effects on body length and head trunk angle when compared with deionized water. Furthermore, SAS and LAS increased the phenomena of malformations, such as pericardial edema (PE), yolk sac edema (YSE), opaque yolk (OY), axial curvature (AC), and jaw (J) malformation (Figure 6A,B). The most commonly observed malformations include notochord malformation, yolk edema, axis malformation, and heart malformation. Quantification of the total specific malformation rate of LAS was higher than that of SAS (Figure 6B). Taken together, both SAS and LAS resulted in zebrafish embryo developmental toxicity, increased severe malformation phenotypes and stronger toxic effects were observed when embryos were exposed to concentrations of 10 and 100 µg/mL LAS and SAS. Moreover, LAS induced more severe malformations than SAS.

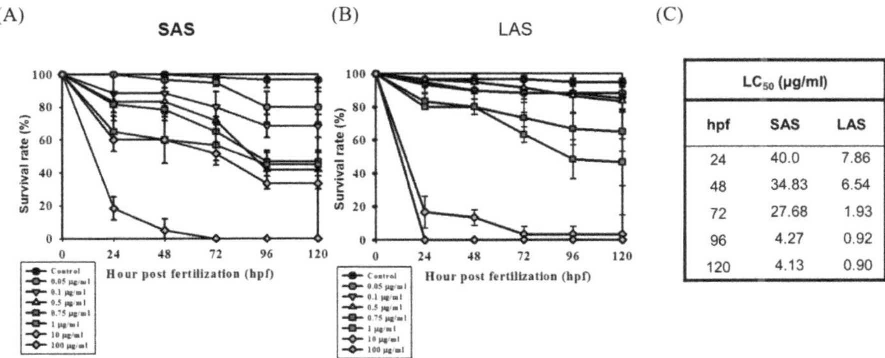

Figure 4. The toxic effects of SAS/LAS. The mortality of embryos treated with 0. 0.05, 0.1, 0.5, 0.75, 1, 10, or 100 µg/mL (**A**) SAS or (**B**) LAS at 24, 48, 72, 96, 120 hpf. (**C**) LC$_{50}$ of SAS/LAS-treated embryos. The lower LC$_{50}$ of LAS indicates that the toxicity of LAS is higher than that of SAS at 24, 48, 72, 96, and 120 hpf. The LC$_{50}$ of SAS/LAS was calculated by IBM SPSS 22 statistics The mortality was performed from 3 replicate trials using 30 embryos at each dose. Error bars present the SEM of the mean.

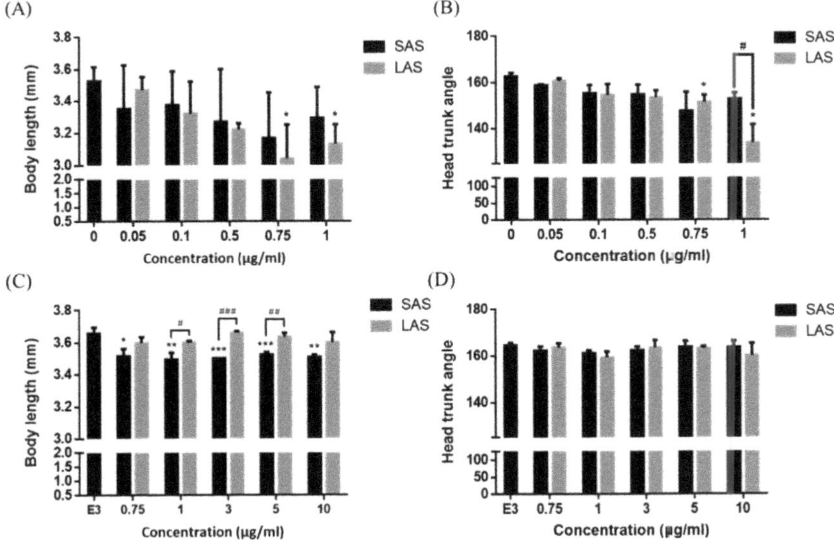

Figure 5. The developmental toxicity of zebrafish embryos. The developmental toxicity of SAS/LAS suspension in deionized water (**A,B**) and suspension in E3 embryo medium (**C,D**) were evaluated via body length and head-trunk angle. Zebrafish embryos were treated with SAS and LAS suspension in deionized water (0, 0.05, 0.1, 0.5, 0.75, 1 µg/mL) and E3 medium (0, 0.75, 1, 3, 5, 10 µg/mL) for 72 hpf to measure the body length and head-trunk angle. The values are presented as the mean ± SEM. Values that are significantly different from the control are indicated by asterisks. (one-way ANOVA, followed by a *t*-test: * $p < 0.05$, ** $p < 0.01$ and *** $p < 0.001$). LAS control versus exposed LAS groups). Values that are significantly different between SAS and LAS is indicated by the pound sign (one-way ANOVA, followed by a *t*-test: # $p < 0.05$, ## $p < 0.01$ and ### $p < 0.001$).

2.5. Quantification of ROS Expression and Damage to the Intestines

Overproduction of ROS induces multiple adverse effects and activates various cellular toxic pathways, including lipid and protein peroxidation, DNA damage, activation of autophagy and cell death [42]. Evidence abounds that excessive ROS generation is an important mechanism induced by metallic NPs [3]. Metallic NP-induced ROS are correlated with physicochemical properties, such as particle size and shape [3]. However, the role of AgNP size on ROS production remains elusive. Therefore, we next determined whether SAS and LAS caused different levels of ROS in zebrafish embryos. As demonstrated in Figure 7A,B, the signals from ROS were mainly located in the intestines. In addition, both SAS- and LAS-treated Tg(IFABP:dsRed) fish embryos induced excessive ROS, and LAS induced higher ROS levels than SAS (Figure 7C,D). These results suggested that SAS and LAS increased ROS levels in the intestines and that LAS triggered greater ROS generation than SAS.

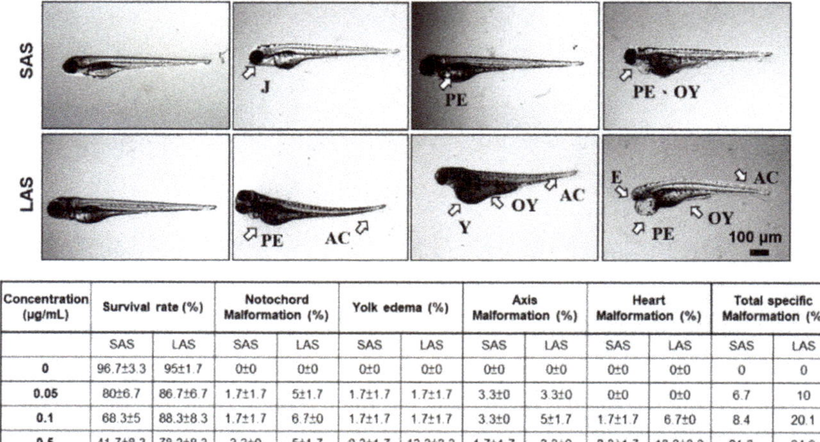

(B)

Concentration (µg/mL)	Survival rate (%)		Notochord Malformation (%)		Yolk edema (%)		Axis Malformation (%)		Heart Malformation (%)		Total specific Malformation (%)	
	SAS	LAS	SAS	LAS	SAS	LAS	SAS	LAS	SAS	LAS	SAS	LAS
0	96.7±3.3	95±1.7	0±0	0±0	0±0	0±0	0±0	0±0	0±0	0±0	0	0
0.05	80±6.7	86.7±6.7	1.7±1.7	5±1.7	1.7±1.7	1.7±1.7	3.3±0	3.3±0	0±0	0±0	6.7	10
0.1	68.3±5	88.3±8.3	1.7±1.7	6.7±0	1.7±1.7	1.7±1.7	3.3±0	5±1.7	1.7±1.7	6.7±0	8.4	20.1
0.5	41.7±8.3	78.3±8.3	3.3±0	5±1.7	8.3±1.7	13.3±3.3	1.7±1.7	3.3±0	8.3±1.7	13.3±3.3	21.6	34.9
0.75	46.7±0	71.7±1.7	0±0	11.7±1.7	3.3±3.3	11.7±1.7	1.7±1.7	8.3±5	3.3±3.3	6.7±6.7	8.3	38.4
1	45±5	40±16.7	1.7±1.7	5±1.7	6.7±0	5±1.7	3.3±0	5±1.7	5±1.7	10±3.3	16.7	25
10	33.3±0	15±8.3	5±1.7	8.3±5	5±1.7	13.3±3.3	8.3±5	8.3±5	5±1.7	15±5	23.3	44.9
100	0±0	0±0	0±0	0±0	0±0	0±0	0±0	0±0	0±0	0±0	0	0

Figure 6. Malformation of zebrafish embryos exposed to LAS and SAS. (**A**) The SAS/LAS-treated embryos (0, 0.1, 1, 10 µg/mL) revealed different malformed phenotypes at 96 hpf. J, jaw malformation; PE, pericardial edema; OY, opaque yolk; AC, axial curvature; Y, yolk sac edema; E, eye malformation. (**B**) Quantification of the survival rate and specific malformation rate, including notochord malformation, yolk edema, axis malformation, and heart malformation. The total malformation indicates the sum of the specific malformation rate. The values are presented as the mean ± SEM.

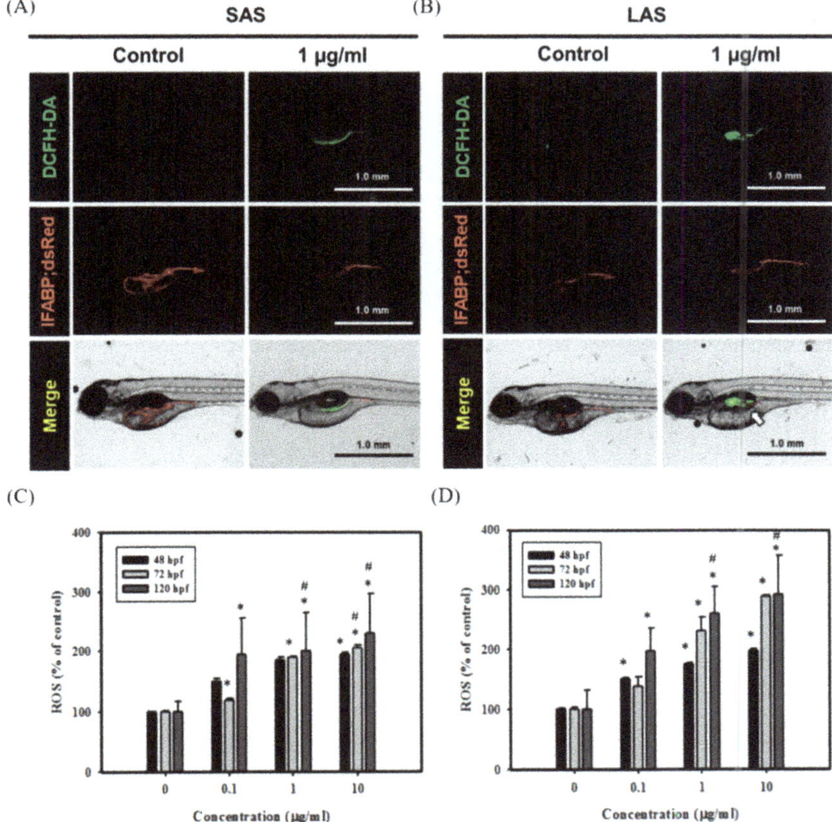

Figure 7. SAS/LAS induced excessive ROS in zebrafish embryos and damaged the intestinal region. Transgenic zebrafish (Tg(IFABP;dsRed)) embryos were treated with 1 µg/mL of (**A**) SAS and (**B**) LAS for observing the intestinal region and ROS production at 120 hpf. The ROS levels were analyzed via DCFH-DA assay. GFP fluorescence signals indicate ROS. RFP fluorescence signals represent the intestinal region. The white arrow indicates the excessive ROS in LAS group. ROS production of (**C**) SAS- (**D**) LAS-treated groups (0, 0.1, 1, and 10 µg/mL) were quantified at 48, 72 and 120 hpf. Quantification of ROS levels was performed with TissueQuest software. The values are presented as the means ± SEM. Values that are significantly different from the control are indicated by asterisks (one-way ANOVA, followed by a *t*-test: *, $p < 0.05$, control versus exposed groups. #, $p < 0.05$).

2.6. AgNP-Induced Zebrafish Embryo Lysosomal Activity and Apoptosis

The lysosome is a critical catabolic and anabolic mediator that can coordinate signals for the degradation of cellular components and cellular stressors [3,43]. Among the multiple functions of the lysosome, lysosomes are involved in the terminal degradation of organelles for autophagy. Therefore, lysosomal activity alterations are tightly related to normal autophagy functions. Recently, several studies have indicated that damaged lysosomes can be an emerging mechanism of nanotoxicity [44]. Therefore, we determined the effects of particle size on lysosomal activity and further conducted a TUNEL assay to detect apoptosis and the main AgNP-induced development defects. To investigate lysosomal activity, zebrafish embryos were treated with SAS and LAS until 48 hpf and were stained with Lysosensor. The results indicated that both SAS and LAS increased lysosomal activity, especially after the treatment with LAS at higher concentrations (Figure 8A,B). In addition, SAS and LAS increased TUNEL$^+$ cells (Red) (Figure 8C,D). Most interestingly, the LAS caused a greater number of TUNEL$^+$-stained cells than the SAS in the notochord of zebrafish. Our results suggest that SAS and LAS induced lysosomal activity as well as apoptosis in the stages of zebrafish development.

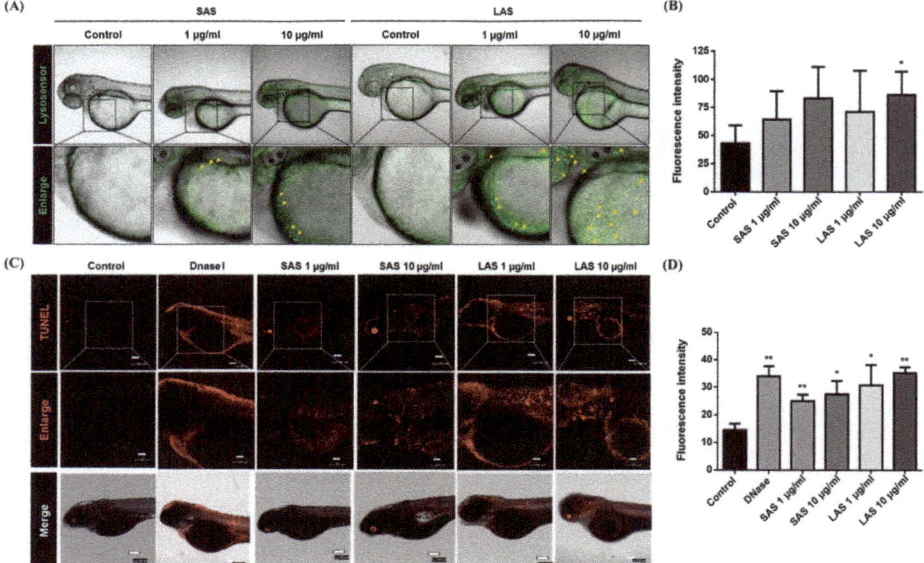

Figure 8. Effect of SAS/LAS on lysosomal activity and apoptosis. (**A**) 4 hpf zebrafish embryos were treated with SAS, LAS (0, 1, 10 µg/mL) and stained by Lysosensor, at 48 hpf. The green color and arrows represent upregulation of lysosomal activity. (**B**) The quantification of lysosensor fluorescence intensity (**C**) The zebrafish embryos treated SAS/LAS (0, 1, 10 µg/mL) were conducted at 72 hpf. Dnase I is the positive control of TUNEL assay. Red color represents apoptotic cell. Apoptotic cell mainly distributed to axis, yolk sac and head region. The scale bars in top and bottom panel (TUNEL and Merge) and middle panel (Enlarge) represents 100 µm and 200 µm respectively. (**D**) The quantification of TUNEL fluorescence intensity. The values are presented as the mean ± SEM. Values that are significantly different from the control are indicated by asterisks (one-way ANOVA, followed by a *t*-test: $* p < 0.05$ and $** p < 0.01$).

3. Discussion

Nanotoxicology is an interdisciplinary field that must bridge both the complex physical and chemical properties of nanoparticles and their interaction with biological systems. The extraordinary properties of nanomaterials are primarily attributed to their nanoscale structure, size, and shape [43]. However, these good benefits may become a serious issue in the application of nanomaterials and bring unknown safety considerations. In the present study, we explored and answered some questions regarding different sizes of nanoparticle-induced nanotoxicity. To evaluate the size-dependent toxic effects of AgNPs, many studies have recently observed the survival rate, malformation, hatching rate, and heart rate of zebrafish embryos after exposure to AgNPs [11,45]. First, we analyzed the survival rate of zebrafish embryos after exposure to SAS and LAS. The results of the LC$_{50}$ calculated by the statistical software indicated that the LAS have the potential to cause higher toxicity (Figure 4C). Most studies have suggested that a smaller particle size creates a higher specific surface area, resulting in increased bioavailability or surface activity of the particles, which in turn leads to increased toxicity [46]. However, some studies have indicated that a larger particle size is highly toxic to zebrafish embryos. Tracking the distribution of AgNPs in zebrafish embryos, the larger-sized AgNPs (41.6 ± 9.1 nm) caused more chorion penetration through passive diffusion and Brownian motion than the smaller particle size (11.6 ± 3.5 nm) [25,46]. Like the human gastric/intestinal mucosal barrier, the chorion offers effective protection and acts as the first barrier against microorganisms, toxins, and other environmental stresses during zebrafish embryo development. Previous studies have shown that nanoparticles may adhere to mucins, causing an enlargement of the pore size with increased susceptibility to penetration from microorganisms [47,48]. However, smaller, charged nanoparticles may be repelled by the hydrophilic domains and will not be able to penetrate the mucosal layer [47,49,50]. Scanning electron microscopy revealed that both the SAS and LAS passed through the pores into the zebrafish embryo and adhered to the inner chorion (Figure 3A). Interestingly, the groups exposed to the LAS had more aggregation in the inner membrane. The results of the BSE experiment also clearly indicated LAS accumulation on the inner and outer layers of the chorion, and microorganisms were observed on both the inside and outside of the chorion (Figure 3B,C). It is possible that we observed more adverse effects from the LAS-exposed zebrafish embryos due to the enlarged pore size of the chorion, which resulted in the penetration of microorganisms and caused infection or toxicity to zebrafish embryos by interfering with the chorion protection. More interestingly, we found the increasing ROS level and lysosomal activity in zebrafish embryo exposed to AgNPs (Figures 7 and 8A). Lysosome is the definitive antimicrobial organelle [51]. It stands in a crucial position in host–pathogen interactions, by being both targeted by pathogens and serving as a major mechanism for killing intracellular invaders [52]. Previous study indicated that the change of lysosomal enzyme activity may control microbial invaders in infected prawns [53]. The induction of ROS and lysosomal activity in zebrafish embryo could be attributed, on one hand, by the increasing pore size caused by AgNPs, leading to decrease of the protection against infection and consequent immune response. On the other hand, AgNPs may also interfere with lysosomal pH and cause excessive ROS production and impair the antimicrobial capabilities of zebrafish embryos.

Several studies have suggested that particle size strongly contributes to internalization and cell responses [3,54]. In the present study, it is possible that the AgNPs of ~50 nm in diameter exhibit both higher Ag deposition (Figure 2) and toxic effects than the AgNPs ~10 nm in diameter (Figure 4) because the specific size of the nanoparticles may incite particular cellular responses. For example, various studies have suggested that the highest cellular uptake occurs for nanoparticles that are approximately 30-50 nm in size due to the membrane-wrapping process [1,55]. Thermodynamically, particle sizes ranging between 30-50 nm are suitable for recruiting and binding to receptors, resulting in the successful activation of membrane wrapping. In contrast, the small-sized nanoparticles exhibit fewer interactions between the ligand and receptor. However, once the nanoparticle size is above 50 nm, it will possess a high affinity for a specific receptor and limit the binding of the receptor to other nanoparticles [1,55]. Therefore, the optimal particle size for internalization and the driving of dramatically adverse effects of nanoparticles could be 30–50 nm in diameter.

Previous studies have shown that AgNPs cause developmental delays and increase the occurrence of malformation [56,57]. In our results, we found that AgNPs increased developmental toxicity (Figure 5) and malformation phenotypes (Figure 6). We observed that the AgNP-induced malformation phenotypes were mainly yolk sac edema and heart failure. Interestingly, the larger size of AgNPs led to a higher percentage of yolk sac edema than the smaller size. Some studies have shown that AgNPs promote the occurrence of malformation via the influences of oxidative stress, cathepsin L, metallothionein, and endoplasmic reticulum calcium ATPase 1, which are malformation-related factors [57]. Moreover, other studies have found that AgNPs disrupt embryogenesis in medaka via the dysregulation of teratogenicity-related genes, including ctsL, tpm1, rbp, mt, and atp2a1 [57]. In the Medaka study, AgNPs significantly downregulated ctsL gene expression levels, which are known as lysosomal cysteine proteinases and are responsible for tumor promotion, bone resorption and growth regulation. In lower vertebrates, ctsL is expressed in the embryonic stages and participates in yolk proteolysis [58]. Therefore, AgNP-decreased ctsL expression levels may lead to teratogenicity and cause zebrafish embryo developmental toxicity.

In our study, we showed that the zebrafish yolk sac was stained with a higher intensity of Lysosensor in both the SAS and LAS groups (Figure 8A). Nanoparticles typically accumulate in lysosomes, and their impact on lysosomal function and autophagy have recently been proposed as an emerging mechanism of nanotoxicity [43,44,59]. Autophagy plays a critical physiological role in mitigating stress-mediated protein aggregation and clearance of damaged organelles [37]. However, growing evidence suggests that various nanoparticles activate autophagy, which may serve as a cellular protective mechanism against toxicity. In our previous study, we proved that AgNPs can disrupt lysosomal activity and cause autophagy dysfunction and apoptosis in mouse fibroblast cells [37,38]. Recent studies have also shown that AgNPs block the process of autophagy due to impairment of autophagosome-lysosome fusion, which results in autophagic defects and consequently triggers the induction of multiple cytotoxic pathways [39].

AgNPs are widely used in many industrial and daily sections and thus facilitate their release into the environment via wastewater. It has been estimated by simulative modeling studies that the concentrations of AgNPs in U.S. surface waters are between 0.09 and 0.43 ng/L, those in European surface waters are between 0.59 and 2.16 ng/L, and those in the river Rhine are between 40 and 320 ng/L [9]. Moreover, the behaviors of nanoparticles can be changed by environmental conditions. It has also been reported that organic matrixes such as humic acid in natural water could attenuate AgNPs induced toxicity [60]. We found both SAS and LAS suspended in E3 embryo medium presented visible precipitation when it lasted for more than 12 h (data not shown).

An interesting finding on the developmental toxicity in embryos showed that AgNPs suspended in E3 medium elicited mild or none toxic effects on body length and head trunk angle when compared with deionized water. One possible explanation is that nanoparticles in the E3 medium could form aggregates. In a previous study, the authors indicated that the dispersion of zinc oxide nanoparticle in E3 medium was significantly reduced and formed aggregates, leading to a size of particles that exceeded the defined nanoscale range (1–100 nm), which may have affected the penetration through the chorion [61]. E3 medium, a multivalent inorganic salt solution, has been reported to trigger the instability of nanoparticles [62]; thus, the high ionic strength solution could contribute to the aggregation of AgNPs as demonstrated previously [63] and in our current study.

Our experimental exposure concentration of AgNPs in this study may be much higher than actual environmental AgNP concentrations. Nonetheless, from the toxicology/ecotoxicology point of view, determination of the dose response relationship is a central dogma for the risk assessment and hazard classification. Our experimental designs are based on "OECD Test No. 236: Fish Embryo Acute Toxicity". Determination of LC_{50} and the toxic mechanisms of AgNPs can provide essential information further for both environmental and human health risk assessment.

In summary, our results suggest that zebrafish embryos take up a certain amount of AgNPs. Interestingly, AgNPs ~50 nm in diameter showed a different potency from that of AgNPs ~10 nm in diameter. The LAS increased the pore size and penetrated the chorion, leading to an abundance of AgNP deposition. The accumulated AgNPs induced adverse toxic effects and developmental defects and increased the potential of specific malformations. AgNPs activate multiple cellular toxic pathways, including excessive ROS, activated lysosomal activity and even increased cell death. Taken together, we determined that AgNPs triggered size- and dose-dependent adverse effects in a zebrafish embryo model (Figure 9).

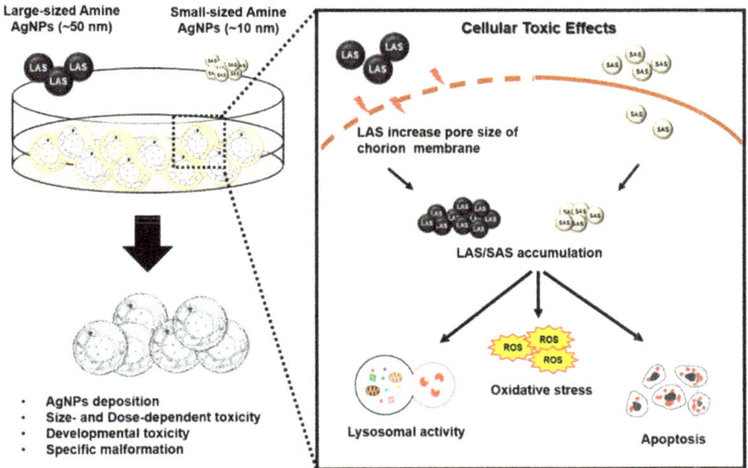

Figure 9. Illustration of SAS-/LAS-induced toxic effects in zebrafish embryos. AgNPs triggered size- as well as dose-dependent toxicity and accumulated in the embryos. The deposition of AgNPs increased the number of developmental defects and the induction of specific malformations. Mechanically, AgNPs activate various cellular toxic signaling pathways, including increased ROS, lysosomal activity and apoptosis. Interestingly, LAS showed different behavior from SAS. LAS increased the pore size and penetrated the chorion, resulting in an abundance of AgNP deposition.

4. Materials and Methods

4.1. SAS/LAS Synthesis and Characterization

SAS and LAS were prepared by surface functional substitution from 20 nm and 50 nm citrate-coated AgNPs, respectively. The AgNPs were prepared by the $NaBH_4$ reduction of $AgNO_3$. Briefly, aqueous silver nitrate solution (1.5 mL, 20 mM) and sodium citrate solution (1.2 mL, 80 mM) were mixed with H_2O (26.7 mL). Ice-cold aqueous $NaBH_4$ solution (0.6 mL, 100 mM) was added dropwise under vigorous stirring. Following this process, the resultant colloidal AgNP suspension was stored at 4 °C and kept from light. The required size of the NPs was separated by centrifugation. Then, 40 mL 40 ppm prepared large (83 nm) and small (26 nm) silver nanoparticles were mixed with 100 mL of cysteamine, and added with 1 mL of sodium hydroxide (NaOH) to neutralize the ions released by cysteamine (the exact volume required depending on the concentration of AgNPs); if the added of cysteamine causes AgNPs to be reduced, the volume needs to be reduced. Then, the mixture was vigorously stirred at room temperature. After reacting for 2 h, the silver nanoparticles were purified by centrifugation and resuspend with deionized water. Characterization of the SAS and LAS was performed using TEM (JEOL Co., Akishima, Tokyo, Japan). SAS and LAS were examined after suspension in M.Q. water and deposition onto copper-coated carbon grids. TEM software was calibrated to measure the sizes of the SAS and LAS. The compositions of SAS and LAS were determined by EDX analysis (JEOL Co., Akishima, Tokyo, Japan). The hydrodynamic sizes, zeta potential, and polydispersity index (PDI) of SAS and LAS examined by DLS (Delsa™ Nano C, Beckman Coulter, Inc., Brea, CA, USA). The zeta potential of the SAS and LAS was analyzed in aqueous dispersion using a phase analysis light scattering (PALS) (Delsa™ Nano C, Beckman Coulter, Inc., Brea, CA, USA).

4.2. Preparation of Rhodamine 6G Conjugated SAS/LAS

The final product cysteamine AgNPs were further synthesized and mixed with Rhodamine B isothiocyanate (R6G) (Sigma-Aldrich, St. Louis, MO, USA). To 20 mL, 800 ppm AgNPs was added

2 mL of 1 mM R6G. The solution was stirred vigorously at room temperature for 2 h. The AgNPs were purified by centrifugation (depending on the size of the desired nanoparticle). The supernatant was removed and deionized water was used to redissolve the precipitate. The centrifugation step was repeated until the supernatant was clarified. Finally, the precipitate was dissolved back into deionized water to obtain R6G-conjugated SAS/LAS. R6G-conjugated SAS/LAS exhibited red fluorescence and we used them to observe the locations of the SAS and LAS in zebrafish embryo.

4.3. Atomic Absorption Spectrometry

To determine the concentration of the synthesized silver nanoparticles, atomic absorption spectrometry (AAS) was used to examine the amount of silver. A standard solution of $AgNO_3$ was prepared and diluted to a proper concentration to make a standard curve. The sample was diluted into the range of the standard curve and then digested at 95 °C for 10 min. Then, the concentration was analyzed using a graphite furnace atomic absorption spectrometer (Perkin Elmer AAnalystTM 600, Waltham, MA, USA).

To quantify the distribution concentration of SAS and LAS retention in the embryos, after zebrafish embryo exposure to AgNPs (1 and 10 µg/mL), embryos were collected at 48, 72, and 96 hpf, and AAS was used to examine the cumulative amount of silver in the embryos. The collected embryos of the experimental and untreated groups were treated with 1.5 mL of nitric acid (69%) for 4 h at 50 °C, followed by another 4 h at 100 °C. Once fully digested, the samples were diluted with 2% nitric acid to an acceptable concentration (approximately 200- to 300-fold dilutions depending on the exposure dose) and then subjected to analysis using a graphite furnace atomic absorption spectrometer. Triplicate measurements of the absorption for the sample were taken for each exposure dose.

4.4. Scanning Electron Microscopy (SEM)

To analyze the chorion, we used SEM with the principle of BSE. The sample chorions were rinsed with PBS and then 20–25 µL of sample was siphoned to load onto the center of the glasses. Fifty microliters of 2.5% glutaraldehyde was added to every sample and followed by sealing. The glasses were swayed for 20 min at 4 °C. The glutaraldehyde was removed and the sample was rinsed with PBS for 3–5 min three times. To each plate was added 1.0% OsO_4 (osmium tetroxide) (Sigma-Aldrich, St. Louis, MO, USA) to start the reaction for 40 min. Dehydration was operated by the following concentrations: 50%, 70%, 80%, 90%, 95% and 100% three times (using 0.5 mL of acetone and 3 min for every glass). Every glass was packed with filtrate paper and the samples were placed on the upper side. The samples were dried at the temperature critical point and coated with gold. Finally, the sample as well as the glass were loaded on the field-emission scanning electron microscope instrument (JSM-7001F, JEOL Co., Tokyo, Japan) at 28–30 kV to scope.

4.5. Zebrafish Maintenance and Egg Spawning

Zebrafish (Danio rerio) (Gendanio, AB line wild-type, New Taipei, R.O.C) in adulthood were maintained at 28.0 °C with a light cycle of 14 h light/10 h dark and fed brine shrimp twice per day and feed (containing 60% rough protein) once per day. All experiments on zebrafish were performed according to the guidelines of our institute (Guide for Care and Use of Laboratory Animals, National Cheng Kung University Medical College) and were approved by the Institutional Animal Care and Use Committee of National Cheng Kung University, Taiwan (Approval No.: 107133; Date: February 27, 2018). To obtain zebrafish embryos, male and female zebrafish were set up as pairs the night before mating in breeding tanks with a divider. The next morning, the zebrafish were stimulated with light, and the dropped eggs were collected, pooled, and rinsed with tap water and then deionized water several times. For the experiments, fertilized eggs were chosen and collected under a microscope, and the dead/unfertilized embryos were removed to avoid test solution pollution. All zebrafish experiments were performed in compliance with guidelines from the OECD Test Guidelines (TG203 and TG236) [64].

4.6. Zebrafish Embryo Acute Toxicity Test

The fish embryo acute toxicity (FET) test using zebrafish is used to examine toxicity in 96 h [65]. In our research, we modified some testing conditions to meet our endpoint. Fish embryos were collected and exposed to the test solution at 4 hpf, 28.0 °C ($n = 30$). For the embryo acute toxicity test, all the development takes place in milli Q water or AgNPs solution. The embryos were maintained in milli-Q water with AgNPs (LAS/SAS groups) or milli-Q water alone (Control group) during 4 hpf to the specific endpoint. The exposure solutions were renewed every 24 h to avoid aggregation of AgNPs. Then, the numbers of alive and dead embryos at 24, 48, 72, 96, and 120 hpf were counted, and photos were taken using a microscope with a mounted CCD camera (SMZ800, Nikon Instruments Inc., Melville, NY, USA).

4.7. Developmental Toxicity

We used healthy zebrafish to spawn embryos. We then washed the embryos with E3-media and egg water after collection. The used criteria came from prior study [2]. For developmental toxicity assay, the zebrafish embryos were exposed to AgNPs suspended both in deionized water and E3 medium, a culture medium usually applied to developmental study. The development of zebrafish embryos could be measured by their body length and head-trunk angle. The head-trunk angle is a mathematical function of zebrafish embryo development at 28.5 °C. The main purpose is to measure the angle between the head and trunk and then determine whether the development is normal, stimulated or inhibited. We then calculated the malformation rate and the kind of malformation, including some criteria indicated malformations, such as yolk sac edema, opaque yolk, pericardial edema and axial curvature. Additionally, we observed the malformations of bent axial, tail damage and jaw bulge.

4.8. ROS Measurement

ROS generation induced by different doses (0, 0.1, 1, 10 µg/mL) of LAS and SAS was measured. For sample preparation, 0.003% 1-phenyl-2-thiourea (PTU) (Sigma-Aldrich, St. Louis, MO, USA) was added to decrease the rate of pigment formation. Then, 48, 72, or 120 hpf, the reactive oxygen species (ROS) in zebrafish embryos was measured. Before measuring, 1 µg/mL DCFH-DA dye (Sigma-Aldrich, St. Louis, MO, USA) was added to each group (0, 0.1, 1, 10 µg/mL). After cultivating and avoiding light for 30 min under an ambient environment, the embryos were rinsed with deionized water several times and fixed on glass with 3% methylcellulose. Then, we observed the ROS generation with a confocal microscope (excitation/scattering = 488 nm/543 nm) and captured photos. An apical point was to analyze AgNPs-induced ROS in zebrafish embryos. Finally, the quantitative fluorescence was measured using the panoramic tissue somatic cell quantitative software TissueQuest (TissueGnostics, Vienna, Austria).

4.9. Lysosomal Activity Assay

LysoSensor™ (Thermo Fisher Scientific Inc., Waltham, MA, USA) induces green florescence in a pH < 5 environment and distinguishes the activity of the cellular lysosome. Embryos at 4 hpf were exposed to AgNPs and 0.03% PTU was added to avoid fish pigment precipitation. At 48 hpf, the AgNP solution was removed and the embryos were rinsed with Milli-Q water 3 times. Finally, 10 µM Lysosensor kit was added, and the samples were incubated for 45 min at 28 °C. After cultivation, the solution was removed and the samples were rinsed with 1 mL of Milli-Q water 3 times. The fish were fixed with 3% methylcellulose and observed under a vertical fluorescence microscope (BX51, Olympus Co., Tokyo, Japan). The phenotypes of the lysosomes are red dots.

4.10. TUNEL Assay

To detect apoptotic cells in live embryos, we conducted a whole-mount TUNEL assay (Sigma-Aldrich, St. Louis, MO, USA). Embryos were exposed to different doses (0, 0.1, 1, 10 µg/mL) of LAS and SAS at 4 hpf, and the exposure method was the same as the zebrafish embryo acute toxicity test. The endpoint of the experiment was at 72 hpf. The exposure solution containing AgNPs was placed in a 12-well plate and then washed three times with deionized water, and the embryos were collected in an Eppendorf microcentrifuge tube in the test concentration group. Embryos were fixed with 4% paraformaldehyde at room temperature for 1 h. After fixation for 1 h, we washed the embryos with PBS on a shaker (40 rpm) for 5 min, which was repeated twice. Then, we incubated the embryos with blocking buffer on a shaker for 30 min. After 30 min, the embryos were washed with PBS on a shaker (40 rpm) for 5 min, which was repeated twice. Then, the embryos were incubated with permeabilization solution (0.1% Triton X-100 in 0.1%) on ice for 30 min. After that, we washed the embryos with PBS on a shaker (40 rpm) for 5 min twice. Finally, the TUNEL reaction mixture (labeling solution:enzyme solution = 9:1) was incubated at 37 °C for 1 h in a water bath and finally observed with a fluorescence microscope.

4.11. Statistics

All experiments were conducted in triplicate. We used statistical software to analyze the data, such as IBM SPSS statistics 22 and Excel. Statistical analyses were carried out using the two-tailed Student's *t*-test for comparisons between the means or using one-way ANOVA. Differences were considered to be significant when $p < 0.05$. All data are presented as the means± SD.

Author Contributions: Z.-Y.C., Y.-H.L., N.-J.L. and Y.-J.W. conceptualized the ideas. Z.-Y.C., Y.-H.L., N.-J.L. and C.-C.H. performed the experiments and conducted data analyzed. F.-Y.C. and Y.-H.L. participated characterization and synthesis of AgNPs. J.-F.H., F.-I.L., T.-F.F. and S.-J.Y. interpreted the results of the experiment. All authors participated in writing, discussion and approving the final manuscript. All authors have read and agreed to the published version of the manuscript.

Funding: This work was supported by the Ministry of Science and Technology, Taiwan (MOST 106-2314-B-006-029-MY3, MOST 108-2638-B-006 -001 -MY2 and MOST 108-2314-B-039-061-MY3), China Medical University, Taichung, Taiwan (CMU108-N-09), Ministry of Labor, Taiwan (Q105-P005), Toxic and Chemical Substances Bureau, Environmental Protection Administration, Executive Yuan, Taiwan (107A024 and 109A012) and Environmental Analysis Laboratory, Environmental Protection Administration, Executive Yuan, Taiwan(Q106-P014).

Acknowledgments: We thank Taiwan Zebrafish Technology and Resource Center and we also thank Center for Micro/Nano Science and Technology, National Cheng Kung University for technical supporting. Experiments and data analysis were performed in part through the use of the Medical Research Core Facilities Center, Office of Research & Development at China medical University, Taichung, Taiwan, R.O.C. We are appreciative all members in the lab to help and advice.

Conflicts of Interest: The authors declare no conflict of interest.

Abbreviations

AgNPs	Silver nanoparticles
LAS	larger-sized AgNPs
SAS	small-sized AgNPs
NPs	Nanoparticles
ZFE	Zebrafish embryo
ROS	Reactive oxygen species
ER	Endoplasmic reticulum
TEM	Transmission electron microscopy
R6G	Rhodamine B isothiocyanate
AAS	Atomic absorption spectroscopy
SEM	Scanning electron microscopy
BSE	Backscattered electrons
PE	Pericardial edema
YSE	Yolk sac edema
OY	Opaque yolk
AC	Axial curvature
J	Jaw

LC$_{50}$	Lethal concentration 50
FET	Fish embryo acute toxicity
PTU	1-phenyl-2-thiourea
PDI	Polydispersity index
PALS	Phase analysis light scattering

References

1. Hoshyar, N.; Gray, S.; Han, H.; Bao, G. The effect of nanoparticle size on in vivo pharmacokinetics and cellular interaction. *Nanomedicine* **2016**, *11*, 673–692. [CrossRef] [PubMed]
2. Kimmel, C.B.; Ballard, W.W.; Kimmel, S.R.; Ullmann, B.; Schilling, T.F. Stages of embryonic development of the zebrafish. *Dev. Dyn.* **1995**, *203*, 253–310. [CrossRef] [PubMed]
3. Abdal Dayem, A.; Hossain, M.K.; Lee, S.B.; Kim, K.; Saha, S.K.; Yang, G.M.; Choi, H.Y.; Cho, S.G. The role of reactive oxygen species (ROS) in the biological activities of metallic nanoparticles. *Int. J. Mol. Sci.* **2017**, *18*, 120. [CrossRef] [PubMed]
4. Chen, X.; Schluesener, H.J. Nanosilver: A nanoproduct in medical application. *Toxicol. Lett.* **2008**, *176*, 1–12. [CrossRef] [PubMed]
5. Ahamed, M.; Alsalhi, M.S.; Siddiqui, M.K. Silver nanoparticle applications and human health. *Clin. Chim. Acta* **2010**, *411*, 1841–1848. [CrossRef] [PubMed]
6. Cohen, M.S.; Stern, J.M.; Vanni, A.J.; Kelley, R.S.; Baumgart, E.; Field, D.; Libertino, J.A.; Summerhayes, I.C. In vitro analysis of a nanocrystalline silver-coated surgical mesh. *Surg. Infect.* **2007**, *8*, 397–403. [CrossRef]
7. Lee, H.Y.; Park, H.K.; Lee, Y.M.; Kim, K.; Park, S.B. A practical procedure for producing silver nanocoated fabric and its antibacterial evaluation for biomedical applications. *Chem. Commun.* **2007**, 2959–2961. [CrossRef]
8. Vigneshwaran, N.; Kathe, A.A.; Varadarajan, P.V.; Nachane, R.P.; Balasubramanya, R.H. Functional finishing of cotton fabrics using silver nanoparticles. *J. Nanosci. Nanotechnol.* **2007**, *7*, 1893–1897. [CrossRef]
9. McGillicuddy, E.; Murray, I.; Kavanagh, S.; Morrison, L.; Fogarty, A.; Cormican, M.; Dockery, P.; Prendergast, M.; Rowan, N.; Morris, D. Silver nanoparticles in the environment: Sources, detection and ecotoxicology. *Sci. Total Environ.* **2017**, *575*, 231–246. [CrossRef]
10. Massarsky, A.; Trudeau, V.L.; Moon, T.W. Predicting the environmental impact of nanosilver. *Environ. Toxicol. Pharmacol.* **2014**, *38*, 861–873. [CrossRef]
11. Khan, I.; Bahuguna, A.; Krishnan, M.; Shukla, S.; Lee, H.; Min, S.H.; Choi, D.K.; Cho, Y.; Bajpai, V.K.; Huh, Y.S.; et al. The effect of biogenic manufactured silver nanoparticles on human endothelial cells and zebrafish model. *Sci. Total Environ.* **2019**, *679*, 365–377. [CrossRef] [PubMed]
12. Mao, B.H.; Chen, Z.Y.; Wang, Y.J.; Yan, S.J. Silver nanoparticles have lethal and sublethal adverse effects on development and longevity by inducing ROS-mediated stress responses. *Sci. Rep.* **2018**, *8*, 2445. [CrossRef] [PubMed]
13. Haque, E.; Ward, A.C. Zebrafish as a Model to Evaluate Nanoparticle Toxicity. *Nanomaterials* **2018**, *8*, 561. [CrossRef] [PubMed]
14. Hassanen, E.I.; Khalaf, A.A.; Tohamy, A.F.; Mohammed, E.R.; Farroh, K.Y. Toxicopathological and immunological studies on different concentrations of chitosan-coated silver nanoparticles in rats. *Int. J. Nanomed.* **2019**, *14*, 4723–4739. [CrossRef]
15. Ribeiro, F.; Van Gestel, C.A.M.; Pavlaki, M.D.; Azevedo, S.; Soares, A.; Loureiro, S. Bioaccumulation of silver in Daphnia magna: Waterborne and dietary exposure to nanoparticles and dissolved silver. *Sci. Total Environ.* **2017**, *574*, 1633–1639. [CrossRef]
16. Barker, L.K.; Giska, J.R.; Radniecki, T.S.; Semprini, L. Effects of short- and long-term exposure of silver nanoparticles and silver ions to Nitrosomonas europaea biofilms and planktonic cells. *Chemosphere* **2018**, *206*, 606–614. [CrossRef]
17. Hauck, T.S.; Ghazani, A.A.; Chan, W.C. Assessing the effect of surface chemistry on gold nanorod uptake, toxicity, and gene expression in mammalian cells. *Small* **2008**, *4*, 153–159. [CrossRef]
18. Zhao, F.; Zhao, Y.; Liu, Y.; Chang, X.; Chen, C.; Zhao, Y. Cellular uptake, intracellular trafficking, and cytotoxicity of nanomaterials. *Small* **2011**, *7*, 1322–1337. [CrossRef]

19. Poynton, H.C.; Lazorchak, J.M.; Impellitteri, C.A.; Blalock, B.J.; Rogers, K.; Allen, H.J.; Loguinov, A.; Heckman, J.L.; Govindasmawy, S. Toxicogenomic responses of nanotoxicity in Daphnia magna exposed to silver nitrate and coated silver nanoparticles. *Environ. Sci. Technol.* **2012**, *46*, 6288–6296. [CrossRef]

20. Hou, J.; Liu, H.; Wang, L.; Duan, L.; Li, S.; Wang, X. Molecular Toxicity of Metal Oxide Nanoparticles in Danio rerio. *Environ. Sci. Technol.* **2018**, *52*, 7996–8004. [CrossRef]

21. Gliga, A.R.; Skoglund, S.; Wallinder, I.O.; Fadeel, B.; Karlsson, H.L. Size-dependent cytotoxicity of silver nanoparticles in human lung cells: The role of cellular uptake, agglomeration and Ag release. *Part. Fibre Toxicol.* **2014**, *11*, 11. [CrossRef] [PubMed]

22. Kim, T.H.; Kim, M.; Park, H.S.; Shin, U.S.; Gong, M.S.; Kim, H.W. Size-dependent cellular toxicity of silver nanoparticles. *J. Biomed. Mater. Res. Part A* **2012**, *100*, 1033–1043. [CrossRef] [PubMed]

23. Kim, K.T.; Tanguay, R.L. The role of chorion on toxicity of silver nanoparticles in the embryonic zebrafish assay. *Environ. Health Toxicol.* **2014**, *29*, e2014021. [CrossRef]

24. Browning, L.M.; Lee, K.J.; Nallathamby, P.D.; Xu, X.H. Silver nanoparticles incite size- and dose-dependent developmental phenotypes and nanotoxicity in zebrafish embryos. *Chem. Res. Toxicol.* **2013**, *26*, 1503–1513. [CrossRef] [PubMed]

25. Lee, K.J.; Browning, L.M.; Nallathamby, P.D.; Desai, T.; Cherukuri, P.K.; Xu, X.H. In vivo quantitative study of sized-dependent transport and toxicity of single silver nanoparticles using zebrafish embryos. *Chem. Res. Toxicol.* **2012**, *25*, 1029–1046. [CrossRef]

26. Lee, K.J.; Browning, L.M.; Nallathamby, P.D.; Osgood, C.J.; Xu, X.H. Silver nanoparticles induce developmental stage-specific embryonic phenotypes in zebrafish. *Nanoscale* **2013**, *5*, 11625–11636. [CrossRef]

27. Hill, A.J.; Teraoka, H.; Heideman, W.; Peterson, R.E. Zebrafish as a model vertebrate for investigating chemical toxicity. *Toxicol. Sci.* **2005**, *86*, 6–19. [CrossRef]

28. Asharani, P.V.; Lianwu, Y.; Gong, Z.; Valiyaveettil, S. Comparison of the toxicity of silver, gold and platinum nanoparticles in developing zebrafish embryos. *Nanotoxicology* **2011**, *5*, 43–54. [CrossRef]

29. Brun, N.R.; Lenz, M.; Wehrli, B.; Fent, K. Comparative effects of zinc oxide nanoparticles and dissolved zinc on zebrafish embryos and eleuthero-embryos: Importance of zinc ions. *Sci. Total Environ.* **2014**, *476–477*, 657–666. [CrossRef]

30. Chen, D.; Zhang, D.; Yu, J.C.; Chan, K.M. Effects of Cu2O nanoparticle and CuCl2 on zebrafish larvae and a liver cell-line. *Aquat. Toxicol.* **2011**, *105*, 344–354. [CrossRef]

31. Weil, M.; Meissner, T.; Busch, W.; Springer, A.; Kuhnel, D.; Schulz, R.; Duis, K. The oxidized state of the nanocomposite Carbo-Iron(R) causes no adverse effects on growth, survival and differential gene expression in zebrafish. *Sci. Total Environ.* **2015**, *530*, 198–208. [CrossRef] [PubMed]

32. Praetorius, A.; Tufenkji, N.; Goss, K.-U.; Scheringer, M.; von der Kammer, F.; Elimelech, M. The road to nowhere: Equilibrium partition coefficients for nanoparticles. *Environ. Sci. Nano.* **2014**, *1*. [CrossRef]

33. Scholz, S.; Fischer, S.; Gundel, U.; Kuster, E.; Luckenbach, T.; Voelker, D. The zebrafish embryo model in environmental risk assessment–applications beyond acute toxicity testing. *Environ. Sci. Pollut. Res. Int.* **2008**, *15*, 394–404. [CrossRef] [PubMed]

34. Wang, J.; Wang, W.X. Significance of physicochemical and uptake kinetics in controlling the toxicity of metallic nanomaterials to aquatic organisms. *J. Zhejiang Univ. Sci. A* **2014**, *15*, 573–592. [CrossRef]

35. Lee, K.J.; Nallathamby, P.D.; Browning, L.M.; Osgood, C.J.; Xu, X.H. In vivo imaging of transport and biocompatibility of single silver nanoparticles in early development of zebrafish embryos. *Acs Nano* **2007**, *1*, 133–143. [CrossRef] [PubMed]

36. Duan, J.; Yu, Y.; Shi, H.; Tian, L.; Guo, C.; Huang, P.; Zhou, X.; Peng, S.; Sun, Z. Toxic effects of silica nanoparticles on zebrafish embryos and larvae. *PLoS ONE* **2013**, *8*, e74606. [CrossRef]

37. Lee, Y.H.; Fang, C.Y.; Chiu, H.W.; Cheng, F.Y.; Tsai, J.C.; Chen, C.W.; Wang, Y.J. Endoplasmic Reticulum Stress-Triggered Autophagy and Lysosomal Dysfunction Contribute to the Cytotoxicity of Amine-Modified Silver Nanoparticles in NIH 3T3 Cells. *J. Biomed. Nanotechnol.* **2017**, *13*, 778–794. [CrossRef]

38. Lee, Y.H.; Cheng, F.Y.; Chiu, H.W.; Tsai, J.C.; Fang, C.Y.; Chen, C.W.; Wang, Y.J. Cytotoxicity, oxidative stress, apoptosis and the autophagic effects of silver nanoparticles in mouse embryonic fibroblasts. *Biomaterials* **2014**, *35*, 4706–4715. [CrossRef]

39. Mao, B.H.; Tsai, J.C.; Chen, C.W.; Yan, S.J.; Wang, Y.J. Mechanisms of silver nanoparticle-induced toxicity and important role of autophagy. *Nanotoxicology* **2016**, *10*, 1021–1040. [CrossRef]

40. Parng, C. In vivo zebrafish assays for toxicity testing. *Curr. Opin. Drug Discov. Dev.* **2005**, *8*, 100–106.

41. Howe, K.; Clark, M.D.; Torroja, C.F.; Torrance, J.; Berthelot, C.; Muffato, M.; Collins, J.E.; Humphray, S.; McLaren, K.; Matthews, L.; et al. The zebrafish reference genome sequence and its relationship to the human genome. *Nature* **2013**, *496*, 498–503. [CrossRef] [PubMed]

42. Zhang, J.; Wang, X.; Vikash, V.; Ye, Q.; Wu, D.; Liu, Y.; Dong, W. ROS and ROS-Mediated Cellular Signaling. *Oxidative Med. Cell. Longev.* **2016**, *2016*, 4350965. [CrossRef] [PubMed]

43. Wang, F.; Salvati, A.; Boya, P. Lysosome-dependent cell death and deregulated autophagy induced by amine-modified polystyrene nanoparticles. *Open Biol.* **2018**, *8*, 170271. [CrossRef] [PubMed]

44. Stern, S.T.; Adiseshaiah, P.P.; Crist, R.M. Autophagy and lysosomal dysfunction as emerging mechanisms of nanomaterial toxicity. *Part. Fibre Toxicol.* **2012**, *9*, 20. [CrossRef] [PubMed]

45. Xia, G.; Liu, T.; Wang, Z.; Hou, Y.; Dong, L.; Zhu, J.; Qi, J. The effect of silver nanoparticles on zebrafish embryonic development and toxicology. *Artif. CellsNanomed. Biotechnol.* **2016**, *44*, 1116–1121. [CrossRef] [PubMed]

46. Lin, S.; Zhao, Y.; Nel, A.E.; Lin, S. Zebrafish: An in vivo model for nano EHS studies. *Small* **2013**, *9*, 1608–1618. [CrossRef]

47. Wang, Y.Y.; Lai, S.K.; So, C.; Schneider, C.; Cone, R.; Hanes, J. Mucoadhesive nanoparticles may disrupt the protective human mucus barrier by altering its microstructure. *PLoS ONE* **2011**, *6*, e21547. [CrossRef]

48. Shvedova, A.A.; Sager, T.; Murray, A.R.; Kisin, E.; Porter, D.W.; Leonard, S.S.; Schwegler-Berry, D.; Robinson, V.A.; Castranova, V. Critical issues in the evaluation of possible adverse pulmonary effects from airborne nanoparticles. In *Nanotoxicology, Characterization, Dosing and Health Effects*; Informa Healthcare, USA Inc.: New York, NY, USA, 2007.

49. Cone, R.A. Barrier properties of mucus. *Adv. Drug Deliv. Rev.* **2009**, *61*, 75–85. [CrossRef]

50. Lai, S.K.; Wang, Y.Y.; Hanes, J. Mucus-penetrating nanoparticles for drug and gene delivery to mucosal tissues. *Adv. Drug Deliv. Rev.* **2009**, *61*, 158–171. [CrossRef]

51. Uribe-Querol, E.; Rosales, C. Control of Phagocytosis by Microbial Pathogens. *Front. Immunol.* **2017**, *8*. [CrossRef]

52. Vural, A.; Al-Khodor, S.; Cheung, G.Y.C.; Shi, C.S.; Srinivasan, L.; McQuiston, T.J.; Hwang, I.Y.; Yeh, A.J.; Blumer, J.B.; Briken, V.; et al. Activator of G-Protein Signaling 3–Induced Lysosomal Biogenesis Limits Macrophage Intracellular Bacterial Infection. *J. Immunol.* **2016**, *196*, 846. [CrossRef] [PubMed]

53. Misra, C.K.; Das, B.K.; Pradhan, J.; Pattnaik, P.; Sethi, S.; Mukherjee, S.C. Changes in lysosomal enzyme activity and protection against Vibrio infection in Macrobrachium rosenbergii (De Man) post larvae after bath immunostimulation with β-glucan. *Fish Shellfish Immunol.* **2004**, *17*, 389–395. [CrossRef] [PubMed]

54. Foroozandeh, P.; Aziz, A.A. Insight into Cellular Uptake and Intracellular Trafficking of Nanoparticles. *Nanoscale Res. Lett.* **2018**, *13*, 339. [CrossRef] [PubMed]

55. Zhang, S.; Li, J.; Lykotrafitis, G.; Bao, G.; Suresh, S. Size-Dependent Endocytosis of Nanoparticles. *Adv. Mater.* **2009**, *21*, 419–424. [CrossRef]

56. Ema, M.; Okuda, H.; Gamo, M.; Honda, K. A review of reproductive and developmental toxicity of silver nanoparticles in laboratory animals. *Reprod. Toxicol.* **2017**, *67*, 149–164. [CrossRef]

57. Kashiwada, S.; Ariza, M.E.; Kawaguchi, T.; Nakagame, Y.; Jayasinghe, B.S.; Gartner, K.; Nakamura, H.; Kagami, Y.; Sabo-Attwood, T.; Ferguson, P.L.; et al. Silver nanocolloids disrupt medaka embryogenesis through vital gene expressions. *Environ. Sci. Technol.* **2012**, *46*, 6278–6287. [CrossRef]

58. Tingaud-Sequeira, A.; Cerda, J. Phylogenetic relationships and gene expression pattern of three different cathepsin L (Ctsl) isoforms in zebrafish: Ctsla is the putative yolk processing enzyme. *Gene* **2007**, *386*, 98–106. [CrossRef]

59. Panzarini, E.; Inguscio, V.; Tenuzzo, B.A.; Carata, E.; Dini, L. Nanomaterials and Autophagy: New Insights in Cancer Treatment. *Cancers* **2013**, *5*, 296–319. [CrossRef]

60. Caceres-Velez, P.R.; Fascineli, M.L.; Sousa, M.H.; Grisolia, C.K.; Yate, L.; de Souza, P.E.N.; Estrela-Lopis, I.; Moya, S.; Azevedo, R.B. Humic acid attenuation of silver nanoparticle toxicity by ion complexation and the formation of a Ag(3+) coating. *J. Hazard. Mater.* **2018**, *353*, 173–181. [CrossRef]

61. Bai, W.; Zhang, Z.; Tian, W.; He, X.; Ma, Y.; Zhao, Y.; Chai, Z. Toxicity of zinc oxide nanoparticles to zebrafish embryo: A physicochemical study of toxicity mechanism. *J. Nanoparticle Res.* **2010**, *12*, 1645–1654. [CrossRef]

62. Zhang, Y.; Chen, Y.; Westerhoff, P.; Hristovski, K.; Crittenden, J.C. Stability of commercial metal oxide nanoparticles in water. *Water Res.* **2008**, *42*, 2204–2212. [CrossRef] [PubMed]

63. Anna, B.; Barbara, K.; Magdalena, O. How the surface properties affect the nanocytotoxicity of silver? Study of the influence of three types of nanosilver on two wheat varieties. *Acta Physiol. Plant.* **2018**, *40*, 31. [CrossRef]

64. Lee, W.S.; Cho, H.J.; Kim, E.; Huh, Y.H.; Kim, H.J.; Kim, B.; Kang, T.; Lee, J.S.; Jeong, J. Bioaccumulation of polystyrene nanoplastics and their effect on the toxicity of Au ions in zebrafish embryos. *Nanoscale* **2019**. [CrossRef]

65. OECD, Test No. 236: Fish Embryo Acute Toxicity (FET) Test. 2013. Available online: www.oecd.org (accessed on 19 April 2020).

International Journal of
Molecular Sciences

Article

Nanoplastics Cause Neurobehavioral Impairments, Reproductive and Oxidative Damages, and Biomarker Responses in Zebrafish: Throwing up Alarms of Wide Spread Health Risk of Exposure

Sreeja Sarasamma [1,2,†], Gilbert Audira [1,2,†], Petrus Siregar [2], Nemi Malhotra [3], Yu-Heng Lai [4], Sung-Tzu Liang [2], Jung-Ren Chen [5,*], Kelvin H.-C. Chen [6,*] and Chung-Der Hsiao [1,2,7,*]

[1] Department of Chemistry, Chung Yuan Christian University, Chung-Li 32023, Taiwan; sreejakarthik@hotmail.com (S.S.); gilbertaudira@yahoo.com (G.A.)
[2] Department of Bioscience Technology, Chung Yuan Christian University, Chung-Li 32023, Taiwan; siregar.petrus27@gmail.com (P.S.); stliang3@gmail.com (S.-T.L.)
[3] Department of Biomedical Engineering, Chung Yuan Christian University, Chung-Li 32023, Taiwan; nemi.malhotra@gmail.com
[4] Department of Chemistry, Chinese Culture University, Taipei 11114, Taiwan; lyh21@ulive.pccu.edu.tw
[5] Department of Biological Science & Technology, College of Medicine, I-Shou University, Kaohsiung 82445, Taiwan
[6] Department of Applied Chemistry, National Pingtung University, Pingtung 90003, Taiwan
[7] Center for Nanotechnology, Chung Yuan Christian University, Chung-Li 32023, Taiwan
* Correspondence: jrchen@isu.edu.tw (J.-R.C.); kelvin@mail.nptu.edu.tw (K.H.-C.C.); cdhsiao@cycu.edu.tw (C.-D.H.); Tel.: +886-7-6151100#7320 (J.-R.C.); Tel.: +886-8-7663800#33254 (K.H.-C.C.); +886-3-2653545 (C.-D.H.)
† These authors contributed equally to this work.

Received: 19 January 2020; Accepted: 16 February 2020; Published: 19 February 2020

Abstract: Plastic pollution is a growing global emergency and it could serve as a geological indicator of the Anthropocene era. Microplastics are potentially more hazardous than macroplastics, as the former can permeate biological membranes. The toxicity of microplastic exposure on humans and aquatic organisms has been documented, but the toxicity and behavioral changes of nanoplastics (NPs) in mammals are scarce. In spite of their small size, nanoplastics have an enormous surface area, which bears the potential to bind even bigger amounts of toxic compounds in comparison to microplastics. Here, we used polystyrene nanoplastics (PS-NPs) (diameter size at ~70 nm) to investigate the neurobehavioral alterations, tissue distribution, accumulation, and specific health risk of nanoplastics in adult zebrafish. The results demonstrated that PS-NPs accumulated in gonads, intestine, liver, and brain with a tissue distribution pattern that was greatly dependent on the size and shape of the NPs particle. Importantly, an analysis of multiple behavior endpoints and different biochemical biomarkers evidenced that PS-NPs exposure induced disturbance of lipid and energy metabolism as well as oxidative stress and tissue accumulation. Pronounced behavior alterations in their locomotion activity, aggressiveness, shoal formation, and predator avoidance behavior were exhibited by the high concentration of the PS-NPs group, along with the dysregulated circadian rhythm locomotion activity after its chronic exposure. Moreover, several important neurotransmitter biomarkers for neurotoxicity investigation were significantly altered after one week of PS-NPs exposure and these significant changes may indicate the potential toxicity from PS-NPs exposure. In addition, after ~1-month incubation, the fluorescence spectroscopy results revealed the accumulation and distribution of PS-NPs across zebrafish tissues, especially in gonads, which would possibly further affect fish reproductive function. Overall, our results provided new evidence for the adverse consequences of PS-NPs-induced behavioral dysregulation and changes at the molecular level that eventually reduce the survival fitness of zebrafish in the ecosystem.

Int. J. Mol. Sci. **2020**, *21*, 1410

Keywords: polystyrene; nanoplastics; neurotoxicity; behavior test; ecotoxicity; oxidative stress; zebrafish

---

1. Introduction

Approximately 300 million tons of plastic waste is being produced every year, which is equivalent to the weight of the entire human population. If the current trend continues, by 2050 the plastic industry could account for 20% of the world's total oil consumption. Only 9% of plastic waste ever produced was recycled and 12% was incinerated, while the remaining 79% accumulated in landfills, dumps, or the natural environment [1]. However, plastic waste whether in a river, an ocean, or on land, can persist in the environment for centuries. Plastic debris can come in all shapes and sizes. Macroplastic is relatively large particles of plastic (typically more than about 5 mm) found, especially in the marine environment. Those that are less than five millimeters in length are called microplastics. Macroplastic is clearly visible plastic that can be caught and generally will not have a direct impact on the food chain. Farm animals or fish that mistake them for food swallow the majority of these tiny plastic particles and, thus, can find their way into the human body [2].

However, the potential risks that are associated with microplastics exposure are unknown for both wildlife and humans. Critical information—long/short mode of exposure and effects data needed for the risk assessments—are still lacking. Human populations could be exposed to microplastics directly from the food or the environment. Multiple studies have documented that microplastics are globally dispersed in marine sediments, oceans, and shorelines in the form of debris that is generated from the degradation/hydrolysis or buoyant Styrofoam of plastics [3,4]. Microplastics can disrupt the food biota and environmental health due to their poor degradation capability and small size [5,6]. Likewise, the toxic effects of smaller sized nanoplastics are also very plausible, however, there are limited relevant data and their research is in its infancy [7,8]. Thus far, it has been demonstrated that microplastics can be ingested and accumulated by small aquatic organisms, such as zooplankton (*Daphnia magna, Centropages typicus*) and shrimp (*Crangon cargon*), subsequently being transferred to the large organism, such as sea turtles [9,10]. Another recent study on water flea (*Ceriodaphinia dubia*) reported that the acute and chronic exposure of polyester fibers resulted in significant effects on reproduction and survival [11]. In fish, microplastics have been found to cause major adverse effects, including hepatic stress, intestinal damages, and oxidative stress, among others [12–14]. Moreover, Karami et al. also found that the potential presence of microplastics in dried fish tissue: viscera and gills [15]. In embryos and adult fish, microplastics have been reported translocating from the digestive tract to gills and liver of zebrafish [12]. Together, these studies suggested that microplastics could exhibit toxicity in living organisms, the challenge could be extensive due to ubiquity in the environment and the translocation of potentially moving particles to animal body parts, usually consumed by humans.

The most abundant microplastics that are found in marine waters are low-density polyolefins and polystyrene [16,17]. Microplastics can be ingested by organisms and accumulated for a long time because of their tiny size and poor biodegradability. It was reported that microplastics can induce reactive oxygen species (ROS) formation and oxidative stress in mussels and lugworms [18,19]. In addition, microplastics can also induce the regulation of phagocytic activity of immune cells in worms [19]. However, the mechanism of the toxicity of microplastics is unclear and it needs further investigation. Enzymatic analysis and histology are widely used to determine the toxic effects of microplastics [20]. In all of these studies, the major factor was set on oral ingestion of microplastics, and only a few pieces of researches included other modalities for microplastic exposure. A study reported that, upon exposure, clean microspheres were not only found in the intestinal tract but also in the gills of the shore crab (*Carcinus maenas*) after 21 days of exposure [21]. Nanoplastics have an enormous surface area, bearing the potential to bind even bigger amounts of toxic compounds than microplastics.

Currently, very little data are available regarding the effects of nanoplastics on the development of aquatic organisms, whereas most of the studies have focused on the chemical aspects of nanoplastics. Indeed, the toxic effects of leachate from particles have been analyzed in different organisms, such as sea urchins [22] and mussels [22], whereas the behavioral and physical effects are still poorly explored [23]. To fill this gap, we evaluated the effects of environmentally relevant concentrations of nanoplastics on adult zebrafish behavior and biochemistry to assess the ecological toxicity of polystyrene nanoplastics. Polystyrene is a major kind of plastic of high production volume as well as the primary components of plastic debris contaminating the environment [24]. So far, no evidence exists regarding the potential effect of polystyrene nanoplastics on adult zebrafish behavior. In the present study, adult zebrafish were used to investigate the neurobehavioral changes that are induced by polystyrene nanoplastics and their distribution and accumulation in various tissues. Neurotransmitter amounts and enzyme activities for wither dopamine (DA), acetylcholine (ACh), serotonin (5-HT), melatonin (MT), γ-aminobutyric acid (GABA), Cytochrome P450, family 1, subfamily A, polypeptide 1 (CYP1A1), reactive oxygen species (ROS), or creatinine kinase (CK) contents were quantified to explore the toxicity mechanism.

2. Results

2.1. Overview of Experimental Design

A schematic representation of biological responses in zebrafish in response to polystyrene nanoplastics (PS-NPs) exposure is presented in Figure 1 based on data from this study. Figure 1 shows the shape and size of PS-NPs, as confirmed by transmission electron microscopy (upper panel). The diameter of PS-NPs used in this study was estimated to be around 70 nm. In addition, the Fluorescein isothiocyanate (FITC)-conjugated PS-NPs was used to trace the tissue distribution after waterborne exposure. The adult zebrafish aged six to seven months old were incubated with PS-NPs at doses of 0.5 and 1.5 ppm for around seven consecutive days and five behavioral tests (novel tank, mirror biting, predator avoidance, social interaction, and shoaling) were performed to measure its consequent behavioral alterations (lower left panel, highlighted in red). In addition, circadian rhythm locomotion activity test was performed after ~7 weeks incubation of 5 ppm PS-NPs to observe the long-term effect of high dose PS-NPs on zebrafish sleep/wake behavior. Finally, after all of the behavioral tests, excluding circadian rhythm locomotion activity test, were done, zebrafish were sacrificed and the brain, the liver, and the muscle tissues were dissected for biochemical marker measurement (lower right panel, highlighted in blue). For PS-NPs distribution, the brain, liver, male gonad, and intestine were collected and subjected to indirect PS-NPs quantification based on green fluorescence signal intensity.

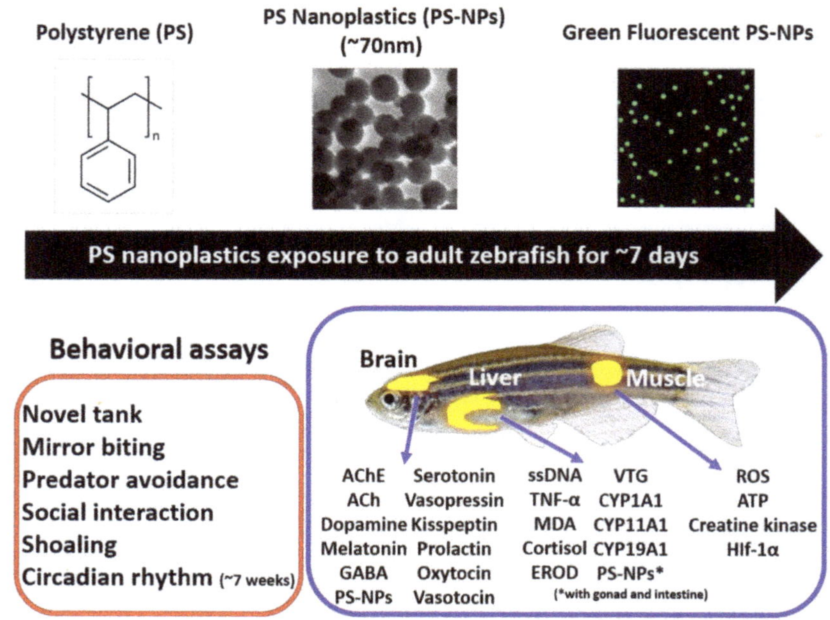

Figure 1. The experimental design to evaluate the ecotoxicity of polystyrene nanoplastics (PS-NPs). The chemical structure and particle size of polystyrene used in this study were summarized in the upper panel. The behavioral toxicity assays for PS-NPs were summarized in the left-right panel (red color). The biochemical endpoints for PS-NPs toxicity were summarized in the right lower panel (blue color).

2.2. PS-NPs Exposure Reduced Average Speed and Exploration Behavior in Zebrafish

The locomotion activity and exploratory behavior were evaluated while using a novel tank test in PS-NPs-treated fishes at two different concentrations. Novel tank test is a behavior test to assess the zebrafish exploratory nature. This test exploits the natural tendency of zebrafish to initially diving to the bottom part of a novel tank, with a gradual increase in vertical activity over time. Typically, when zebrafishes are moved to a new environment, they display high anxiety and bottom-dwelling behavior, but they will start to explore the tank and move towards the upper area of the tank with anxiety relief once they adapt to the new environment (Figure 2G,J) [25]. In this test, we measured six important zebrafish behavior endpoints representing their swimming activity and exploratory behavior. The result showed that adult fish that were exposed to 0.5 ppm of PS-NPs showed no difference in their swimming activity when compared to the controls while hyperactivity-like behavior shown by the 1.5 ppm concentration group, supported by the high level of average speed and low level of freezing time movement ratio (Figure 2A,B). Furthermore, alteration in their exploratory behavior was observed in the high concentration of the PS-NPs-treated group. This abnormality was indicated by an increase in the number of entries in the top and total distance traveled in the top (Figure 2C,D,I,L), even though there was no significant difference shown in the time in top duration and latency to enter the top endpoints (Figure 2E,F). On another hand, this behavior alteration was not exhibited by the low concentration group fish (Figure 2C,D,H,K). Figure 2G–L and Supplementary Video S1 display the locomotion trajectories and behavioral changes for the control, 0.5 and 1.5 ppm PS-NPs exposed fish in the novel tank test.

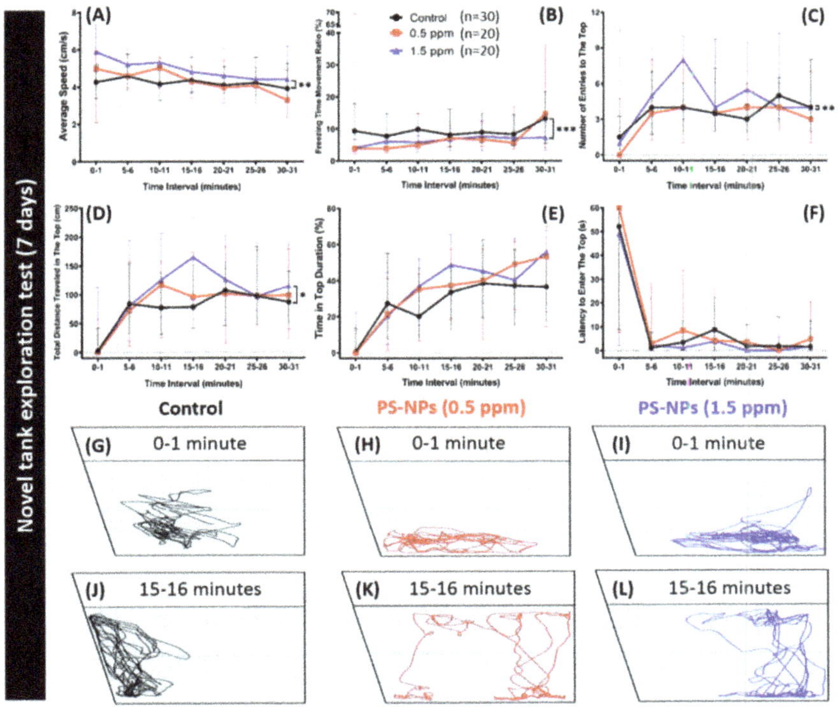

Figure 2. Novel tank behavior endpoints comparisons between control and polystyrene nanoplastics (PS-NPs)-exposed zebrafish groups after a ~7-day exposure. (**A**) Average speed, (**B**) freezing time movement ratio, (**C**) number of entries to the top, (**D**) total distance traveled in the top, (**E**) time in top duration, and (**F**) latency to enter the top were analyzed. The 1 min. locomotion trajectories for the control, 0.5 and 1.5 ppm PS-NPs exposed fish in the novel tank test were presented in (**G** to **L**), respectively. The black line represents the control group, the red line represents the low concentration PS-NPs group (0.5 ppm), and the blue line represents the high concentration PS-NPs group (1.5 ppm). The data are expressed as the median with interquartile range and were analyzed by a Kruskal–Wallis test, which continued with Dunn's multiple comparisons test as a follow-up test (n = 30 for control; n = 20 for each PS-NPs-exposed group; * p < 0.05, ** p < 0.01, *** p < 0.001).

2.3. PS-NPs Exposure Reduced Aggression and Predator Avoidance in Zebrafish

A mirror biting test was conducted to determine the aggressive nature of zebrafish altered by the exposure of PS-NPs. The mirror biting assay is a simple and efficient method to test fish aggressiveness in terms of the frequency of the tested fish to bite their mirror images [26]. This test might also indicate a more general measure of social motivation or the intent to interact with a social partner [27]. In this test, chronic exposure of PS-NPs in both concentrations significantly reduced zebrafish aggressiveness. This phenomenon is shown by a decrease in the mirror biting time percentage and in the longest duration in the mirror side (Figure 3A,B). Interestingly, both concentrations of PS-NPs exposure reduced their locomotion behavior, as indicated by a decrease in the average speed, swimming time movement ratio, and rapid movement ratio, and an increase in the freezing time movement ratio as compared to the control group (Figure 3C–F). This result suggests that both 0.5 and 1.5 ppm of PS-NPs exposure reduce the aggressiveness in zebrafish. Figure 3G–I and Supplementary Video S2 display the locomotion trajectories and behavioral changes for the control, 0.5 and 1.5 ppm PS-NPs-exposed fish in the mirror biting test.

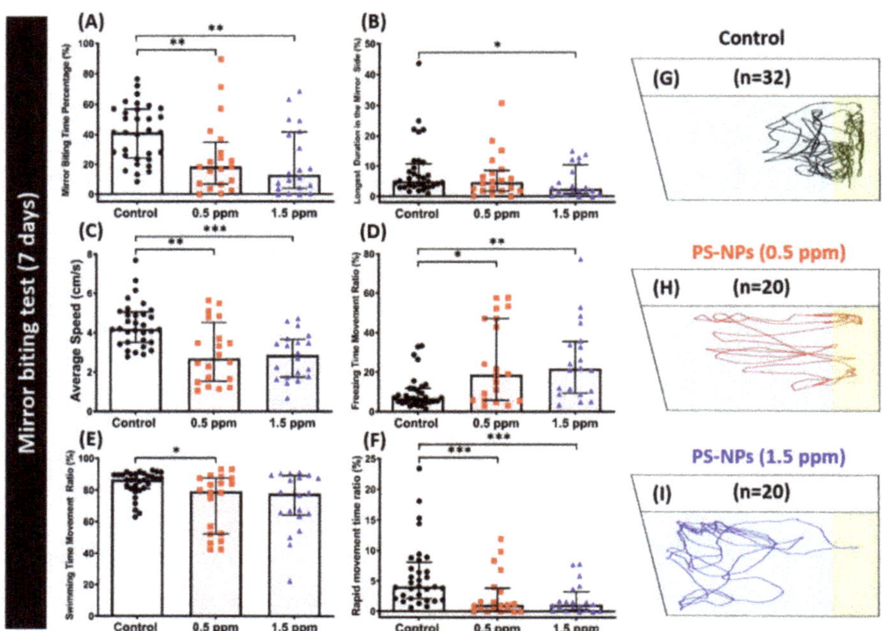

Figure 3. Mirror biting behavior endpoints comparisons between control and polystyrene nanoplastics (PS-NPs)-exposed zebrafish groups after a ~7-day exposure. (**A**) Mirror biting time percentage, (**B**) longest duration in the mirror side, (**C**) average speed, (**D**) freezing time movement ratio, (**E**) swimming time movement ratio, and (**F**) rapid movement time ratio were analyzed for mirror biting assay. The 1 min. locomotion trajectories for the control, 0.5 and 1.5 ppm PS-NPs-exposed fish in mirror biting tests were presented in **G** to **I**, respectively, with the yellow-colored zone as the mirror biting region. The data are expressed as the median with interquartile range and were analyzed by a Kruskal–Wallis test, which continued with Dunn's multiple comparisons test as a follow-up test ($n = 32$ for control; $n = 20$ for each PS-NPs-exposed group; * $p < 0.05$, ** $p < 0.01$, *** $p < 0.001$).

Fear is a collection of behavioral responses that are elicited by negative stimuli that are associated with imminent danger, such as the presence of a predator. Predator avoidance is an innate response for fish when facing their natural predator by showing high anxiety or even freezing behavior. Zebrafish have an innate response as freezing or anxiety when exposed to the sight of a natural predator. Therefore, this response is helpful in examining certain behavior alterations during the course of revelation to the predator [28]. In the predator avoidance test, we assessed zebrafish fear reactions, including predator avoidance behavior. We exposed the different groups of zebrafish to the predator fish convict cichlid (*Amatitlania nigrofasciata*). From the results, we found that exposure of PS-NPs at 0.5 ppm did not alter zebrafish fear response to the predator, as shown in Figure 4. No significant difference was detected in all of the endpoints measured during the test between control and 0.5 ppm PS-NPs-exposed fish group (Figure 4A–H). However, exposure of PS-NPs at a higher concentration (1.5 ppm) did alter their fear response to the convict cichlid. This difference was indicated by the low level of the average distance to separator between zebrafish and predator fish exhibited by high concentration PS-NPs-exposed fish (Figure 4B). In addition, there was also a slight increment in their predator approaching time, even though it did not reach a statistically significant difference (Figure 4A). Furthermore, we found a difference in 1.5 ppm PS-NPs-exposed fish movement types with the control fish, even though there was no significant difference in the average speed (Figure 4C). This difference was indicated by a high level of swimming time movement ratio, while there were no significant differences in their freezing and rapid movement time ratios (Figure 4D–F). These results suggest that

1.5 ppm of PS-NPs exposure reduces the predator avoidance behavior performance in zebrafish. In addition, Figure 4G–I and Supplementary Video S3 illustrate the locomotion trajectories and behavioral changes for the control, 0.5 and 1.5 ppm PS-NPs-exposed fish in the predator avoidance test.

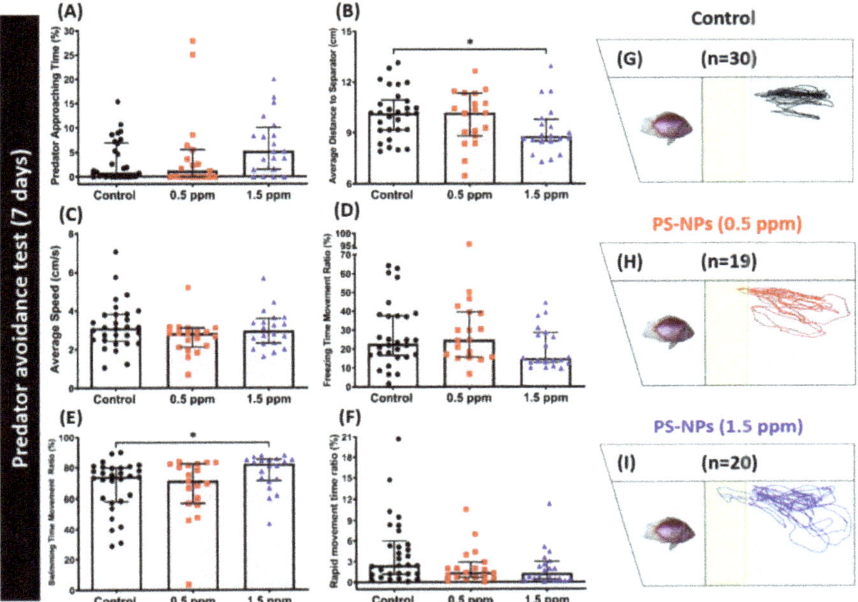

Figure 4. Predator avoidance behavior endpoints comparisons between control and polystyrene nanoplastics (PS-NPs)-exposed zebrafish groups after a ~7-day exposure. (**A**) Predator approaching time percentage, (**B**) average distance to the separator, (**C**) average speed, (**D**) freezing time movement ratio, (**E**) swimming time movement ratio, and (**F**) rapid movement time ratio were analyzed. The 1 min. locomotion trajectories for the control, 0.5 and 1.5 ppm PS-NPs exposed fish in the predator avoidance test were presented in **G** to **I**, respectively, with the yellow-colored zone as the predator approaching region. The data are expressed as the median with interquartile range and were analyzed by a Kruskal–Wallis test, which continued with Dunn's multiple comparisons test as a follow-up test (n = 30 for control; n = 19 for 0.5 ppm PS-NPs-exposure fish; n = 20 for 1.5 ppm MP-exposure fish; p < 0.05).

2.4. PS-NPs Exposure Did Not Alter Conspecific Social Behavior in Zebrafish

We assessed zebrafish social nature by conspecific interactions by performing social interaction tests. We tested their social interest to another conspecific after a one-week exposure of PS-NPs. Here, we did not find any significant change between the experimental and untreated groups in terms of interaction time percentage (Figure 5A), longest duration in separator side (Figure 5B), and average distance to the separator (Figure 5D). However, the average speed was increased in the experimental groups when compared to untreated controls (Figure 5C), which is consistent with the novel tank test result. This result suggests that both 0.5 and 1.5 ppm of PS-NPs exposure does not change the conspecific interaction in zebrafish. Figure 5E–G and Supplementary Video S4 illustrate the locomotion trajectories and behavioral changes for the control, 0.5 and 1.5 ppm PS-NPs-exposed fish in the social interaction test.

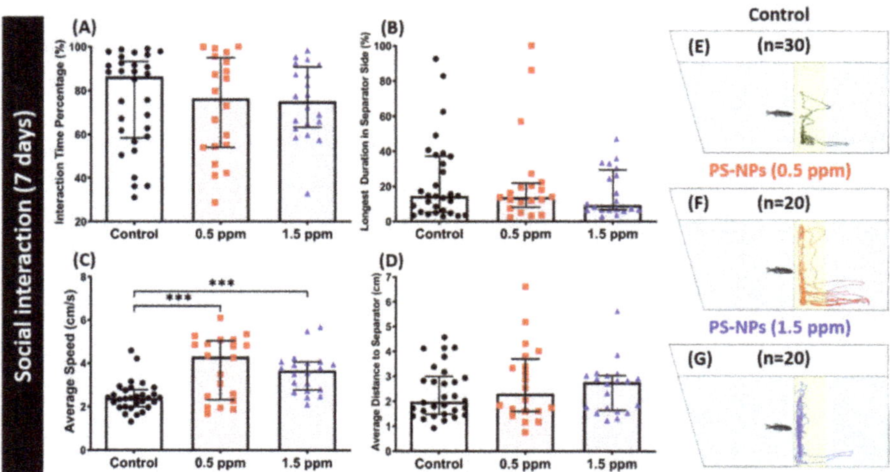

Figure 5. Social interaction behavior endpoints comparisons between control and polystyrene nanoplastics (PS-NPs)-exposed zebrafish groups after a ~7-day exposure. (**A**) Interaction time percentage, (**B**) longest duration in the separator, (**C**) average speed, and (**D**) average distance to separator side were analyzed for social interaction assay. The 1 min. locomotion trajectories for the control, 0.5 and 1.5 ppm PS-NPs exposed fish in mirror biting tests were presented in **E**, **F**, **G**, respectively with the yellow-colored zone as the conspecific interaction region. The data are expressed as the median with interquartile range and were analyzed by a Kruskal-Wallis test, which continued with Dunn's multiple comparisons test as a follow-up test (*n* = 30 for control; *n* = 20 for each PS-NPs-exposed group).

2.5. PS-NPs Exposure Tighten the Shoal in Zebrafish

Shoaling is an inherently social behavior of zebrafish. Shoaling is an innate behavior for fish to swim together to reduce anxiety and the risk of being captured by predators. When zebrafish sense a certain kind of threat or when challenged in a situation they tend to avoid, they usually swim into very tight groups together. Thus, the shoaling test can specify certain neurological behaviors of zebrafish after PS-NPs exposure. Shoaling nature provides the individual fish with multiple benefits, including efficient foraging, defense against predators, and access to mates [29,30]. From the result, we found tight shoal were formed by the PS-NPs-treated fish. This phenomenon was supported by a decrease in average inter fish distance, average nearest neighbor area, and average farthest neighbor distance showed in both the 0.5 ppm and 1.5 ppm PS-NPs-treated groups (Figure 6C,E,F). In addition, there was a slight decrement in their average shoal area, even though it did not reach a statistically significant difference (Figure 6D). Furthermore, 1.5 ppm of PS-NPs exposure reduced fish group exploratory behavior, which was indicated by a decrease in time spent at the top portion of the tank (Figure 6B). However, there was no significant difference regarding their locomotion activity observed during this test (Figure 6A). This result suggests both 0.5 and 1.5 ppm of PS-NPs exposure trigger a tight shoaling behavior in zebrafish. Figure 5G–I and Supplementary Video S5 illustrate the locomotion trajectories and behavioral changes for the control, 0.5 and 1.5 ppm PS-NPs-exposed fish in the shoaling test.

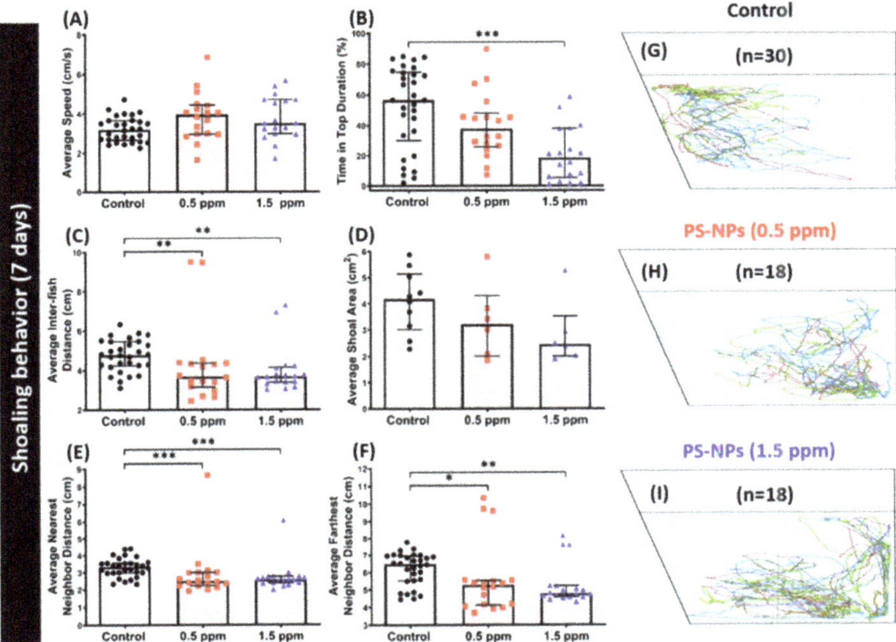

Figure 6. Shoaling behavior endpoint comparisons between the control and polystyrene nanoplastics (PS-NPs)-exposed zebrafish groups after a ~7-day exposure. (**A**) Average speed, (**B**) time in top duration, (**C**) average inter-fish distance, (**D**) average shoal area, (**E**) average nearest neighbor distance, and (**F**) average farthest neighbor distance, were analyzed. Groups of three fish were tested for shoaling behavior. The 1 min. locomotion trajectories for the control, 0.5 and 1.5 ppm PS-NPs exposed fish in shoaling tests were presented in **G**, **H**, **I**, respectively. The data are expressed as the median with interquartile range were analyzed by a Kruskal–Wallis test, which continued with Dunn's multiple comparisons test as a follow-up test ($n = 30$ for control; $n = 18$ for PS-NP-exposed fish; * $p < 0.05$, ** $p < 0.01$, *** $p < 0.005$).

2.6. High Dose of PS-NPs Exposure Dysregulated the Circadian Rhythm

Circadian rhythm genes and their regulation are well documented, particularly in the zebrafish brain and in pineal independently in several vertebrates [31,32]. In this test, we assessed zebrafish circadian rhythm locomotion activity after a high dose of PS-NPs was exposed chronically. The result showed a chronic exposure of 5 ppm PS-NPs dysregulated their circadian rhythm locomotion activity (Figure 7A). In the light cycle, the reduction of locomotion activity and abnormal movement orientation were observed in the treated fish, which was indicated by the significant reduction in average speed and rapid movement time ratio of PS-NPs-exposed fishes (Figure 7B,G), followed by the increment in meandering and freezing movement time ratio (Figure 7D,E). However, there were no significant differences found in the average angular velocity and swimming movement time ratio between the control and treated groups (Figure 7C,F). Furthermore, a similar phenomenon was also shown during the dark cycle. In the dark cycle, more robust hypoactivity behavior was exhibited by the treated fish, supported by the reduction of average speed, average angular velocity, swimming movement ratio, and rapid movement ratio of the PS-NPs-treated fish as compared to the controls (Figure 7H–I,L–M). In addition, the high level of freezing movement time ratio might also support this behavior alteration (Figure 7K). However, there was no statistical difference in their meandering during the dark cycle (Figure 7J). This result suggests that the chronic exposure of 5 ppm PS-NPs dysregulates circadian rhythm locomotion activity in zebrafish.

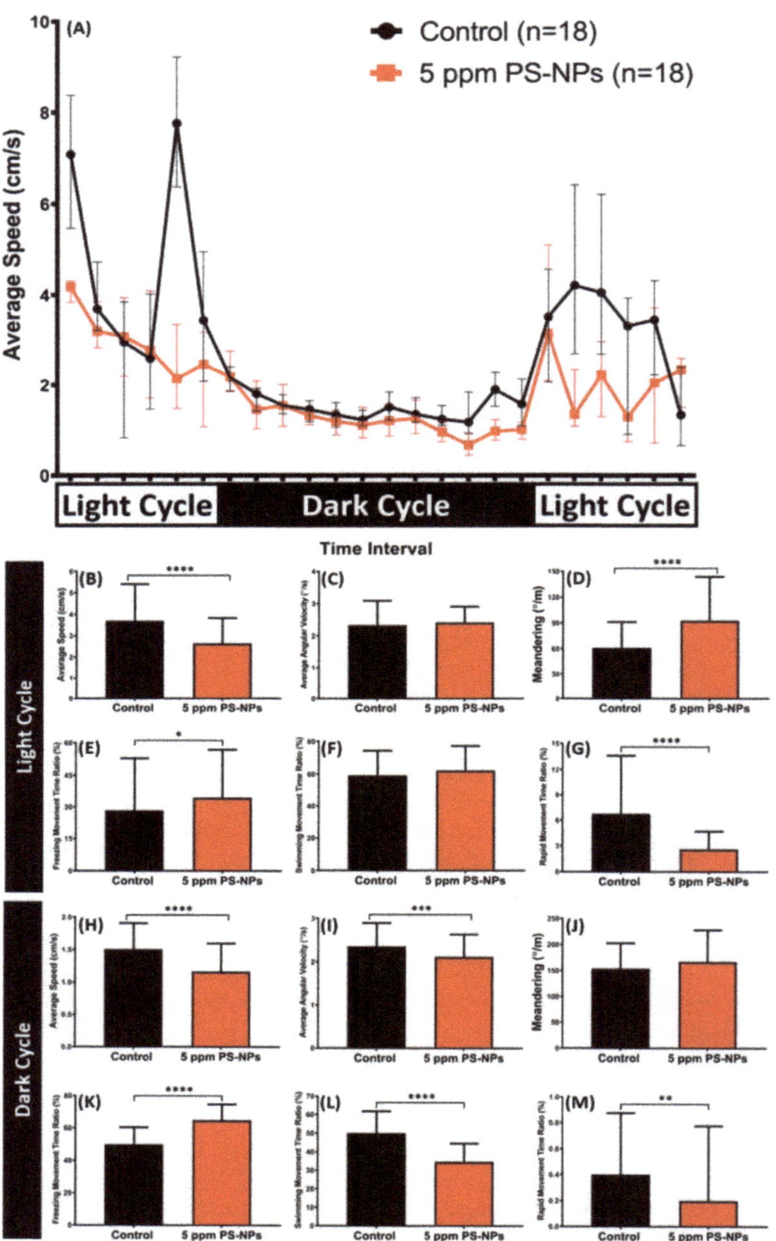

Figure 7. The circadian rhythm locomotion activity assay for control and 5 ppm polystyrene nanoplastics (PS-NPs)-exposed zebrafish groups after a seven-week exposure. (**A**) Comparison of the time chronology changes of the average speed between the control and the PS-NPs-exposed fish in light and dark cycles. The white area shows the light period and the black area shows the dark period. Comparison of the average speed (**B,H**), average angular velocity (**C,I**), and meandering (**D,J**), freezing movement time ratio (**E,K**), swimming movement time ratio (**F,L**), and rapid movement ratio (**G,M**) during the light and dark cycles, respectively. The data are expressed as the median with interquartile range and were analyzed by Mann–Whitney test ($n = 18$ for control; $n = 18$ for PS-NPs-exposed fish; * $p < 0.05$, ** $p < 0.01$, *** $p < 0.001$, ****, $p < 0.0001$).

2.7. Tissue Distribution of Fluorescent PS-NPs in Zebrafish

We performed fluorescence spectroscopy analysis on the treated tissues, which were collected from the emission spectra 489 nm to 590 nm, at an excitation wavelength of 485 nm. We found strong green fluorescence in the PS-NPs-treated tissues as compared to the controls (Figure 8). The green fluorescence was predominantly seen in gonads (can reach around 15 ug/mg total protein) and intestine, and other tissues, including liver and brains (around five times less than those in the gonads). This result demonstrates that the green-labeled PS-NPs can accumulate in various tissues, including the nervous, digestive, and reproductive organs after long term exposure.

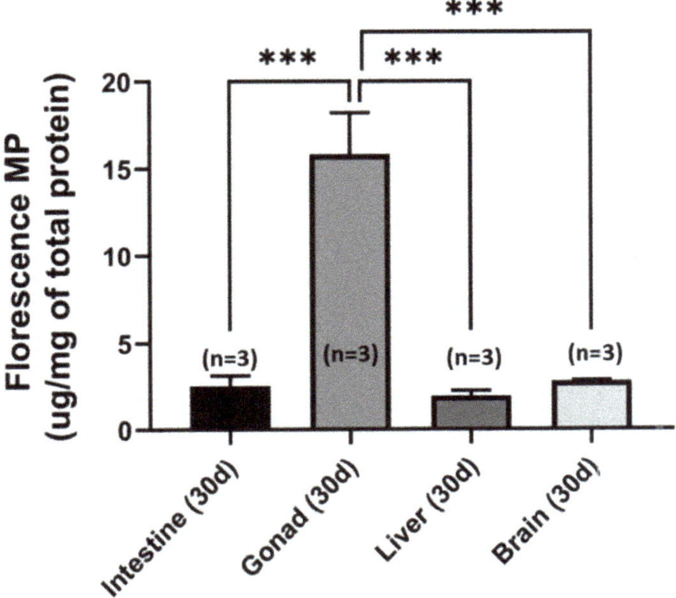

Figure 8. Comparison of the tissue distribution of green fluorescence-labeled nanoplastics among different tissues in zebrafish after ~30 days of polystyrene nanoplastics (PS-NPs) exposure. The data are expressed as the mean ± SEM and they were analyzed by One-way ANOVA, which continued with post hoc analysis ($n = 3$; *** $p < 0.001$).

2.8. Measurement of Marker Expression in Muscle and Liver after PS-NPs Exposure

Several biochemical parameters that were related to oxidative stress, energy and lipid metabolisms, oxygen uptake, DNA damage, inflammatory response, and environmental toxic responses were examined in muscle and liver tissues of the PS-NPs-exposed fish to better understand the corresponding toxic mechanism induced by PS-NPs (summarized in Table 1). From the muscle tissues, we examined oxidative stress, energy metabolism, and oxygen uptake changes after PS-NPs exposure. Regarding oxidative stress, we found that the relative amount of reactive oxygen species (ROS) level increased in the fish exposed to the high concentration of PS-NPs while we found that a high concentration of PS-NPs induced a significant decrease in ATP level for energy metabolism. On another hand, a significant change was absent in both treated groups regarding their level of creatine kinase, a muscle energy marker. Furthermore, from this tissue, we also found that Hif-1α, a key marker for hypoxia, was not changed after the exposure.

Table 1. Various biomarker analyses in the muscle, liver, and brain tissues for adult zebrafish exposed to low and high doses of polystyrene nanoparticles (PS-NPs). The data are expressed as the mean ± SEM and analyzed by One-way ANOVA continued with post hoc analysis.

Biomarker	WT	PS-NPs (0.5 ppm)		PS-NPs (1.5 ppm)		Unit	Significance	ANOVA F Value	p Value
Muscle									
ROS	4.53 ± 0.3	5.27 ± 0.9	NS	7.31 ± 0.32	*	U/ug of total protein	YES	F (2, 6) =5.862	$p = 0.0388$
ATP	296.4 ± 5.80	290.3 ± 4.36	NS	242.4 ± 3.78	***	ng/ug of total protein	YES	F (2,6) = 39.27	$p = 0.0004$
Creatine kinase	6.70 ± 0.11	7.50 ± 0.78	NS	6.88 ± 0.39	NS	pg/ug of total protein	NO	F (2,6) = 0.468	$p = 0.646$
Hif-1α	18.77 ± 0.55	21.07 ± 1.6	NS	22.26 ± 0.30	NS	pg/ug of total protein	NO	F (2,6) = 2.974	$p = 0.126$
Liver									
ssDNA	2.63 ± 0.29	2.28 ± 0.61	NS	5.13 ± 0.58	*	U/ug of total protein	YES	F (2,6) = 9.0	$p = 0.01$
TNF-α	21.71 ± 1.67	17.59 ± 3.37	NS	45.26 ± 8.20	*	pg/ug of total protein	YES	F (2,6) = 8.20	$p = 0.01$
MDA	0.32 ± 0.08	0.20 ± 0.06	NS	0.27 ± 0.06	NS	pg/ug of total protein	NO	F (2,6) = 0.84	$p = 0.47$
Cortisol	64.76 ± 5.50	57.35 ± 5.50	NS	88.5 ± 6.93	*	pg/ug of total protein	YES	F (2,6) = 9.60	$p = 0.01$
EROD	362.5 ± 45.89	315.2 ± 29.00	NS	319.7 ± 8.40	NS	U/ug of total protein	NO	F (2,6) = 0.67	$p = 0.5$
VTG	32.14 ± 2.5	54.19 ± 3.8	**	68.34 ± 2.2	***	ng/ug of total protein	YES	F (2,6) = 37.82	$p = 0.0004$
CYP1A1	2.4 ± 0.62	1.01 ± 0.24	NS	0.77 ± 0.03	*	ng/ug of total protein	YES	F (2,6) = 5.42	$p = 0.045$
CYP11A1	0.30 ± 0.03	0.27 ± 0.02	NS	0.50 ± 0.05	*	ng/ug of total protein	YES	F (2,6) = 9.64	$p = 0.013$
CYP19A1	0.42 ± 0.07	0.28 ± 0.03	NS	0.77 ± 0.01	**	ng/ug of total protein	YES	F (2,6) = 25.38	$p = 0.001$
Brain									
AChE	19.48 ± 0.90	16.73 ± 0.73	*	13.5 ± 0.02	**	U/ug of total protein	YES	F (2,6) = 20.86	$p = 0.002$
Acetylcholine	32.24 ± 1.80	30.02 ± 1.90	NS	29.54 ± 1.20	NS	U/ug of total protein	NO	F (2,6) = 14.16	$p = 0.005$

Table 1. *Cont.*

Biomarker	WT	PS-NPs (0.5 ppm)		PS-NPs (1.5 ppm)		Unit	Significance	ANOVA *F* Value	*p* Value
Dopamine	66.80 ± 2.80	60.57 ± 2.25	NS	50.42 ± 2.80	**	pg/ug of total protein	YES	F (2,6) = 9.10	$p = 0.01$
Melatonin	7.91 ± 0.12	6.64 ± 0.06	**	5.64 ± 0.23	***	pg/ug of total protein	YES	F (2,6) = 52.11	$p = 0.0002$
GABA	0.26 ± 0.01	0.23 ± 0.00	NS	0.19 ± 0.00	**	U/ug of total protein	YES	F (2,6) = 19.48	$p = 0.002$
5-HT	0.85 ± 0.05	0.72 ± 0.01	NS	0.58 ± 0.03	**	ng/ug of total protein	YES	F (2,6) = 14.16	$p = 0.005$
Vassopressin	4.18 ± 0.23	3.28 ± 0.21	*	2.47 ± 0.13	**	ng/ug of total protein	YES	F (2,6) = 18.58	$p = 0.002$
Kisspeptin	11.72 ± 0.91	9.53 ± 0.18	NS	7.49 ± 0.31	**	ng/ug of total protein	YES	F (2,6) = 13.91	$p = 0.005$
PRL	60.45 ± 6.9	39.46 ± 6.9	NS	49.01 ± 1.9	NS	ng/ug of total protein	NO	F (2,6) = 3.26	$p = 0.10$
Oxytocin	24.86 ± 1.7	19.85 ± 0.17	NS	15.9 ± 1.43	**	pg/ug of total protein	YES	F (2,6) = 12.04	$p = 0.007$
Vasotocin	381 ± 15.29	364 ± 13.37	NS	419 ± 30.58	NS	pg/ug of total protein	NO	F (2,6) = 1.78	$p = 0.24$

NS, no significant; significant level was test by one-way ANOVA and post hoc analysis.

Next, the effects of PS-NPs exposure on DNA damage, inflammation, and lipid peroxidation were evaluated by measuring several important biomarkers, such as single-stranded DNA (ssDNA), TNF-α, and malondialdehyde (MDA) from the liver tissue. Regarding the DNA damage, the relative amount of ssDNA was increased in the high concentration group. Furthermore, in the 0.5 ppm PS-NPs-treated group, the relative amount of TNF-α, an inflammatory marker, was not significantly different from those of the control. However, the relative amount of those markers in the liver of 1.5 ppm PS-NPs-treated group were significantly higher than those of the control. In addition, MDA, which is a marker for lipid peroxidation, the relative amount from the liver tissue was unchanged in both groups. Several biomarkers for chemical exposure-response were also measured from the liver tissues, including cortisol and ethoxyresorufin-O-deethylase (EROD). From the result, we found a significant increment in the cortisol level exhibited by the high concentration group only. However, there were no significant differences regarding the EROD level that was observed in all of the groups. In addition, we also found increments in the vitellogenin (VTG), a biomarker for environmental estrogens, levels in both treated groups. Later, we analyzed the regulation of three isoforms of cytochrome P450, CYP enzymes (CYP1A1, CYP11A1, and CYP19A1) in zebrafish liver after the PS-NPs exposure. Interestingly, all three CYP isoenzyme expressions were significantly elevated in 1.5 ppm PS-NPs-exposed zebrafish when compared to the control.

2.9. Measurement of Neurotransmitter Expression in Brain after PS-NPs Exposure

Neurotransmitters affect a wide variety of both physical and psychological functions, including heart rate, sleep, appetite, and behaviors, such as mood and fear. The expression of neurotransmitters in the brain was measured biochemically while using enzyme-linked immunosorbent assay (ELISA) to investigate PS-NPs exposure to neurotransmitters. Fixed amount of the total soluble protein in brain was subjected to ELISA to determine the expression level of neurotransmitters, such as acetylcholine esterase (AChE), acetylcholine (ACh), dopamine (DA), melatonin, γ-aminobutyric acid (GABA), serotonin (5-HT), vasopressin, kisspeptin, prolactin (PRL), oxytocin, and vasotocin. From the result, the activity of AChE was significantly inhibited in the 1.5 ppm PS-NPs group (Table 1). Interestingly, the relative amount of ACh did not significantly change in both of the treated groups. Furthermore, the other neurotransmitters, such as DA, melatonin, GABA, 5-HT, vasopressin, kisspeptin, and oxytocin, were also significantly decreased after exposure to PS-NPs, especially in the high concentration group. However, the relative amount of PRL and vasotocin did not show any appreciable change in the treatment groups.

3. Discussion

3.1. Microplastic Pollutions

Over the past few years, very high concentrations of microplastics have been detected in freshwater bodies (0–1×10^6 items/m^3) [33] and oceans (0–1×10^4 items/m^3) [34]. It was calculated that between 4.8 and 12.7 million metric tons of plastic waste disposed into the ocean in 2010 and this mass could drastically increase by one order of magnitude by 2025 [1]. Microplastics enter the ecosystem from many sources, including clothing, industrial processes in the form of microbeads, plastic pellets, and microfibers that degraded from plastic bags and fishing nets. Plastics take hundreds to thousands of years for degradation, thus increasing their probability of being ingested and accumulated in bodies and tissues of water organisms. The evidence of this incorporation of microplastics in the organism bodies has already been observed in various species of animals, particularly fishes and mussels that are commonly used for human consumption. It is becoming increasingly evident that microplastics can be transmitted through the aquatic food web, which leads to biological accumulation. The entire movement and life cycle of microplastics in the environment is still under research, in a critical review and assessment of data quality of the occurrence studies of microplastics, Dr. Koelmans and colleagues at the Wageningen University, the Netherlands assessed the quality of fifty studies researching microplastics in drinking water and its major freshwater sources. They concluded that more high-quality data are needed on the occurrence of microplastics in drinking water, to better understand potential exposure and inform human health risk assessments [35]. Additionally, in a study in *Mytilus edulis*, the accumulation of microplastics in the form of microfibers varied from 0.9 to 4.6 items/g, where more microplastics were detected in the wild groups than in the farmer's group [36]. In the current study, we evaluated the effects of several different concentrations of PS-NPs on the adult zebrafish behavior and physiological aspects with an environmentally relevant concentration after chronic incubation. The obtained results added new important information regarding the neurotoxic effects of this emerging pollutant on adult zebrafish.

3.2. Inflammatory Protein Expression and Oxidative Stress

We investigated the inflammatory protein expression and measured oxidative stress to understand the exacerbation of PS-NPs induced toxicity in zebrafish. First, we evaluated the level of TNF-α marker as a representative protein of inflammations by ELISA. The 1.5 ppm treatment induced the marker level to rise in liver tissues and triggered a significant immune response, thus suggesting a synergistic effect on inflammation. Furthermore, the lack of anti-inflammatory function observed from liver tissue when fish were exposed to a high concentration of PS-NPs with elevated ssDNA, suggesting an oxidative stress mechanism of toxicity [37]. Reactive oxygen species (ROS) are one of the important features that result from toxicant-induced cell death and they are implicated in the inflammatory response. ROS generation by PS-NPs was examined using the ROS ELISA kit showing ROS production was dramatically increased in the PS-NPs 1.5 ppm treated group, indicating that the presence of PS-NPs synergistically aggravated ROS production. In a prior study, when 0.05, 0.5, & 6 μm polystyrene beads were exposed to rotifers, different sizes of microbeads induced ROS in a size-dependent manner [38]. Besides, the size dependency of microbeads side effects was proved as a consequence of oxidative stress, which is P-JNK, P-p38 activation with increased ROS level [38]. Smaller sized PS-NPs had been shown to induce the activities of ROS and Hif-1α biomarkers. These studies demonstrated that smaller sized particles are more toxic than the larger particles due to their specific surface area [39,40]. Moreover, a high level of ROS might also be due to insufficient nutrition or the inhibition of fish food digestion that is caused by aggregated PS-NPs in fish to a larger extent, although they were not aggregated in the water. In addition, even though the ROS levels were investigated after ~7 days of PS-NPs treatment, the aggregation and accumulation of fluorescence-labeled PS-NPs in fish determined after ~7 weeks of accumulation may also be related to this phenomenon. More future experiments are considered to be conducted to validate this hypothesis. Additionally, polyethylene and polystyrene microplastics were shown to adsorb pyrene in time and dose-dependent manner in *Mytilus galloprovincialis*, depicting cellular deformities with an alteration in oxidative stress, neurotoxic effect, and antioxidant system in a previous study [18]. Therefore, this result might be an important consideration for future research efforts. To conclude, the high concentration and size of nanoplastics appeared to be one of the key determinants of such toxicity, leading to elevated ROS production and cell death, as well as pro-inflammatory responses.

3.3. Energy Metabolism

The toxic effects of PS-NPs in the liver of zebrafish were investigated after ~7 days of exposure. In our study, the altered level of ATP marker in a biochemical assay confirmed the disruption of energy metabolism. Similar studies have reported that in marine worms and copepods, ingested NPs depleted energy reserves [24,41] and particularly affected the feeding behavior of fish [42]. A possible explanation is that the large quantity of ingested PS-NPs without any nutritional value hindered the normal absorption of food. In addition, the dysregulation of food intake can also lead to severe changes in energy and lipid metabolism.

3.4. Reproductive Toxicity of PS-NPs

In a previous study, the 2 and 6 µm diameter and 0.023 mg/L of micro-PS significantly decreased the oocyte number, diameter, and sperm velocity in oysters when exposed for two months, after assessing the reproductive cycle on different eco-physiological parameters [43]. Based on the biochemical analysis of differentially expressed CYP proteins, PS-NPs may directly affect the ovary. Moreover, nanoplastics may cause oxidative stress to the developing follicles and interfere with the expression of key regulatory genes, impairing the function and growth of stage 1/11 follicles. The aborted follicular development, as well as the inhibition of E2 synthesis, could be a consequence of cyp1a1 down-regulation [44–46]. Generally, the accumulation of yolk in oocytes during oocyte development after fertilization is a key success in zebrafish embryonic development [47]. Therefore, vitellogenin (VTG), a female-specific protein that is responsible as a precursor of egg yolk proteins in vertebrates, plays an important role in zebrafish reproduction [48]. This protein is synthesized in the liver, secreted to the bloodstream, and then transported to the developing oocytes for vitellogenesis process [49]. Thus, VTG has been used as a biomarker of estrogenic pollution or as a biomarker of an endocrine disruptor in vertebrates [50–52]. In addition, male zebrafish also can synthesize and secrete VTG when exposed to estrogen mimic pollutants, even though VTG is a female-specific protein [53,54]. In this experiment, the increased level of VTG in the treated fish suggests that PS-NPs caused endocrine disruption pollution in the environment.

Later, a fluorescence spectrophotometer was used to quantify fluorescence coated PS-NPs that accumulated in tissues of zebrafish after one-month incubation. Even though this technique relies on the fluorescent dye encapsulated in plastic beads, it is very useful to quantify the accumulation of PS-NPs in different tissues after the exposure. It is noted that, in zebrafish, the gills are considered as the initial site for uptake and elimination of nanoparticles, while the brain, gonads, and liver connected with the gills via arterial blood. In addition, blood circulating in veins from the gonads will also reach the liver [55]. We found that 70 nm PS-NPs in low concentration preferentially accumulated in the gonad as compared to other tissues/organs, in accordance with previous studies in another nanoparticle [56]. This result indicated that PS-NPs could pass through the gonad blood barriers and accumulate in gonad tissues. These gonadal alterations would possibly further affect reproductive activity by inducing germ cells apoptosis in gonad tissues [57]. In another prior study, this issue was shown in the zebrafish when their parents were exposed to the poly-N-vynil-2-pirrolidone and polyethylenimine (PVP/PEI) coated silver-nanoparticles. Parental exposure to Ag-NPs increased embryo malformation prevalence, suggesting that silver is able to reach the gonads and accumulate in the oocytes causing a disturbance in developing embryos in fish waterborne exposed to the Ag-NPs suspension [58]. Furthermore, nanoparticles were also found to invade the protective barrier and disrupt the oocytes in the mice study. In their study, NPs crossed the blood-brain barrier and accumulated in the central nervous system (CNS) [59]. Later, NPs disrupted hormone secretion, such as gonadotropin-releasing hormone (GnRH), which is responsible for oogenesis, through the hypothalamic-pituitary-gonadal axis [60–62]. Moreover, NPs could also cross through the placenta into the fetus, causing treated mice to likely exhibit fetal inflammation, genotoxicity, apoptosis, reproductive deficiency, and immunodeficiency [60]. In addition, another study in *Bombyx mori*, an invertebrate model organism, also proved that Ag-NPs could induce reproductive toxicity. In this study, the possible mechanism is Ag-NPs passed through the gonads and generated ROS in spermatocytes and internal germ cells, inducing the early germ cell death that led to the cells and DNA damage. The genetic integrity of the gonads is an essential aspect of reproductive success [63]. Thus, as ROS can harm the DNA and cells, this damaged genetic will be inherited to the next generation as the defective genetic [64,65]. In summary, even though the direct evidence that the PS-NPs passed through the bio-barrier and moved to further tissues in fish, such as gonads and brain, could not be shown; the fact that these organs were affected become an indicator from the perspective of biological effects. The alterations that were observed in several tissues demonstrate that PS-NPs do exert potential toxic effects on both the reproductive and nervous systems.

3.5. Behavioral and Neurotransmitters Alterations Caused by PS-NPs

The novel tank test has emerged as a potentially useful behavioral measure of anxiety in zebrafish. It exploits the tendency of zebrafish to initially dive to the bottom of a novel experimental tank, which has been compared to thigmotaxis in rodents, with a gradual increase in vertical activity over time [25]. Similar to the rodent open-field test, the novel tank test endpoints can be applied to zebrafish models of anxiety [66]. During the novel tank test, similar locomotion activity and exploratory behavior between control and low concentration of PS-NPs-exposed groups were observed. On another hand, the high concentration of the PS-NPs-exposed group exhibited hyperactivity and abnormal exploratory behaviors. Various important neurotransmitters in the brain were measured by performing ELISA with antigen-specific antibodies to understand the molecular mechanisms involved in this behavioral impairment of PS-NPs exposed to zebrafish. Abnormal behavior showed by the treated group might be associated with the deregulation of several important biomarkers, including oxytocin, vasopressin, and kisspeptin. Oxytocin, which is known to buffer the stress response, in the brain modulates a broad variety of behaviors, as well as anxiety-related behavior and stress coping. Together with vasopressin, oxytocin is an essential part of the hypothalamo-neurohypophysial system that appears to activate bond-relevant behaviors, such as the exploration of a novel environment [67,68]. In humans, oxytocin projects into the amygdala, hippocampus, and regions of the spinal cord that regulate the parasymphatic branch of the autonomic nervous system, which attenuates stress responses [69]. In addition, another study also found the antidepressant-like effects that are caused by kisspeptin-13 in mice modified forced swimming test via adrenergic and serotonergic receptors [70]. Further, recent research suggests that anxiety-like behavior is also directly associated with the acetylcholinesterase (AChE) activity of the hippocampus since AChE knockdown in the hippocampus promotes anxiety-like behavior in mice [71]. AChE is an enzyme that is responsible for the hydrolysis of the neurotransmitter acetylcholine, and it has been implicated in several non-cholinergic actions, including acute stress response and neurite outgrowth [72]. Several studies have shown that nanoplastics exposure could dysregulate AChE activity in aquatic organisms [18,73,74]. In the current study, the AChE level found to be deregulated in the treated fish that may contribute to the anxiety-like behavior that was exhibited by these fish. The previous study in mice has found that anxiety was linked to AChE activity by demonstrating that pubertal BPA (Bisphenol A) exposure increased anxiety-like behavior and decreases AChE activity in the hippocampus [75]. Another prior study also reported that exposure to high doses of thiamethoxam, a neonicotinoid insecticide, in rats produces AChE inhibition that persists some days after exposure accompanied by deficits in behavioral performance [76]. In reptiles and amphibians, exposure to AChE-inhibiting pesticides, such as carbaryl compromised their locomotion activity, which is such a critical process as a predator avoidance and prey capture [77]. Equally important, abnormal levels of several neurotransmitters, including γ-Aminobutyric acid (GABA), dopamine, and serotonin, may also play a role in this abnormal behavior. Extensive evidence indicated that GABA transmission plays a primary role in the modulation of behavioral sequelae resulting from stress, while serotonin (5-hydroxytryptamine) is implicated in the regulation of several developmental, behavioral, and physiologic processes, including anxiety and affective states in human and nonhuman species [78]. Serotonin is a key modulatory neurotransmitter in the central nervous system [79]. In humans, dysfunction in the neurotransmission of serotonin is implicated in a variety of psychiatric disorders, including major depression and anxiety [80]. As an addition, cortisol, a steroid hormone, might also contribute to this abnormal behavior. It is considered as the principal corticosteroid that is secreted by the teleost fish adrenal system in response to acute and chronic stress, which might explain the elevated level of cortisol that was observed in this study [81,82].

Aggressiveness, an ancestral behavior that is common to all animal species at least from fishes onwards, is one of the behavioral traits that is often closely linked to fitness [83]. It can be defined as the execution of action against animals belonging to the same or different species [84]. In zebrafish, mirror-image stimulation is traditionally used for studying zebrafish aggressive behavior [26]. Our findings demonstrated that both concentrations (0.5 and 1.5 ppm) of PS-NPs significantly impaired zebrafish aggressiveness in the mirror-biting test. The activity of several important biomarkers was also evaluated to investigate the mechanism related to this behavior alteration. From the result, significant reduced oxytocin and vasopressin activities observed in the brain tissue may be related to the aggressiveness reduction. Supporting the current result, previous studies showed that oxytocin increased the sexual and aggressive behavior of dominant monkeys and modulated a broad variety of behaviors, including pair bonding and aggression in rats [68,85]. Furthermore, aggressive behavior is partially influenced by vasopressin, a neuropeptide hormone implicated in the regulation of several social behaviors, mostly through its receptors in the brain [69,86,87]. In mice, a lower binding of vasopressin to its receptor 1a is connected with lower aggression, and knocking out the vasopressin receptor 1b significantly reduces its aggressive behavior. In addition, the injection of arginine vasopressin into the ventrolateral hypothalamus was found to stimulate aggressive behavior in gonadally intact male mice [88]. Generally, in mammals, oxytocin and vasopressin neurons are both involved in the control of social behaviors, such as aggression and reproduction [89,90]. The dysregulation of several important brain neurotransmitters, including dopamine and serotonin, might also be related to the less aggressive behavior observed in treated fish since these transmitters are critically involved in the neural circuits for many types of human and animal aggression [91]. In preclinical studies, the role of dopamine D1, D2, and D3 receptors in the modulation of aggression has been documented. Moreover, more persuasive evidence for a significant role of dopamine D2 receptors was mentioned from studies that focus on defensive-aggressive behavior in cats [92]. Traditionally, several studies have shown that elevated serotonin and GABA levels lead to decreased aggression in many different species, including humans [92]. However, the aggression level decreased in the treated fish along with downregulated serotonin activities in the present study. One possible explanation is that the majority (95%) of total body serotonin is released into the gut by intestinal enterochromaffin cells, which might compensate for the downregulated serotonin in the brain that is caused by the microplastic NP exposure [93].

As one of the important behavioral reactions, fear responses may have a significant fitness component, since it might allow the animal (and human) to avoid predation or other forms of danger in nature [94]. Based on the previous method, fear responses were induced in zebrafish by presenting a convict cichlid (*Amatitlania nigrofasciata*) during the predator avoidance test [95]. In the present study, alteration of predator avoidance behavior was shown by the high concentration of the PS-NPs-treated group, while this phenomenon was absent in the low concentration of the PS-NPs-treated group. Later, we quantitatively compared the neurotransmitters and other biochemical markers between the control and PS-NPs-treated fish. From the result, we found a significant decrease in kisspeptin levels after the PS-NPs treatments to the zebrafish. Kisspeptin is a hypothalamic neuropeptide that is derived from the *Kiss1* and it has proven to play a major role in vertebrate reproduction. However, several studies demonstrated a unique role for the kisspeptin system in altering the fear response in several animal models. In mice, the effects of kisspeptin-13 on mice passive avoidance learning were reported. Kisspeptin-Kiss-R signaling could be involved in the contextualization of fear in mammals since the expression of the *Kiss1* gene and Kiss-R was shown in the hypothalamus and the medial amygdala, which is a fear-regulating region in rodent brain [96]. In addition, very recent findings showed a unique role of kisspeptin in inhibiting fear response in zebrafish. Thus, it suggests the interaction between the vHb-expressing *Kiss1* and the serotonin system in the modulation of alarm substance-evoked fear responses [70]. Furthermore, the decreased activities of vasopressin and oxytocin may also related to this behavioral impairment, because these hormones are well known as modulators of a variety of cognitive and emotional processes, most notably learning and memory, trust and selective affiliation, and fear [90]. A pioneering study also found that oxytocin reduced amygdala activation and its coupling to brainstem centers responsible for autonomic and behavioral components of fear [69]. The low level of GABA was also observed in the treated fish brain. GABA is the most abundant inhibitory neurotransmitter in the mammalian brain and is known to be related to the anxiety/fear response in human and animal models [97,98]. In the rat study, it has been found that GABA neurotransmission in the nucleus accumbens shell played a role in defensive or fear-related behavior, which might contribute to the diminished predator avoidance behavior exhibited by the treated fish during the assay [99].

Social interactions are an important domain of animal and human behavior since they play a major role in the acquisition and development of learned behaviors [100]. Social interaction and shoaling behavior tests are very commonly used behavior assessments in fish models to study their social behaviors. In zebrafish, these tests assess their sociability by observing the interactions between several fishes [26]. Even though social behavior deficit was not shown by the PS-NPs-treated fish during the social interaction test, tighten shoals were formed by both of the treated groups, which might indicate stress responses in zebrafish [101]. We found that the reduction of oxytocin and vasopressin levels might play roles in this phenomenon after several major brain neurotransmitters were measured. As mentioned earlier, oxytocin and vasopressin are both best known for their contribution to the regulation of social behavior; moreover, a plethora of studies in humans already found that both of these neuropeptides modulate human social behavior and cognition [68,102]. In the previous study, the oxytocin effect in the social behavior modulation was studied in the oxytocin knockout mice. The mutant mice failed to recognize familiar conspecifics after repeated social encounters while it can be restored with central oxytocin administration into the amygdala [103]. Thus, the effects of oxytocin on partner preferences strongly suggest a causal connection between oxytocin and the formation of social bonds [67]. Furthermore, vasopressin was found to modulate social behaviors and motor activity in both dominant and subordinate monkeys [85]. Finally, comparisons across fishes, amphibians, and mammals also indicate that oxytocin and vasopressin and their homologs both regulate many of the same types of social behaviors throughout the vertebrate lineage [90]. In addition, the development of shoaling has also been found to be associated with whole-brain GABA and dopamine levels [104]. Previously, a study demonstrated that, in response to social stimuli, dopamine and 3,4-Dihydroxyphenylacetic acid (DOPAC) levels rapidly rise in the brain of adult zebrafish [105]. Dopamine D1-receptor antagonism was also found to impair shoaling in adult zebrafish in a dose-dependent manner [106]. In goldfish, the tendency to shoal has been demonstrated to increase when they were administered with anxiogenic drugs-GABA agonists [107]. Further, the study in medaka (*Oryzias latipes*) was also shown that an anti-anxiety drug—the positive allosteric modulator of GABA diazepam—selectively modified their shoaling behavior [98]. Taken together, the low level of dopamine and GABA that was observed in this research might contribute to the abnormal shoal that formed by the PS-NPs-treated fish group.

Circadian rhythms play a central role in adapting the physiology and behavior of living organisms to anticipate daily environmental changes [108]. After seven weeks of PS-NPs exposure, the dysregulation of circadian rhythm locomotion activity was shown by the treated fish group and might be related to the melatonin, a key hormone controlling the circadian rhythm. Generally, this physiological hormone gets involved in sleep timing and currently used as a primary treatment for sleep disorders in humans [109]. Melatonin has also been reported to promote sleep-like behavior in diurnal vertebrates, such as zebrafish, and is responsible for other physiological processes, including immune function, blood pressure, and retinal physiology [110,111]. However, in this experiment, PS-NPs-treated groups both showed a significantly low level of melatonin, while sleep-like behaviors were observed during the circadian rhythm locomotion activity test in light and dark cycles. One possible explanation is that the sleep-like behavior showed by the treated fish was likely caused by anxiety or depression, not by the melatonin level [112]. The downregulated vasopressin level might also contribute to this behavior alteration since vasopressin is the product of one such clock-controlled gene that evokes circadian rhythms. The previous study reported that the circadian rhythmicity of locomotor activities was significantly reduced in V1a (one of the vasopressin receptor)-deficient mice [113]. In addition, there is a possibility that serotonin also plays a role in this phenomenon. This is supported by another prior study in mice lacking the serotonin transporter. The abnormality in REM sleep was shown by the mutant mice, followed with other abnormalities, including reward-related and locomotor responses to psychostimulants and analgesic responses [114]. Besides, the hypoactivity behavior of fish, in this case, might also be attributed to an increase in oxidative damage [2]. However, this hypoactivity behavior that was observed during this test was in contrast with the hyperactivity behavior shown during the novel tank test. A possible explanation is that higher concentration (5 ppm) and longer exposure of PS-NPs were applied in the circadian rhythm locomotion activity test, while, in the novel tank test, only lower concentrations (0.5 and 1.5 ppm) and shorter exposures of microplastic PS-NPs were used.

Taken together, we found that PS-NPs treatments inhibited the activity of several important neurotransmitters at the highest concentration of PS-NPs, which might lead to cholinergic neurotransmission insufficiency. These dysregulations raise the potentiality that PS-NPs exposure could induce adverse effects on neurotransmission in zebrafish. This investigation is a step in the direction of understanding the adverse effect of PS-NPs on freshwater or marine organisms to understand the phenomenon at a very basic level.

4. Conclusions

To the best of our knowledge, the behavior impairments that result from the exposure to 70 nm PS-NPs have not been investigated before in fish. Our study confirmed that acute (~7 days) and chronic (~7 weeks) exposure of 70 nm PS-NPs could alter the neurobehavior and accumulated in at least four tissues (brain, liver, gonads, and intestine). Moreover, PS-NPs accumulation induced several effects on behavioral profiles and biochemical biomarkers predicting the potential health risk to mammals (as summarized in Figure 9). We provide evidence that PS-NPs exposure causes disruptions to behavior, induces oxidative stress, and elicits neurotoxic responses based on a comprehensive analysis of multiple parameters. Further studies are needed to better understand the mechanism underlying the biological effects of PS-NPs on the aquatic biosystem. Other factors, such as shape, size, and composition of PS-NPs, exposure period, and physiological characteristics of the exposed organisms may be closely related to the toxic effects of PS-NPs and interlacing PS-NPs and aquatic biota.

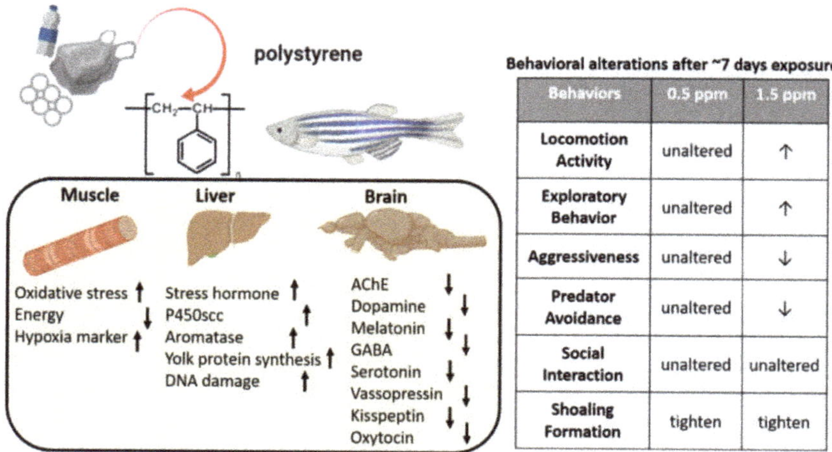

Behaviors	0.5 ppm	1.5 ppm
Locomotion Activity	unaltered	↑
Exploratory Behavior	unaltered	↑
Aggressiveness	unaltered	↓
Predator Avoidance	unaltered	↓
Social Interaction	unaltered	unaltered
Shoaling Formation	tighten	tighten

Figure 9. Schematic diagram of the biochemical and behavioral changes after polystyrene nanoplastics (PS-NPs) exposure in zebrafish. Left panel showing the biomarker expression alteration (↑: up regulated, ↓: down regulated) in the muscle, liver, and brain after PS-NPs exposure. Right panel showing the behavioral alteration after PS-NPs exposure at either 0.5 or 1.5 ppm (↑: higher behavior level, ↓: lower behavior level).

5. Materials and Methods

5.1. Particle Characterization

Polystyrene (PS) particles (nominal size ~70 nm) were purchased from Tianjin BaseLine ChromTech Research Centre (Tianjin, China). According to the manufacturer, the particles are spherical, white opaque, and can be detected by Transmission Electron Microscopy. The PS-NPs solutions were prepared with purified water (Milli-Q, Darmstadt, Germany) and sonicated before every usage. Lumisphere uniform green fluorescent polystyrene microspheres (~70 nm) were purchased from Tianjin BaseLine ChromTech Research Centre (Tianjin, China) to study the tissue distribution after waterborne exposing to zebrafish. The manufacturer performed transmission electron microscopy to verify the particle size.

5.2. Zebrafish Husbandry

Healthy adult zebrafish (~6 months old and 0.30 ± 0.022 g in wet weight) were maintained at 25 ± 1 °C with a 14/10 h light/dark cycles in culture water (UV sterilized and well-aerated water, pH 7.2 + 0.4, dissolved oxygen, 6.5 ± 0.2 mg/L, electrical conductivity, 0.254 ± 0.004 mS/cm, water hardness, 183 ± 5 mg of $CaCO_3$/L). The Committee for Animal Experimentation of the Chung Yuan Christian University approved all of the experimental protocols and procedures involving zebrafish (Number: CYCU107030, issue date 19 Dec. 2018). All of the experiments were performed in accordance with the guidelines for laboratory animals.

5.3. Exposure of PS-NPs

Adult zebrafish were randomly placed into three different glass tanks, each tank containing 20 fishes and the 2 L test solutions. During the entire experiment period, the glass tanks were continuously aerated to maintain the complete dispersion of the particles in the test solution and make sure that no aggregation was observed. For the toxicity test, fishes were randomly assigned to control and PS-NPs treated group (n = 20 for each group). For the treatment groups, 70 nm virgin PS-NPs were used and the exposure concentrations were 0.5 ppm, 1.5 ppm, and 5 ppm, which were chosen in accordance with the previous study about the toxicity of NPs on aquatic organisms [38]. The exposure protocol includes acute (~7 days) and chronic (~30 days and ~7 weeks) validation for the PS-NPs toxicity. For the experiment, the test solution in each tank was refreshed every two days.

5.4. Adult Behavior Test Battery

8 ± 1 days exposed adult zebrafish were tested mostly within the morning until afternoon (10.00 to 16:00) and start with a ~3–5 min. pre-acclimation in the tank, excluding the novel tank test, in a series of five behavioral assays: novel tank, mirror biting, predator avoidance, social interaction, and shoaling tests, as referenced from the previous publication [95]. The whole experiment was conducted in a temperature-controlled room (25 ± 1 °C). Novel tank and shoaling tests were performed on the seven days exposure, while mirror biting and social interaction tests were carried out on the following day. Lastly, a predator avoidance test was conducted on the nine days of exposure. Every day, the experiment began after the routine morning brine shrimp feeding. Fish tanks designated for conducting the experiment were transferred to the behavior testing room and freshly made system water was used in all of the testing apparatus. In these assays, untreated adult zebrafish and PS-NPs-exposed fishes were placed in an experimental tank filled with ~1.25 L of fish water. Canon EOS 600D camera with a long-range zoom lens (Canon Inc., Tokyo, Japan) was used for video recording in all of the tests mentioned above and, later, the videos were loaded by the idTracker [115] software for fish movement tracking and activity analysis based on the previous method [95,116]. In the novel tank test, the behavioral responses were recorded at a 1-min. interval of 0, 5, 10, 15, 20, 25, and 30 min. The behavioral endpoints were followed: time in top duration, average speed, freezing time movement ratio, number of entries to the top, latency to enter the top, and total distance traveled at the top. Next, the experimental tank with a mirror that was placed vertically to one side of the wall was used

in the mirror biting test. After acclimation, their behavior was recorded for 5 min. The measured behavior endpoints were mirror biting time percentage, longest duration in the mirror side, average speed, freezing time movement ratio, swimming time movement ratio, and rapid movement time ratio. The tested fish were placed in the tank with a transparent glass separator placed at ~15 cm away from the vertical side of the tank wall to test the predator avoidance behavior. After acclimation, *Amatitlania nigrofasciata*, a convict cichlid, was placed on the empty side of the test tank and their behavior was recorded for 5 min. For this test, several important endpoints were calculated, including predator approaching time percentage, the average distance to the separator, average speed, freezing time movement ratio, swimming time movement ratio, and rapid movement time ratio. Later, a social interaction test was performed with a transparent glass separator that was placed at ~11 cm away from the vertical side of the tank wall. A conspecific was placed on the empty side of the test tank after acclimation. Predator approaching time percentage, average distance to separator, average speed, freezing time movement ratio, swimming time movement ratio, and rapid movement time ratio were calculated after the videos were recorded for 5 min. Finally, the shoaling assay was conducted with three fish in each test tank. After 5 min. videos were recorded, several important endpoints including average speed, time in top duration, average inter-fish distance, average shoal area, average nearest neighbor distance, and average farthest neighbor distance were calculated.

5.5. Circadian Rhythm Locomotion Activity Assay

A circadian rhythm locomotion activity test was performed to analyze the zebrafish locomotion activity during the light and dark cycle after ~7 weeks exposure of 5 ppm PS-NPs. This test was conducted based on a previously published method [117]. The dark/cycle test apparatus consisted of six fish tanks (20 × 10 × 5 cm), with three fishes for each tank. As a light source below the tanks, a lightbox consisting of a light-emitting diode (LED) and an infrared light-emitting diode (IR-LED) were used in light and dark cycles, respectively. One infrared-sensitive charge-coupled device (CCD; detection window: 700–1000 nm) with a maximum resolution of 1920 × 1080 pixels and a 30-fps frame rate was used for video recording (3206_1080P module, Shenzhen, China). We recorded the zebrafish locomotion activity (average speed, average angular velocity, meandering, freezing time movement ratio, swimming time movement ratio, and rapid movement time ratio) for 1 min. every hour and, later, idTracker was used to track fish movement trajectories [115].

5.6. Biochemical Analysis of Biomarkers

After all of the behavioral analyses, three fishes were randomly collected from each tank (nine fishes/treatment) to evaluate the toxic effects of PS-NPs, alterations of biomarkers in different tissues due to PS-NPs exposure were determined. The fish were rinsed with water to remove the microplastics from the skin. Afterward, they were anesthetized by immersing in Tricaine (MS-222) where they were sacrificed and examined later. Different tissues (muscle, liver, and brain) were collected and a pool of three zebrafish tissues was used for homogenate preparation. The tissue samples were dissected and immediately frozen in liquid nitrogen and stored at −80 °C for transcriptomic analysis and biochemical validation or fixed in 10% formalin for histopathological examinations. For transcriptomic analysis and biochemical validation, the tissues were homogenized at medium speed with a Bullet blender with 50 volumes of (*v*/*w*) ice-cold phosphate saline buffer adjusted to pH 7.2. The samples were further centrifuged at 12,000 rpm for 20 min. and the crude homogenate was stored in 100 µl aliquots at −80 °C until required. Tissues were analyzed at the end of all behavioral experiments, excluding the circadian rhythm locomotion activity, to determine the possible effects of PS-NPs exposure in the following tissues: for the muscle, oxidative stress (ROS), energy (ATP and creatine kinase), and lipid metabolisms (MDA); for liver, oxygen uptake (Hif-1α), DNA damage (ssDNA), inflammation (TNF-α), stress (cortisol), yolk protein precursor (VTG), and detoxification enzyme (EROD, CYPs); for brain, AChE, acetylcholine, dopamine, melatonin, GABA, 5-HT, vasopressin, kisspeptin, prolactin, oxytocin, and vasotocin. All of these markers were detected by using commercial ELISA kits purchased from Zgenebio Inc. (Taipei, Taiwan) and all of the measurements were conducted according to the manufacture protocols.

5.7. Florescence Analysis of Lumisphere Microplastics on Adult Zebrafish

Following the behavior analysis mentioned above, a one-month exposure to lumisphere microplastics on adult zebrafish was conducted for the fluorescence analysis. After ~30 days incubation with 1.5 ppm florescence PS-NPs, the fishes were anesthetized by Tricain methanesulfonate (MS222), sacrificed, and then dissected to obtain several different tissues, such as intestine, gonads, liver, and brain. Each type of tissue was used for homogenate preparation by using phosphate buffers saline. The concentrations of PS-NPs in different tissues of treated and untreated samples were measured while using a fluorescence spectrophotometer (SYNERGY-HT, Bio-Tek) with excitation 485 nm and emission at 590 nm. The standard curve was generated by using serial dilutions of lumisphere green florescence PS-NP suspensions. The background luminescence of the tissues of untreated fish was detected and then subtracted from that of the PS-NPs-exposed samples. Each detection was run in triplicate.

5.8. Statistical Analysis

All of the statistical analyses were performed in GraphPad Prism (GraphPad Software, Inc., version 7.01, La Jolla, CA, USA). Non-parametric analysis was used to analyze all of the behavioral and biochemical results. Post hoc comparisons were done while using Dunnett's test unless stated otherwise. All the data are presented neither as median with interquartile range or mean ± standard error of the mean (S.E.M).

Supplementary Materials: Supplementary Materials can be found at http://www.mdpi.com/1422-0067/21/4/1410/s1.

Author Contributions: Conceptualization, C.-D.H.; Data curation, S.S., G.A., S.-T.L. and C.-D.H.; Formal analysis, S.S., G.A., P.S., N.M., Y.-H.L., S.-T.L. and C.-D.H.; Funding acquisition, K.H.-C.C. and C.-D.H.; Investigation, Y.-H.L.; Methodology, G.A. and P.S.; Software, C.-D.H.; Supervision, J.-R.C., K.H.-C.C. and C.-D.H.; Writing—original draft, S.S., J.-R.C., K.H.-C.C. and C.-D.H. All authors have read and agreed to the published version of the manuscript.

Funding: This study was funded by the grants sponsored by the Ministry of Science and Technology, MOST 105-2313-B-033-001-MY3 and MOST 107-2622-B-033-001-CC2 to C.-D.H. and MOST 106-2113-M-153-002 and MOST 108-2113-M-153-003 to K.H.-C.C.

Acknowledgments: We appreciated two anonymous reviewers on reviewing and providing professional comments on improving this paper quality.

Conflicts of Interest: The authors declare no conflicts of interest.

References

1. Geyer, R.; Jambeck, J.R.; Law, K.L. Production, use, and fate of all plastics ever made. *Sci. Adv.* **2017**, *3*, e1700782. [CrossRef] [PubMed]
2. Chen, Q.; Gundlach, M.; Yang, S.; Jiang, J.; Velki, M.; Yin, D.; Hollert, H. Quantitative investigation of the mechanisms of microplastics and nanoplastics toward zebrafish larvae locomotor activity. *Sci. Total. Environ.* **2017**, *584*, 1022–1031. [CrossRef] [PubMed]
3. Martin, J.; Lusher, A.; Thompson, R.C.; Morley, A. The deposition and accumulation of microplastics in marine sediments and bottom water from the irish continental shelf. *Sci. Rep.* **2017**, *7*, 10772. [CrossRef] [PubMed]
4. Booth, A.; Kubowicz, S.; Beegle-Krause, C.; Skancke, J.; Nordam, T.; Landsem, E.; Throne-Holst, M.; Jahren, S. Microplastic in global and norwegian marine environments: Distributions, degradation mechanisms and transport. In *Miljødirektoratet M-918*; SINTEF: Trondheim, Norway, 2017.
5. Auta, H.; Emenike, C.; Fauziah, S. Distribution and importance of microplastics in the marine environment: A review of the sources, fate, effects, and potential solutions. *Environ. Int.* **2017**, *102*, 165–176. [CrossRef]
6. Law, K.L.; Thompson, R.C. Microplastics in the seas. *Science* **2014**, *345*, 144–145. [CrossRef]

7. Galloway, T.S.; Cole, M.; Lewis, C. Interactions of microplastic debris throughout the marine ecosystem. *Nat. Ecol. Evol.* **2017**, *1*, 116. [CrossRef]

8. Koelmans, A.A.; Besseling, E.; Shim, W.J. Nanoplastics in the aquatic environment. Critical review; In *Marine Anthropogenic Litter*; Springer: Cham, Switzerland, 2015; pp. 325–340.

9. Cole, M.; Lindeque, P.; Fileman, E.; Halsband, C.; Goodhead, R.; Moger, J.; Galloway, T.S. Microplastic ingestion by zooplankton. *Environ. Sci. Technol.* **2013**, *47*, 6646–6655. [CrossRef]

10. Devriese, L.I.; van der Meulen, M.D.; Maes, T.; Bekaert, K.; Paul-Pont, I.; Frère, L.; Robbens, J.; Vethaak, A.D. Microplastic contamination in brown shrimp (crangon crangon, linnaeus 1758) from coastal waters of the southern north sea and channel area. *Mar. Pollut. Bull.* **2015**, *98*, 179–187. [CrossRef]

11. Ziajahromi, S.; Kumar, A.; Neale, P.A.; Leusch, F.D. Impact of microplastic beads and fibers on waterflea (ceriodaphnia dubia) survival, growth, and reproduction: Implications of single and mixture exposures. *Environ. Sci. Technol.* **2017**, *51*, 13397–13406. [CrossRef]

12. Ferreira, P.; Fonte, E.; Soares, M.E.; Carvalho, F.; Guilhermino, L. Effects of multi-stressors on juveniles of the marine fish pomatoschistus microps: Gold nanoparticles, microplastics and temperature. *Aquat. Toxicol.* **2016**, *170*, 89–103. [CrossRef]

13. Pedà, C.; Caccamo, L.; Fossi, M.C.; Gai, F.; Andaloro, F.; Genovese, L.; Perdichizzi, A.; Romeo, T.; Maricchiolo, G. Intestinal alterations in european sea bass dicentrarchus labrax (linnaeus, 1758) exposed to microplastics: Preliminary results. *Environ. Pollut.* **2016**, *212*, 251–256. [CrossRef]

14. Rochman, C.M.; Hoh, E.; Kurobe, T.; Teh, S.J. Ingested plastic transfers hazardous chemicals to fish and induces hepatic stress. *Sci. Rep.* **2013**, *3*, 3263. [CrossRef]

15. Karami, A.; Golieskardi, A.; Ho, Y.B.; Larat, V.; Salamatinia, B. Microplastics in eviscerated flesh and excised organs of dried fish. *Sci. Rep.* **2017**, *7*, 5473. [CrossRef] [PubMed]

16. Song, Y.K.; Hong, S.H.; Jang, M.; Han, G.M.; Shim, W.J. Occurrence and distribution of microplastics in the sea surface microlayer in jinhae bay, south korea. *Arch. Environ. Contam. Toxicol.* **2015**, *69*, 279–287. [CrossRef] [PubMed]

17. Song, Y.K.; Hong, S.H.; Jang, M.; Kang, J.-H.; Kwon, O.Y.; Han, G.M.; Shim, W.J. Large accumulation of micro-sized synthetic polymer particles in the sea surface microlayer. *Environ. Sci. Technol.* **2014**, *48*, 9014–9021. [CrossRef] [PubMed]

18. Avio, C.G.; Gorbi, S.; Milan, M.; Benedetti, M.; Fattorini, D.; d'Errico, G.; Pauletto, M.; Bargelloni, L.; Regoli, F. Pollutants bioavailability and toxicological risk from microplastics to marine mussels. *Environ. Pollut.* **2015**, *198*, 211–222. [CrossRef] [PubMed]

19. Browne, M.A.; Niven, S.J.; Galloway, T.S.; Rowland, S.J.; Thompson, R.C. Microplastic moves pollutants and additives to worms, reducing functions linked to health and biodiversity. *Curr. Biol.* **2013**, *23*, 2388–2392. [CrossRef]

20. Von Moos, N.; Burkhardt-Holm, P.; Köhler, A. Uptake and effects of microplastics on cells and tissue of the blue mussel mytilus edulis l. After an experimental exposure. *Environ. Sci. Technol.* **2012**, *46*, 11327–11335. [CrossRef]

21. Watts, A.J.; Lewis, C.; Goodhead, R.M.; Beckett, S.J.; Moger, J.; Tyler, C.R.; Galloway, T.S. Uptake and retention of microplastics by the shore crab carcinus maenas. *Environ. Sci. Technol.* **2014**, *48*, 8823–8830. [CrossRef]

22. E Silva, P.P.G.; Nobre, C.R.; Resaffe, P.; Pereira, C.D.S.; Gusmão, F. Leachate from microplastics impairs larval development in brown mussels. *Water Res.* **2016**, *106*, 364–370. [CrossRef]

23. Kaposi, K.L.; Mos, B.; Kelaher, B.P.; Dworjanyn, S.A. Ingestion of microplastic has limited impact on a marine larva. *Environ. Sci. Technol.* **2014**, *48*, 1638–1645. [CrossRef] [PubMed]

24. Sadri, S.S.; Thompson, R.C. On the quantity and composition of floating plastic debris entering and leaving the tamar estuary, southwest england. *Mar. Pollut. Bull.* **2014**, *81*, 55–60. [CrossRef]

25. Blaser, R.E.; Rosemberg, D.B. Measures of anxiety in zebrafish (danio rerio): Dissociation of black/white preference and novel tank test. *PLoS ONE* **2012**, *7*, e36931. [CrossRef] [PubMed]

26. Pham, M.; Raymond, J.; Hester, J.; Kyzar, E.; Gaikwad, S.; Bruce, I.; Fryar, C.; Chanin, S.; Enriquez, J.; Bagawandoss, S. Assessing social behavior phenotypes in adult zebrafish: Shoaling, social preference, and mirror biting tests. In *Zebrafish Protocols for Neurobehavioral Research*; Springer: Berlin, Germany, 2012; pp. 231–246.

27. Moretz, J.A.; Martins, E.P.; Robison, B.D. Behavioral syndromes and the evolution of correlated behavior in zebrafish. *Behav. Ecol.* **2007**, *18*, 556–562. [CrossRef]

28. Speedie, N.; Gerlai, R. Alarm substance induced behavioral responses in zebrafish (danio rerio). *Behav. Brain Res.* **2008**, *188*, 168–177. [CrossRef] [PubMed]

29. Ledesma, J.M.; McRobert, S.P. Innate and learned shoaling preferences based on body coloration in juvenile mollies, poecilia latipinna. *Ethology* **2008**, *114*, 1044–1048. [CrossRef]

30. Morrell, L.J.; James, R. Mechanisms for aggregation in animals: Rule success depends on ecological variables. *Behav. Ecol.* **2007**, *19*, 193–201. [CrossRef]

31. Bégay, V.R.; Falcón, J.; Cahill, G.M.; Klein, D.C.; Coon, S.L. Transcripts encoding two melatonin synthesis enzymes in the teleost pineal organ: Circadian regulation in pike and zebrafish, but not in trout. *Endocrinology* **1998**, *139*, 905–912. [CrossRef]

32. Reppert, S.M.; Weaver, D.R. Molecular analysis of mammalian circadian rhythms. *Annu. Rev. Physiol.* **2001**, *63*, 647–676. [CrossRef]

33. Eerkes-Medrano, D.; Thompson, R.C.; Aldridge, D.C. Microplastics in freshwater systems: A review of the emerging threats, identification of knowledge gaps and prioritisation of research needs. *Water Res.* **2015**, *75*, 63–82. [CrossRef]

34. Desforges, J.-P.W.; Galbraith, M.; Dangerfield, N.; Ross, P.S. Widespread distribution of microplastics in subsurface seawater in the ne pacific ocean. *Mar. Pollut. Bull.* **2014**, *79*, 94–99. [CrossRef] [PubMed]

35. Koelmans, A.A.; Nor, N.H.M.; Hermsen, E.; Kooi, M.; Mintenig, S.M.; De France, J. Microplastics in freshwaters and drinking water: Critical review and assessment of data quality. *Water Res.* **2019**. [CrossRef] [PubMed]

36. Li, J.; Qu, X.; Su, L.; Zhang, W.; Yang, D.; Kolandhasamy, P.; Li, D.; Shi, H. Microplastics in mussels along the coastal waters of china. *Environ. Pollut.* **2016**, *214*, 177–184. [CrossRef] [PubMed]

37. Xu, X.; Wang, X.; Li, Y.; Wang, Y.; Yang, L. A large-scale association study for nanoparticle c60 uncovers mechanisms of nanotoxicity disrupting the native conformations of DNA/rna. *Nucleic Acids Res.* **2012**, *40*, 7622–7632. [CrossRef] [PubMed]

38. Kang, H.-M.; Jeong, C.-B.; Lee, Y.H.; Cui, Y.-H.; Kim, D.-H.; Lee, M.-C.; Kim, H.-S.; Han, J.; Hwang, D.-S.; Lee, S.-J. Cross-reactivities of mammalian mapks antibodies in rotifer and copepod: Application in mechanistic studies in aquatic ecotoxicology. *Mar. Pollut. Bull.* **2017**, *124*, 614–623. [CrossRef] [PubMed]

39. Yamamoto, A.; Honma, R.; Sumita, M.; Hanawa, T. Cytotoxicity evaluation of ceramic particles of different sizes and shapes. *J. Biomed. Mater. Res. Part A* **2004**, *68*, 244–256. [CrossRef] [PubMed]

40. Yang, L.; Watts, D.J. Particle surface characteristics may play an important role in phytotoxicity of alumina nanoparticles. *Toxicol. Lett.* **2005**, *158*, 122–132. [CrossRef]

41. Cole, M.; Lindeque, P.; Fileman, E.; Halsband, C.; Galloway, T.S. The impact of polystyrene microplastics on feeding, function and fecundity in the marine copepod calanus helgolandicus. *Environ. Sci. Technol.* **2015**, *49*, 1130–1137. [CrossRef]

42. Cedervall, T.; Hansson, L.-A.; Lard, M.; Frohm, B.; Linse, S. Food chain transport of nanoparticles affects behaviour and fat metabolism in fish. *PLoS ONE* **2012**, *7*, e32254. [CrossRef]

43. Sussarellu, R.; Suquet, M.; Thomas, Y.; Lambert, C.; Fabioux, C.; Pernet, M.E.J.; Le Goïc, N.; Quillien, V.; Mingant, C.; Epelboin, Y. Oyster reproduction is affected by exposure to polystyrene microplastics. *Proc. Natl. Acad. Sci. USA* **2016**, *113*, 2430–2435. [CrossRef]

44. Monteiro, P.R.R.; Reis-Henriques, M.A.; Coimbra, J. Polycyclic aromatic hydrocarbons inhibit in vitro ovarian steroidogenesis in the flounder (platichthys flesus l.). *Aquat. Toxicol.* **2000**, *48*, 549–559. [CrossRef]

45. Britt, K.L.; Saunders, P.K.; McPherson, S.J.; Misso, M.L.; Simpson, E.R.; Findlay, J.K. Estrogen actions on follicle formation and early follicle development. *Biol. Reprod.* **2004**, *71*, 1712–1723. [CrossRef] [PubMed]

46. Gupta, R.K.; Singh, J.M.; Leslie, T.C.; Meachum, S.; Flaws, J.A.; Yao, H.H. Di-(2-ethylhexyl) phthalate and mono-(2-ethylhexyl) phthalate inhibit growth and reduce estradiol levels of antral follicles in vitro. *Toxicol. Appl. Pharmacol.* **2010**, *242*, 224–230. [CrossRef] [PubMed]

47. Wiegand, M.D. Composition, accumulation and utilization of yolk lipids in teleost fish. *Rev. Fish Biol. Fish.* **1996**, *6*, 259–286. [CrossRef]

48. Lazier, C. Vitellogenin gene expression in teleost fish. *Biochem. Mol. Biol. Fishes* **1993**, 391–405.

49. Arukwe, A.; Goksøyr, A. Eggshell and egg yolk proteins in fish: Hepatic proteins for the next generation: Oogenetic, population, and evolutionary implications of endocrine disruption. *Comp. Hepatol.* **2003**, *2*, 4. [CrossRef]

50. Denslow, N.D.; Chow, M.C.; Kroll, K.J.; Green, L. Vitellogenin as a biomarker of exposure for estrogen or estrogen mimics. *Ecotoxicology* **1999**, *8*, 385–398. [CrossRef]

51. Zhong, L.; Yuan, L.; Rao, Y.; Li, Z.; Zhang, X.; Liao, T.; Xu, Y.; Dai, H. Distribution of vitellogenin in zebrafish (danio rerio) tissues for biomarker analysis. *Aquat. Toxicol.* **2014**, *149*, 1–7. [CrossRef]

52. Ota, M.; Saito, T.; Yoshizaki, G.; Otsuki, A. Vitellogenin-like gene expression in liver of male zebrafish (danio rerio) by 1 μm 3-methylcholanthrene treatment: The possibility of a rapid and sensitive in vivo bioassay. *Water Res.* **2000**, *34*, 2400–2403. [CrossRef]

53. Marin, M.G.; Matozzo, V. Vitellogenin induction as a biomarker of exposure to estrogenic compounds in aquatic environments. *Mar. Pollut. Bull.* **2004**, *48*, 835–839. [CrossRef]

54. Sumpter, J.P.; Jobling, S. Vitellogenesis as a biomarker for estrogenic contamination of the aquatic environment. *Environ. Health Perspect.* **1995**, *103*, 173–178. [PubMed]

55. Peéry, A.R.; Devillers, J.; Brochot, C.; Mombelli, E.; Palluel, O.; Piccini, B.; Brion, F.; Beaudouin, R.M. A physiologically based toxicokinetic model for the zebrafish danio rerio. *Environ. Sci. Technol.* **2013**, *48*, 781–790. [CrossRef] [PubMed]

56. Li, J.; Ying, G.-G.; Jones, K.C.; Martin, F.L. Real-world carbon nanoparticle exposures induce brain and gonadal alterations in zebrafish (danio rerio) as determined by biospectroscopy techniques. *Analyst* **2015**, *140*, 2687–2695. [CrossRef] [PubMed]

57. Ma, Y.-B.; Lu, C.-J.; Junaid, M.; Jia, P.-P.; Yang, L.; Zhang, J.-H.; Pei, D.-S. Potential adverse outcome pathway (aop) of silver nanoparticles mediated reproductive toxicity in zebrafish. *Chemosphere* **2018**, *207*, 320–328. [CrossRef]

58. Orbea, A.; González-Soto, N.; Lacave, J.M.; Barrio, I.; Cajaraville, M.P. Developmental and reproductive toxicity of pvp/pei-coated silver nanoparticles to zebrafish. *Comp. Biochem. Physiol. Part C Toxicol. Pharmacol.* **2017**, *199*, 59–68. [CrossRef]

59. Oberdörster, G.; Sharp, Z.; Atudorei, V.; Elder, A.; Gelein, R.; Kreyling, W.; Cox, C. Translocation of inhaled ultrafine particles to the brain. *Inhal. Toxicol.* **2004**, *16*, 437–445. [CrossRef]

60. Hou, C.-C.; Zhu, J.-Q. Nanoparticles and female reproductive system: How do nanoparticles affect oogenesis and embryonic development. *Oncotarget* **2017**, *8*, 109799. [CrossRef]

61. Hsueh, A.J.; Billig, H.; Tsafriri, A. Ovarian follicle atresia: A hormonally controlled apoptotic process. *Endocr. Rev.* **1994**, *15*, 707–724.

62. Rollerova, E.; Jurcovicova, J.; Mlynarcikova, A.; Sadlonova, I.; Bilanicova, D.; Wsolova, L.; Kiss, A.; Kovriznych, J.; Kronek, J.; Ciampor, F. Delayed adverse effects of neonatal exposure to polymeric nanoparticle poly (ethylene glycol)-block-polylactide methyl ether on hypothalamic–pituitary–ovarian axis development and function in wistar rats. *Reprod. Toxicol.* **2015**, *57*, 165–175. [CrossRef]

63. Dayal, N.; Thakur, M.; Patil, P.; Singh, D.; Vanage, G.; Joshi, D. Histological and genotoxic evaluation of gold nanoparticles in ovarian cells of zebrafish (danio rerio). *J. Nanoparticle Res.* **2016**, *18*, 291. [CrossRef]

64. Yan, S.-Q.; Xing, R.; Zhou, Y.-F.; Li, K.-L.; Su, Y.-Y.; Qiu, J.-F.; Zhang, Y.-H.; Zhang, K.-Q.; He, Y.; Lu, X.-P. Reproductive toxicity and gender differences induced by cadmium telluride quantum dots in an invertebrate model organism. *Sci. Rep.* **2016**, *6*, 34182. [CrossRef] [PubMed]

65. Wang, R.; Song, B.; Wu, J.; Zhang, Y.; Chen, A.; Shao, L. Potential adverse effects of nanoparticles on the reproductive system. *Int. J. Nanomed.* **2018**, *13*, 8487. [CrossRef] [PubMed]

66. Cachat, J.; Stewart, A.; Grossman, L.; Gaikwad, S.; Kadri, F.; Chung, K.M.; Wu, N.; Wong, K.; Roy, S.; Suciu, C.; et al. Measuring behavioral and endocrine responses to novelty stress in adult zebrafish. *Nat. Protoc.* **2010**, *5*, 1786. [CrossRef] [PubMed]

67. Brown, S.L.; Brown, R.M. Selective investment theory: Recasting the functional significance of close relationships. *Psychol. Inq.* **2006**, *17*, 1–29. [CrossRef]

68. Neumann, I.D. Brain oxytocin: A key regulator of emotional and social behaviours in both females and males. *J. Neuroendocrinol.* **2008**, *20*, 858–865. [CrossRef] [PubMed]

69. De Dreu, C.K.; Shalvi, S.; Greer, L.L.; Van Kleef, G.A.; Handgraaf, M.J. Oxytocin motivates non-cooperation in intergroup conflict to protect vulnerable in-group members. *PLoS ONE* **2012**, *7*, e46751. [CrossRef]

70. Parhar, I.S.; Ogawa, S.; Ubuka, T. Reproductive neuroendocrine pathways of social behavior. *Front. Endocrinol.* **2016**, *7*, 28. [CrossRef]

71. Mineur, Y.S.; Obayemi, A.; Wigestrand, M.B.; Fote, G.M.; Calarco, C.A.; Li, A.M.; Picciotto, M.R. Cholinergic signaling in the hippocampus regulates social stress resilience and anxiety-and depression-like behavior. *Proc. Natl. Acad. Sci. USA* **2013**, *110*, 3573–3578. [CrossRef]

72. Chacón, M.A.; Reyes, A.E.; Inestrosa, N.C. Acetylcholinesterase induces neuronal cell loss, astrocyte hypertrophy and behavioral deficits in mammalian hippocampus. *J. Neurochem.* **2003**, *87*, 195–204. [CrossRef]

73. Oliveira, M.; Ribeiro, A.; Hylland, K.; Guilhermino, L. Single and combined effects of microplastics and pyrene on juveniles (0+ group) of the common goby pomatoschistus microps (teleostei, gobiidae). *Ecol. Indic.* **2013**, *34*, 641–647. [CrossRef]

74. Ribeiro, F.; Garcia, A.R.; Pereira, B.P.; Fonseca, M.; Mestre, N.C.; Fonseca, T.G.; Ilharco, L.M.; Bebianno, M.J. Microplastics effects in scrobicularia plana. *Mar. Pollut. Bull.* **2017**, *122*, 379–391. [CrossRef] [PubMed]

75. Luo, G.; Wei, R.; Niu, R.; Wang, C.; Wang, J. Pubertal exposure to bisphenol a increases anxiety-like behavior and decreases acetylcholinesterase activity of hippocampus in adult male mice. *Food Chem. Toxicol.* **2013**, *60*, 177–180. [CrossRef] [PubMed]

76. Rodrigues, K.; Santana, M.; Do Nascimento, J.; Picanço-Diniz, D.; Maués, L.; Santos, S.; Ferreira, V.; Alfonso, M.; Duran, R.; Faro, L. Behavioral and biochemical effects of neonicotinoid thiamethoxam on the cholinergic system in rats. *Ecotoxicol. Environ. Saf.* **2010**, *73*, 101–107. [CrossRef] [PubMed]

77. DuRant, S.E.; Hopkins, W.A.; Talent, L.G. Impaired terrestrial and arboreal locomotor performance in the western fence lizard (sceloporus occidentalis) after exposure to an ache-inhibiting pesticide. *Environ. Pollut.* **2007**, *149*, 18–24. [CrossRef]

78. Manzanares, P.A.R.; Isoardi, N.A.; Carrer, H.F.; Molina, V.A. Previous stress facilitates fear memory, attenuates gabaergic inhibition, and increases synaptic plasticity in the rat basolateral amygdala. *J. Neurosci.* **2005**, *25*, 8725–8734. [CrossRef]

79. Lira, A.; Zhou, M.; Castanon, N.; Ansorge, M.S.; Gordon, J.A.; Francis, J.H.; Bradley-Moore, M.; Lira, J.; Underwood, M.D.; Arango, V. Altered depression-related behaviors and functional changes in the dorsal raphe nucleus of serotonin transporter-deficient mice. *Biol. Psychiatry* **2003**, *54*, 960–971. [CrossRef]

80. Lee, J.-H.; Kim, H.J.; Kim, J.G.; Ryu, V.; Kim, B.-T.; Kang, D.-W.; Jahng, J.W. Depressive behaviors and decreased expression of serotonin reuptake transporter in rats that experienced neonatal maternal separation. *Neurosci. Res.* **2007**, *58*, 32–39. [CrossRef]

81. Wysocki, L.E.; Dittami, J.P.; Ladich, F. Ship noise and cortisol secretion in european freshwater fishes. *Biol. Conserv.* **2006**, *128*, 501–508. [CrossRef]

82. Øverli, Ø.; Kotzian, S.; Winberg, S. Effects of cortisol on aggression and locomotor activity in rainbow trout. *Horm. Behav.* **2002**, *42*, 53–61. [CrossRef]

83. Ariyomo, T.O.; Carter, M.; Watt, P.J. Heritability of boldness and aggressiveness in the zebrafish. *Behav. Genet.* **2013**, *43*, 161–167. [CrossRef]

84. Giammanco, M.; Tabacchi, G.; Giammanco, S.; Di Majo, D.; La Guardia, M. Testosterone and aggressiveness. *Med. Sci. Monit.* **2005**, *11*, RA136–RA145. [PubMed]

85. Winslow, J.T.; Insel, T.R. Social status in pairs of male squirrel monkeys determines the behavioral response to central oxytocin administration. *J. Neurosci.* **1991**, *11*, 2032–2038. [CrossRef] [PubMed]

86. Oldfield, R.G.; Hofmann, H.A. Neuropeptide regulation of social behavior in a monogamous cichlid fish. *Physiol. Behav.* **2011**, *102*, 296–303. [CrossRef] [PubMed]

87. Coverdill, A.; McCarthy, M.; Bridges, R.; Nephew, B. Effects of chronic central arginine vasopressin (avp) on maternal behavior in chronically stressed rat dams. *Brain Sci.* **2012**, *2*, 589–604. [CrossRef]

88. Ogrizek, M.; Grgurevič, N.; Snoj, T.; Majdič, G. Injections to pregnant mice produce prenatal stress that affects aggressive behavior in their adult male offspring. *Horm. Behav.* **2018**, *106*, 35–43. [CrossRef]

89. Kanda, S.; Akazome, Y.; Mitani, Y.; Okubo, K.; Oka, Y. Neuroanatomical evidence that kisspeptin directly regulates isotocin and vasotocin neurons. *PLoS ONE* **2013**, *8*, e62776. [CrossRef]

90. Campbell, P.; Ophir, A.G.; Phelps, S.M. Central vasopressin and oxytocin receptor distributions in two species of singing mice. *J. Comp. Neurol.* **2009**, *516*, 321–333. [CrossRef]

91. Miczek, K.A.; Fish, E.W.; Joseph, F.; de Almeida, R.M. Social and neural determinants of aggressive behavior: Pharmacotherapeutic targets at serotonin, dopamine and γ-aminobutyric acid systems. *Psychopharmacology* **2002**, *163*, 434–458. [CrossRef]

92. De Almeida, R.M.; Ferrari, P.F.; Parmigiani, S.; Miczek, K.A. Escalated aggressive behavior: Dopamine, serotonin and gaba. *Eur. J. Pharmacol.* **2005**, *526*, 51–64. [CrossRef]

93. Berger, M.; Gray, J.A.; Roth, B.L. The expanded biology of serotonin. *Annu. Rev. Med.* **2009**, *60*, 355–366. [CrossRef]

94. Ahmed, O.; Seguin, D.; Gerlai, R. An automated predator avoidance task in zebrafish. *Behav. Brain Res.* **2011**, *216*, 166–171. [CrossRef] [PubMed]

95. Audira, G.; Sampurna, B.; Juniardi, S.; Liang, S.-T.; Lai, Y.-H.; Hsiao, C.-D. A versatile setup for measuring multiple behavior endpoints in zebrafish. *Inventions* **2018**, *3*, 75. [CrossRef]

96. Ogawa, S.; Nathan, F.M.; Parhar, I.S. Habenular kisspeptin modulates fear in the zebrafish. *Proc. Natl. Acad. Sci. USA* **2014**, *111*, 3841–3846. [CrossRef]

97. Brambilla, P.; Perez, J.; Barale, F.; Schettini, G.; Soares, J. Gabaergic dysfunction in mood disorders. *Mol. Psychiatry* **2003**, *8*, 721. [CrossRef]

98. Tsubokawa, T.; Saito, K.; Kawano, H.; Kawamura, K.; Shinozuka, K.; Watanabe, S. Pharmacological effects on mirror approaching behavior and neurochemical aspects of the telencephalon in the fish, medaka (oryzias latipes). *Soc. Neurosci.* **2009**, *4*, 276–286. [CrossRef]

99. Reynolds, S.M.; Berridge, K.C. Fear and feeding in the nucleus accumbens shell: Rostrocaudal segregation of gaba-elicited defensive behavior versus eating behavior. *J. Neurosci.* **2001**, *21*, 3261–3270. [CrossRef]

100. Brown, C.; Laland, K. Social learning and life skills training for hatchery reared fish. *J. Fish Biol.* **2001**, *59*, 471–493. [CrossRef]

101. Green, J.; Collins, C.; Kyzar, E.J.; Pham, M.; Roth, A.; Gaikwad, S.; Cachat, J.; Stewart, A.M.; Landsman, S.; Grieco, F.; et al. Automated high-throughput neurophenotyping of zebrafish social behavior. *J. Neurosci. Methods* **2012**, *210*, 266–271. [CrossRef]

102. Young, L.J.; Flanagan-Cato, L.M. Editorial comment: Oxytocin, vasopressin and social behavior. *Horm. Behav.* **2012**, *61*, 227. [CrossRef]

103. Winslow, J.T.; Insel, T.R. The social deficits of the oxytocin knockout mouse. *Neuropeptides* **2002**, *36*, 221–229. [CrossRef]

104. Shams, S.; Amlani, S.; Buske, C.; Chatterjee, D.; Gerlai, R. Developmental social isolation affects adult behavior, social interaction, and dopamine metabolite levels in zebrafish. *Dev. Psychobiol.* **2018**, *60*, 43–56. [CrossRef]

105. Saif, M.; Chatterjee, D.; Buske, C.; Gerlai, R. Sight of conspecific images induces changes in neurochemistry in zebrafish. *Behav. Brain Res.* **2013**, *243*, 294–299. [CrossRef]

106. Scerbina, T.; Chatterjee, D.; Gerlai, R. Dopamine receptor antagonism disrupts social preference in zebrafish: A strain comparison study. *Amino Acids* **2012**, *43*, 2059–2072. [CrossRef]

107. Nishimura, Y.; Yoshida, M.; Watanabe, S. The effect on other individual presentations of the goldfish by fg7142 injection. *Nihon shinkei seishin yakurigaku zasshi. Jpn. J. Psychopharmacol.* **2002**, *22*, 55–59.

108. Lahiri, K.; Vallone, D.; Gondi, S.B.; Santoriello, C.; Dickmeis, T.; Foulkes, N.S. Temperature regulates transcription in the zebrafish circadian clock. *PLoS Biol.* **2005**, *3*, e351. [CrossRef]

109. Auld, F.; Maschauer, E.L.; Morrison, I.; Skene, D.J.; Riha, R.L. Evidence for the efficacy of melatonin in the treatment of primary adult sleep disorders. *Sleep Med. Rev.* **2017**, *34*, 10–22. [CrossRef]

110. Zhdanova, I.V.; Wang, S.Y.; Leclair, O.U.; Danilova, N.P. Melatonin promotes sleep-like state in zebrafish. *Brain Res.* **2001**, *903*, 263–268. [CrossRef]

111. Altun, A.; Ugur-Altun, B. Melatonin: Therapeutic and clinical utilization. *Int. J. Clin. Pract.* **2007**, *61*, 835–845. [CrossRef]

112. Li, X.; Liu, X.; Li, T.; Li, X.; Feng, D.; Kuang, X.; Xu, J.; Zhao, X.; Sun, M.; Chen, D. Sio 2 nanoparticles cause depression and anxiety-like behavior in adult zebrafish. *RSC Adv.* **2017**, *7*, 2953–2963. [CrossRef]

113. Li, J.-D.; Burton, K.J.; Zhang, C.; Hu, S.-B.; Zhou, Q.-Y. Vasopressin receptor v1a regulates circadian rhythms of locomotor activity and expression of clock-controlled genes in the suprachiasmatic nuclei. *Am. J. Physiol. Regul. Integr. Comp. Physiol.* **2009**, *296*, R824–R830. [CrossRef]

114. Wisor, J.; Wurts, S.; Hall, F.; Lesch, K.; Murphy, D.; Uhl, G.; Edgar, D. Altered rapid eye movement sleep timing in serotonin transporter knockout mice. *Neuroreport* **2003**, *14*, 233–238. [CrossRef] [PubMed]

115. Pérez-Escudero, A.; Vicente-Page, J.; Hinz, R.C.; Arganda, S.; De Polavieja, G.G. Idtracker: Tracking individuals in a group by automatic identification of unmarked animals. *Nat. Methods* **2014**, *11*, 743. [CrossRef] [PubMed]

116. Audira, G.; Sampurna, B.; Juniardi, S.; Liang, S.-T.; Lai, Y.-H.; Hsiao, C.-D. A simple setup to perform 3d locomotion tracking in zebrafish by using a single camera. *Inventions* **2018**, *3*, 11. [CrossRef]
117. Audira, G.; Sampurna, B.P.; Juniardi, S.; Liang, S.-T.; Lai, Y.-H.; Han, L.; Hsiao, C.-D. Establishing simple image-based methods and a cost-effective instrument for toxicity assessment on circadian rhythm dysregulation in fish. *Biol. Open* **2019**, *8*, bio041871. [CrossRef] [PubMed]

International Journal of
Molecular Sciences

Article

Behavioral Impairments and Oxidative Stress in the Brain, Muscle, and Gill Caused by Chronic Exposure of C_{70} Nanoparticles on Adult Zebrafish

Sreeja Sarasamma [1,2,†], Gilbert Audira [1,2,†], Prabu Samikannu [1], Stevhen Juniardi [2], Petrus Siregar [2], Erwei Hao [3,4,*], Jung-Ren Chen [5,*] and Chung-Der Hsiao [1,2,3,4,6,7,*]

1   Department of Chemistry, Chung Yuan Christian University, Chung-Li 32023, Taiwan;
    sreejakarthik@hotmail.com (S.S.); gilbertaudira@yahoo.com (G.A.); chemistprabu5590@gmail.com (P.S.)
2   Department of Bioscience Technology, Chung Yuan Christian University, Chung-Li 32023, Taiwan;
    stvn.jun@gmail.com (S.J.); siregar.petrus27@gmail.com (P.S.)
3   Guangxi Key Laboratory of Efficacy Study on Chinese Materia Medica, Guangxi University of Chinese
    Medicine, Nanning 530200, China
4   Guangxi Collaborative Innovation Center for Research on Functional Ingredients of Agricultural Residues,
    Guangxi University of Chinese Medicine, Nanning 530200, China
5   Department of Biological Science & Technology College of Medicine, I-Shou University,
    Kaohsiung 82445, Taiwan
6   Center for Biomedical Technology, Chung Yuan Christian University, Chung-Li 32023, Taiwan
7   Center for Nanotechnology, Chung Yuan Christian University, Chung-Li 32023, Taiwan
*   Correspondence: ewhao@163.com (E.H.); jrchen@isu.edu.tw (J.-R.C.); cdhsiao@cycu.edu.tw (C.-D.H.);
    Tel.: +86-0771-473-3831 (E.H.); +886-7-615-1100-7320 (J.-R.C.); +886-3-265-3545 (C.-D.H.)
†   These authors contributed equally to this work.

Received: 11 October 2019; Accepted: 13 November 2019; Published: 18 November 2019

Abstract: There is an imperative need to develop efficient whole-animal-based testing assays to determine the potential toxicity of engineered nanomaterials. While previous studies have demonstrated toxicity in lung and skin cells after C_{70} nanoparticles (NPs) exposure, the potential detrimental role of C_{70} NPs in neurobehavior is largely unaddressed. Here, we evaluated the chronic effects of C_{70} NPs exposure on behavior and alterations in biochemical responses in adult zebrafish. Two different exposure doses were used for this experiment: low dose (0.5 ppm) and high dose (1.5 ppm). Behavioral tests were performed after two weeks of exposure of C_{70} NPs. We found decreased locomotion, exploration, mirror biting, social interaction, and shoaling activities, as well as anxiety elevation and circadian rhythm locomotor activity impairment after ~2 weeks in the C_{70} NP-exposed fish. The results of biochemical assays reveal that following exposure of zebrafish to 1.5 ppm of C_{70} NPs, the activity of superoxide dismutase (SOD) in the brain and muscle tissues increased significantly. In addition, the concentration of reactive oxygen species (ROS) also increased from 2.95 ± 0.12 U/ug to 8.46 ± 0.25 U/ug and from 0.90 ± 0.03 U/ug to 3.53 ± 0.64 U/ug in the muscle and brain tissues, respectively. Furthermore, an increased level of cortisol was also observed in muscle and brain tissues, ranging from 17.95 ± 0.90 pg/ug to 23.95 ± 0.66 pg/ug and from 3.47 ± 0.13 pg/ug to 4.91 ± 0.51 pg/ug, respectively. Increment of Hif1-α level was also observed in both tissues. The elevation was ranging from 11.65 ± 0.54 pg/ug to 18.45 ± 1.00 pg/ug in the muscle tissue and from 4.26 ± 0.11 pg/ug to 6.86 ± 0.37 pg/ug in the brain tissue. Moreover, the content of DNA damage and inflammatory markers such as ssDNA, TNF-α, and IL-1β were also increased substantially in the brain tissues. Significant changes in several biomarker levels, including catalase and malondialdehyde (MDA), were also observed in the gill tissues. Finally, we used a neurophenomic approach with a particular focus on environmental influences, which can also be easily adapted for other aquatic fish species, to assess the toxicity of metal and carbon-based nanoparticles. In summary, this is the first study to illustrate the adult zebrafish toxicity and the alterations in several neurobehavior parameters after zebrafish exposure to environmentally relevant amounts of C_{70} NPs.

Keywords: C_{70}; reactive oxygen species; nanoparticle toxicity; behavior tests; phenomics

---

1. Introduction

As the industrial applications of stabilized nanoparticles continue to expand, it becomes crucial to understand their potential environmental implications. The potential application of nanomaterials includes drug delivery, medical equipment, biosensors, and personal care products [1]. The projected widespread use and large-scale production volume have led to growing concern over the potential for most engineered nanomaterials to adversely affect human health and the environment [2,3].

Nowadays, among carbon-based nanomaterials, fullerenes have received much attention with many potential applications, including in electronics, site-specific drug delivery, and pharmaceutical nanocarriers [4–6]. In fact, in the pharmaceutical industry, the potential antioxidant nature of fullerenes could be identified as a therapeutic intervention for the nervous system and neurodegenerative diseases, diabetes, pancreatic disease, skin damage, hearing loss, septic shock, and kidney diseases. Fullerenes are being marketed to consumers as a therapeutic agent in cosmetics and creams [7]. In addition, fullerenes can serve as major components in a variety of plastics, including filtration membranes [8]. However, many studies are raising safety concerns by demonstrating possible cytotoxic effects of fullerenes and their derivatives [9–13]. With increasing large production volume and use of fullerene, it is imperative to determine/monitor the possible human health and environmental implications [14] of these nanomaterials since adverse effects as well as protective effects are reported upon exposure [15–17]. Lack of toxicological data on nanomaterials makes it difficult to determine if there is a risk connected with nanoparticle exposure. Timely assessment of nanoparticle toxicity would provide this critical data, enhance the public trust of the nanotechnology industry, and aid regulators in deciding the environmental and health risks of commercial nanomaterials [18]. Hence, there is an urgent need to develop efficient, rapid, and appropriate testing strategies to assess toxicity and potential risks posed by fullerenes and their derivatives.

Various biological models have been used for the toxicological assessment of nanomaterials. In vitro techniques, such as cell culture, are often used because they are efficient, rapid, and less expensive [15]. While these in vitro studies are useful, a direct translation to human health risk is often difficult to understand [19]. In vivo studies, on the other hand, can provide improved prediction of the biological response in an intact system. Since in vivo studies often employ rodent models, assessments are generally time-consuming, expensive, and require extensive facilities for housing experimental animals. Time, labor, infrastructure, and cost can be significantly reduced by replacing the traditional rodent model with the zebrafish model [20]. Zebrafish are a well-established model for studying toxicological assessments and basic developmental biological processes [21–24]. Currently, only a handful of studies have investigated the adverse effects of C_{70} NPs on *Daphnia* [25] but there is a paucity of studies on adult zebrafish. Furthermore, the cellular and biochemical mechanisms underlying C_{70} NP toxicity are still poorly understood.

A number of diverse platforms are available to assess toxicity, ranging from in vitro studies to basic model organisms, such as *Daphnia* or sea urchins, to higher vertebrate models, such as rodents and primates [26,27]. Recent studies have begun to apply "big data" approaches to aid in data analysis and interpretation for validation of drugs and behaviors in zebrafish [28,29]. In this context, zebrafish behavioral phenomics are emerging as a new platform directed towards assessing various behavioral phenotypes by means of high-throughput screening and test batteries [30]. This new area of zebrafish phenomics-based biology is gaining importance in aquatic toxicology and neuropharmacology, in addition to the search for genes and pathways that can serve as biomarkers or targets for drug exposure.

In this continuum, a small number of reports deal with possible toxicities of C_{70} NPs with the aid of either in vitro or in vivo studies [31,32], but their neurobehavior impairments were not definitively determined. No evidence was currently available to allow for predictions of behavioral functions that would most likely be affected by these nanomaterials. There is an urgent need for a molecular biomarker that would be used as an endpoint to evaluate neurobehavior toxicities. To this end, this study aimed to investigate the effects of toxicity level and stress response of adult zebrafish to fullerene C_{70} NPs. To understand the mechanism underlying the abnormal neurobehavior and the oxidative inflammation in the brain caused by nanoparticulate C_{70}, we investigate the different endpoints of behavior analysis and the pathological changes in various tissues following exposure to zebrafish, assess the oxidative stress markers, and examine the effects of neurotransmitters including γ-aminobutyric acid (GABA), acetylcholinesterase (AChE) activity, and levels of dopamine (DA), serotonin (5-HT), and melatonin in the zebrafish brain. The experimental design and times for behavioral endpoint measurement were summarized in Figure 1A.

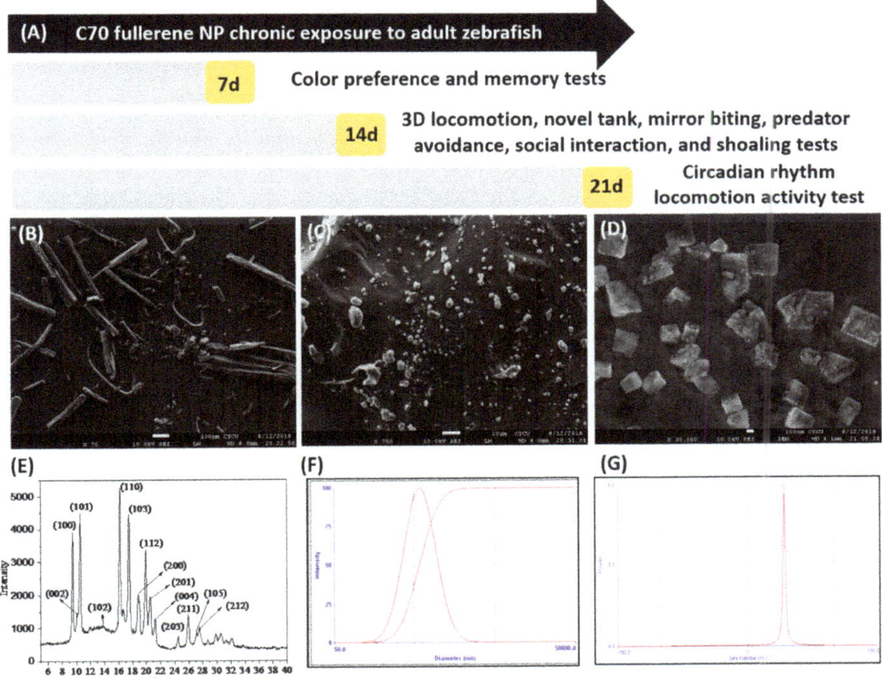

Figure 1. (**A**) Overview of the experimental design and time points for chronic exposure of C_{70} fullerene nanoparticles (NPs) to adult zebrafish. For chronic toxicity, we measured color preference and short-term memory at 7 days post-exposure (dpe). 3D locomotion, novel tank, mirror biting, predator avoidance, social interaction, and shoaling tests were given at 14 dpe. The circadian rhythm test was given at 21 dpe. After all behavior tests, fish were dissected and subjected to biochemical assays by 22 dpe. Characterization of the C_{70} NPs used in this study: (**B**) SEM micrograph of C_{70} NPs stock solution in the absence of solvents, (**C**) C_{70} NPs dissolved in DMSO showing wide disparity in aggregation, (**D**) high magnification scanning electron micrograph showing the size of C_{70} NPs used in this study, and (**E**) X-ray diffraction patterns of the crystal quality of the C_{70} NPs. (**F**) The particle size distribution of 0.5 ppm C_{70} NPs in DMSO was measured by dynamic light scattering. C_{70} NP suspensions were sonicated prior to measurement to resuspend the large particles and assess changes in large aggregate status. (**G**) The zeta potential value of C_{70} NPs is estimated at −34.0 mV.

2. Results

2.1. Physical Property Characterization of C_{70} NPs (Nanoparticles)

The size of C_{70} NPs was determined by X-ray diffraction (XRD) analysis and scanning electron microscopy (SEM). As shown in Figure 1B, the nanoparticles appear mainly spherical. Due to the insolubility of C_{70} NPs in water, it was sonicated to form a uniform suspension in 0.1% of dimethyl sulfoxide (DMSO). Previous evaluations in our laboratory have demonstrated no adverse biological effect of DMSO at this concentration [33]. The SEM data showed crystalline particles of C_{70} NPs with a diameter of 95.02 ± 0.25 nm (Figure 1D). The crystal structure of C_{70} NPs was analyzed by XRD and showed intense diffraction peaks. In Figure 1E, the XRD patterns of C_{70} NPs peaks at $2\theta = 9.5°$, $11.9°$, $14.3°$, $16.2°$, $18.5°$, $19.6°$, $21.5°$ were assigned to (100), (101), (102), (110), (103) respectively (Figure 1E). The following suspension at DMSO C_{70} nanoparticles size distribution was measured by dynamic light scattering (DLS) (Figure 1F). Zeta potential measurements, evaluated by electrophoretic mobility of C_{70} nanoparticle in Figure 1G, indicated that it acquired a negative surface charge of −34.0 mV.

2.2. Low-Dose Exposure of C_{70} NPs Reduced Locomotion and Exploration Behaviors

The locomotor activity was assessed using a three-dimensional (3D) locomotion test assay in zebrafish after chronic exposure to two different concentrations of C_{70} NPs (0.5 and 1.5 ppm) for two weeks. In this test, we measured six important zebrafish behavior endpoints representative of swimming activity and orientation. From the results, we found that 1.5 ppm of C_{70} NP-treated fish showed a significant reduction in their locomotor activity. On another hand, a lower concentration of C_{70} NP_S (0.5 ppm) did not show any behavioral alteration in swimming activity. These phenomena were indicated by a lower average speed, a rapid movement ratio, and a higher freezing time ratio exhibited by zebrafish treated with 1.5 ppm C_{70} $NP_{S.}$ Furthermore, there were no significant differences in the average speed, freezing, and rapid movement time ratio between 0.5 ppm of C_{70} NP-treated fish and the control group (Figure 2A,E–F). In addition, similar average angular velocity and meandering were recorded in both treated fish and the control groups suggesting swimming orientation was not affected (Figure 2B,D). Interestingly, it was observed that both 0.5 and 1.5 ppm C_{70} NP-treated groups showed a significant increase in time of top duration compared to the control group (Figure 2C).

The novel tank test, another experiment to assess zebrafish locomotor activity, exploratory behavior, and anxiety level was conducted after two weeks of C_{70} NPs exposure. This test exploits the natural tendency of zebrafish to initially dive to the bottom part of a novel tank, with a gradual increase in activity in the vertical axis over time [34]. This test revealed adult fish exposed to both concentrations of C_{70} NPs showed a significant decrease in their swimming activity compared to the controls. This finding was shown by a lower average speed and a higher freezing time movement ratio of treated fish compared with the control group in most of the experiment time (Figure 2G,H). Furthermore, 0.5 ppm of C_{70} NP_s altered the exploratory behavior, indicated by a lower average of time in top duration, a lower level of entries to the top, and a lower total distance traveled in the top, and a higher level of latency to enter the top portion of the test tank (Figure 2I–L). Surprisingly, higher concentration of C_{70} NP_s did not cause a more severe effect in the exploratory behavior, as indicated by the similar average of time in top duration, number of entries to the top, total distance traveled in the top, and latency to enter the top portion of the test tank in most of the test time interval seen in the treated fish and the control group (Figure 2I–L).

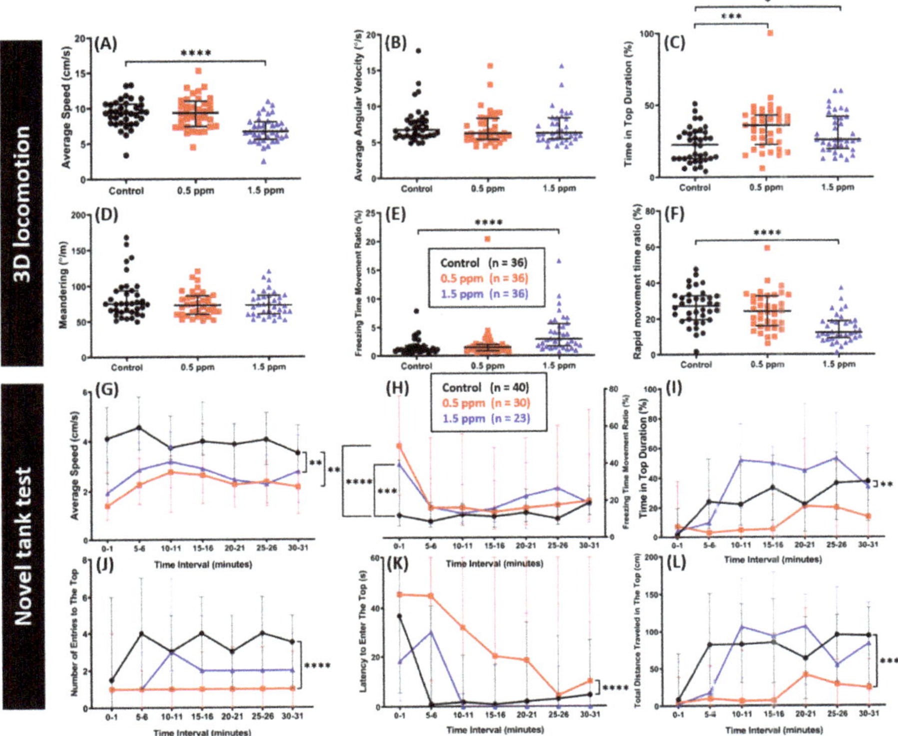

Figure 2. Comparison of behavior endpoints between the untreated control and C_{70} NP-exposed zebrafish in 3D locomotion and novel tank tests after 14-day exposure. (**A**) Average speed, (**B**) average angular velocity, (**C**) time in top duration, (**D**) meandering, (**E**) freezing time movement ratio, and (**F**) rapid movement time ratio were analyzed for the 3D locomotion test. For the novel tank test, (**G**) average speed, (**H**) freezing time movement ratio, (**I**) time in top duration, (**J**) number of entries to the top, (**K**) latency to enter the top, and (**L**) total distance traveled in the top were analyzed. The data are expressed as the median with interquartile range. The 3D locomotion test data were analyzed by the Kruskal–Wallis test, with Dunn's multiple comparisons test as a follow-up test (n = 36 for both control and treatment groups). The novel tank test data were analyzed by two-way ANOVA with Geisser–Greenhouse correction (n = 40 for the untreated control; n = 30 for the 0.5 ppm C_{70} NP-exposed fish; n = 23 for the 1.5 ppm C_{70} NPs-exposed fish; * $p < 0.05$, ** $p < 0.01$, *** $p < 0.001$, **** $p < 0.0001$).

2.3. Low Doses of C_{70} NPs Exposure Reduced Aggression and Predator Avoidance

To measure the aggressiveness of the fish, the mirror biting test was conducted after two weeks of C_{70} NPs exposure. Biting the mirror may also indicate, more generally, social motivation or the intent to interact with a social partner [15]. In this test, chronic exposure of C_{70} NPs in both concentrations significantly reduced zebrafish aggressiveness, as indicated by a lower mirror biting time percentage and the longest duration in the mirror side (Figure 3B–C). Furthermore, in line with our previous C_{60} NPs studies [15], 1.5 ppm of C_{70} NPs exposure reduced their locomotion behavior, displayed by a lower average speed, swimming and rapid movement time ratios, and a higher freezing time movement ratio compared to the control fish (Figure 3A,D–F). In addition, a slight decrement in locomotor activity was also detected in 0.5 ppm of C_{70} NP-treated fish, as shown by a lower rapid movement time ratio (Figure 3F). Meanwhile, there was no significant difference found in other types of fish movement and average speed between 0.5 ppm of C_{70} NPs treated and the control group (Figure 3A,D–E).

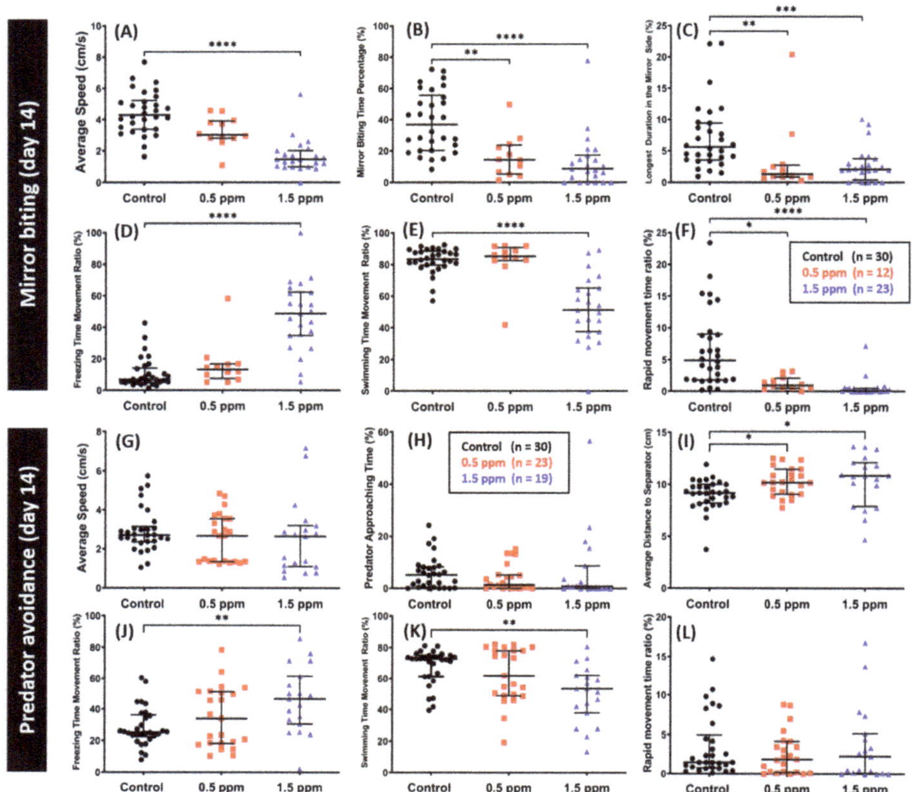

Figure 3. Comparison of mirror biting and predator avoidance behavior endpoints between the untreated control and C_{70}-exposed fish after 14-day exposure. (**A**) Average speed, (**B**) mirror biting time percentage, (**C**) longest duration in the mirror side, (**D**) freezing time movement ratio, (**E**) swimming time movement ratio, and (**F**) rapid movement time ratio were analyzed for the mirror biting assay (n = 30 for the untreated control; n = 12 for the 0.5 ppm C_{70} NP-exposed fish; n = 23 for the 1.5 ppm C_{70} NP-exposed fish). For predator avoidance test, the (**G**) average speed, (**H**) predator approaching time ratio, (**I**) average distance to separator, (**J**) freezing time movement ratio, (**K**) swimming time movement ratio, and (**L**) rapid movement time ratio were analyzed (n = 30 for the untreated control; n = 23 for the 0.5 ppm C_{70} NP-exposed fish; n = 19 for the 1.5 ppm C_{70} NP-exposed fish). The data are expressed as the median with interquartile range and were analyzed by the Kruskal–Wallis test continued with Dunn's multiple comparisons test as a follow-up test (* $p < 0.05$, ** $p < 0.01$, *** $p < 0.001$, **** $p < 0.0001$).

Fear is a collection of behavioral responses that are elicited by negative stimuli associated with imminent danger such as the presence of a predator [35]. In the predator avoidance test, we assessed zebrafish fear reactions, including anti-predatory behavior after two weeks of C_{70} NPs exposure. Predator avoidance analysis was conducted by confronting zebrafish with a predator, convict cichlid (*Amatitlania nigrofasciata*). We found that exposure of C_{70} NPs in zebrafish did not alter their fear response behavior to their predator, showed by similar predator approaching time between the treated zebrafish and the control (Figure 3H). Meanwhile, there were slight increments in the average distance to separator between zebrafish and the predator fish exhibited by both the 0.5 ppm and the 1.5 ppm C_{70} NP-treated zebrafish (Figure 3I). Furthermore, we found some differences in the types of movement in the 1.5 ppm of C_{70} NP-treated zebrafish compared with the control fish, even though their average speed was similar (Figure 3G). These differences were indicated by a higher freezing time movement ratio and a lower swimming time movement ratio, while there was no significant difference in their rapid movement time ratios (Figure 3J–L).

2.4. Low Doses of C$_{70}$ NPs Exposure Dysregulated Social Interaction and Shoaling Behavior

The social interaction test is another useful model to study fish social phenotypes. It encompasses more than simply preferring to be near members of the same species as humans [15]. We assessed zebrafish sociability by observing zebrafish interactions with conspecific after two weeks of C$_{70}$ NPs exposure [36]. Based on the results, we found that zebrafish chronically exposed to 1.5 ppm of C$_{70}$ NPs had reduced conspecific social interaction, as shown by a lower interaction time percentage and a higher average distance to the separator and the longest duration in the separator side (Figure 4B–D). On another hand, low concentration of C$_{70}$ NPs did not alter zebrafish conspecific interaction behavior, as shown by similar interaction time percentage and the longest duration in the separator side between the C$_{70}$ NPs treated fish and untreated control, even though there was an increment in the average distance to the separator was noted in the 0.5 ppm C$_{70}$ NPs treated fish (Figure 4B–D). Furthermore, locomotor activity alteration was detected in the high concentration of C$_{70}$ NP-treated fish, which was shown by the low level of average speed while there was no similar phenomenon displayed by the low concentration of C$_{70}$ NPs group, as shown by similar average speed compared to the control group (Figure 4A).

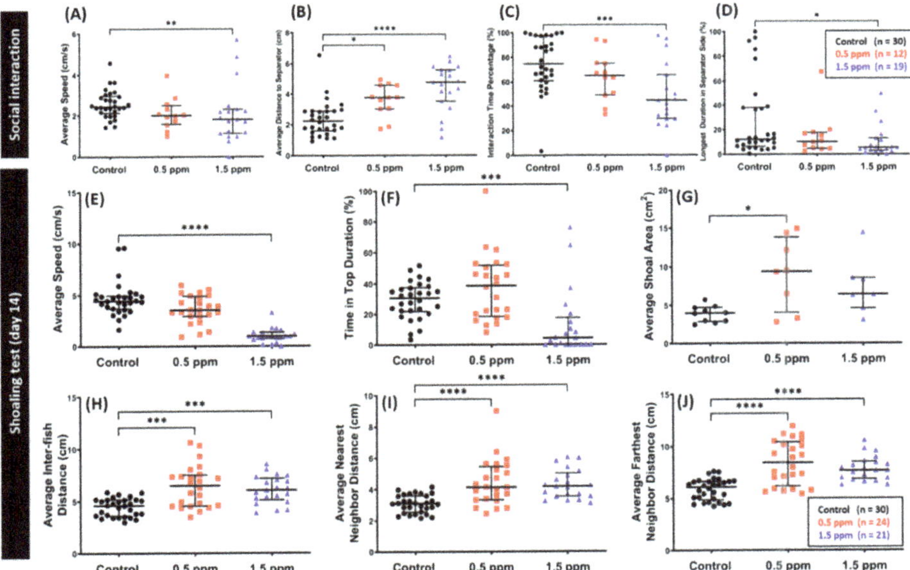

Figure 4. Comparison of social interaction and shoaling behavior endpoints between the untreated control and C$_{70}$ NP-exposed fish after 14-day exposure. (**A**) Average speed, (**B**) average distance to separator side, (**C**) interaction time percentage, and (**D**) the longest duration in the separator side were analyzed for the social interaction test ($n = 30$ for the untreated control; $n = 12$ for the 0.5 ppm C$_{70}$ NP-exposed fish; $n = 19$ for the 1.5 ppm C$_{70}$ NP-exposed fish). For the shoaling test: (**E**) average speed, (**F**) time in top duration, (**G**) average shoal area, (**H**) average inter-fish distance, (**I**) average nearest neighbor distance, and (**J**) average farthest neighbor distance were analyzed ($n = 30$ for the untreated control; $n = 24$ for the 0.5 ppm C$_{70}$ NPs-exposed fish; $n = 21$ for the 1.5 ppm C$_{70}$ NPs-exposed fish, with 3 fish for each shoal). The data are expressed as the median with interquartile range and were analyzed by the Kruskal–Wallis test, with Dunn's multiple comparisons test as a follow-up test (* $p < 0.05$, ** $p < 0.01$, *** $p < 0.001$, **** $p < 0.0001$).

Shoaling behavior represents the complex interaction of animals moving together in coordinated movements which is very common in fish models [37]. We tested the social behavior of C_{70} NP-exposed fish using a shoaling assay after two weeks of C_{70} NPs exposure. In this test, loose shoals formed by C_{70} NP-treated fish were observed, as indicated by increased average shoal area, average inter-fish distance, average nearest neighbor distance, and average farthest neighbor distance in both C_{70} NPs treated groups (Figure 4G–J). Furthermore, in line with our previous C_{60} NPs studies [38], a significant decrease in locomotor activity was observed in the high concentration of C_{70} NPs compared to untreated control (Figure 4E). In addition, a high concentration of C_{70} NPs exposure also reduced the exploratory behavior as shown by decreased time in the top duration (Figure 4F).

2.5. Low Doses of C_{70} NPs Exposure Dysregulated the Color Preference

Color vision is one of the most prominent modalities to recognize biologically-important stimulation and it plays a significant role in visual perception. Zebrafish possess eyes and retinas that are very similar to those of other vertebrates includes humans [38]. The color preference test has been used to assess phenotypical and behavioral changes in zebrafish [15,39,40]. Our previous color preference study showed that adult zebrafish exposed to C_{60} NPs have a strong aversion towards green/blue and red/blue compared to other colors [41]. The color preference patterns in C_{70} NP-treated fish showed changes in blue–red color combination after one-week exposure (Figure 5C), with overall preference pattern shifted from a red > blue > green > yellow preference in the untreated control to a blue > red > green > yellow preference in the C_{70} NP-treated fish. Other color combinations for either blue–green (Figure 5A), green–yellow (Figure 5B), red–green (Figure 5D), red–yellow (Figure 5E), or blue–yellow (Figure 5F) showed non-significant differences between the untreated control and the C_{70} NP-treated fish.

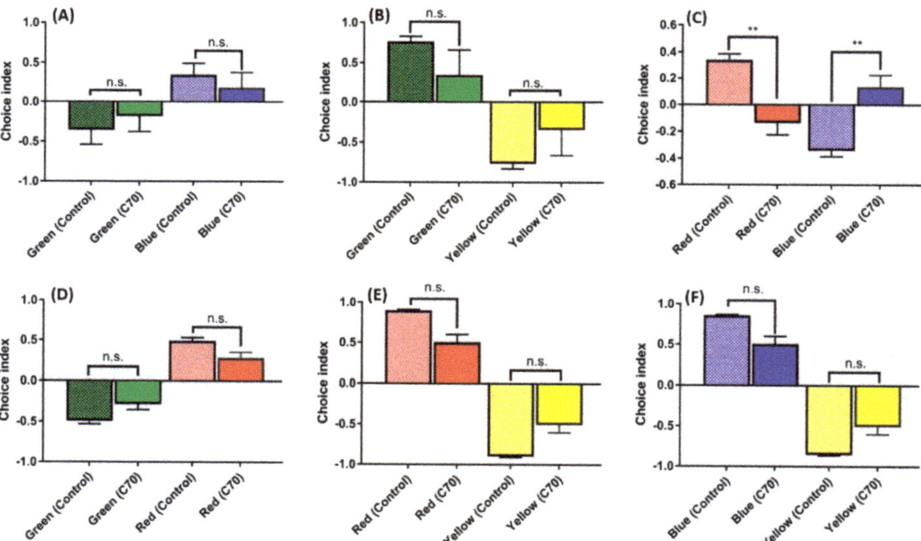

Figure 5. Comparison of color preferences between the untreated control and 1.5 ppm of C_{70} NP-exposed fish: (**A**) green vs. blue combination; (**B**) green vs. yellow combination; (**C**) red vs. blue combination; (**D**) green vs. red combination; (**E**) red vs. yellow combination; and (**F**) blue vs. yellow combination. Since all of the data were not normally distributed, they were analyzed using non-parametric Kruskal–Wallis followed by Dunn's post-hoc test, and $p < 0.05$ was considered significantly different. The data are presented with mean ± SEM with $n = 24$, n.s. = not significant, ** $p < 0.01$.

2.6. Low Doses of C_{70} NPs Exposure Dysregulated the Circadian Rhythm but Did Not Alter Short-Term Memory

The zebrafish represents a potential model to study memory function and impairment in vertebrates. Here, we used a passive avoidance test to explore the potential short-term memory deficiency in zebrafish after three weeks exposed to environmental-level C_{70} NPs. This passive avoidance test was conducted by using a white–black shuttle box equipped with an electric shock following our previously published protocol (Figure 6A) [42]. However, no significant alteration on a short-term memory test in terms of either training-phase latency (Figure 6B) or testing-phase latency (Figure 6C) was found in the C_{70} NP-treated fish. This result demonstrated that chronic exposure to C_{70} NPs does not cause short-term memory alteration in zebrafish.

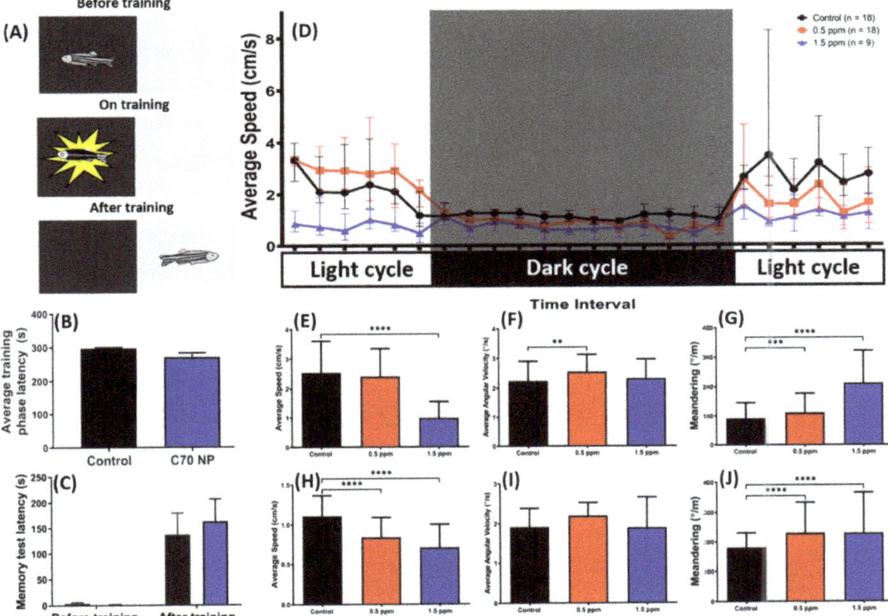

Figure 6. The short-term memory and circadian rhythm assay for the untreated control and C_{70} NP-exposed fish after 7- and 21-day exposure, respectively. (**A**) Schematic showing the experimental protocol for the passive avoidance test. (**B**) The average training phase latency on zebrafish learning. (**C**) The memory retention latency for the memory test. The data are expressed as the mean ± SEM and were analyzed by two-way ANOVA with Sidak's multiple comparisons test as a follow-up test ($n = 10$ for the untreated control; $n = 6$ for the C_{70} NP-exposed fish). (**D**) Comparison of time chronological changes of the average speed between wild-type and C_{70} NP-exposed fish in the day and night cycle. The grey area shows the dark period and the unshaded area is the light period. Comparison of (**E**) average speed, (**F**) average angular velocity, and (**G**) meandering at light period. Comparison of (**H**) average speed, (**I**) average angular velocity and (**J**) meandering at the dark period. The data are expressed as the median with interquartile range and were analyzed by Kruskal–Wallis test, with Dunn's multiple comparisons test as a follow-up test ($n = 18$ for the untreated control; $n = 18$ for 0.5 ppm C_{70} NP-exposed fish; $n = 9$ for the 1.5 ppm C_{70} NP-exposed fish; *** $p < 0.001$, **** $p < 0.0001$).

Light cycles are the most important synchronizers of biological rhythms in nature [15]. We assessed zebrafish circadian rhythms and the effects of light to dark photoperiods on zebrafish locomotor activity after three weeks of C_{70} NPs exposure (Figure 6D), represented as average swimming speed over time. In agreement with our other locomotor activity test results done in C_{60} NPs [42], high concentration of C_{70} NPs exposure-treated fish showed lower circadian locomotor activities than the control group during the light and dark periods, while low concentration of C_{70} NPs exposure only affected their locomotor activity during the dark cycle as indicated by abnormal level of average speed during both periods (Figure 6E,H). Furthermore, zig-zag-like movement behavior was also detected in C_{70} NP-treated groups, indicated by a higher level of meandering compared to the control group in both of the light and dark periods (Figure 6G,J). Meanwhile, there was no irregular swimming orientation exhibited by both treated fish groups, except for a high average angular velocity in the 0.5 ppm C_{70} NP-treated group during the light period (Figure 6F,I).

2.7. Impact of C_{70} NPs Exposure on Biomarker Expression in the Muscle, Brain, and Gill

By the end of all behavioral tests on day 22, we sacrificed the fish, dissected the muscle, brain, and gill tissues, extracted total proteins and subjected to perform an enzyme-linked immunosorbent assay (ELISA) to measure biomarker expression after C_{70} NP_S exposure. In muscle tissue, the effects of C_{70} NPs exposure on several important biomarkers such as reactive oxygen species (ROS) generation, lipid peroxidation (MDA and TBARS) production, and anti-oxidative stress enzyme (CAT and SOD) activities are shown in Table 1. The significant increases of ROS, TBARS and MDA contents were observed in the muscle after treatment of C_{70} NPs. While the increment of the MDA level was also shown in the gill tissue, there was no significant change in the treated fish gill regarding the TBARS level. The C_{70} NPs exposure also triggered the activation of antioxidative enzymes since we found the relative activities of CAT and SOD were significantly elevated in the muscle after C_{70} NPs exposure. This similar phenomenon was also shown in the gill tissues after the CAT activity was measured. By measuring the stress hormone, we provided evidence to show the strong anxiety behavior induced by C_{70} NP_S exposure was well agreed with the high elevation of cortisol level in the muscle. In addition, the hypoactivity in locomotion after C_{70} NP_S exposure led us to ask whether it was correlated to the muscle energy or oxygen supplement deficiency. We addressed this question by measuring the creatine kinase (CK), ATP and Hif1-α levels. Results showed the relative activity or content of CK and ATP was reduced in the muscle after C_{70} NP_S expose. The hypoxia marker of Hif1-α, on the contrary, displayed significantly elevation after C_{70} NP_S expose.

From the results, we also found that the activities of antioxidative enzymes in the brains from the 1.5 ppm and 0.5 ppm group were significantly lower than the control. Furthermore, the reduction of enzymatic activities in the brain caused by 1.5 ppm C_{70} NPs was greater than that by the 0.5 ppm and untreated controls. Combined with those findings, we concluded the oxidative stress in the muscle induced by the C_{70} NPs treatment was apparently more severe than the control group. The effects of C_{70} NPs exposure on DNA damage, hypoxia and inflammation were evaluated by measuring biomarkers such as ssDNA, Hif-1α, IL-1β, and TNF-α activity. In 0.5 ppm C_{70} NP-treated groups, the activities of inflammatory markers were not significantly different from those of the control. However, the activities of inflammatory markers in the brains and gills of 1.5 ppm C_{70} NP-treated groups were significantly higher than those of the control. Taken together, 1.5 ppm C_{70} NP-exposed zebrafish muscle and gill tissues showed a higher inflammatory marker response compared to the 0.5 ppm C_{70} NP-exposed groups and those of the untreated controls.

Considering that behavioral changes induced by C_{70} NPs exposure may be related to alterations on the cholinergic system, the effect of C_{70} NPs exposure on AChE, ACh, serotonin, and melatonin activity from the brain of zebrafish were evaluated. Data in Table 1 showed a significant inhibition of ACh, melatonin, and serotonin activities in the brain of the 1.5 ppm C_{70} NP-treated group. Moreover, AChE activity was significantly increased by C_{70} NPs treatment after three weeks of chronic exposure.

Later, we analyzed the regulation of two isoforms of cyclooxygenase (COX-1 and COX-2) in the zebrafish brain after the C_{70} NPs exposure. Cyclooxygenase plays a role to produce prostanoids like prostaglandins and thromboxanes that are all responsible for the inflammatory response. Interestingly, COX-1 expression was significantly elevated in the 1.5 ppm C_{70} NP-treated group compared to the untreated control and 0.5 ppm C_{70} NPs treated. The COX-2 protein was expressed at a very low level in the brain tissue of C_{70} NPs exposed to adult zebrafish compared to untreated control.

Table 1. Comparison of biomarker expression in the muscle, brain, and gill tissues in C_{70} NP-exposed zebrafish. Data are expressed as the mean ± SEM.

Biomarker	WT ($n = 10$)	C70 (0.5 ppm)	C70 (1.5 ppm)	Unit	Significance	ANOVA F Value	p Value
				Muscle			
ROS	2.95 ± 0.12	5.62 ± 0.26 ***	8.46 ± 0.25 ****	U/ug of total protein	YES	$F_{(2, 6)} = 157.2$	$p < 0.0001$
CAT	3.03 ± 0.13	5.01 ± 0.44 **	3.82 ± 0.01 NS	U/ug of total protein	YES	$F_{(2, 6)} = 14.10$	$p = 0.0054$
SOD	6.19 ± 0.78	14.38 ± 0.34 ***	14.34 ± 0.54 ***	U/ug of total protein	YES	$F_{(2, 6)} = 66.33$	$p < 0.0001$
TBARS	5.60 ± 0.34	7.39 ± 0.36 *	8.37 ± 0.29 **	ng/ug of total protein	YES	$F_{(2, 6)} = 17.92$	$p = 0.0029$
MDA	0.15 ± 0.00	0.20 ± 0.01 *	0.22 ± 0.01 *	ng/ug of total protein	YES	$F_{(2, 6)} = 8.826$	$p = 0.0163$
Cortisol	17.95 ± 0.90	20.87 ± 1.17 NS	23.95 ± 0.66 **	pg/ug of total protein	YES	$F_{(2, 6)} = 10.3$	$p = 0.0114$
Hif1-α	11.65 ± 0.54	17.60 ± 0.87 **	18.45 ± 1.00 **	pg/ug of total protein	YES	$F_{(2, 6)} = 20.05$	$p = 0.0022$
ssDNA	0.61 ± 0.01	0.81 ± 0.07 *	0.89 ± 0.05 **	U/ug of total protein	YES	$F_{(2, 6)} = 9.408$	$p = 0.0141$
TNF-α	6.86 ± 0.41	8.90 ± 0.33 **	11.76 ± 0.26 ***	pg/ug of total protein	YES	$F_{(2, 6)} = 52.24$	$p = 0.0002$
IL1-β	0.42 ± 0.05	0.47 ± 0.02 NS	0.57 ± 0.03 *	ng/ug of total protein	YES	$F_{(2, 6)} = 5.946$	$p = 0.0377$
ATP	363.60 ± 9.02	343.60 ± 15.23 NS	282.20 ± 9.30 **	ng/ug of total protein	YES	$F_{(2, 6)} = 13.5$	$p = 0.0060$
CK	3.06 ± 0.14	4.78 ± 0.34 **	2.79 ± 0.14 NS	pg/ug of total protein	YES	$F_{(2, 6)} = 22.17$	$p = 0.0017$
				Brain			
ROS	0.90 ± 0.03	1.61 ± 0.19 NS	3.53 ± 0.64 **	U/ug of total protein	YES	$F_{(2, 6)} = 12.38$	$p = 0.0074$
SOD	2.20 ± 0.06	3.44 ± 0.08 ****	3.28 ± 0.09 ***	U/ug of total protein	YES	$F_{(2, 6)} = 73.1$	$p < 0.0001$
Cortisol	3.47 ± 0.13	4.76 ± 0.08 *	4.91 ± 0.51 *	pg/ug of total protein	YES	$F_{(2, 6)} = 6.561$	$p = 0.0309$
Hif1-α	4.26 ± 0.11	4.91 ± 0.16 NS	6.86 ± 0.37 ***	pg/ug of total protein	YES	$F_{(2, 6)} = 32.09$	$p = 0.0006$
ssDNA	0.13 ± 0.01	0.28 ± 0.09 NS	0.37 ± 0.02 *	U/ug of total protein	YES	$F_{(2, 6)} = 5.274$	$p = 0.0477$
ACh	4.44 ± 0.15	4.61 ± 0.08 NS	2.93 ± 0.31 **	U/ug of total protein	YES	$F_{(2, 6)} = 20.92$	$p = 0.0020$
AChE	0.70 ± 0.10	0.77 ± 0.02 NS	1.03 ± 0.08 **	U/ug of total protein	YES	$F_{(2, 6)} = 12.75$	$p = 0.0069$
Melatonin	1.02 ± 0.03	0.94 ± 0.04 NS	0.68 ± 0.03 ***	pg/ug of total protein	YES	$F_{(2, 6)} = 31.66$	$p = 0.0006$
Serotonin	0.20 ± 0.01	0.22 ± 0.01 NS	0.14 ± 0.00 **	ng/ug of total protein	YES	$F_{(2, 6)} = 31.05$	$p = 0.0007$
Dopamine	9.04 ± 0.08	9.30 ± 0.20 NS	6.92 ± 0.09 ****	pg/ug of total protein	YES	$F_{(2, 6)} = 95.27$	$p < 0.0001$
COX-1	0.16 ± 0.00	0.18 ± 0.00 NS	0.38 ± 0.01 ****	U/pg of total protein	YES	$F_{(2, 6)} = 258.2$	$p < 0.0001$
COX-2	0.74 ± 0.01	0.94 ± 0.12 NS	0.64 ± 0.03 NS	U/pg of total protein	YES	$F_{(2, 6)} = 4.32$	$p = 0.0688$
				Gills			
CAT	1.44 ± 0.03	1.64 ± 0.02 **	1.30 ± 0.03 *	U/ug of total protein	YES	$F_{(2, 6)} = 44.21$	$p = 0.0003$
TBARS	1.71 ± 0.14	1.64 ± 0.06 NS	2.30 ± 0.68 NS	ng/ug of total protein	NO	$F_{(2, 6)} = 0.821$	$p = 0.4842$
MDA	0.08 ± 0.00	0.09 ± 0.00 NS	0.11 ± 0.00 **	ng/ug of total protein	YES	$F_{(2, 6)} = 21.29$	$p = 0.0019$
TNF-α	3.06 ± 0.10	2.98 ± 0.03 NS	8.68 ± 0.74 ***	pg/ug of total protein	YES	$F_{(2, 6)} = 56.94$	$p = 0.0001$

N.S. = not significant, * $p < 0.05$, ** $p < 0.01$, *** $p < 0.001$, **** $p < 0.0001$.

2.8. Clustering of Zebrafish Behavior Distinguished Carbon NPs (C_{60} and C_{70}) Based on Their Exposure Concentration

Next, we would like to explore the difference of behavioral alteration between C_{70} NPs and other chemical toxicity by using a novel phenomic approach. In addition to the high and low doses of C_{70} fullerene NPs treatment, we also included our previous published C_{60} NPs [43] and $ZnCl_2$ [15] data for cluster comparison. Initially, we transformed the behavioral endpoints for novel tank exploration, mirror biting, predator avoidance, social interaction, and shoaling into a scoring matrix. Later, this scoring matrix was subjected to principal component analysis (PCA) to elucidate the relationship between each experimental group. Both PCA (Figure 7A) and hierarchical clustering analysis (Figure 7B) demonstrate the behavioral alteration patterns between C_{60} and C_{70} NPs were close to each other and could be grouped into a single clade. The $ZnCl_2$-exposed fish, on the contrary, displayed a distinct behavioral alteration pattern from those for the fullerene-exposed fish. For instance, $ZnCl_2$ exposure can strongly increase the freezing behavior (behavioral endpoint 3-4) and reduce the average swimming speed (behavioral endpoint 3-1), mirror biting time percentage (behavioral endpoint 3-2), longest duration in the mirror side (behavioral endpoint 3-3), and swimming time movement ratio (behavioral endpoint 3-5) of the treated zebrafish when compared to C_{60}/C_{70} NPs exposure. Based on behavioral phenomic evidence collected here, we concluded each chemical can induce unique, fingerprint-like behavioral alteration patterns in zebrafish.

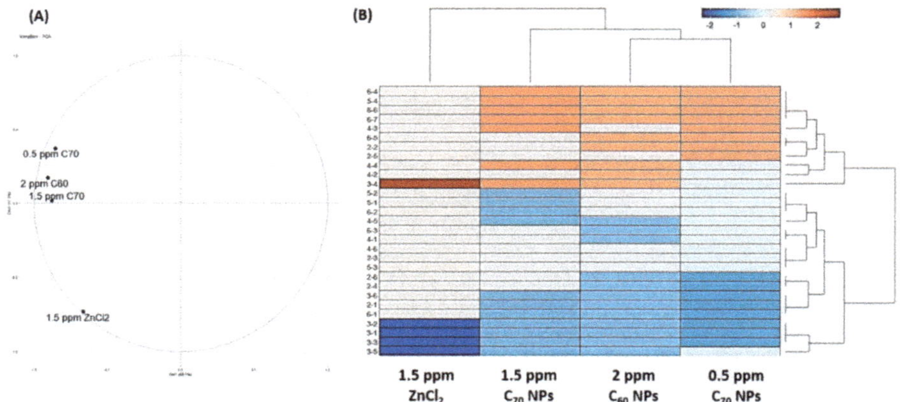

Figure 7. Comparison of behavioral alterations in zebrafish after exposing to either C_{60} NPs, C_{70} NPs or $ZnCl_2$. (**A**) Results obtained from principal component analysis (PCA). (**B**) Results obtained from hierarchical clustering and heat map analysis.

3. Discussion

The biochemical and behavioral effects of fullerene C_{70} NPs were assessed in vivo using the adult zebrafish as a model organism. The result presented herein clearly demonstrates the usefulness of this model as an effective platform to rapidly assess manufactured fullerene nanomaterial toxicity. Thus far, much of the data on the effects of fullerene exposure had been obtained from in vitro methods [15,37,42,44,45] with few exceptions [37]. Research on C_{70} NPs toxicity is scarce. In vitro data may be of lacking the predictive ability of in vivo responses, particularly since those results might be dependent on the cell culture system selected for the experiment. For instance, the cytotoxicity of fullerene C_{60} to human liver carcinoma cells (HepG2), neuronal human astrocytes, and dermal fibroblasts were found to be dependent on cell types [46].

Recently, several reports indicated that nano-sized particulate matters can reach the brain and may be related to neurodegenerative diseases [47,48]. Another study in mice revealed that nanoparticles might be taken up to the brain from olfactory epithelium to the various parts of the rat brain through olfactory nerves [49]. Our results on adult zebrafish exposure to C_{70} NPs were similar to those observations and caused behavioral impairments, the deregulated levels of various biomarkers. These observations indicated that C_{70} NPs produced direct or indirect inflammations to the fish brain.

The present research is the first to demonstrate an environmentally relevant carbon-based nanoparticulate induces behavioral dysfunction in zebrafish. Here, we use the zebrafish model to show the effect of two different doses of C_{70} NPs on different behavior parameters including locomotion, exploration, shoaling, circadian rhythm, and social interaction, important stress-related biomarkers and neurotransmitters. Since animal behavior is considered as the result of complex interactions between a species and the environment, the pattern of the behavioral repertoire of a species could be used as an indicator of the health status of an organism [50]. Behavioral responses can have ecological effects at community and population levels [16]. Therefore, behavioral data could provide valuable information as early reporters of toxicity at higher levels of biological organization [51]. In this experiment, novel tank test results showed zebrafish chronically exposed to 1.5 ppm of C_{70} NPs presented significant alterations on total distance traveled, suggesting both motor and locomotor behavior patterns are impaired compared with the 0.5 ppm and untreated controls. In addition, in a circadian rhythm test, both of the C_{70} NP-treated groups also showed a significant decrease in locomotor activity in both light and dark cycles. Altered activity patterns and locomotion can lead to an increased vulnerability to predators [52]. Another changed parameter was aggressive behavior. Fish exposed to C_{70} NPs revealed a significant reduction in the aggressive nature of zebrafish, which plays a crucial role in the behavior and ecology of adult fish. Furthermore, C_{70} NPs were also found to dysregulate zebrafish social interaction behavior. Since this behavior is related to foraging, mating, fear response, and defense against predators, these behavioral deficits may also be related to the loose shoals formed observed during the shoaling test. These results showed that alterations in normal zebrafish behavior and impaired mobility in fish exposed to C_{70} NPs, which could compromise the survival of a population in natural environments. In addition, C_{70} NPs exposure was also found to change color preferences pattern. Together, behavioral response assessment may serve as a discerning tool for quantitative monitoring of toxicological effects of nanoparticle-based water contaminants in fish species [53].

Among the biomarkers of toxicity, our data showed a distinct production of ROS and lipid peroxidation (MDA content increased) in the muscle and gills of the fish treated with a higher dose of nanoparticles (Table 1), indicating that these C_{70} NP-treated fish experienced severe oxidative stress. Similarly, in another rodent study reported particulate matters in polluted air caused oxidative stress in the mouse brain [54]. The over-accumulation of ROS would tip the balance of the antioxidative/oxidative system in the brain, resulting in the significant reduction of the antioxidative enzymes such as SOD and CAT (Table 1). Nanoparticles were no longer freely circulated in the cytoplasm after being internalized by cells but were preferentially located in mitochondria [15]. However, when the mitochondria were invaded by the nanomaterials, the antioxidant defense capacity could be compromised [55]. Our study showed that the total anti-oxidation capacity decreased with increasing C_{70} NP doses. Muscle, gills, brains, and liver are organs with an active metabolism, responsible for vital functions of the body such as respiration, motion, behavior, excretion, and accumulation of xenobiotics [56]. Finally, it is noteworthy that due to its smaller size, and larger surface per mass and high reactivity, our data showed that C_{70} NPs were able to migrate into the brain more readily, were absorbed more from the circulation, and thus led to more severe toxicity in adult zebrafish than the bulk fullerene.

Recent trends in information technology have seen great momentum in biomedical research that helped usher in a new generation of approaches to understanding and sharing knowledge in both disciplines. Today, systems biology utilizes large-scale datasets to investigate molecular signaling networks from an integrative and comprehensive standpoint. Since the advent of proteomics, transcriptomics, and genomics, various data-rich fields of biology such as metabolomics and glycomics have emerged based on the compilation and validation of large-scale datasets. It is in this regard that the massive amount of phenotypic datasets generated by high-throughput behavioral screens has given rise to a new and vibrant field of a new subject neuro-phenomics: the integrative analysis of neural phenotypes and their regulation by various environmental and genetic factors [57]. With the wide application of zebrafish in neuroscience, a better understanding of the role of environmental factors in aquatic models facilitates further in-depth neuro-phenotyping studies by this small animal model [58]. The influence of various environmental pollutants/modifiers on zebrafish behavioral phenotypes is scarce [59]. To this end, we applied the phenomics' approach; a cluster vector-based method was used following cross-calculated PCA and heatmap to highlight the main behavioral alterations induced by fullerene C_{70} NPs. In essence, this phenomics approach combines fast and simple data acquisition with complex and extensive behavioral analysis (enabled by behavior recognition, movement pattern, and video tracking), with ease in studies with a small sample size.

Most relevant to the present investigation is research in zebrafish embryos, which indicated a significant increase in pericardial edema, malformations, and mortality resulted after exposure to C_{70} NPs [60]. They directly determined in vivo fullerene exposures induced cellular death by two independent cellular death assays. We did not investigate any cell death assays but analyzed the behavioral changes and significant changes in biochemical assays that can provide important insights into the physiology of the organism. Up-regulation of proteins with antioxidant activity (including reactive oxygen species, MDA and TBARS) after fullerene C_{70} NPs treatments in the present study was consistent with the notion that the fish were responding to oxidative injury by activating a defense mechanism following exposure to nanoparticles. Oxidative stress is a crucial subject in aquatic toxicology. Damage to mitochondrial structure and function accelerates ROS production and causes oxidative stress [61]. The present study showed that C_{70} NPs could inhibit the activities of antioxidant enzymes, including CAT and SOD as confirmed by the biochemical assay in adult zebrafish tissues after exposure. These results demonstrated that C_{70} NPs elicited oxidative damages. In addition, the immune system plays an important role when assessing chemical toxicity. Previous reports have shown that chemicals can dysregulate the immune system and exert immunotoxicity on animals [34,62]. In the present study, the expression of TNF-α and IL-1β, representative proteins of inflammations, were obviously up-regulated at 1.5 ppm C_{70} NPs exposure. This phenomenon indicates that C_{70} NPs triggered a significant immune response suggesting a synergistic effect on inflammation. Inflammation and oxidative stress are concatenated processes that are usually activated in cells due to stress [63]. Consistent with the ROS measurement result, over-accumulation of inflammation is well known to contribute to the high level of ROS in the brain. ROS is produced through the Fenton reaction of amyloid Aβ with metal ions and causes the accumulation of inflammatory cytokines, including TNF-α and IL-1β that attract active plaques [64]. In addition, this result is in line with another prior study by Zhang and colleagues. In their study, it was found that the bare and starch-coated NPs displayed different tissue toxicity and both types of NP could induce inflammation and oxidative stress [65]. Moreover, the lack of anti-inflammatory function observed in the tissues was also related to the up-regulated ssDNA level. The ssDNA content in all treatment groups was higher than that of the untreated group. These results indicated that the DNA damage response was also activated after C_{70} NPS treatment. It might be interpreted as an adaptation during fullerene intoxication.

Adult zebrafish exposed to C_{70} NPs were not lethal, but behavioral changes and biochemical responses were obvious compared to untreated controls. Of all the biochemical markers tested, the majority of them showed alterations at the protein level and gills pathology. Further investigation is required to determine if these changes relate to corresponding gene expression or signal transduction pathways indicative of exposure to C_{70} NPs aggregates and changes in proteomic profiles. The behavioral impairments observed in zebrafish exposed to C_{70} NPs indicated that the exposure scenario used in this study had significant effects on the fish through ecologically-relevant doses used. Longer exposure to C_{70} NPs or other exposure scenarios (including different species, mode of exposure) may result in different effects. A fine but the significant point is that, in this study, changes in biochemical results were investigated after 21 days, presumably the expression of these proteins/enzymes could have been affected differently after different exposure durations and at different developmental stages of zebrafish.

4. Summary and Conclusions

To our knowledge, this is the first study to show behavioral alterations induced by fullerene C_{70} NPs chronic exposure in adult zebrafish at environmentally-relevant concentrations. These effects seem to be correlated with changes in oxidative stress, inflammation, hypoxia, and imbalance of neurotransmitters in the brain (summarized in Figure 8). By utilizing this multiple behavioral test approach, our dynamic whole-animal model can be used to reveal the toxic potential of novel manufactured nanomaterials at the behavioral, physiological, and cellular levels. Furthermore, information obtained using this animal model system could be used as rapid feedback for engineers manufacturing novel nanomaterials, such that they can consider potential toxicity to favor the development of engineered nanoparticles with minimal toxicity.

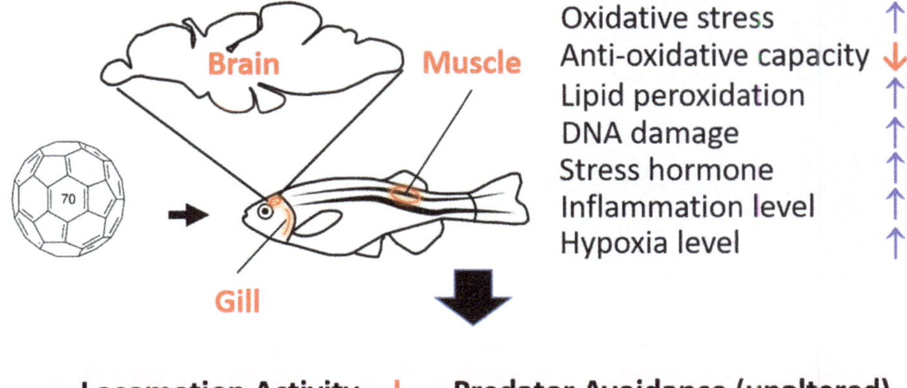

Figure 8. Schematic diagram showing the detrimental effects of chronic exposure of C_{70} nanoparticles on adult zebrafish. The corresponding behavioral alterations (pink color) and biochemical alterations in the brain tissue (blue color) after C_{70} nanoparticles exposure are summarized.

5. Materials and Methods

5.1. Chemicals

C_{70} Fullerene (C_{70}, 99.5% purity) was purchased from Sigma-Aldrich (St. Louis, MO, USA) (Cat no. 482994) and DMSO was purchased from Fisher scientific (Waltham, MA, USA). All the reagents used were analytical grade.

5.2. Animals

Adult wild-type zebrafish AB strain (*Danio rerio*) of both sexes (6–7 months old) were used in this study. Animals were acclimatized for at least one week before the experiments and were fed by lab-grown brine shrimps two times a day. The fish were maintained in a healthy condition and free of any signs of infections, and were used according to the guidelines for the care and use of laboratory animals by CYCU. All procedures in the present study were approved by the Animal Ethics Committee of the Chung Yuan Christian University (Approval ID 107030; approval date: 19/12/2018).

5.3. C_{70} NP_S Suspensions

Suspensions of C_{70} NPs in DMSO were prepared as in the previously described protocol with some modifications [66,67]. C_{70} NPs suspensions were further sonicated for one hour prior to use. To avoid the photoexcitation of C_{70} NPs, the whole exposure procedure was in the absence of light. However, overnight sonication was necessary to uniformly distribute in DMSO.

5.4. Characterization of C_{70} NPs

Prior to the exposure, C_{70} NPs were characterized for size, surface area, and structural properties by scanning electron microscope and X-ray diffraction methods. The stock solution dispersion was confirmed by SEM. However, without the use of SDS the C_{70} NPs were not dispersed and clearly visible as polymerized long strands about 50–70 nm thick (Figure 1B). The C_{70} NPs were suspended in 0.1% DMSO, stirred well and then sonicated overnight, pipetted in 20ul droplets, deposited on a copper grid, and the sample grid was dried in a microwave oven for about 4 h without vacuum. Then the copper grid was directly inserted into a FESEM machine after it was completely dried. The images were taken with a 10k magnification CCD camera. The fullerene C_{70} NPs were further analyzed by the X-ray diffraction method and determined to be of high purity. C_{70} NPs dispersed in DMSO and stock solutions were made at concentrations of 0.5 mg/L and 1.5 mg/L. Samples of these suspensions were taken for analysis of particle size distribution, zeta potential, and particle dissolution. Particle size distributions were obtained using a Zetasizer Nano ZS with a 633 nm red laser and were capable of both particle size analysis (using dynamic light scattering as the basic principle of operation) and zeta potential measurement (Zeta Potential Instruments Inc., Long Island, NY, USA).

5.5. Zebrafish Exposure Protocol

The experimental adult zebrafish of males and females with an average age group of 6 months, an average weight of 0.60 ± 0.10 g, and an average length of 43.25 ± 2.76 mm, were selected for the study. Two different concentrations of C_{70} NPs suspensions (0.5 and 1.5 mg/L) were prepared prior to use. The fish were separated into three groups, each consisting of 20 adult zebra fish. The three groups were exposed to tank water (as a control), 0.5 ppm C_{70} NPs, and 1.5 ppm C_{70} NPs for a test period in 20 L glass tanks containing 15 L of a test solution. In order to maintain the constant concentration of the C_{70} NPs, the test suspensions were replaced every 24 h. For the behavior toxicity test, the control fish were not exposed to any nanoparticle or solvent, while the experimental group was treated with C_{70} NPs at different doses and about 70 % water was changed every 24h with refilling after each change. All fish were terminated at the end of the particle exposure and were sacrificed within minutes by immersed in the working concentration of tricaine (160 ppm) prior to body weight measurement. Once all the behavioral tests were done, the fish were anesthetized and immediately euthanized by immersion in high-dose tricaine solution at 1600 ppm (Sigma) and their tissues were removed for further biochemical assays. From each fish, brain, muscle, and gill tissues were independently harvested and all biochemical assays were performed. Exposure experiments were conducted three times.

5.6. Behavioral Endpoints

The behavioral endpoints being measured were the following: 3D locomotion, novel tank exploration, aggression, predator avoidance test, shoaling, conspecific social interaction, circadian rhythm, color preferences, and short-term memory tests. For 3D locomotion, novel tank exploration, aggression, predator avoidance test, shoaling, conspecific social interaction, and shoaling behaviors, the camera was placed around 5 m in front of the zebrafish tower which is described in our previously published protocol [68,69]. The video was recorded in black and white mode with a frame rate of 50 fps (frames per second).

5.6.1. 3D Locomotion Test

3D locomotion test was performed on the 14th day of C_{70} NPs exposure to the zebrafish. The tracking procedure of C_{70} NP-treated fish was followed by our previously published method [70]. Two different concentrations of fullerene C_{70} NPs were used for this study (0.5 and 1.5 mg/L). Three separate experiments were performed using the same batch of C_{70} NP-exposed fish.

5.6.2. Novel Tank Test

The novel tank test was defined by our previous publication [70] which reflected the congenital characteristics in the swimming behavior of zebrafish. Normally, zebrafish have two behavioral phenomena: freezing, which was defined as a total absence of movement, except for the gills and eyes for 1s or longer; and erratic movements, which was defined as sharp changes in direction or velocity and repeated darting behaviors. In this experiment, zebrafish were placed individually in the tank. Their behaviors were recorded for 1 min at intervals of 0, 5, 10, 15, 20, 25, and 30 min. A video camera with a long-range zoom lens feature was positioned in front of the test tank. The novel tank test parameters of behaviors were the following: average speed (cm/s), freezing time movement ratio (%), time in top duration (%), number of entries to the top, latency to enter the top (s), and total distance traveled in the top (cm). Later, the recorded videos were analyzed by idTracker [70] and data tracking was calculated using Microsoft Excel.

5.6.3. Aggression Test

The aggression test was referenced from the previous study [70–72] with some modifications. C_{70} NP-treated adult fish and untreated fish were placed into the test tank containing a mirror placed on vertically to one side of the wall. The aggressive test parameters were the following: average speed (cm/s), mirror biting time percentage (%), longest duration in the mirror side (%), freezing, swimming, and rapid time movement ratio (%).

5.6.4. Predator-Avoidance Test

The predator-avoidance behavior test of zebrafish was referenced from our previous publication [70]. The fear and escape behaviors as a response for the predator presence were determined by measuring average speed (cm/s), predator approaching time (%), the average distance to a separator (cm), freezing, swimming, and rapid movement ratio (%).

5.6.5. Social Interaction Test

The social interaction test which was based on our previous publication was conducted to assess zebrafish social interaction behavior with their conspecifics [70]. In this assay, we measured interaction time percentage (%), longest duration in separator side (%), average speed (cm/s), and average distance to a separator (cm).

5.6.6. Shoaling Test

The sixth zebrafish behavioral test in this experiment was the shoaling test, an assessment of group affiliation behavior. This test was conducted based on our previous publication [70]. In this test, average speed (cm/s), time in top duration (%), average shoal area (cm^2), average inter-fish distance (cm), and average nearest and farthest neighbor distance (cm) were calculated.

5.6.7. Circadian Rhythm Test

The circadian rhythm test was carried out to evaluate zebrafish sleep/wake behaviors on the 21st day of C_{70} NPs exposure and the test was based on previous publications [70]. In this experiment, we recorded zebrafish locomotor activity (average speed (cm/s), average angular velocity (°/s), and meandering (°/m)) for 60 s every 60 min in 24 h. Later, idTracker software was used to track fish movement trajectories.

5.6.8. Color Preferences Test

The zebrafish color preferences were assessed using a 20 × 20 × 10 cm tank divided into a two-color partition (blue–red, blue–yellow, blue–green, red–yellow, red–green, and yellow–green). The experiment was recorded using a CCD camera (ONTOP, M2 module, China) for 30 min. The video was analyzed using idTracker to determine the position of the zebrafish.

5.6.9. Short-Term Memory Test

We performed a short-term memory test by using a passive avoidance setting according to our previous publication [73,74]. Initially, 20 fish were randomly grouped into control and C_{70} NPs exposed groups with 10 fish each. Later, experimental group fish exposed to 1.5 ppm C_{70} NPs were exposed to a shuttle box to perform short-term memory tests. The learning latency, total number of electric shocks given for training, and memory latency were recorded for comparison between control and 1.5 C_{70} NP-exposed fish.

5.7. Total Protein Extraction

After behavioral tests, nine fish per treatment were collected from each tank for biochemical analysis. Gills, brain, and red muscle tissues were carefully removed and a pool of three fish was used for homogenate preparation. Tissues were homogenized at 8000 rpm in a bullet blender with 50 volumes of (v/w) ice-cold phosphate saline buffer with a pH of 7.2. Samples were further centrifuged at 4000 rpm for 20 min and the crude homogenate was stored in 100 µL aliquots at −20 °C until required. Total protein concentration in tissues was determined using a Pierce BCA Protein Assay kit (23225, Thermo Fisher Scientific, Massachusetts, MA, USA). The color formation was analyzed at 562 nm using a microplate reader (Multiskan GO, Thermo Fisher Scientific). Exposed fish tissues were analyzed to determine the possible effects of lipid peroxidation, oxidative stress, neurotransmitter changes, and antioxidant activity.

5.8. Biochemical Parameter Assay

To evaluate the toxic effects of C_{70} NPs, alterations of biomarkers in zebrafish tissues due to NPs exposure were determined. The tissue oxidative stress marker like reactive oxygen species was measured by ELISA kits purchased from a commercial company (ZGB-E1561, Zgenebio Inc., Taipei, Taiwan). The DNA damage marker (ssDNA) and lipid peroxidation (malondialdehyde, MDA, and thiobarbituric acid reactive substances—TBARS) markers were measured by target-specific ELISA kits purchased from a commercial company (ZGB-E1595, ZGB-E1592, and ZGB-E1605 Zgenebio Inc., Taipei, Taiwan). Cortisol, a stress hormone, and two inflammation markers of TNF-α and IL-1β were measured by using commercial target-specific ELISA kits (ZGB-E1560, ZGB-E1612, and ZGB-E1609, Zgenebio Inc., Taipei, Taiwan). Creatine kinase (CK) and adenosine-5′–triphosphate (ATP), keys marker for energy metabolism, and hypoxia-inducible factor 1-alpha (Hif1-α), a key marker for hypoxia, were measured by using target-specific ELISA kits purchased from a commercial company (ZGB-E1646, ZGB-E1645, and ZGB-E1643, Zgenebio Inc., Taipei, Taiwan). Cyclooxygenase (COX-1 and COX-2), enzymes for prostanoids production, were measured by target-specific ELISA kits purchased from a commercial company (ZGB-E1655 and ZGB-E1656, Zgenebio Inc., Taipei, Taiwan). The antioxidant enzymes-related biomarkers, like superoxide dismutase (SOD, ZGB-E1604) and catalase (CAT, ZGB-E1598), and neurotoxic responses in terms of acetylcholine esterase (AChE, ZGB-E1637), dopamine (DA, ZGB-E1573), acetylcholine (ACh, ZGB-E1585), melatonin (ZGB-E1597), and serotonin (ZGB-E1572) activities were measured by target-specific ELISA kits purchased from commercial company (Zgenebio Inc., Taipei, Taiwan) according to the manufacture's protocol. These assay kits are based on the sandwich ELISA method that involves a specific antibody for the detection of the molecules of interest. Briefly, 10 µL of brain homogenate was placed onto the well. Then, 100 ul of horseradish peroxidase (HRP)-conjugate reagent was added into each well, covered with an adhesive strip and incubated for 60 min at 37 °C. The well was washed with wash solution (400 µL) and chromogen A 50 ul and chromogen B 50 µL were added to each well. The mixture was incubated for 15 min at 37 °C. Then the reaction was stopped by adding 50 µL of stop solution to each well. The absorbance was analyzed at 450 nm using a microplate reader (Multiskan GO, Thermo Fisher Scientific) within 15 min. The data were expressed as either U/µg, ng/µg, or pg/µg of total protein.

5.9. Statistical Analysis

The zebrafish behavioral data were analyzed by different types of statistical analysis. In the novel tank test, two-way ANOVA with Geisser–Greenhouse correction was conducted and the significant differences between the control group and C_{70} NP-treated groups during the whole 30 min test were described by the "*" symbol. For the 3D locomotion, mirror biting, predator avoidance, social interaction, shoaling, and circadian rhythm tests, the Kruskal–Wallis test with Dunn's multiple comparisons test as a follow-up test was used. The color preference data were analyzed using two-way ANOVA followed by a Tukey post-hoc test. If the data were not normally distributed, they were analyzed using non-parametric Kruskal–Wallis followed by Dunn's post-hoc test. The biochemical data were analyzed individually ($n = 20$ for both the control fish group and C_{70} nanoparticle-treated fish group. Biomarker responses of exposed fish were compared with control fish by a one-way ANOVA test followed by the post-hoc test of Tukey, depending upon the data normality for significant data. The level of significance was set at a p value <0.05. All statistics were plotted and compiled by using GraphPad Prism (GraphPad Software version 7 Inc., La Jolla, CA, USA).

5.10. PCA, Heatmap, and Clustering Analysis

All behavior results data were converted in to excel file using Microsoft Excel. Every endpoint alteration was defined as a number, ranging from -4 to 4, where -4 means the value of the endpoint was significantly lower than the control group (**** $p < 0.0001$) and 4 means the endpoints values was significantly higher than the control group (**** $p < 0.0001$). If the treated fish behavioral endpoint was not significantly different from the control fish, the number was defined as 0 ($p > 0.05$). All of the important behavioral endpoints in each test were listed in the previous study [42]. Next, the excel file was converted to a comma delimited type file (.csv) in order to be readable by R software. PCA, heatmap, and clustering analysis were carried out by R software (https://www.r-project.org/).

Author Contributions: For research articles with several authors, a short paragraph specifying their individual contributions must be provided. The following statements should be used "conceptualization, S.S. and G.A.; methodology, G.A. and P.S.; validation, S.J., P.S. and J.-R.C.; formal analysis, P.S.; investigation, S.S.; resources, E.H.; data curation, S.J. and P.S.; writing—original draft preparation, C.-D.H.; supervision, J.-R.C. and C.-D.H.; project administration, C.-D.H.; funding acquisition, C.-D.H."

Funding: This study was funded by the grants sponsored by the Ministry of Science Technology (MOST 105-2313-B-033-001-MY3 and MOST107-2622-B-033-001-CC2) to C.D.H.

Conflicts of Interest: The authors declare no conflicts of interest.

References

1. Afreen, S.; Muthoosamy, K.; Manickam, S.; Hashim, U. Functionalized fullerene (C60) as a potential nanomediator in the fabrication of highly sensitive biosensors. *Biosens. Bioelectron.* **2015**, *63*, 354–364. [CrossRef]

2. Bogdanović, G.; Đorđević, A. Carbon nanomaterials: Biologically active fullerene derivatives. *Srpski Arhiv Celokupno Lekarstvo* **2016**, *144*, 222–231. [CrossRef]

3. Liu, F.-f.; Zhao, J.; Wang, S.; Du, P.; Xing, B. Effects of solution chemistry on adsorption of selected pharmaceuticals and personal care products (PPCPs) by graphenes and carbon nanotubes. *Environ. Sci. Technol.* **2014**, *48*, 13197–13206. [CrossRef] [PubMed]

4. Ju-Nam, Y.; Lead, J.R. Manufactured nanoparticles: An overview of their chemistry, interactions and potential environmental implications. *Sci. Total Environ.* **2008**, *400*, 396–414. [CrossRef] [PubMed]

5. Usepa, U. *Nanotechnology white papper*; U.S. Environmental Protection Agency: Washington, DC, USA, 2007; p. 20460.

6. Innocenzi, P.; Brusatin, G. Fullerene-based organi—Inorganic nanocomposites and their applications. *Chem. Mater.* **2001**, *13*, 3126–3139. [CrossRef]

7. Zhu, X.; Sollogoub, M.; Zhang, Y. Biological applications of hydrophilic C60 derivatives (hC60s)–A structural perspective. *Eur. J. Med. Chem.* **2016**, *115*, 438–452. [CrossRef]

8. Soleyman, R.; Hirbod, S.; Adeli, M. Advances in the biomedical application of polymer-functionalized carbon nanotubes. *Biomater. Sci.* **2015**, *3*, 695–711. [CrossRef]

9. Bakry, R.; Vallant, R.M.; Najam-ul-Haq, M.; Rainer, M.; Szabo, Z.; Huck, C.W.; Bonn, G.K. Medicinal applications of fullerenes. *Int. J. Nanomed.* **2007**, *2*, 639.

10. Harrison, B.S.; Atala, A. Carbon nanotube applications for tissue engineering. *Biomaterials* **2007**, *28*, 344–353. [CrossRef]

11. Kamat, J.P.; Devasagayam, T.P.; Priyadarsini, K.; Mohan, H.; Mittal, J.P. Oxidative damage induced by the fullerene C60 on photosensitization in rat liver microsomes. *Chem. Biol. Interact.* **1998**, *114*, 145–159. [CrossRef]

12. Wolff, D.J.; Papoiu, A.D.; Mialkowski, K.; Richardson, C.F.; Schuster, D.I.; Wilson, S.R. Inhibition of nitric oxide synthase isoforms by tris-malonyl-C60-fullerene adducts. *Arch. Biochem. Biophys.* **2000**, *378*, 216–223. [CrossRef]

13. Oberdörster, E. Manufactured nanomaterials (fullerenes, C60) induce oxidative stress in the brain of juvenile largemouth bass. *Environ. Health Perspect.* **2004**, *112*, 1058–1062. [CrossRef]

14. Zhu, S.; Oberdörster, E.; Haasch, M.L. Toxicity of an engineered nanoparticle (fullerene, C60) in two aquatic species, Daphnia and fathead minnow. *Marine Environ. Res.* **2006**, *62*, S5–S9. [CrossRef]

15. Sarasamma, S.; Audira, G.; Juniardi, S.; Sampurna, B.; Lai, Y.-H.; Hao, E.; Chen, J.-R.; Hsiao, C.-D. Evaluation of the Effects of Carbon 60 Nanoparticle Exposure to Adult Zebrafish: A Behavioral and Biochemical Approach to Elucidate the Mechanism of Toxicity. *Int. J. Mol. Sci.* **2018**, *19*, 3853. [CrossRef] [PubMed]

16. Semmler, M.; Seitz, J.; Erbe, F.; Mayer, P.; Heyder, J.; Oberdörster, G.; Kreyling, W. Long-term clearance kinetics of inhaled ultrafine insoluble iridium particles from the rat lung, including transient translocation into secondary organs. *Inhalation Toxicol.* **2004**, *16*, 453–459. [CrossRef] [PubMed]

17. John, A.A.; Subramanian, A.P.; Vellayappan, M.V.; Balaji, A.; Mohandas, H.; Jaganathan, S.K. Carbon nanotubes and graphene as emerging candidates in neuroregeneration and neurodrug delivery. *Int. J. Nanomed.* **2015**, *10*, 4267.

18. Yang, X.; Ebrahimi, A.; Li, J.; Cui, Q. Fullerene–biomolecule conjugates and their biomedicinal applications. *Int. J. Nanomed.* **2014**, *9*, 77. [CrossRef]

19. Hurt, R.H.; Monthioux, M.; Kane, A. Toxicology of carbon nanomaterials: Status, trends, and perspectives on the special issue. *Carbon* **2006**, *44*, 1028–1033. [CrossRef]

20. Monteiro-Riviere, N.; Inman, A.; Zhang, L. Limitations and relative utility of screening assays to assess engineered nanoparticle toxicity in a human cell line. *Toxicol. Appl. Pharmacol.* **2009**, *234*, 222–235. [CrossRef]

21. Kroll, A.; Pillukat, M.H.; Hahn, D.; Schnekenburger, J. Current in vitro methods in nanoparticle risk assessment: Limitations and challenges. *Eur. J. Pharm.* **2009**, *72*, 370–377. [CrossRef]

22. Chakraborty, C.; Sharma, A.R.; Sharma, G.; Lee, S.-S. Zebrafish: A complete animal model to enumerate the nanoparticle toxicity. *J. Nanobiotechnol.* **2016**, *14*, 65. [CrossRef] [PubMed]

23. Rubinstein, A.L. Zebrafish assays for drug toxicity screening. *Exp. Opin. Drug Metabol. Toxicol.* **2006**, *2*, 231–240. [CrossRef] [PubMed]

24. Hill, A.J.; Teraoka, H.; Heideman, W.; Peterson, R.E. Zebrafish as a model vertebrate for investigating chemical toxicity. *Toxicol. Sci.* **2005**, *86*, 6–19. [CrossRef] [PubMed]

25. De Esch, C.; Slieker, R.; Wolterbeek, A.; Woutersen, R.; de Groot, D. Zebrafish as potential model for developmental neurotoxicity testing: A mini review. *Neurotoxicology* **2012**, *34*, 545–553. [CrossRef]

26. Scholz, S.; Fischer, S.; Gündel, U.; Küster, E.; Luckenbach, T.; Voelker, D. The zebrafish embryo model in environmental risk assessment—Applications beyond acute toxicity testing. *Environ. Sci. Pollut. Res.* **2008**, *15*, 394–404. [CrossRef]

27. Seda, B.C.; Ke, P.C.; Mount, A.S.; Klaine, S.J. Toxicity of aqueous C70-gallic acid suspension in Daphnia magna. *Environ. Toxicol.* **2012**, *31*, 215–220. [CrossRef]

28. Tchounwou, P.B.; Yedjou, C.G.; Patlolla, A.K.; Sutton, D.J. Heavy metal toxicity and the environment. In *Molecular, Clinical and Environmental Toxicology*; Springer: Cham, Switzerland, 2012; pp. 133–164.

29. Karen, S.; Brown, T.M. *Principles of Toxicology*; CRC Press: Boca Raton, FL, USA, 2006.

30. Herculano, A.M.; Maximino, C. Serotonergic modulation of zebrafish behavior: Towards a paradox. *Progress Neuro-Psychopharmacol. Biol. Psychiatry* **2014**, *55*, 50–66. [CrossRef]

31. Kokel, D.; Peterson, R.T. Using the zebrafish photomotor response for psychotropic drug screening. In *Methods in Cell Biology*; Elsevier: Amsterdam, The Netherlands, 2011; Volume 105, pp. 517–524.

32. Stewart, A.M.; Gerlai, R.; Kalueff, A.V. Developing highER-throughput zebrafish screens for in-vivo CNS drug discovery. *Front. Behav. Neurosci.* **2015**, *9*, 14. [CrossRef]

33. Horie, M.; Nishio, K.; Kato, H.; Shinohara, N.; Nakamura, A.; Fujita, K.; Kinugasa, S.; Endoh, S.; Yoshida, Y.; Hagihara, Y. In vitro evaluation of cellular influences induced by stable fullerene C70 medium dispersion: Induction of cellular oxidative stress. *Chemosphere* **2013**, *93*, 1182–1188. [CrossRef]

34. Usenko, C.Y.; Harper, S.L.; Tanguay, R.L. In vivo evaluation of carbon fullerene toxicity using embryonic zebrafish. *Carbon* **2007**, *45*, 1891–1898. [CrossRef]

35. Moretz, J.A.; Martins, E.P.; Robison, B.D. Behavioral syndromes and the evolution of correlated behavior in zebrafish. *Behav. Ecol.* **2007**, *18*, 556–562. [CrossRef]

36. Speedie, N.; Gerlai, R. Alarm substance induced behavioral responses in zebrafish (Danio rerio). *Behav. Brain Res.* **2008**, *188*, 168–177. [CrossRef] [PubMed]

37. Wang, X.; Zheng, Y.; Zhang, Y.; Li, J.; Zhang, H.; Wang, H. Effects of β-diketone antibiotic mixtures on behavior of zebrafish (Danio rerio). *Chemosphere* **2016**, *144*, 2195–2205. [CrossRef] [PubMed]

38. Pham, M.; Raymond, J.; Hester, J.; Kyzar, E.; Gaikwad, S.; Bruce, I.; Fryar, C.; Chanin, S.; Enriquez, J.; Bagawandoss, S. Assessing social behavior phenotypes in adult zebrafish: Shoaling, social preference, and mirror biting tests. In *Zebrafish Protocols for Neurobehavioral Research*; Springer: Cham, Switzerland, 2012; pp. 231–246.

39. Glass, A.S.; Dahm, R. The zebrafish as a model organism for eye development. *Ophthalmic Res.* **2004**, *36*, 4–24. [CrossRef]

40. Bault, Z.A.; Peterson, S.M.; Freeman, J.L. Directional and color preference in adult zebrafish: Implications in behavioral and learning assays in neurotoxicology studies. *J. Appl. Toxicol. JAT* **2015**, *35*, 1502–1510. [CrossRef]

41. Park, J.S.; Ryu, J.H.; Choi, T.I.; Bae, Y.K.; Lee, S.; Kang, H.J.; Kim, C.H. Innate color preference of zebrafish and its use in behavioral analyses. *Mol. Cells* **2016**, *39*, 750–755. [CrossRef]

42. Sarasamma, S.; Audira, G.; Juniardi, S.; Sampurna, B.; Liang, S.-T.; Hao, E.; Lai, Y.-H.; Hsiao, C.-D. Zinc Chloride exposure inhibits brain acetylcholine levels, produces neurotoxic signatures, and diminishes memory and motor activities in adult zebrafish. *Int. J. Mol. Sci.* **2018**, *19*, 3195. [CrossRef]

43. López-Olmeda, J.F.; Madrid, J.A.; Sánchez-Vázquez, F.J. Light and temperature cycles as zeitgebers of zebrafish (Danio rerio) circadian activity rhythms. *Chronobiol. Int.* **2006**, *23*, 537–550. [CrossRef]

44. Larner, S.F.; Wang, J.; Goodman, J.; O'Donoghue Altman, M.B.; Xin, M.; Wang, K.K. In vitro neurotoxicity resulting from exposure of cultured neural cells to several types of nanoparticles. *J. Cell Death* **2017**, *10*, 1179670717694523. [CrossRef]

45. Ershova, E.; Sergeeva, V.; Chausheva, A.; Zheglo, D.; Nikitina, V.; Smirnova, T.; Kameneva, L.; Porokhovnik, L.; Kutsev, S.; Troshin, P. Toxic and DNA damaging effects of a functionalized fullerene in human embryonic lung fibroblasts. *Mutat. Res. Genetic Toxicol. Environ. Mutagen.* **2016**, *805*, 46–57. [CrossRef]

46. Nakagawa, Y.; Inomata, A.; Ogata, A.; Nakae, D. Comparative effects of sulfhydryl compounds on target organellae, nuclei and mitochondria, of hydroxylated fullerene-induced cytotoxicity in isolated rat hepatocytes. *J. Appl. Toxicol.* **2015**, *35*, 1465–1472. [CrossRef] [PubMed]

47. Usenko, C.Y.; Harper, S.L.; Tanguay, R.L. Fullerene C60 exposure elicits an oxidative stress response in embryonic zebrafish. *Toxicol. Appl. Pharmacol.* **2008**, *229*, 44–55. [CrossRef] [PubMed]

48. Sayes, C.M.; Gobin, A.M.; Ausman, K.D.; Mendez, J.; West, J.L.; Colvin, V.L. Nano-C60 cytotoxicity is due to lipid peroxidation. *Biomaterials* **2005**, *26*, 7587–7595. [CrossRef] [PubMed]

49. Block, M.; Wu, X.; Pei, Z.; Li, G.; Wang, T.; Qin, L.; Wilson, B.; Yang, J.; Hong, J.; Veronesi, B. Nanometer size diesel exhaust particles are selectively toxic to dopaminergic neurons: The role of microglia, phagocytosis, and NADPH oxidase. *FASEB J.* **2004**, *18*, 1618–1620. [CrossRef] [PubMed]

50. Hawkins, S.J.; Crompton, L.A.; Sood, A.; Saunders, M.; Boyle, N.T.; Buckley, A.; Minogue, A.M.; McComish, S.F.; Jiménez-Moreno, N.; Cordero-Llana, O. Nanoparticle-induced neuronal toxicity across placental barriers is mediated by autophagy and dependent on astrocytes. *Nat. Nanotechnol.* **2018**, *13*, 427. [CrossRef] [PubMed]

51. Schmidel, A.J.; Assmann, K.L.; Werlang, C.C.; Bertoncello, K.T.; Francescon, F.; Rambo, C.L.; Beltrame, G.M.; Calegari, D.; Batista, C.B.; Blaser, R.E. Subchronic atrazine exposure changes defensive behaviour profile and disrupts brain acetylcholinesterase activity of zebrafish. *Neurotoxicol. Teratol.* **2014**, *44*, 62–69. [CrossRef]

52. Weis, J.S.; Candelmo, A. Pollutants and fish predator/prey behavior: A review of laboratory and field approaches. *Curr. Zool.* **2012**, *58*, 9–20. [CrossRef]

53. Kalueff, A.V.; Gebhardt, M.; Stewart, A.M.; Cachat, J.M.; Brimmer, M.; Chawla, J.S.; Craddock, C.; Kyzar, E.J.; Roth, A.; Landsman, S. Towards a comprehensive catalog of zebrafish behavior 1.0 and beyond. *Zebrafish* **2013**, *10*, 70–86. [CrossRef]

54. Scott, G.R.; Sloman, K.A. The effects of environmental pollutants on complex fish behaviour: Integrating behavioural and physiological indicators of toxicity. *Aquat. Toxicol.* **2004**, *68*, 369–392. [CrossRef]

55. MohanKumar, S.M.; Campbell, A.; Block, M.; Veronesi, B. Particulate matter, oxidative stress and neurotoxicity. *Neurotoxicology* **2008**, *29*, 479–488. [CrossRef]

56. De Lorenzo, A.D. The olfactory neuron and the blood-brain barrier. In *Ciba Foundation Symposium-Internal Secretions of the Pancreas (Colloquia on Endocrinology)*; John Wiley and Sons: Chichester, UK, 1970; pp. 151–176.

57. Long, T.C.; Saleh, N.; Tilton, R.D.; Lowry, G.V.; Veronesi, B. Titanium dioxide (P25) produces reactive oxygen species in immortalized brain microglia (BV2): Implications for nanoparticle neurotoxicity. *Environ. Sci. Technol.* **2006**, *40*, 4346–4352. [CrossRef] [PubMed]

58. Gernhöfer, M.; Pawert, M.; Schramm, M.; Müller, E.; Triebskorn, R. Ultrastructural biomarkers as tools to characterize the health status of fish in contaminated streams. *J. Aquat. Ecosyst. Stress Recovery* **2001**, *8*, 241–260. [CrossRef]

59. Gerlai, R. Phenomics: Fiction or the future? *Trends Neurosci.* **2002**, *25*, 506–509. [CrossRef]

60. Stewart, A.M.; Kaluyeva, A.A.; Poudel, M.K.; Nguyen, M.; Song, C.; Kalueff, A.V. Building zebrafish neurobehavioral phenomics: Effects of common environmental factors on anxiety and locomotor activity. *Zebrafish* **2015**, *12*, 339–348. [CrossRef] [PubMed]

61. Collymore, C.; Tolwani, R.J.; Rasmussen, S. The behavioral effects of single housing and environmental enrichment on adult zebrafish (Danio rerio). *J. Am. Assoc. Lab. Animal Sci.* **2015**, *54*, 280–285.

62. James, A.M.; Yau-Huei, W.; Cheng-Yoong, P.; MURPHY, M.P. Altered mitochondrial function in fibroblasts containing MELAS or MERRF mitochondrial DNA mutations. *Biochem. J.* **1996**, *318*, 401–407. [CrossRef]

63. Eder, K.J.; Clifford, M.A.; Hedrick, R.P.; Köhler, H.-R.; Werner, I. Expression of immune-regulatory genes in juvenile Chinook salmon following exposure to pesticides and infectious hematopoietic necrosis virus (IHNV). *Fish Shellfish Immunol.* **2008**, *25*, 508–516. [CrossRef]

64. Jin, Y.; Pan, X.; Cao, L.; Ma, B.; Fu, Z. Embryonic exposure to cis-bifenthrin enantioselectively induces the transcription of genes related to oxidative stress, apoptosis and immunotoxicity in zebrafish (Danio rerio). *Fish Shellfish Immunol.* **2013**, *34*, 717–723. [CrossRef]

65. Dandekar, A.; Mendez, R.; Zhang, K. Cross talk between ER stress, oxidative stress, and inflammation in health and disease. In *Stress Responses*; Springer: Cham, Switzerland, 2015; pp. 205–214.

66. Condello, C.; Yuan, P.; Schain, A.; Grutzendler, J. Microglia constitute a barrier that prevents neurotoxic protofibrillar $A\beta42$ hotspots around plaques. *Nat. Commun.* **2015**, *6*, 6176. [CrossRef]

67. Zheng, M.; Lu, J.; Zhao, D. Effects of starch-coating of magnetite nanoparticles on cellular uptake, toxicity and gene expression profiles in adult zebrafish. *Sci. Total Environ.* **2018**, *622*, 930–941. [CrossRef]

68. Isaacson, C.W.; Usenko, C.Y.; Tanguay, R.L.; Field, J.A. Quantification of fullerenes by LC/ESI-MS and its application to in vivo toxicity assays. *Anal. Chem.* **2007**, *79*, 9091–9097. [CrossRef] [PubMed]

69. Kim, K.-T.; Jang, M.-H.; Kim, J.-Y.; Kim, S.D. Effect of preparation methods on toxicity of fullerene water suspensions to Japanese medaka embryos. *Sci. Total Environ.* **2010**, *408*, 5606–5612. [CrossRef] [PubMed]

70. Audira, G.; Sampurna, B.; Juniardi, S.; Liang, S.-T.; Lai, Y.-H.; Hsiao, C.-D. A versatile setup for measuring multiple behavior endpoints in zebrafish. *Inventions* **2018**, *3*, 75. [CrossRef]

71. Pérez-Escudero, A.; Vicente-Page, J.; Hinz, R.C.; Arganda, S.; De Polavieja, G.G. idTracker: Tracking individuals in a group by automatic identification of unmarked animals. *Nat. Methods* **2014**, *11*, 743. [CrossRef]

72. Weber, D.N.; Hoffmann, R.G.; Hoke, E.S.; Tanguay, R.L. Bisphenol A exposure during early development induces sex-specific changes in adult zebrafish social interactions. *J. Toxicol. Environ. Health Part A* **2015**, *78*, 50–66. [CrossRef]

73. Audira, G.; Sarasamma, S.; Chen, J.-R.; Juniardi, S.; Sampurna, B.; Liang, S.-T.; Lai, Y.-H.; Lin, G.-M.; Hsieh, M.-C.; Hsiao, C.-D. Zebrafish Mutants carrying leptin a (lepa) gene deficiency display obesity, anxiety, less aggression and fear, and circadian rhythm and color preference dysregulation. *Int. J. Mol. Sci.* **2018**, *19*, 4038. [CrossRef]
74. Audira, G.; Sampurna, B.P.; Juniardi, S.; Liang, S.-T.; Lai, Y.-H.; Han, L.; Hsiao, C.-D. Establishing simple image-based methods and a cost-effective instrument for toxicity assessment on circadian rhythm dysregulation in fish. *Biol. Open* **2019**, *8*, bio041871. [CrossRef]

International Journal of
Molecular Sciences

Review

Potent Impact of Plastic Nanomaterials and Micromaterials on the Food Chain and Human Health

Yung-Li Wang [1], Yu-Hsuan Lee [2], I-Jen Chiu [1,3], Yuh-Feng Lin [1,3,*] and Hui-Wen Chiu [1,3,*]

[1] Graduate Institute of Clinical Medicine, College of Medicine, Taipei Medical University, Taipei 11031, Taiwan; cetuspower@gmail.com (Y.-L.W.); stirbar2000@yahoo.com.tw (I.-J.C.)

[2] Department of Cosmeceutics, China Medical University, Taichung 40402, Taiwan; bmm175@hotmail.com

[3] Division of Nephrology, Department of Internal Medicine, Shuang Ho Hospital, Taipei Medical University, New Taipei City 23561, Taiwan

* Correspondence: linyf@s.tmu.edu.tw (Y.-F.L.); leu3@tmu.edu.tw (H.-W.C.); Tel.: +886-2-22490088 (Y.-F.L.); +886-2-22490088 (H.-W.C.)

Received: 4 February 2020; Accepted: 1 March 2020; Published: 3 March 2020

Abstract: Plastic products are inexpensive, convenient, and are have many applications in daily life. We overuse plastic-related products and ineffectively recycle plastic that is difficult to degrade. Plastic debris can be fragmented into smaller pieces by many physical and chemical processes. Plastic debris that is fragmented into microplastics or nanoplastics has unclear effects on organismal systems. Recently, this debris was shown to affect biota and to be gradually spreading through the food chain. In addition, studies have indicated that workers in plastic-related industries develop many kinds of cancer because of chronic exposure to high levels of airborne microplastics. Microplastics and nanoplastics are everywhere now, contaminating our water, air, and food chain. In this review, we introduce a classification of plastic polymers, define microplastics and nanoplastics, identify plastics that contaminate food, describe the damage and diseases caused by microplastics and nanoplastics, and the molecular and cellular mechanisms of this damage and disease as well as solutions for their amelioration. Thus, we expect to contribute to the understanding of the effects of microplastics and nanoplastics on cellular and molecular mechanisms and the ways that the uptake of microplastics and nanoplastics are potentially dangerous to our biota. After understanding the issues, we can focus on how to handle the problems caused by plastic overuse.

Keywords: plastic products; food chain; microplastics; nanoplastics

1. Introduction

Recently, plastic products have become inexpensive and convenient and are used in all aspects of daily life, such as food and product packaging, clothing, construction and car materials, household goods, medical devices, personal care products, and, toys. [1,2]. Although plastic products are relatively convenient, the negative influences of the "plastic era" caused by these inexpensive and convenient products include high levels of plastic production coupled with a slow biodegradation rate, uncontrolled use, and ineffective and irresponsible waste recycling, leading to plastic accumulation in our global environment, particularly in freshwater and marine environments [3–10]. The use of plastic products has increased rapidly, and 33 billion tons of plastic will likely be produced by 2050 [11], making the Pacific Ocean a giant garbage dump [12]. The plastic debris in aquatic environments is fragmented into smaller pieces by ultraviolet light and biodegraded plastic forms microplastics and nanoplastics [13–15]. However, the largest proportion of microplastics and nanoplastics is generated from the laundering of textiles with mixed synthetic fibers [16] and the friction of the tires of moving cars [17,18]. These microplastics and nanoplastics have unclear effects on organismal systems. Recently, evidence has been presented indicating that plastics significantly affect the growth and oxygen production of *Prochlorococcus* and microalgae. *Prochlorococcus*, especially, is the ocean's most abundant photosynthetic bacteria and produces 10% of global oxygen. [19,20]. However, the growth of earthworms is meaningfully different in soil ecosystems, particularly agricultural land, contaminated with microplastics [21]. In addition, it is noteworthy that microplastics are widespread in naturally-occurring Arctic deep-sea sediments [22] and in snow ranging from the Alps to the Arctic [23]. Therefore, microplastic and nanoplastic contamination is everywhere [24]. Microplastics have been found in human stool [25] and humans can consume microplastics and nanoplastics through seafood [26–30] and water [31–36], etc. Whether plastics will harm our health is unclear; however, the potential consequences may affect the ecological functioning of the globe and future generations of organisms (Figure 1). In this review, we briefly introduce a classification of plastic materials and describe the origin of microplastics and nanoplastics, the food contaminated by microplastics and nanoplastics, the damage and diseases caused by microplastics and nanoplastics, the molecular and cellular mechanisms of the damage and diseases caused by microplastics and nanoplastics, and solutions to mediate the problems caused by plastic overuse.

Figure 1. Some sources and deposits of microplastic and nanoplastic are the result of human needs. The potential impacts of these plastics on air, water, and many foods finally returns to affect humans. All pictures come from Pixabay (https://pixabay.com/).

2. Classification of Plastic Materials and Related Product Applications

Plastic production has gradually increased every year, from 1.5 million tons in the 1950s [37] to an estimated 33 billion tons in 2050 [11]. Plastics are specifically derived from synthetic polymers generated by the polymerization of many monomers and mixtures of a range of materials [38]. Therefore, plastic is predominantly generated into polyethylene (PE), polyester (PES), polyethylene terephthalate (PET), polyetherimide (PEI) (Ultem), polystyrene (PS), polypropylene (PP), low-density polyethylene (LDPE) high-density polyethylene (HDPE), polyvinyl chloride (PVC), polyvinylidene chloride (PVDC) (Saran), polycarbonate (PC), polycarbonate/acrylonitrile butadiene styrene (PC/ABS), high-impact polystyrene (HIPS), polyamides (PA) (nylon), acrylonitrile butadiene styrene (ABS), polyurethanes (PU), urea–formaldehyde (UF), melamine formaldehyde (MF), polymethyl methacrylate (PMMA), polytetrafluoroethylene (PTFE), and polylactic acid (PLA), etc. [39] (Figure 2). The highest percentages of plastics produced worldwide meet the definition of thermoplastic: PP (21%), LDPE (18%), PVC (17%), and HDPE (15%) [40]. Plastic polymers have numerous applications in daily life [41]. PP is usually used in pots for plants, bags, industrial fibers, netting, medical masks, bottle caps, ropes, straws, containers, tanks and jugs, appliances, car fenders, plastic pressure pipe systems, and centrifuge tubes. LDPE is usually used in outdoor furniture, siding, wire cable, floor tiles, plastic bags, shower curtains, buckets, clamshell packaging, and soap dispenser bottles. PVC is usually used in plumbing pipes and guttering, siding, shower curtains, blood bags, window frames, and flooring. HDPE is usually used in detergent bottles, plastic bottles, plastic bags, bottle caps, and milk jugs [37,39,42].

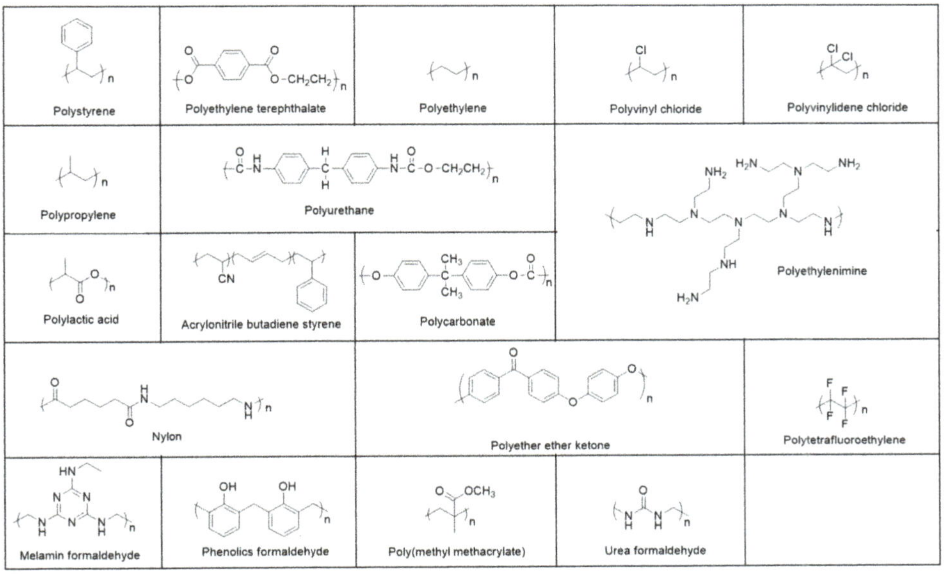

Figure 2. Typical polymer types and their chemical structures. Chemical structures are shown for polyethylene (PE), polyester (PES), polyethylene terephthalate (PET), polyethylenimine (PEI) (Ultem), polystyrene (PS), polylactic acid (PLA), polypropylene (PP), polyvinyl chloride (PVC), polyvinylidene chloride (PVDC) (Saran), polycarbonate (PC), polycarbonate/acrylonitrile butadiene styrene (PC/ABS), polyamides (PA) (nylon), acrylonitrile butadiene styrene (ABS), polyurethanes (PU), urea–formaldehyde (UF), melamine formaldehyde (MF), polymethyl methacrylate (PMMA), and polytetrafluoroethylene (PTFE).

3. Routes of Plastic Micromaterial and Nanomaterial Pollution

Plastic fragments can be generally divided into several types: macroplastics and mesoplastics are greater than 5 mm in size [43], microplastic particles are smaller than 5 mm [44], and nanoplastics are less than 1000 nm or 100 nm [45]. Currently, microplastics are found worldwide in freshwater and marine systems [46], in sediment [47], in soil [48], and within biota [49]. Nanoplastics are generated by the abiotic and biotic degradation of microplastics. For example, UV degradation of microplastics has been shown to generate nanoplastics [13,50], and digestive fragmentation has been proposed as a means by which nanoplastics can be generated from microplastics [51]. However, two types of microplastics, primary and secondary microplastics, are categorized by the form in which they are released. Primary microplastics are directly released into the environment as small particles. Secondary microplastics are derived from large plastic items being degraded into small plastic fragments upon exposure to the environment [52]. The largest proportion of these particles is derived from laundering textiles with mixed synthetic fibers [16] and the friction of car tires [17,18]. Secondary microplastics from synthetic textiles in garments are the major type of microplastics [53,54]. Each textile may release approximately 1900 fibers per washing [55]. Other sources of microplastics and nanoplastics are urban dust [56], road markings [17], and personal care products [57].

4. Presence of Plastic Micromaterials and Nanomaterials in Food and Food Products

Microplastics and nanoplastics are currently everywhere. In marine environments, seabirds and marine mammals ingest microplastics at low trophic levels [52,58]. Microplastics and nanoplastics have been detected at the base of the food web, specifically, in zooplankton, such as chaetognaths [59]. Crustaceans, such as the Japanese shore crab [60] and North Pacific krill [61], contain microplastics and nanoplastics. Sea fish, such as the northeastern Pacific Ocean forage fishes [62], areolate grouper and goldbanded jobfish, are contaminated with microplastics [63]. Oysters ingest polystyrene microplastics, which affect their reproduction [64]. The mussel *Mytilus edulis* ingests microplastics that translocate to the circulatory system [65]. Many kinds of mussels are contaminated, including blue mussels [26,37,66], Mediterranean mussels [61,67], brown mussels [66], and northern horse mussels [68]. Plastic additives and hydrophobic organic compounds (HOCs) are also found in mussels [69]. In summary, many kinds of seafood are potentially contaminated by microplastics and/or nanoplastics [30]. In addition, in our daily life, many consumables, such as tap water [70], bottled water [34,71], beer [70,72], sea salt [70,73], sugar [74], honey [74] and plastic teabags [75] have also been found to contain microplastics or nanoplastics. Even air [76,77] and unprocessed water [78] have been contaminated with microplastics. Sooner or later, the entire food chain will be contaminated with plastic. Some statistics from studies published by PubMed on animals contaminated by microplastics and/or nanoplastics are presented in Table 1.

Table 1. Published studies from the National Center for Biotechnology Information (NCBI) on microplastic and/or nanoplastic contamination in different animal species.

Animal Species	Number of Published Studies from NCBI
Human	2
Bear	1
Mouse	5
Birds	5
Seabirds	8
Crustaceans	68
Bivalves	79
Fish and sea mammals	161
Insects	5
Turtles	5
Amphipods	2
Seaplants	3

5. Damage and Diseases Caused by Plastic Micromaterials and Nanomaterials

Many marine animals suffer from ingesting high amounts of plastic debris [79,80]. Fragments of this plastic debris, such as microplastics, accumulate in the gut and cause obstruction and inflammation in many organs in a wide range of living creatures [52,81]. Microplastics reduce photosynthesis in microalgae [20] and have a negative influence on the feeding behavior of zooplankton [58] and lugworms [82]. They also accumulate in and probably negatively influence the gills, stomach, and hepatopancreas of crabs [33], and they change the biomarkers and histology of fish tissues [84]. Evidence indicates that PS microplastics decrease the number of eggs and larvae produced and sperm velocity of oysters [64]. PS microplastics may also transport contaminants to microorganisms [81]. Studies have described the influence of microplastics on the digestive system [85]. The aquatic ecosystem has accommodated the ingestion of contaminated organisms [86]. Finally, this leads to the uptake of microplastics in the human intestine [26]. Several studies have indicated that PS microplastics can cause metabolic disorders of amino acids, bile acids [87], and liver lipids [88] in mice. Microplastics change gut microbiota dysbiosis and decrease gut mucin secretion in mice [87,88]. However, these microplastics or nanoplastics are also released to the atmosphere, becoming airborne contaminants [55,86,89]. Indeed, a study shows contamination in working environments [33]. Workers in the synthetic textiles, flock and vinyl chloride (VC), or polyvinyl chloride (PVC) industries are potentially exposed to high concentrations of microplastics in the air during work [76]. Synthetic textile workers potentially suffer higher rates of lung-cancer-related mortality [90] or stomach and esophageal cancers [91]. Flock workers have a high incidence of interstitial and lung diseases [92,93]. VC has been considered a carcinogenic factor and mostly causes angiosarcoma of the liver [94–96]. Microplastics or nanoplastics disrupt the endocrine system [97], induce neurotoxicity [98], and produce reproductive abnormities with trans-generational effects [99–103]. In addition, food and drink are a major vehicles of microplastic and nanoplastic exposure through which polymer elements and additives are potentially transported [104]. Risk assessments on using food packaging nanomaterials with antimicrobial activity, including titanium dioxide [105] and carbon nanotubes [106], have shown that they present risks comparable to those of using nanopolymers. The complex mixtures of plastic additives can dissolve in the polymer and leak into the surrounding environment [107]. The physical–chemical characteristics of these particles, such as the size, external charge, length:width ratio, porosity, surface corona, and hydrophilicity, cause different circulation times [108]. In addition, microplastics can absorb persistent organic pollutants (POPs) such as polycyclic aromatic hydrocarbons (PAHs), polychlorinated biphenyls (PCBs), and pesticides, including dichlorodiphenyltrichloroethane (DDT) and hexachlorobenzene (HCB), in the ocean [109,110]. These compounds have a higher affinity for plastic than for water [110,111]. Microplastics and/or nanoplastics are taken up into the gut and lungs, and enter many organs, where they potentially cause damage and result in disease.

6. Molecular and Cellular Mechanisms of Plastic Micromaterial and Nanomaterial Damage and Disease

Microplastics and/or nanoplastics can enter the circulation from the gut via trophic transfer [30] and from air [76,77]. Microplastics or nanoplastics inhibit the efflux pump and induce cytotoxicity in human intestinal cells [112]. The cytotoxicity induced by microplastics and/or nanoplastics stimulates oxidative stress via free radical generation originating from reactive oxygen species (ROS) [98,101,103,113–115]. Several studies have shown this connection in monogonont rotifer [116], *Caenorhabditis elegans* [117], *Danio rerio* [118], mouse liver [119], and human intestine cell lines [112]. Overproduced ROS can alter the homeostasis of cells by mediating antioxidant systems. ROS overwhelm the antioxidants produced in response to damage to cellular components, including DNA, carbohydrates, lipids, and proteins. This damage is associated with gene instability, physiological alterations, and carcinogenesis [120,121]. For example, scleractinian coral, *Pocillopora damicornis*, exposed to microplastics have increased superoxide dismutase (SOD) and catalase (CAT) activity and glutathione S-transferase (GST) and alkaline phosphatase (ALP) loss of function. SOD and CAT are antioxidant enzymes, GST is a detoxifying enzyme and ALP is an immune enzyme in coral. In addition, in coral, microplastics regulate genes that are related to the stress response, zymogen granules, c-Jun N-terminal kinase (JNK) signaling pathways, sterol transport, and the epidermal growth factor–extracellular signal-regulated kinase 1/2 (EGF-ERK1/2) pathway (Figure 3) [115]. In addition, PS nanoplastics increase oxidative stress, activate the expression of genes in the nuclear factor E2-related factor (Nrf) signaling pathway (Figure 3) [114], and increase expression of glutathione S-transferase 4 (GST-4) enzyme in *Caenorhabditis elegans* [117]. Additionally, a previous report showed that PS microplastics also induce inflammation and activate SOD and CAT activity in the livers of *Danio rerio* [118] and mice [119,122]. These findings indicate that microplastics induce oxidative stress as the main mechanism of toxicity induction in these organisms. PS microplastics can affect amino acid metabolism by increasing arginine and tyrosine and affect bile acid metabolism by mediating the levels of taurocholic acid (TCA), β-muricholic acid (βMCA), adenosine triphosphate (ATP)-binding cassette, subfamily B, member 11 (*Abcb11*) and cholesterol 7a-hydroxylase (*Cyp7a1*) [87]. They also affect liver lipid metabolism by changing the triglyceride (TG), total cholesterol (TCH), and pyruvate levels (Figure 3) [88]. PS microplastics also increased the acetylcholinesterase (AChE) activity and the related neurotransmitters such as threonine, aspartate, and taurine in a mouse model [122]. In addition, microplastics and nanoplastics elicit immunological responses [115,123,124], alter gene expression [88,103,113,114,125] and induce genotoxicity [113,126]. In kidney cells, VC stimulates the expression of fibrosis-related proteins, such as CTGF, PAI-1, and collagen 1, and autophagy-related proteins, such as Beclin 1 and LC3-II [127]. VC is also a carcinogenic factor and results in angiosarcoma of the liver [94–96]. Studies have indicated that VC causes several DNA mutations, such as *Ras* mutations [128], *K-ras-2* mutations [129], *p53* mutations [130,131], and *p21* mutations [132].

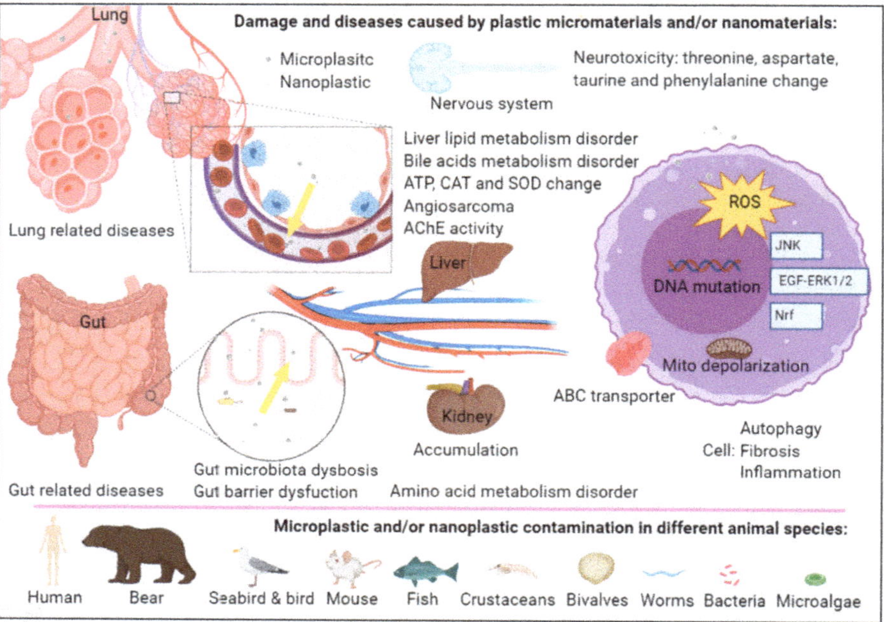

Figure 3. Impact of plastic micromaterials and nanomaterials in organisms. Microplastics and/or nanoplastics can enter the circulation from the gut and lungs and accumulate in the gut, liver, and kidney resulting in several diseases. At the cell level, microplastics or nanoplastics can inhibit the efflux pump and mitochondria depolarization, induce reactive oxygen species (ROS). They also affect several signaling pathways, cause fibrosis, autophagy, and even DNA mutations. Many animal species have been contaminated by microplastics and/or nanoplastics. The figure was created with BioRender.com.

7. Solutions for Reducing Plastic Micromaterials and Nanomaterials

A large area of accumulated garbage is adrift in the ocean [133]. Prevention and clean-up proposals have been made by political bodies around the world [134], such as the plastic reduction policy of Africa, which ranks first in the world [135]. A Netherlands-based organization, the Ocean Cleanup, uses massive drift nets to reduce the size of the Great Pacific Garbage Patch [136]. Wastewater treatment plants (WWTPs) in several countries have found microplastic particles [49,137–139]. Australia uses filters in large drains to stop garbage from entering the ocean [140]. Single-use plastic items are one of the components in this large area of plastic waste. In India, single-use plastic items, such as plastic bags, plastic spoons, plastic cups, plastic drinking straws, plastic jars, and plastic bottles, have been banned since October 2, 2019. The European Union has set a target to eliminate some single-use plastic items by 2021 [141]. Single-use plastic items, such as plastic straws, are being replaced—a Vietnamese company has developed a reed pipe to replace plastic straws [142], and a Taiwanese company has developed a straw with sugar cane [143]. In addition, facial cleansers containing plastic particles [57] have been banned in many countries [144]. On the one hand, plastic waste has been turned into resources. For example, a company in the Netherlands uses plastic to replace traditional road materials [145], and it is better than asphalt, with 60% greater strength [146]. In India, abandoned fishing nets have been turned into surfboards [147], in UK, students have successfully used fish skin and red algae as raw materials to develop plastic substitutes [148], and in Mexico, scientists have used cactus fruit to make nontoxic edible plastic [149]. On the other hand, due to physical and chemical changes, plastics become microplastics and nanoplastics. Therefore, some microbial biodegradation can be used to depolymerize those polymers into smaller monomers. Biodegradation is ultimately successful when plastics degrade monomers into CO_2 and water. Marine bacteria are potential candidates for use in the biodegradation of plastic wastes [150]. PS is known to biodegrade in the gut of yellow mealworms [151]. Many fungal strains can also degrade several plastics, such as PVC, PHB, and PLA [39]. Recently, several enzymes have been identified as capable of degrading PET plastics [152].

8. Conclusions

We over-use plastics because they are inexpensive and convenient, and worldwide, ecosystems are suffering, and contaminating-levels of plastic debris is a concern that has been reported. Plastic must be managed (especially in single-use items) and recycled such that it is finally fragmented to small plastics. Most of the contamination by microplastics and nanoplastics is derived from laundering synthetic textiles and the friction from the tires of driven cars. Currently, there is no effective way to reduce the amount of microplastics and nanoplastics in the food chain. Furthermore, it is unclear how the mixture of different sized groups and material types interacts with living creatures. Previous studies have indicated that workers in plastic-related industries suffer many kinds of cancer by being exposed to high levels of airborne microplastics over many years. In addition, it is important to characterize the microplastics and nanoplastics that have accumulated in the food chain and to gain a clearer understanding of their negative impact on our bodies. Finally, the degradation of microplastics and nanoparticles from environmental bacteria and fungi remains a challenge for the scientific community.

Author Contributions: Conceptualization, Y.-L.W., Y.-F.L., and H.-W.C.; writing—original draft preparation, Y.-L.W., Y.-F.L., and H.-W.C.; writing—review and editing, Y.-L.W., Y.-H.L., I.-J.C., Y.-F.L., and H.-W.C. All authors have read and agreed to the published version of the manuscript.

Funding: This study was supported by the Ministry of Science and Technology, Taiwan (MOST 106-2314-B-038-069-MY3, MOST 108-2314-B-039-061-MY3 and MOST 108-2314-B-038-044) and China Medical University, Taichung, Taiwan (CMU108-N-09).

Conflicts of Interest: The authors declare no conflict of interest. The funders had no role in the design of the study; in the collection, analyses, or interpretation of data; in the writing of the manuscript, or in the decision to publish the results.

References

1. Geyer, R.; Jambeck, J.R.; Law, K.L. Production, use, and fate of all plastics ever made. *Sci. Adv.* **2017**, *3*, e1700782. [CrossRef] [PubMed]
2. Rhodes, C.J. Plastic pollution and potential solutions. *Sci. Prog.* **2018**, *101*, 207–260. [CrossRef] [PubMed]
3. Diaz-Torres, E.R.; Ortega-Ortiz, C.D.; Silva-Iniguez, L.; Nene-Preciado, A.; Orozco, E.T. Floating marine debris in waters of the Mexican Central Pacific. *Mar. Pollut. Bull.* **2017**, *115*, 225–232. [CrossRef] [PubMed]
4. Ryan, P.G.; Musker, S.; Rink, A. Low densities of drifting litter in the African sector of the Southern Ocean. *Mar. Pollut. Bull.* **2014**, *89*, 16–19. [CrossRef]
5. Ryan, P.G. Litter survey detects the South Atlantic 'garbage patch'. *Mar. Pollut. Bull.* **2014**, *79*, 220–224. [CrossRef]
6. Eriksen, M.; Maximenko, N.; Thiel, M.; Cummins, A.; Lattin, G.; Wilson, S.; Hafner, J.; Zellers, A.; Rifman, S. Plastic pollution in the South Pacific subtropical gyre. *Mar. Pollut. Bull.* **2013**, *68*, 71–76. [CrossRef]
7. Cozar, A.; Echevarria, F.; Gonzalez-Gordillo, J.I.; Irigoien, X.; Ubeda, B.; Hernandez-Leon, S.; Palma, A.T.; Navarro, S.; Garcia-de-Lomas, J.; Ruiz, A.; et al. Plastic debris in the open ocean. *Proc. Natl. Acad. Sci. USA* **2014**, *111*, 10239–10244. [CrossRef]
8. Derraik, J.G. The pollution of the marine environment by plastic debris: A review. *Mar. Pollut. Bull.* **2002**, *44*, 842–852. [CrossRef]

9. Jambeck, J.R.; Geyer, R.; Wilcox, C.; Siegler, T.R.; Perryman, M.; Andrady, A.; Narayan, R.; Law, K.L. Marine pollution. Plastic waste inputs from land into the ocean. *Science* **2015**, *347*, 768–771. [CrossRef]

10. Lamb, J.B.; Willis, B.L.; Fiorenza, E.A.; Couch, C.S.; Howard, R.; Rader, D.N.; True, J.D.; Kelly, L.A.; Ahmad, A.; Jompa, J.; et al. Plastic waste associated with disease on coral reefs. *Science* **2018**, *359*, 460–462. [CrossRef]

11. Rochman, C.M.; Browne, M.A.; Halpern, B.S.; Hentschel, B.T.; Hoh, E.; Karapanagioti, H.K.; Rios-Mendoza, L.M.; Takada, H.; Teh, S.; Thompson, R.C. Policy: Classify plastic waste as hazardous. *Nature* **2013**, *494*, 169–171. [CrossRef]

12. Lebreton, L.; Slat, B.; Ferrari, F.; Sainte-Rose, B.; Aitken, J.; Marthouse, R.; Hajbane, S.; Cunsolo, S.; Schwarz, A.; Levivier, A.; et al. Evidence that the Great Pacific Garbage Patch is rapidly accumulating plastic. *Sci. Rep.* **2018**, *8*, 4666. [CrossRef] [PubMed]

13. Yousif, E.; Haddad, R. Photodegradation and photostabilization of polymers, especially polystyrene: Review. *Springerplus* **2013**, *2*, 398. [CrossRef] [PubMed]

14. Gewert, B.; Plassmann, M.M.; MacLeod, M. Pathways for degradation of plastic polymers floating in the marine environment. *Environ. Sci. Process. Impacts* **2015**, *17*, 1513–1521. [CrossRef] [PubMed]

15. Song, Y.K.; Hong, S.H.; Jang, M.; Han, G.M.; Jung, S.W.; Shim, W.J. Combined effects of UV exposure duration and mechanical abrasion on microplastic fragmentation by polymer type. *Environ. Sci. Technol.* **2017**, *51*, 4368–4376. [CrossRef] [PubMed]

16. Carney Almroth, B.M.; Astrom, L.; Roslund, S.; Petersson, H.; Johansson, M.; Persson, N.K. Quantifying shedding of synthetic fibers from textiles; a source of microplastics released into the environment. *Environ. Sci. Pollut. Res. Int.* **2018**, *25*, 1191–1199. [CrossRef]

17. Kole, P.J.; Lohr, A.J.; Van Belleghem, F.; Ragas, A.M.J. Wear and tear of tyres: A stealthy source of microplastics in the environment. *Int. J. Environ. Res. Public Health* **2017**, *14*, 1265. [CrossRef]

18. Redondo-Hasselerharm, P.E.; De Ruijter, V.N.; Mintenig, S.M.; Verschoor, A.; Koelmans, A.A. Ingestion and chronic effects of car tire tread particles on freshwater benthic macroinvertebrates. *Environ. Sci. Technol.* **2018**, *52*, 13986–13994. [CrossRef]

19. Tetu, S.G.; Sarker, I.; Schrameyer, V.; Pickford, R.; Elbourne, L.D.H.; Moore, L.R.; Paulsen, I.T. Plastic leachates impair growth and oxygen production in Prochlorococcus, the ocean's most abundant photosynthetic bacteria. *Commun. Biol.* **2019**, *2*, 184. [CrossRef]

20. Sjollema, S.B.; Redondo-Hasselerharm, P.; Leslie, H.A.; Kraak, M.H.S.; Vethaak, A.D. Do plastic particles affect microalgal photosynthesis and growth? *Aquat. Toxicol.* **2016**, *170*, 259–261. [CrossRef]

21. Boots, B.; Russell, C.W.; Green, D.S. Effects of microplastics in soil ecosystems: Above and below ground. *Environ. Sci. Technol.* **2019**. [CrossRef] [PubMed]

22. Bergmann, M.; Wirzberger, V.; Krumpen, T.; Lorenz, C.; Primpke, S.; Tekman, M.B.; Gerdts, G. High quantities of microplastic in arctic deep-sea sediments from the HAUSGARTEN observatory. *Environ. Sci. Technol.* **2017**, *51*, 11000–11010. [CrossRef] [PubMed]

23. Bergmann, M.; Mutzel, S.; Primpke, S.; Tekman, M.B.; Trachsel, J.; Gerdts, G. White and wonderful? Microplastics prevail in snow from the Alps to the Arctic. *Sci. Adv.* **2019**, *5*. [CrossRef] [PubMed]

24. Shahul Hamid, F.; Bhatti, M.S.; Anuar, N.; Anuar, N.; Mohan, P.; Periathamby, A. Worldwide distribution and abundance of microplastic: How dire is the situation? *Waste Manag. Res.* **2018**, *36*, 873–897. [CrossRef]

25. Schwabl, P.; Koppel, S.; Konigshofer, P.; Bucsics, T.; Trauner, M.; Reiberger, T.; Liebmann, B. Detection of various microplastics in human stool: A prospective case series. *Ann. Intern. Med.* **2019**. [CrossRef]

26. Van Cauwenberghe, L.; Janssen, C.R. Microplastics in bivalves cultured for human consumption. *Environ. Pollut.* **2014**, *193*, 65–70. [CrossRef]

27. Cho, Y.; Shim, W.J.; Jang, M.; Han, G.M.; Hong, S.H. Abundance and characteristics of microplastics in market bivalves from South Korea. *Environ. Pollut.* **2019**, *245*, 1107–1116. [CrossRef]

28. Li, J.; Qu, X.; Su, L.; Zhang, W.; Yang, D.; Kolandhasamy, P.; Li, D.; Shi, H. Microplastics in mussels along the coastal waters of China. *Environ. Pollut.* **2016**, *214*, 177–184. [CrossRef]

29. Li, J.; Green, C.; Reynolds, A.; Shi, H.; Rotchell, J.M. Microplastics in mussels sampled from coastal waters and supermarkets in the United Kingdom. *Environ. Pollut.* **2018**, *241*, 35–44. [CrossRef]

30. Smith, M.; Love, D.C.; Rochman, C.M.; Neff, R.A. Microplastics in seafood and the implications for human health. *Curr. Environ. Health Rep.* **2018**, *5*, 375–386. [CrossRef]

31. Cox, K.D.; Covernton, G.A.; Davies, H.L.; Dower, J.F.; Juanes, F.; Dudas, S.E. Human consumption of microplastics. *Environ. Sci. Technol.* **2019**, *53*, 7068–7074. [CrossRef] [PubMed]

32. Kniggendorf, A.K.; Wetzel, C.; Roth, B. Microplastics detection in streaming tap water with raman spectroscopy. *Sensors* **2019**, *19*, 1839. [CrossRef] [PubMed]

33. Panno, S.V.; Kelly, W.R.; Scott, J.; Zheng, W.; McNeish, R.E.; Holm, N.; Hoellein, T.J.; Baranski, E.L. Microplastic contamination in karst groundwater systems. *Ground Water* **2019**, *57*, 189–196. [CrossRef] [PubMed]

34. Welle, F.; Franz, R. Microplastic in bottled natural mineral water—Literature review and considerations on exposure and risk assessment. *Food Addit. Contam. Part A Chem. Anal. Control Expo. Risk Assess* **2018**, *35*, 2482–2492. [CrossRef]

35. Mason, S.A.; Welch, V.G.; Neratko, J. Synthetic polymer contamination in bottled water. *Front. Chem.* **2018**, *6*, 407. [CrossRef]

36. Koelmans, A.A.; Mohamed Nor, N.H.; Hermsen, E.; Kooi, M.; Mintenig, S.M.; De France, J. Microplastics in freshwaters and drinking water: Critical review and assessment of data quality. *Water Res.* **2019**, *155*, 410–422. [CrossRef]

37. Li, W.C.; Tse, H.F.; Fok, L. Plastic waste in the marine environment: A review of sources, occurrence and effects. *Sci. Total Environ.* **2016**, *566–567*, 333–349. [CrossRef]

38. Thompson, R.C.; Swan, S.H.; Moore, C.J.; Vom Saal, F.S. Our plastic age. *Philos. Trans. R. Soc. Lond. B Biol. Sci.* **2009**, *364*, 1973–1976. [CrossRef]

39. Ghosh, S.K.; Pal, S.; Ray, S. Study of microbes having potentiality for biodegradation of plastics. *Environ. Sci. Pollut. Res. Int.* **2013**, *20*, 4339–4355. [CrossRef]

40. Hahladakis, J.N.; Velis, C.A.; Weber, R.; Iacovidou, E.; Purnell, P. An overview of chemical additives present in plastics: Migration, release, fate and environmental impact during their use, disposal and recycling. *J. Hazard. Mater.* **2018**, *344*, 179–199. [CrossRef]

41. Andrady, A.L.; Neal, M.A. Applications and societal benefits of plastics. *Philos. Trans. R. Soc. Lond. B Biol. Sci.* **2009**, *364*, 1977–1984. [CrossRef] [PubMed]

42. Huang, J.T.; Huang, V.I. Evaluation of the efficiency of medical masks and the creation of new medical masks. *J. Int. Med. Res.* **2007**, *35*, 213–223. [CrossRef] [PubMed]

43. Alimba, C.G.; Faggio, C. Microplastics in the marine environment: Current trends in environmental pollution and mechanisms of toxicological profile. *Environ. Toxicol. Pharmacol.* **2019**, *68*, 61–74. [CrossRef] [PubMed]

44. Schmidt, C.; Krauth, T.; Wagner, S. Export of plastic debris by rivers into the sea. *Environ. Sci. Technol.* **2017**, *51*, 12246–12253. [CrossRef]

45. Gigault, J.; Halle, A.T.; Baudrimont, M.; Pascal, P.Y.; Gauffre, F.; Phi, T.L.; El Hadri, H.; Grassl, B.; Reynaud, S. Current opinion: What is a nanoplastic? *Environ. Pollut.* **2018**, *235*, 1030–1034. [CrossRef] [PubMed]

46. Barnes, D.K.; Galgani, F.; Thompson, R.C.; Barlaz, M. Accumulation and fragmentation of plastic debris in global environments. *Philos. Trans. R. Soc. Lond. B Biol. Sci.* **2009**, *364*, 1985–1998. [CrossRef]

47. Graca, B.; Szewc, K.; Zakrzewska, D.; Dolega, A.; Szczerbowska-Boruchowska, M. Sources and fate of microplastics in marine and beach sediments of the Southern Baltic Sea-a preliminary study. *Environ. Sci. Pollut. Res. Int.* **2017**, *24*, 7650–7661. [CrossRef]

48. Scheurer, M.; Bigalke, M. Microplastics in swiss floodplain soils. *Environ. Sci. Technol.* **2018**, *52*, 3591–3598. [CrossRef]

49. Leslie, H.A.; Brandsma, S.H.; Van Velzen, M.J.; Vethaak, A.D. Microplastics en route: Field measurements in the Dutch river delta and Amsterdam canals, wastewater treatment plants, North Sea sediments and biota. *Environ. Int.* **2017**, *101*, 133–142. [CrossRef]

50. Lambert, S.; Wagner, M. Characterisation of nanoplastics during the degradation of polystyrene. *Chemosphere* **2016**, *145*, 265–268. [CrossRef]

51. Dawson, A.L.; Kawaguchi, S.; King, C.K.; Townsend, K.A.; King, R.; Huston, W.M.; Bengtson Nash, S.M. Turning microplastics into nanoplastics through digestive fragmentation by Antarctic krill. *Nat. Commun.* **2018**, *9*, 1001. [CrossRef] [PubMed]

52. Wright, S.L.; Thompson, R.C.; Galloway, T.S. The physical impacts of microplastics on marine organisms: A review. *Environ. Pollut.* **2013**, *178*, 483–492. [CrossRef] [PubMed]

53. Salvador Cesa, F.; Turra, A.; Baruque-Ramos, J. Synthetic fibers as microplastics in the marine environment: A review from textile perspective with a focus on domestic washings. *Sci. Total Environ.* **2017**, *598*, 1116–1129. [CrossRef] [PubMed]

54. Lares, M.; Ncibi, M.C.; Sillanpaa, M.; Sillanpaa, M. Occurrence, identification and removal of microplastic particles and fibers in conventional activated sludge process and advanced MBR technology. *Water Res.* **2018**, *133*, 236–246. [CrossRef]

55. Browne, M.A.; Crump, P.; Niven, S.J.; Teuten, E.; Tonkin, A.; Galloway, T.; Thompson, R. Accumulation of microplastic on shorelines woldwide: Sources and sinks. *Environ. Sci. Technol.* **2011**, *45*, 9175–9179. [CrossRef]

56. Dehghani, S.; Moore, F.; Akhbarizadeh, R. Microplastic pollution in deposited urban dust, Tehran metropolis, Iran. *Environ. Sci. Pollut. Res. Int.* **2017**, *24*, 20360–20371. [CrossRef]

57. Duis, K.; Coors, A. Microplastics in the aquatic and terrestrial environment: Sources (with a specific focus on personal care products), fate and effects. *Environ. Sci. Eur.* **2016**, *28*, 2. [CrossRef]

58. Setala, O.; Fleming-Lehtinen, V.; Lehtiniemi, M. Ingestion and transfer of microplastics in the planktonic food web. *Environ. Pollut.* **2014**, *185*, 77–83. [CrossRef]

59. Von Moos, N.; Burkhardt-Holm, P.; Kohler, A. Uptake and effects of microplastics on cells and tissue of the blue mussel Mytilus edulis L. After an experimental exposure. *Environ. Sci. Technol.* **2012**, *46*, 11327–11335. [CrossRef]

60. Karlsson, T.M.; Vethaak, A.D.; Almroth, B.C.; Ariese, F.; Van Velzen, M.; Hassellov, M.; Leslie, H.A. Screening for microplastics in sediment, water, marine invertebrates and fish: Method development and microplastic accumulation. *Mar. Pollut. Bull.* **2017**, *122*, 403–408. [CrossRef]

61. Wesch, C.; Bredimus, K.; Paulus, M.; Klein, R. Towards the suitable monitoring of ingestion of microplastics by marine biota: A review. *Environ. Pollut.* **2016**, *218*, 1200–1208. [CrossRef] [PubMed]

62. Hipfner, J.M.; Galbraith, M.; Tucker, S.; Studholme, K.R.; Domalik, A.D.; Pearson, S.F.; Good, T.P.; Ross, P.S.; Hodum, P. Two forage fishes as potential conduits for the vertical transfer of microfibres in Northeastern Pacific Ocean food webs. *Environ. Pollut.* **2018**, *239*, 215–222. [CrossRef] [PubMed]

63. Baalkhuyur, F.M.; Bin Dohaish, E.A.; Elhalwagy, M.E.A.; Alikunhi, N.M.; AlSuwailem, A.M.; Rostad, A.; Coker, D.J.; Berumen, M.L.; Duarte, C.M. Microplastic in the gastrointestinal tract of fishes along the Saudi Arabian Red Sea coast. *Mar. Pollut. Bull.* **2018**, *131*, 407–415. [CrossRef] [PubMed]

64. Sussarellu, R.; Suquet, M.; Thomas, Y.; Lambert, C.; Fabioux, C.; Pernet, M.E.; Le Goic, N.; Quillien, V.; Mingant, C.; Epelboin, Y.; et al. Oyster reproduction is affected by exposure to polystyrene microplastics. *Proc. Natl. Acad. Sci. USA* **2016**, *113*, 2430–2435. [CrossRef]

65. Browne, M.A.; Dissanayake, A.; Galloway, T.S.; Lowe, D.M.; Thompson, R.C. Ingested microscopic plastic translocates to the circulatory system of the mussel, Mytilus edulis (L). *Environ. Sci. Technol.* **2008**, *42*, 5026–5031. [CrossRef]

66. De Witte, B.; Devriese, L.; Bekaert, K.; Hoffman, S.; Vandermeersch, G.; Cooreman, K.; Robbens, J. Quality assessment of the blue mussel (Mytilus edulis): Comparison between commercial and wild types. *Mar. Pollut. Bull.* **2014**, *85*, 146–155. [CrossRef]

67. Li, J.; Yang, D.; Li, L.; Jabeen, K.; Shi, H. Microplastics in commercial bivalves from China. *Environ. Pollut.* **2015**, *207*, 190–195. [CrossRef]

68. Catarino, A.I.; Macchia, V.; Sanderson, W.G.; Thompson, R.C.; Henry, T.B. Low levels of microplastics (MP) in wild mussels indicate that MP ingestion by humans is minimal compared to exposure via household fibres fallout during a meal. *Environ. Pollut.* **2018**, *237*, 675–684. [CrossRef]

69. Hermabessiere, L.; Paul-Pont, I.; Cassone, A.L.; Himber, C.; Receveur, J.; Jezequel, R.; El Rakwe, M.; Rinnert, E.; Riviere, G.; Lambert, C.; et al. Microplastic contamination and pollutant levels in mussels and cockles collected along the channel coasts. *Environ. Pollut.* **2019**, *250*, 807–819. [CrossRef]

70. Kosuth, M.; Mason, S.A.; Wattenberg, E.V. Anthropogenic contamination of tap water, beer, and sea salt. *PLoS ONE* **2018**, *13*, e0194970. [CrossRef]

71. Schymanski, D.; Goldbeck, C.; Humpf, H.U.; Furst, P. Analysis of microplastics in water by micro-Raman spectroscopy: Release of plastic particles from different packaging into mineral water. *Water Res.* **2018**, *129*, 154–162. [CrossRef] [PubMed]

72. Liebezeit, G.; Liebezeit, E. Synthetic particles as contaminants in German beers. *Food Addit. Contam. Part A Chem. Anal. Control Expo. Risk Assess* **2014**, *31*, 1574–1578. [CrossRef] [PubMed]

73. Iniguez, M.E.; Conesa, J.A.; Fullana, A. Microplastics in Spanish table salt. *Sci. Rep.* **2017**, *7*, 8620. [CrossRef]

74. Liebezeit, G.; Liebezeit, E. Non-pollen particulates in honey and sugar. *Food Addit. Contam. Part A Chem. Anal. Control Expo. Risk Assess* **2013**, *30*, 2136–2140. [CrossRef] [PubMed]

75. Hernandez, L.M.; Xu, E.G.; Larsson, H.C.E.; Tahara, R.; Maisuria, V.B.; Tufenkji, N. Plastic teabags release billions of microparticles and nanoparticles into tea. *Environ. Sci. Technol.* **2019**. [CrossRef] [PubMed]

76. Prata, J.C. Airborne microplastics: Consequences to human health? *Environ. Pollut.* **2018**, *234*, 115–126. [CrossRef] [PubMed]

77. Dris, R.; Gasperi, J.; Mirande, C.; Mandin, C.; Guerrouache, M.; Langlois, V.; Tassin, B. A first overview of textile fibers, including microplastics, in indoor and outdoor environments. *Environ. Pollut.* **2017**, *221*, 453–458. [CrossRef]

78. Pivokonsky, M.; Cermakova, L.; Novotna, K.; Peer, P.; Cajthaml, T.; Janda, V. Occurrence of microplastics in raw and treated drinking water. *Sci. Total Environ.* **2018**, *643*, 1644–1651. [CrossRef]

79. Wilcox, C.; Puckridge, M.; Schuyler, Q.A.; Townsend, K.; Hardesty, B.D. A quantitative analysis linking sea turtle mortality and plastic debris ingestion. *Sci. Rep.* **2018**, *8*, 12536. [CrossRef]

80. Van Franeker, J.A.; Bravo Rebolledo, E.L.; Hesse, E.; LL, I.J.; Kuhn, S.; Leopold, M.; Mielke, L. Plastic ingestion by harbour porpoises Phocoena phocoena in the Netherlands: Establishing a standardised method. *Ambio* **2018**, *47*, 387–397. [CrossRef]

81. Wang, J.; Tan, Z.; Peng, J.; Qiu, Q.; Li, M. The behaviors of microplastics in the marine environment. *Mar. Environ. Res.* **2016**, *113*, 7–17. [CrossRef] [PubMed]

82. Besseling, E.; Wegner, A.; Foekema, E.M.; Van den Heuvel-Greve, M.J.; Koelmans, A.A. Effects of microplastic on fitness and PCB bioaccumulation by the lugworm Arenicola marina (L.). *Environ. Sci. Technol.* **2013**, *47*, 593–600. [CrossRef] [PubMed]

83. Brennecke, D.; Ferreira, E.C.; Costa, T.M.; Appel, D.; Da Gama, B.A.; Lenz, M. Ingested microplastics (>100 mum) are translocated to organs of the tropical fiddler crab Uca rapax. *Mar. Pollut. Bull.* **2015**, *96*, 491–495. [CrossRef] [PubMed]

84. Karami, A.; Romano, N.; Galloway, T.; Hamzah, H. Virgin microplastics cause toxicity and modulate the impacts of phenanthrene on biomarker responses in African catfish (Clarias gariepinus). *Environ. Res.* **2016**, *151*, 58–70. [CrossRef] [PubMed]

85. McGoran, A.R.; Clark, P.F.; Morritt, D. Presence of microplastic in the digestive tracts of European flounder, Platichthys flesus, and European smelt, Osmerus eperlanus, from the River Thames. *Environ. Pollut.* **2017**, *220*, 744–751. [CrossRef] [PubMed]

86. Vandermeersch, G.; Van Cauwenberghe, L.; Janssen, C.R.; Marques, A.; Granby, K.; Fait, G.; Kotterman, M.J.; Diogene, J.; Bekaert, K.; Robbens, J.; et al. A critical view on microplastic quantification in aquatic organisms. *Environ. Res.* **2015**, *143*, 46–55. [CrossRef]

87. Jin, Y.; Lu, L.; Tu, W.; Luo, T.; Fu, Z. Impacts of polystyrene microplastic on the gut barrier, microbiota and metabolism of mice. *Sci. Total Environ.* **2019**, *649*, 308–317. [CrossRef]

88. Lu, L.; Wan, Z.; Luo, T.; Fu, Z.; Jin, Y. Polystyrene microplastics induce gut microbiota dysbiosis and hepatic lipid metabolism disorder in mice. *Sci. Total Environ.* **2018**, *631–632*, 449–458. [CrossRef]

89. Frias, J.P.; Gago, J.; Otero, V.; Sobral, P. Microplastics in coastal sediments from Southern Portuguese shelf waters. *Mar. Environ. Res.* **2016**, *114*, 24–30. [CrossRef]

90. Hours, M.; Fevotte, J.; Lafont, S.; Bergeret, A. Cancer mortality in a synthetic spinning plant in Besancon, France. *Occup. Environ. Med.* **2007**, *64*, 575–581. [CrossRef]

91. Gallagher, L.G.; Li, W.; Ray, R.M.; Romano, M.E.; Wernli, K.J.; Gao, D.L.; Thomas, D.B.; Checkoway, H. Occupational exposures and risk of stomach and esophageal cancers: Update of a cohort of female textile workers in Shanghai, China. *Am. J. Ind. Med.* **2015**, *58*, 267–275. [CrossRef] [PubMed]

92. Kern, D.G.; Crausman, R.S.; Durand, K.T.; Nayer, A.; Kuhn, C., III. Flock worker's lung: Chronic interstitial lung disease in the nylon flocking industry. *Ann. Intern. Med.* **1998**, *129*, 261–272. [CrossRef] [PubMed]

93. Turcotte, S.E.; Chee, A.; Walsh, R.; Grant, F.C.; Liss, G.M.; Boag, A.; Forkert, L.; Munt, P.W.; Lougheed, M.D. Flock worker's lung disease: Natural history of cases and exposed workers in Kingston, Ontario. *Chest* **2013**, *143*, 1642–1648. [CrossRef] [PubMed]

94. Huang, N.C.; Wann, S.R.; Chang, H.T.; Lin, S.L.; Wang, J.S.; Guo, H.R. Arsenic, vinyl chloride, viral hepatitis, and hepatic angiosarcoma: A hospital-based study and review of literature in Taiwan. *BMC Gastroenterol.* **2011**, *11*, 142. [CrossRef]

95. Vianna, N.J.; Brady, J.; Harper, P. Angiosarcoma of the liver: A signal lesion of vinyl chloride exposure. *Environ. Health Perspect.* **1981**, *41*, 207–210. [CrossRef]

96. Elliott, P.; Kleinschmidt, I. Angiosarcoma of the liver in Great Britain in proximity to vinyl chloride sites. *Occup. Environ. Med.* **1997**, *54*, 14–18. [CrossRef]

97. Rochman, C.M.; Kurobe, T.; Flores, I.; Teh, S.J. Early warning signs of endocrine disruption in adult fish from the ingestion of polyethylene with and without sorbed chemical pollutants from the marine environment. *Sci. Total Environ.* **2014**, *493*, 656–661. [CrossRef]

98. Barboza, L.G.A.; Vieira, L.R.; Branco, V.; Figueiredo, N.; Carvalho, F.; Carvalho, C.; Guilhermino, L. Microplastics cause neurotoxicity, oxidative damage and energy-related changes and interact with the bioaccumulation of mercury in the European seabass, Dicentrarchus labrax (Linnaeus, 1758). *Aquat. Toxicol.* **2018**, *195*, 49–57. [CrossRef]

99. Gardon, T.; Reisser, C.; Soyez, C.; Quillien, V.; Le Moullac, G. Microplastics affect energy balance and gametogenesis in the pearl oyster pinctada margaritifera. *Environ. Sci. Technol.* **2018**, *52*, 5277–5286. [CrossRef]

100. Tallec, K.; Huvet, A.; Di Poi, C.; Gonzalez-Fernandez, C.; Lambert, C.; Petton, B.; Le Goic, N.; Berchel, M.; Soudant, P.; Paul-Pont, I. Nanoplastics impaired oyster free living stages, gametes and embryos. *Environ. Pollut.* **2018**, *242*, 1226–1235. [CrossRef]

101. Pitt, J.A.; Trevisan, R.; Massarsky, A.; Kozal, J.S.; Levin, E.D.; Di Giulio, R.T. Maternal transfer of nanoplastics to offspring in zebrafish (Danio rerio): A case study with nanopolystyrene. *Sci. Total Environ.* **2018**, *643*, 324–334. [CrossRef] [PubMed]

102. Martins, A.; Guilhermino, L. Transgenerational effects and recovery of microplastics exposure in model populations of the freshwater cladoceran Daphnia magna Straus. *Sci. Total Environ.* **2018**, *631–632*, 421–428. [CrossRef] [PubMed]

103. Liu, Z.; Yu, P.; Cai, M.; Wu, D.; Zhang, M.; Huang, Y.; Zhao, Y. Polystyrene nanoplastic exposure induces immobilization, reproduction, and stress defense in the freshwater cladoceran Daphnia pulex. *Chemosphere* **2019**, *215*, 74–81. [CrossRef] [PubMed]

104. Haldimann, M.; Alt, A.; Blanc, A.; Brunner, K.; Sager, F.; Dudler, V. Migration of antimony from PET trays into food simulant and food: Determination of Arrhenius parameters and comparison of predicted and measured migration data. *Food Addit. Contam. Part A Chem. Anal. Control Expo. Risk Assess* **2013**, *30*, 587–598. [CrossRef] [PubMed]

105. Wang, J.; Zhou, G.; Chen, C.; Yu, H.; Wang, T.; Ma, Y.; Jia, G.; Gao, Y.; Li, B.; Sun, J.; et al. Acute toxicity and biodistribution of different sized titanium dioxide particles in mice after oral administration. *Toxicol. Lett.* **2007**, *168*, 176–185. [CrossRef] [PubMed]

106. Poland, C.A.; Duffin, R.; Kinloch, I.; Maynard, A.; Wallace, W.A.; Seaton, A.; Stone, V.; Brown, S.; Macnee, W.; Donaldson, K. Carbon nanotubes introduced into the abdominal cavity of mice show asbestos-like pathogenicity in a pilot study. *Nat. Nanotechnol.* **2008**, *3*, 423–428. [CrossRef] [PubMed]

107. Engler, R.E. The complex interaction between marine debris and toxic chemicals in the ocean. *Environ. Sci. Technol.* **2012**, *46*, 12302–12315. [CrossRef]

108. Huang, Y.; Mei, L.; Chen, X.; Wang, Q. Recent developments in food packaging based on nanomaterials. *Nanomaterials* **2018**, *8*, 830. [CrossRef]

109. Mato, Y.; Isobe, T.; Takada, H.; Kanehiro, H.; Ohtake, C.; Kaminuma, T. Plastic resin pellets as a transport medium for toxic chemicals in the marine environment. *Environ. Sci. Technol.* **2001**, *35*, 318–324. [CrossRef]

110. Rochman, C.M.; Hoh, E.; Hentschel, B.T.; Kaye, S. Long-term field measurement of sorption of organic contaminants to five types of plastic pellets: Implications for plastic marine debris. *Environ. Sci. Technol.* **2013**, *47*, 1646–1654. [CrossRef]

111. Andrady, A.L. Microplastics in the marine environment. *Mar. Pollut. Bull.* **2011**, *62*, 1596–1605. [CrossRef] [PubMed]

112. Wu, B.; Wu, X.; Liu, S.; Wang, Z.; Chen, L. Size-dependent effects of polystyrene microplastics on cytotoxicity and efflux pump inhibition in human Caco-2cells. *Chemosphere* **2019**, *221*, 333–341. [CrossRef] [PubMed]

113. Brandts, I.; Teles, M.; Goncalves, A.P.; Barreto, A.; Franco-Martinez, L.; Tvarijonaviciute, A.; Martins, M.A.; Soares, A.; Tort, L.; Oliveira, M. Effects of nanoplastics on Mytilus galloprovincialis after individual and combined exposure with carbamazepine. *Sci. Total Environ.* **2018**, *643*, 775–784. [CrossRef] [PubMed]

114. Qu, M.; Xu, K.; Li, Y.; Wong, G.; Wang, D. Using acs-22 mutant Caenorhabditis elegans to detect the toxicity of nanopolystyrene particles. *Sci. Total Environ.* **2018**, *643*, 119–126. [CrossRef]

115. Tang, J.; Ni, X.; Zhou, Z.; Wang, L.; Lin, S. Acute microplastic exposure raises stress response and suppresses detoxification and immune capacities in the scleractinian coral Pocillopora damicornis. *Environ. Pollut.* **2018**, *243*, 66–74. [CrossRef] [PubMed]

116. Jeong, C.B.; Won, E.J.; Kang, H.M.; Lee, M.C.; Hwang, D.S.; Hwang, U.K.; Zhou, B.; Souissi, S.; Lee, S.J.; Lee, J.S. Microplastic size-dependent toxicity, oxidative stress induction, and p-JNK and p-p38 activation in the monogonont rotifer (brachionus koreanus). *Environ. Sci. Technol.* **2016**, *50*, 8849–8857. [CrossRef]

117. Lei, L.; Wu, S.; Lu, S.; Liu, M.; Song, Y.; Fu, Z.; Shi, H.; Raley-Susman, K.M.; He, D. Microplastic particles cause intestinal damage and other adverse effects in zebrafish Danio rerio and nematode Caenorhabditis elegans. *Sci. Total Environ.* **2018**, *619–620*, 1–8. [CrossRef]

118. Lu, Y.; Zhang, Y.; Deng, Y.; Jiang, W.; Zhao, Y.; Geng, J.; Ding, L.; Ren, H. Uptake and accumulation of polystyrene microplastics in zebrafish (danio rerio) and toxic effects in liver. *Environ. Sci. Technol.* **2016**, *50*, 4054–4060. [CrossRef]

119. Yang, Y.F.; Chen, C.Y.; Lu, T.H.; Liao, C.M. Toxicity-based toxicokinetic/toxicodynamic assessment for bioaccumulation of polystyrene microplastics in mice. *J. Hazard. Mater.* **2019**, *366*, 703–713. [CrossRef]

120. Birben, E.; Sahiner, U.M.; Sackesen, C.; Erzurum, S.; Kalayci, O. Oxidative stress and antioxidant defense. *World Allergy Organ. J.* **2012**, *5*, 9–19. [CrossRef]

121. Nita, M.; Grzybowski, A. The role of the reactive oxygen species and oxidative stress in the pathomechanism of the age-related ocular diseases and other pathologies of the anterior and posterior eye segments in adults. *Oxid. Med. Cell. Longev.* **2016**, *2016*, 3164734. [CrossRef] [PubMed]

122. Deng, Y.; Zhang, Y.; Lemos, B.; Ren, H. Tissue accumulation of microplastics in mice and biomarker responses suggest widespread health risks of exposure. *Sci. Rep.* **2017**, *7*, 46687. [CrossRef] [PubMed]

123. Brandts, I.; Teles, M.; Tvarijonaviciute, A.; Pereira, M.L.; Martins, M.A.; Tort, L.; Oliveira, M. Effects of polymethylmethacrylate nanoplastics on Dicentrarchus labrax. *Genomics* **2018**, *110*, 435–441. [CrossRef] [PubMed]

124. Revel, M.; Yakovenko, N.; Caley, T.; Guillet, C.; Chatel, A.; Mouneyrac, C. Accumulation and immunotoxicity of microplastics in the estuarine worm Hediste diversicolor in environmentally relevant conditions of exposure. *Environ. Sci. Pollut. Res. Int.* **2018**. [CrossRef]

125. Sleight, V.A.; Bakir, A.; Thompson, R.C.; Henry, T.B. Assessment of microplastic-sorbed contaminant bioavailability through analysis of biomarker gene expression in larval zebrafish. *Mar. Pollut. Bull.* **2017**, *116*, 291–297. [CrossRef]

126. Jiang, X.; Chen, H.; Liao, Y.; Ye, Z.; Li, M.; Klobucar, G. Ecotoxicity and genotoxicity of polystyrene microplastics on higher plant Vicia faba. *Environ. Pollut.* **2019**, *250*, 831–838. [CrossRef]

127. Hsu, Y.H.; Chuang, H.C.; Lee, Y.H.; Lin, Y.F.; Chiu, Y.J.; Wang, Y.L.; Wu, M.S.; Chiu, H.W. Induction of fibrosis and autophagy in kidney cells by Vinyl chloride. *Cells* **2019**, *8*, 601. [CrossRef]

128. Boivin-Angele, S.; Lefrancois, L.; Froment, O.; Spiethoff, A.; Bogdanffy, M.S.; Wegener, K.; Wesch, H.; Barbin, A.; Bancel, B.; Trepo, C.; et al. Ras gene mutations in vinyl chloride-induced liver tumours are carcinogen-specific but vary with cell type and species. *Int. J. Cancer* **2000**, *85*, 223–227. [CrossRef]

129. Weihrauch, M.; Benick, M.; Lehner, G.; Wittekind, M.; Bader, M.; Wrbitzk, R.; Tannapfel, A. High prevalence of K-ras-2 mutations in hepatocellular carcinomas in workers exposed to vinyl chloride. *Int. Arch. Occup. Environ. Health* **2001**, *74*, 405–410. [CrossRef]

130. Barbin, A.; Froment, O.; Boivin, S.; Marion, M.J.; Belpoggi, F.; Maltoni, C.; Montesano, R. P53 gene mutation pattern in rat liver tumors induced by vinyl chloride. *Cancer Res.* **1997**, *57*, 1695–1698.

131. Hollstein, M.; Marion, M.J.; Lehman, T.; Welsh, J.; Harris, C.C.; Martel-Planche, G.; Kusters, I.; Montesano, R. P53 mutations at A:T base pairs in angiosarcomas of vinyl chloride-exposed factory workers. *Carcinogenesis* **1994**, *15*, 1–3. [CrossRef] [PubMed]

132. De Vivo, I.; Marion, M.J.; Smith, S.J.; Carney, W.P.; Brandt-Rauf, P.W. Mutant c-Ki-ras p21 protein in chemical carcinogenesis in humans exposed to vinyl chloride. *Cancer Causes Control.* **1994**, *5*, 273–278. [CrossRef] [PubMed]

133. Loulad, S.; Houssa, R.; Rhinane, H.; Boumaaz, A.; Benazzouz, A. Spatial distribution of marine debris on the seafloor of Moroccan waters. *Mar. Pollut. Bull.* **2017**, *124*, 303–313. [CrossRef] [PubMed]

134. Xanthos, D.; Walker, T.R. International policies to reduce plastic marine pollution from single-use plastics (plastic bags and microbeads): A review. *Mar. Pollut. Bull.* **2017**, *118*, 17–26. [CrossRef] [PubMed]

135. Africa Is on the Right Path to Eradicate Plastics. Available online: https://www.unenvironment.org/news-and-stories/story/africa-right-path-eradicate-plastics (accessed on 4 February 2020).

136. A Floating Device Created to Clean Up Plastic from the Ocean Is Finally Doing Its Job, Organizers Say. Available online: https://edition.cnn.com/2019/10/02/tech/ocean-cleanup-catching-plastic-scn-trnd/index.html (accessed on 4 February 2020).

137. Ziajahromi, S.; Neale, P.A.; Rintoul, L.; Leusch, F.D. Wastewater treatment plants as a pathway for microplastics: Development of a new approach to sample wastewater-based microplastics. *Water Res.* **2017**, *112*, 93–99. [CrossRef] [PubMed]

138. Mason, S.A.; Garneau, D.; Sutton, R.; Chu, Y.; Ehmann, K.; Barnes, J.; Fink, P.; Papazissimos, D.; Rogers, D.L. Microplastic pollution is widely detected in US municipal wastewater treatment plant effluent. *Environ. Pollut.* **2016**, *218*, 1045–1054. [CrossRef]

139. Murphy, F.; Ewins, C.; Carbonnier, F.; Quinn, B. Wastewater treatment works (WwTW) as a source of microplastics in the aquatic environment. *Environ. Sci. Technol.* **2016**, *50*, 5800–5808. [CrossRef]

140. Cheap Drainage Nets Keep Water Pollution at Bay in Australia. Available online: https://inhabitat.com/cheap-drainage-nets-keep-water-pollution-at-bay-in-australia/ (accessed on 4 February 2020).

141. Single-Use Plastic Ban: What Is Single-Use Plastic, Which Plastic Items Will Be Banned on October 2? Available online: https://www.jagranjosh.com/current-affairs/current-affairs-august-2019-what-is-singleuse-plastic-and-why-is-it-being-banned-1567074897-1 (accessed on 4 February 2020).

142. Vietnam's Wild Grass Straws. Available online: https://theaseanpost.com/article/vietnams-wild-grass-straws (accessed on 4 February 2020).

143. Taiwanese Entrepreneurs Patent Sugarcane Straws. Available online: https://www.taiwannews.com.tw/en/news/3477607 (accessed on 4 February 2020).

144. Microbeads Are Officially Banned in Cosmetics from Today. Here's What to Use Instead. Available online: https://www.telegraph.co.uk/beauty/skin/the-best-face-and-body-scrubs-without-plastic-microbeads/ (accessed on 4 February 2020).

145. A Road Full of Bottlenecks: Dutch Cycle Path Is Made of Plastic Waste. Available online: https://www.theguardian.com/environment/2018/sep/13/a-road-full-of-bottlenecks-dutch-cycle-path-is-made-of-plastic-waste (accessed on 4 February 2020).

146. Lockerbie Plastic Roads Firm MacRebur Opens First Factory. Available online: https://www.bbc.com/news/uk-scotland-south-scotland-47454719 (accessed on 4 February 2020).

147. Project in India Is Turning Discarded Fishing Nets into Surfboards to Reduce Ocean Plastic. Available online: https://www.mentalfloss.com/article/595467/project-india-fishing-nets-surfboards-reduce-ocean-plastic (accessed on 4 February 2020).

148. Student Designs Sustainable Plastic Alternative Made of Fish Skin and Algae. Available online: http://www.ladbible.com/news/uk-student-designs-sustainable-plastic-alternative-made-using-fish-skin-20190605 (accessed on 4 February 2020).

149. Scientist in Mexico Creates Biodegradable Plastic from Prickly Pear Cactus. Available online: https://www.forbes.com/sites/scottsnowden/2019/07/14/scientist-in-mexico-creates-biodegradable-plastic-from-prickly-pear-cactus/#2dca0dc76c49 (accessed on 4 February 2020).

150. Urbanek, A.K.; Rymowicz, W.; Mironczuk, A.M. Degradation of plastics and plastic-degrading bacteria in cold marine habitats. *Appl. Microbiol. Biotechnol.* **2018**, *102*, 7669–7678. [CrossRef]

151. Brandon, A.M.; Gao, S.H.; Tian, R.; Ning, D.; Yang, S.S.; Zhou, J.; Wu, W.M.; Criddle, C.S. Biodegradation of polyethylene and plastic mixtures in mealworms (Larvae of tenebrio molitor) and effects on the gut microbiome. *Environ. Sci. Technol.* **2018**, *52*, 6526–6533. [CrossRef]

152. Papadopoulou, A.; Hecht, K.; Buller, R. Enzymatic PET degradation. *Chimia (Aarau)* **2019**, *73*, 743–749. [CrossRef]

 International Journal of
Molecular Sciences

Article

Neuron-Like Cells Generated from Human Umbilical Cord Lining-Derived Mesenchymal Stem Cells as a New In Vitro Model for Neuronal Toxicity Screening: Using Magnetite Nanoparticles as an Example

Uliana De Simone [1], Arsenio Spinillo [2], Francesca Caloni [3], Laura Gribaldo [4] and Teresa Coccini [1,*]

1. Laboratory of Clinical & Experimental Toxicology, Toxicology Unit, ICS Maugeri SpA-Benefit Corporation, IRCCS Pavia, Via Maugeri 10, 27100 Pavia, Italy; uliana.desimone@icsmaugeri.it
2. Department of Obstetrics and Gynecology, Fondazione IRCCS Policlinico San Matteo and University of Pavia, 27100 Pavia, Italy; spinillo@smatteo.pv.it
3. Dipartimento di Medicina Veterinaria (DIMEVET), Università degli Studi di Milano, 20133 Milano, Italy; francesca.caloni@unimi.it
4. Chemical Safety and Alternative Methods Unit, Directorate F—Health, Consumers and Reference Materials, Directorate General Joint Research Centre, European Commission, 21027 Ispra, Italy; Laura.GRIBALDO@ec.europa.eu
* Correspondence: teresa.coccini@icsmaugeri.it; Tel.: +39-0382-592416

Received: 27 November 2019; Accepted: 29 December 2019; Published: 31 December 2019

Abstract: The wide employment of iron nanoparticles in environmental and occupational settings underlines their potential to enter the brain. Human cell-based systems are recommended as relevant models to reduce uncertainty and to improve prediction of human toxicity. This study aimed at demonstrating the in vitro differentiation of the human umbilical cord lining-derived-mesenchymal stem cells (hCL-MSCs) into neuron-like cells (hNLCs) and the benefit of using them as an ideal primary cell source of human origin for the neuronal toxicity of Fe_3O_4NPs (magnetite-nanoparticles). Neuron-like phenotype was confirmed by: live morphology; Nissl body staining; protein expression of different neuronal-specific markers (immunofluorescent staining), at different maturation stages (i.e., day-3-early and day-8-full differentiated), namely β-tubulin III, MAP-2, enolase (NSE), glial protein, and almost no nestin and SOX-2 expression. Synaptic makers (SYN, GAP43, and PSD95) were also expressed. Fe_3O_4NPs determined a concentration- and time-dependent reduction of hNLCs viability (by ATP and the Trypan Blue test). Cell density decreased (20–50%) and apoptotic effects were detected at ≥10 µg/mL in both types of differentiated hNLCs. Three-day-differentiated hNLCs were more susceptible (toxicity appeared early and lasted for up to 48 h) than 8-day-differentiated cells (delayed effects). The study demonstrated that (i) hCL-MSCs easily differentiated into neuronal-like cells; (ii) the hNCLs susceptibility to Fe_3O_4NPs; and (iii) human primary cultures of neurons are new in vitro model for NP evaluation.

Keywords: Fe_3O_4 nanoparticles; environmental toxicology; alternative methods; safety assessment; cell-based assay; toxicity-testing strategies; human primary cell culture; predictive nanotoxicology

---

1. Introduction

Among the different types of engineered NPs nanoparticles (NPs), the superparamagnetic iron oxide nanoparticles (SPIONs) are particles formed by small crystals of iron oxide commonly called magnetite Fe_3O_4 or maghemite γ-Fe_2O_3. These SPIONs have gained a huge interest due to their use for several biomedical and clinical applications (e.g., MRI contrast agents, treatments for anemia, magnetic sensing probes, and drug delivery agent) [1,2].

SPIONs have also been developed for use as environmental catalysts and for incorporation into thermoplastics nanocomposites due to their pigmented properties. The latter include several market products such as car tires, paints, etc. In addition, SPION-based technologies have emerged as promising alternatives to current water and wastewater treatment of organic pollutants as nanoadsorbents or as core component of core-shell structures, where the SPIONs function as magnetic separation and the shell provides the desired functionality for pollutant adsorption [3,4]. Moreover, iron oxide particulates, both fine/micron- and ultra-fine/nano-sized, are generated during anthropogenic activities related to the iron and steel industries, as well as during the NP manufacturing process, where they may represent a significant portion of the circulating air becoming a source of potential hazard for at-risk workers [5].

Therefore, there is a considerable need to address the biocompatibility and safety issues associated with the use of SPIONs.

The promising use of SPIONs for diagnostic and therapeutic applications and their wide utilization in environmental and occupational setting, underline the potential of such SPIONs to enter the brain, making it mandatory to study their potential neurotoxicity [6,7].

A relevant number of studies have demonstrated that engineered nanomaterials (ENMs) could cross the blood–brain barrier (BBB) via different routes, after intentionally and unintentionally exposure and further access the central nervous system (CNS), where ENMs could cause neurotoxicity [8–13]. After crossing BBB, ENMs can interact with glial cells and neurons, which could potentially induce a series of disrupted outcomes in the neurological system [14]. The capacity to translocate into the brain has also been demonstrated for SPIONs which, after entering the body, through different routes such as intravenous and intraperitoneal injection, oral administration, and intranasal and intratracheal instillation, gradually accumulate in cerebral tissue, due to their limited excretion, causing damage to neuronal cells and function impairments [15,16]. Several in vivo and in vitro studies have demonstrated the SPIONs neurotoxicity [7,17–20]. In vivo experimental studies have indicated that Fe_3O_4NPs, the predominant form chosen among the SPIONs, can reach the CNS independently of the administration route (e.g., inhalation, intravenous, and intraperitoneal) causing adverse effects in CNS [21–23]. In vitro studies have also supported and mechanistically detailed the Fe_3O_4NPs-induced neurotoxic effects [24] depending on cell type and surface coating of the NPs [25–28]. Regarding to the surface coating, different types of natural and synthetic coating materials (such as dextran, pluronic, and polyethylene glycol) and capping agents (such as poly(ethylenimine), and aminosilane) have been used to improve the colloidal stability, interaction, biodistribution, and biocompatibility of SPIONs in biological systems [26,27] The majority of the in vitro investigations have been performed on different CNS cell types using immortalized cell lines and primary cultures derived from animals, such as PC12, cortical neurons, brain-derived endothelial cells, astrocytes from newborns, microglia (primary or Bv2 cells), and oligodendroglial cells (see review [29]). The major limitation with cell lines is that they are genetically transformed and thus may not represent the normal cell types. Primary cultures have a very limited lifespan and easily lose their tissue-specific characteristics over time. Moreover, it is still difficult to extrapolate animal data to humans because of species differences.

Human cell-based systems are strongly recommended as relevant alternative methods to reduce the uncertainty in species-specific extrapolation of results and to improve prediction in toxicology [30,31]. One of the emerging trends in technologies for developing assays and tools for predictive toxicology goals includes the increased use of stem cells (SCs) [32], which have the advantage, over primary and immortalized cells, of being able to form large populations of stably-differentiated cells representative of different target species including humans. Among SCs, adult stem cells, also known as mesenchymal stem cells (MSCs) are a kind of multipotent progenitors, derived from different human tissues that have a remarkable ability for proliferation, self renewal, and transdifferentiation when they are cultured in specific culture conditions [33–35]. Several recent data are reporting their ability to transdifferentiate into the neurogenic lineage [36–39]. MSCs in vitro differentiation to generate human neuron-like cells (hNLCs) may represent a promising source of cells for neurotoxicity studies.

Among the various sources-derived MSCs, those derived from the human umbilical cord (hUC-MSCs) have the advantages of simple convenient preparation, feasible source, non-traumatic risk of infection, more primitive properties, higher proliferation capacity, and their low immunogenicity and immunosuppressive characteristics turn hUC-MSCs to be an ideal source used as engineering cells in studying stem cell differentiation [40]. Recent investigations have demonstrated that MSCs derived from human umbilical cord can be most efficiently differentiated in vitro into cells of nonmesodermal origin including neuronal-like cells using specific induction protocols [40–45]. MSCs can be obtained from different compartments of the umbilical cord (UC) including the UC lining membrane (hCL-MSCs) [45–48]. These latter cells have been recently isolated and characterized in our lab. The findings have demonstrated that hCL-MSCs may represent a new species-specific tool for establishing efficient platforms for primary screening and toxicity/safety testing of magnetite NPs (Fe_3O_4NPs) [49].

According to the development of new strategies for assessing nanomaterial safety, the SC-derived in vitro models may provide more realistic platforms for nontoxicity study [50–52], and the differentiation of hMSC into hNLCs can further support their use for screening evaluation of neuronal toxicity of NPs in humans. Indeed current available cell types represent a model less close to the in vivo human neuronal cells, and therefore they are not specifically indicated to characterize neuronal toxicity. This approach reflects the most updated recommendations on using human based models to investigate human diseases and toxic effects from xenobiotics. Moreover significant differences on toxicological profile of Fe_3O_4NPs were observed in relation to the cell type model used, rodent versus human [53].

The present study aimed at demonstrating the in vitro differentiation of the human umbilical CL-derived MSCs (hCL-MSCs) into hNLCs (2D-monolayer) and the benefit of using these cells as an ideal primary cell source of human origin for studying the neuronal toxicity of Fe_3O_4NPs.

The neuron-like phenotype was confirmed by: (i) live morphological analysis; (ii) Nissl body staining; and (iii) immunofluorescent staining of the expression of different typical neuronal-specific proteins/markers, at different stages of neuron-like cells maturation (i.e., at day 3-early differentiated and day 8-full differentiated), namely β-tubulin III (a microtubule element of the tubulin family, structural marker), MAP-2 (as a mature neuron marker), enolase (NSE, marker characteristic of neural cells), and glial fibrillary acidic protein (GFAP as an astrocyte marker), and almost no expression of nestin (as an immature neuron marker), and SOX-2 (a transcription factor essential for maintaining self-renewal, or pluripotency of undifferentiated stem cells). The synaptic makers such as synaptophysin (SYN, indicator of the synapses density), growth-associated protein 43 (GAP43, a key factor for axonal growth and elongation), and post-synaptic density 95 (PSD95 an important scaffold protein on the post-synaptic membrane, which plays an important role in the process of synapse formation) were also evaluated.

The toxic effects induced by short-term exposure (24–48 h) to increasing Fe_3O_4NP concentrations (10–100 µg/mL) have been evaluated on hNLCs of both stages of maturation (i.e., at day 3 and 8). Live/dead cells assessment was pursued applying three assays namely MTT: for identification of metabolic activity (i.e., mitochondrial dehydrogenases); ATP: indicator of ATP production breakdown; and Trypan Blue exclusion test: for membrane integrity determination. Since each of these three assays is using a different endpoint to assess cell viability it cannot be taken for granted that results from these three assays correlate. Moreover and most importantly, since NPs interaction with in vitro assay components and read-out systems may result in a wide array of false positives and false negatives, the suitable in vitro cytotoxicity bioassays need to be verified on case-by-case basis. For studies of potential toxicity of NPs it is recommended to apply at least two different biocompatibility assays that have independent underlying principles, in order to become aware of potential disturbances by NPs on the test systems used [7].

Quantification of apoptosis, by caspase-3/7 activity and nuclear fluorescence staining, and cell morphology analysis by light microscopy were complementary estimated.

Physico-chemical characteristics (i.e., hydrodynamic size, polydispersity Index, Zeta potential, and pH) of Fe_3O_4NPs dispersed in the appropriate aqueous media for cell dosing were also assessed, by dynamic light scattering, in order to understand their contribution to toxicity.

2. Results

2.1. Characterization of Mesenchymal Stem Cells Derived from Umbilical Cord Lining Membrane (Hcl-Mscs): Morphology, Cell Surface Markers, and Differentiation into Adipocytes and Osteocytes

The set up to obtain and characterize hCL-MSCs and their transdifferentiation into human neuronal-like cells has been performed following the steps reported in Figure 1.

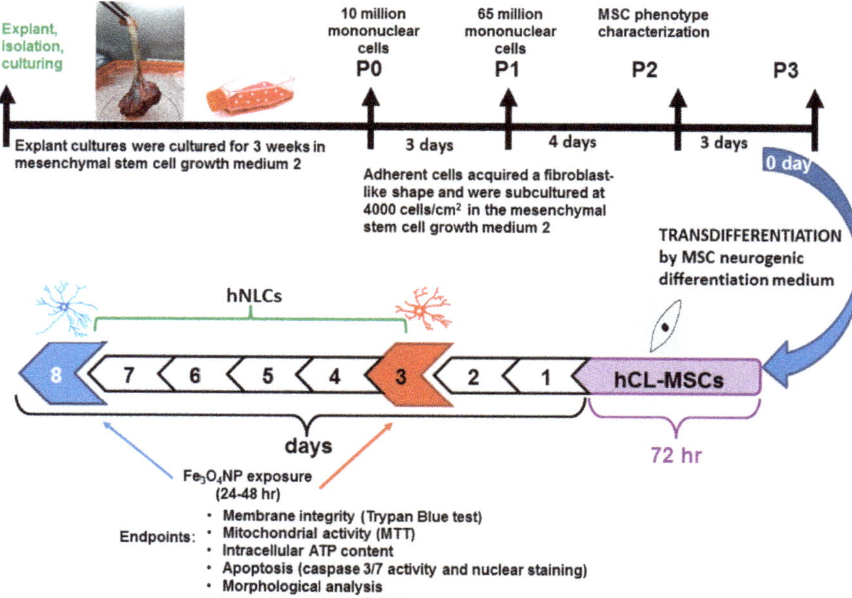

Figure 1. Scheme for the transdifferentiation of hCL-MSCs into neuron-like cells (hNLCs) and the endpoints evaluated at specific timing of Fe_3O_4NPs treatment. P indicates the passage number. The appearance of cell outgrowth from explant cultures was routinely monitored, and roundish or long fusiform cells were observed already after 3 days. The typical fibroblast morphology and homogeneous monolayer were detected between the 10th–17th days and the confluence (80–90%) was reached after approximately 21 days from the seeded cord lining (CL) pieces (Figures 1 and 2A).

After three weeks, when the cells were transferred to subcultures, their morphology appeared rounded (after digestion) and reverted to the fibroblast-like shape after reattachment to the plate and incubation for 24 h. Thereafter, cell growth was rapid, in about 3–4 days after passage adhered completely to plastic, achieved 80% confluency and showed a uniform spindle-shaped exterior with spiral and radial-like growth.

The cells were cultured until the ninth passage and entered into senescence at the twelfth passage.

The immunophenotype was consistent with MSCs: high expression levels of CD73, CD90, and CD105 and HLA-I and negative expression of CD34, CD45, CD14, CD31, and HLA-DR markers (Figure 2B). The hCL-MSCs were capable of differentiating, after the induction in specific media, into adipocytes, as demonstrated by the presence of the intracytoplasmic lipid droplets, and into osteocytes as demonstrated by histological detection of staining for alkaline phosphatase activity and calcium deposition (Figure 2C).

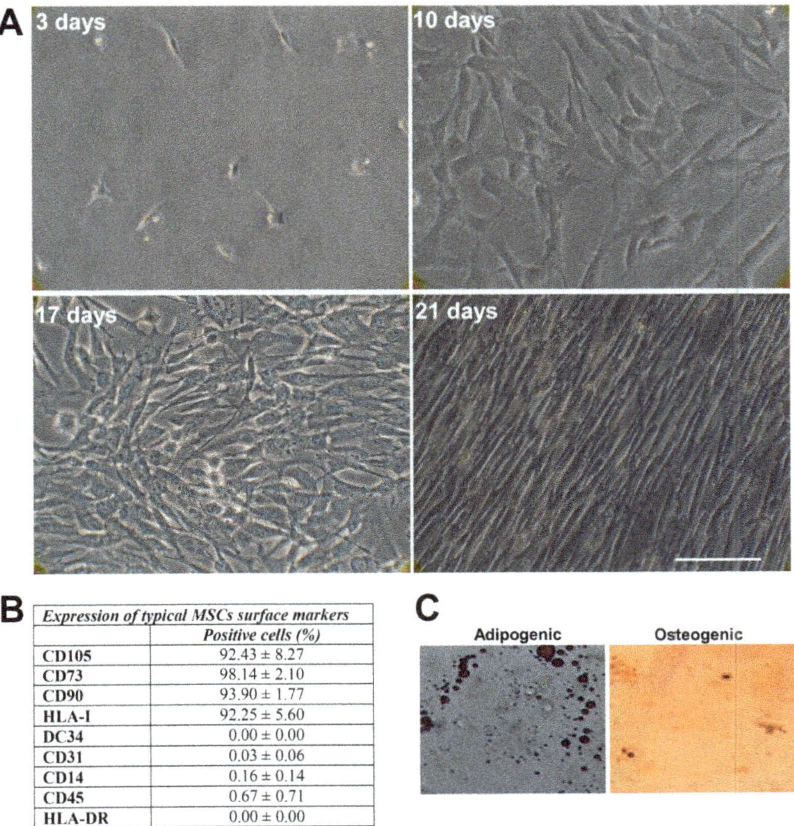

Expression of typical MSCs surface markers	
	Positive cells (%)
CD105	92.43 ± 8.27
CD73	98.14 ± 2.10
CD90	93.90 ± 1.77
HLA-I	92.25 ± 5.60
DC34	0.00 ± 0.00
CD31	0.03 ± 0.06
CD14	0.16 ± 0.14
CD45	0.67 ± 0.71
HLA-DR	0.00 ± 0.00

Figure 2. Characterization of mesenchymal stem cells derived from cord lining membrane (hCL-MSCs). (**A**) Morphology of cell outgrowth from the explants of hCL after 3, 10, 17, and 21 days (until P0) as visualized by phase-contrast microscopy (magnification 32×). hCL-MSCs adhere completely to plastic, exhibit fibroblastic morphology and show a radial or spiral growth (clearly visible at 17 days). Scale bar: 100 µm. (**B**) Typical hCL-MSCs surface markers. Percentage mean ± standard deviation (S.D.) (**C**) Representative images of hCL-MSCs after being induced in adipogenic and osteogenic differentiation (magnification 20×).

Due to the good growth condition at the 3rd passage cells, these hCL-MSCs were selected for the neurogenic transdifferentiation.

2.2. Neurogenic Differentiation

2.2.1. Neuronal-Like Phenotype Characterization

Before assessing the effects of Fe_3O_4 NPs on neuron-like cells (hNLCs) derived from hCL-MSC transdifferentiation, the neuron-like phenotype was confirmed by: (i) live morphological analysis (Figure 3A), (ii) quantitative changes of hNLCs during differentiation time (Figure 3B) and decrease of cell proliferation capacity (Figure 3C), (iii) Nissl body staining (Figure 3D), and (iv) expression of neuronal and synaptic specific proteins/markers (Figure 4A,B and Figure 5).

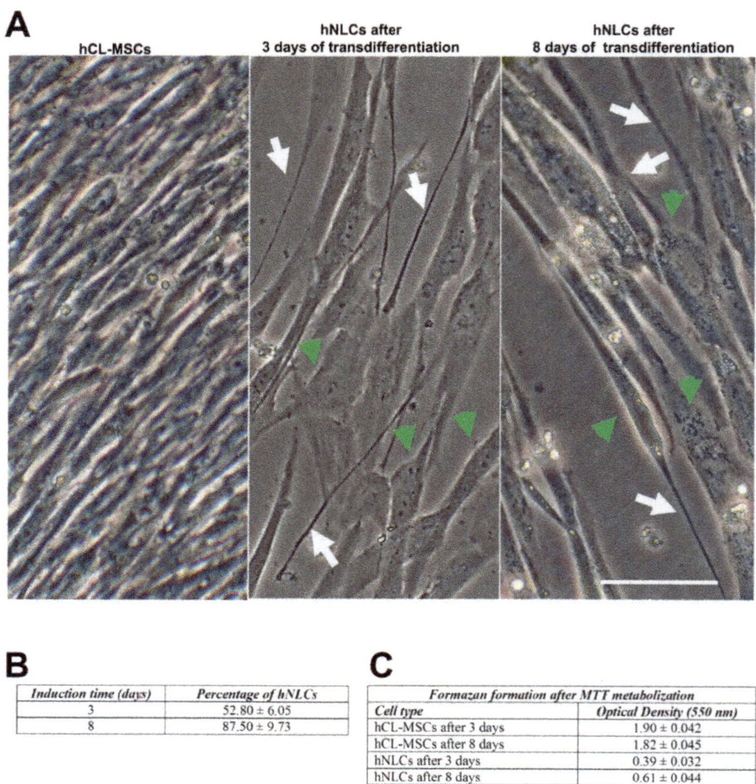

B

Induction time (days)	Percentage of hNLCs
3	52.80 ± 6.05
8	87.50 ± 9.73

C

Formazan formation after MTT metabolization	
Cell type	Optical Density (550 nm)
hCL-MSCs after 3 days	1.90 ± 0.042
hCL-MSCs after 8 days	1.82 ± 0.045
hNLCs after 3 days	0.39 ± 0.032
hNLCs after 8 days	0.61 ± 0.044

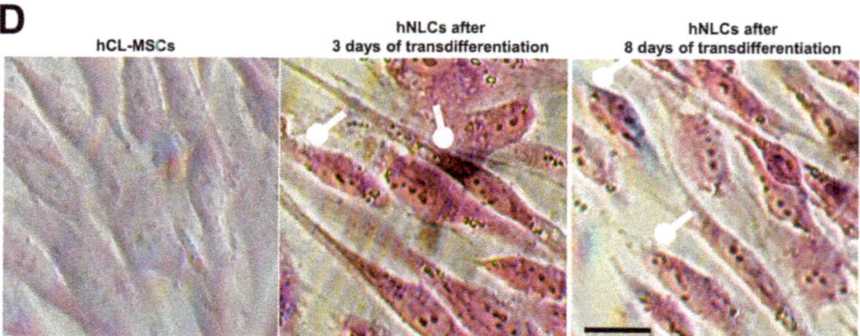

Figure 3. hCL-MSCs transdifferentiated into neuronal lineage at different time points. (**A**) Phase-contrast analysis of hCL-MSCs transdifferentiated into hNLCs at different time points. Cell bodies indicate by green head arrows; elongated structures indicate by white arrows. Scale bar: 100 μm. (**B**) Quantitative changes of hNLCs: the percentage of hNLCs was significantly increased during the induction time. Data are presented as the mean ± S.D. hNLCs at day 3 compared to day 8 of transdifferentiation were statistically different ($p < 0.05$). (**C**) Decrease of cell proliferation capacity during transdifferentiation process into hNLCs (3 and 8 days). Data are presented as the mean ± S.D. (**D**) The Nissl body staining of hCL-MSCs transdifferentiated into neuronal lineage at different time points: differently from the control (hCL-MSCs untransdifferentiated), the hNLCs (after 3 and 8 days) show somata-associated accumulations of the Nissl bodies stained dark black-violet (round-headed white arrows). Scale bar: 100 μm.

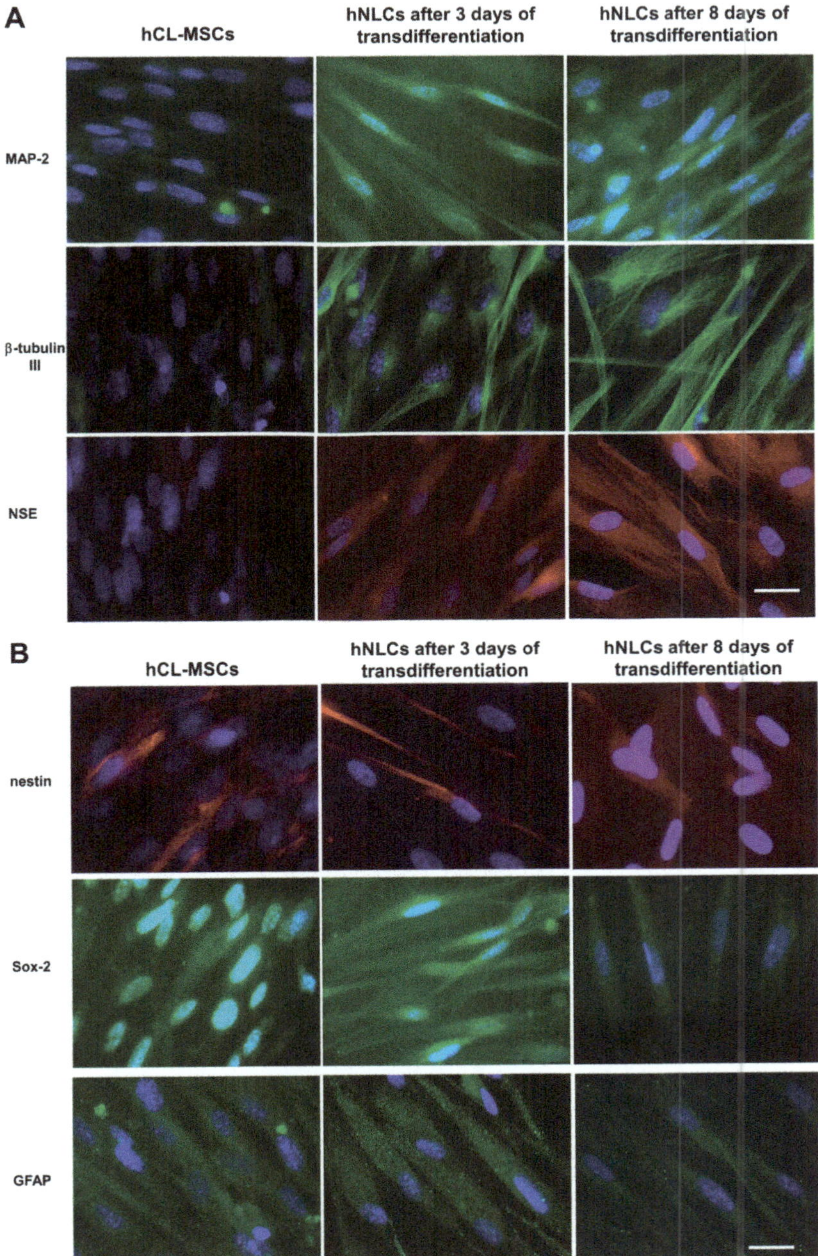

Figure 4. Immunofluorescence characterization of transdifferentiated hNLCs at different time points. (**A**) Representative fluorescence merged microphotographs showing MAP-2- and β-tubulin III-positive (green fluorescence) and enolase-positive (red fluorescence) in hCL-MSCs and transdifferentiated hNLCs at day 3 and 8, (**B**) microphotographs showing nestin-positive (red fluorescence), SOX-2-, and GFAP-positive (green fluorescence) in hCL-MSCs and transdifferentiated hNLCs at day 3 and 8. Nuclei were stained with Hoechst 33258. Scale bar: 100 μm.

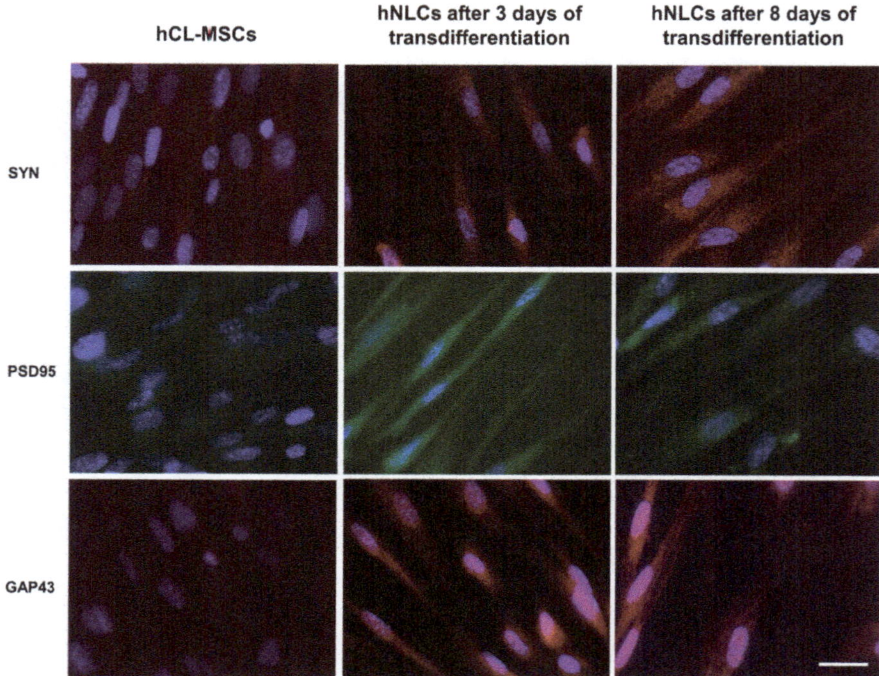

Figure 5. Immunofluorescence of synaptic markers. Representative fluorescence merged microphotographs showing SYN (red fluorescence), PSD95 (green fluorescence), and GAP43 (red fluorescence) positive in hCL-MSCs and transdifferentiated hNLCs at day 3 and 8. Nuclei were stained with Hoechst 33258. Scale bar: 100 μm.

Morphological and Quantitative Changes of hNLCs at Different Time Points (3 and 8 Days)

The images acquired using contrast-phase microscopy showed that hCL-MSCs transdifferentiated towards a neuronal lineage when cultured in mesenchymal stem cell neurogenic differentiation medium: in fact these induced cells exhibited typical neuron-like morphology (Figure 3A).

On day 3 of transdifferentiation, the cells became oval or round with elongated and extended processes (neurite-like); and the total number of cells that changes versus a phenotype neuron-like reached 52.8% ± 6.05% (Figure 3B). The hNLCs appeared more developed on day 8 of transdifferentiation exhibiting a more advanced neuronal appearance: the length of protrusions increased and gradually intertwine connected into an organized network with adjacent cells (Figure 3A); and about 87.50% ± 9.73% appeared as hNLCs (Figure 3B). On the contrary, the hCL-MSCs cultured in mesenchymal stem cell growth medium 2 showed typical spindle-shape morphology with no changes into neuronal morphology (Figure 3A).

The cell proliferative capacity, evaluated by optical density using formazan formation after MTT metabolization, decreased during the transdifferentiation process into hNLCs (3 and 8 days). The cell density was substantially higher in hCL-MSCs even though the same amount of cells (4000 cells/cm^2) was seeded for each group (Figure 3C).

Nissl Body Staining

The cresyl violet staining labeled the Nissl bodies (granular structures of rough endoplasmic reticulum) in the hCL-MSCs undergoing neurogenic transdifferentiation (hNLCs at 3 days and 8 days of transdifferentiation). The Nissl bodies appeared as dark black-violet spot around the nuclei, while, the same were completely absent in hCL-MSCs cultured in classical mesenchymal stem cell growth medium 2 (Figure 3D).

Expression of Neuronal and Synaptic Specific Proteins

The neuronal markers namely MAP-2, β-tubulin III, enolase-NSE, nestin, SOX-2, glial protein-GFAP, and the synaptic makers namely SYN, PSD95, and GAP43, were evaluated after 3 and 8 days of the neurogenic transdifferentiation.

Nuclei were detected using Hoechst 33258 nucleic acid stain, which is a popular nuclear counterstain that emits blue fluorescence when bound to dsDNA.

Figure 4A shows the expression of neuronal markers: MAP-2 and β-tubulin III were visible as green fluorescence around the soma and neurite-like processes in hNLCs at both time points of the neurogenic transdifferentiation, and NSE was visualized as red fluorescent signal into cytoplasm. On the other hand, the MAP-2, β-tubulin III, and NSE antibodies interacted very few with undifferentiated hCL-MSCs. Noteworthy, an improvement of fluorescence intensity of all three of these neuron markers were observed on 8-day hNLCs compared to the 3-day cells (Figure 4A).

Regarding nestin, an early differentiation marker localized in the cytoskeleton, an increase of red fluorescence intensity was observed on 3-day hNLCs, followed by a later decrease of the fluorescence signal after 8 days of transdifferentiation (Figure 4B). In particular, nestin was visible in the vast majority of the hCL-MSCs indicating that these cells were neural stem/progenitor cells (Figure 4B).

By contrast, a weak labeling of SOX-2 (nuclear localization of the fluorescent signal) was detected in both hNLCs when compared to hCL-MSCs: the fluorescence signal decreased during the time of transdifferentiation of these hCL-MSCs into hNLCs (Figure 4B).

Results related to immunofluorescence of GFAP showed the green fluorescence signal in hCL-MSCs and an increase of the fluorescence in hNLCs at 3 days, while hNLCs at 8 days displayed a weak signal, confirming that hCL-MSCs were capable to differentiate not only into hNLCs but also into astrocytes (Figure 4B).

The fluorescence of synaptic makers indicated that the hNLCs (at 3 and 8 days) were able to express both the pre- and post-synaptic proteins such as SYN, PSD95, and GAP43 (Figure 5). Specifically, the anti-SYN staining was exhibited as bright red fluorescence, PSD-95-positive (green fluorescence) was found in proximity to the cell surface, localized to the cell membrane, and GAP43-positive red fluorescence was found in the cell bodies and neurites of hNLCs. Notably, an increase of fluorescence signal for all tested markers were detected on hNLCs at 8 days compared to hNLCs at 3 days and the immunofluorescence was undetectable in hCL-MSC cultures for all markers (Figure 5).

Altogether these findings clearly indicated that hCL-MSCs successfully transdifferentiated into hNLCs. Then this novel in vitro model was used to evaluate the effects induced by short term Fe_3O_4NPs exposure on human hNLCs.

2.3. Physico-Chemical Characterization of Fe_3O_4 Nanoparticles in Mesenchymal Stem Cell Neurogenic Differentiation Medium

The physico-chemical properties of the Fe_3O_4NPs suspension (at 10 and 25 µg/mL) in culture medium used for transdifferentiation were investigated in terms of hydrodynamic size, polydispersity index (pdI), zeta potential (Zp), and pH, since they are parameters that affect the outcomes of cellular uptake, distribution, and reactivity of the nanoparticles in the biological system [26].

Hydrodynamic size measurements (Figure 6A) showed Fe_3O_4NPs agglomeration/aggregation immediately after dispersion in the culture medium (i.e., mesenchymal stem cell neurogenic differentiation medium) exhibiting a size in the micron range: 1213 ± 23.5 and 1368 ± 10 nm at 10 and 25 µg/mL, respectively, after 30 min.

Aggregation still persisted after 24 and 48 h: diameter about 1432 nm. as also observed in phase-contrast micrographs that showed aggregation/agglomeration of the Fe_3O_4NPs as a function of the concentration (Figure 6B,C).

A Size distribution of Fe₃O₄NPs in neurogenic medium

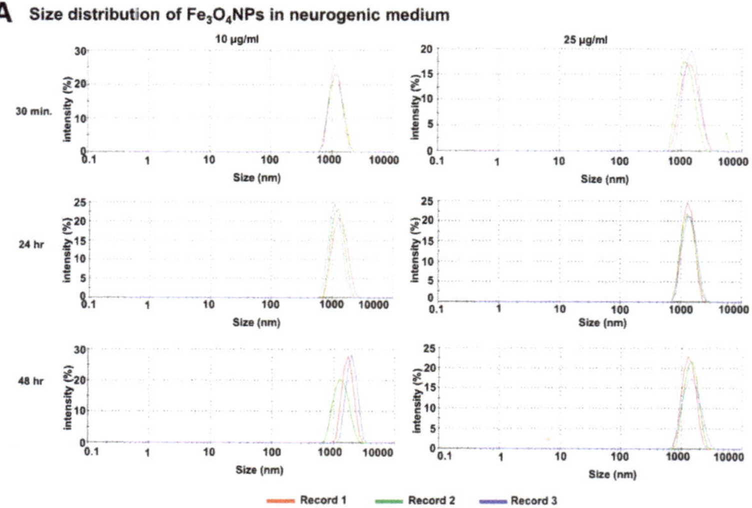

B   Physico-chemical properties of Fe₃O₄NPs in neurogenic medium

	10 µg/ml Fe₃O₄NPs			25 µg/ml Fe₃O₄NPs		
	30 min.	24 hr	48 hr	30 min.	24 hr	48 hr
Mean diameter (nm)	1213 (23.5)	1321 (50.9)	1552 (158.0)	1368 (10.0)	1405 (38.7)	1450 (36.7)
Zp (mV)	-9.50 (0.59)	-8.90 (0.68)	-9.40 (0.34)	-11.30 (0.59)	-10.20 (0.32)	-10.30 (0.25)
PdI	0.264 (0.01)	0.305 (0.09)	0.230 (0.05)	0.264 (0.01)	0.229 (0.01)	0.227 (0.03)
pH	8.09	7.78	7.65	8.09	7.64	7.65

C   Phase-contrast microscopy of Fe₃O₄NPs in neurogenic medium

Figure 6. Physico-chemical characteristics of the Fe₃O₄NPs in mesenchymal stem cell neurogenic differentiation medium. (**A**) Size distribution obtained from dynamic light scattering measurements of Fe₃O₄NPs at concentrations of 10 and 25 µg/mL in mesenchymal stem cell neurogenic differentiation medium after 30 min, 24 and 48 h. (**B**) Physico-chemical properties of the Fe₃O₄NPs in mesenchymal stem cell neurogenic differentiation medium. (**C**) Phase-contrast micrographs of Fe₃O₄NPs in neurogenic medium at 10 and 25 µg/mL after 48 h: aggregations/agglomerations of the Fe₃O₄NPs were observed as brownish sediments, which increased as function of the concentration. Scale bar: 20 µm.

The pdI values (about 0.266 and 0.240 at 10 and 25 µg/mL, respectively) indicated that the Fe_3O_4NPs suspensions were little polydispersed, and also a weak stability in long-term period was evidenced by Zp (around −10 mV) at 30 min, as well as after both 24 and 48 h. The pH values (pH 7.6–8.09) were slightly higher than physiological pH range for each time and concentration point considered (Figure 6B).

2.4. Cytotoxic Effects of Fe₃O₄NPs Exposure on hNLCs at Different Maturation Stages

Cell viability assays are important tools in toxicology studies to assess the cell sensitivity to compounds. In this study, three widely used cell viability assays, namely MTT (activity of mitochondrial dehydrogenases), Trypan blue (TB; membrane integrity) assays, and ATP (breakdown of ATP production) were applied to assess the Fe_3O_4NPs cytotoxicity on neurons. Specifically, hNLCs transdifferentiated at different time points (day 3 and 8) were exposed to increasing Fe_3O_4NPs concentrations (10–100 µg/mL) for 24 and 48 h.

2.4.1. MTT Assay

MTT assay, a standard colorimetric method, has long been regarded as the gold standard of cytotoxicity assays as it is highly sensitive. MTT assay is typically applied to evaluate the mitochondrial function by the activity of mitochondrial dehydrogenases after exposure to Fe_3O_4NPs. When this test was applied in this study, data indicated an unexpected concentration-dependent increase of cell viability after Fe_3O_4NPs treatments in both hNLCs, i.e., early differentiated (at day 3) and fully differentiated stage (at day 8), at both time points considered (24 and 48 h exposure; Figure 7A1,A2).

Cell viability data obtained from MTT assay showed a large difference compared to those obtained from the TB test, ATP assay, and the morphological analysis, which, on the contrary, showed concentration-dependent cell reduction (see below).

The agglomeration/aggregation of Fe_3O_4NPs, as evidenced by its physico-chemical characteristics in culture medium (used for transdifferentiation) and its optical properties (supplementary Figure S1) could be factors responsible of interference with the absorbance readings.

2.4.2. Cell Viability Evaluation by a Trypan Blue Exclusion Test

The membrane integrity evaluation on hNLCs was performed using Trypan blue (TB) exclusion test after 24 and 48 h exposure to increasing concentrations of Fe_3O_4NPs (10–100 µg/mL) (Figure 7B1,B2).

TB data, on 3-day hNLCs, indicated a gradual reduction of viable cells when compared to control after both time points considered. After 24 h, significant cell viability decrease (about 15%) started at 10 µg/mL with a maximum effect (30% cell viability decrease) at 100 µg/mL (Figure 7B1). The cytotoxicity was exacerbated after 48 h exposure: Fe_3O_4NPs treatments induced a significant cell reduction (cell death: 15–45%) at the concentrations ranging from 10 to 100 µg/mL (Figure 7B2).

hNLCs, at day 8-fully differentiated, appeared less susceptible to Fe_3O_4NPs than hNLCs at day 3-early differentiated after 24 h exposure: cell viability was affected at the higher concentrations tested, 50–100 µg/mL, with about 50% cell death (Figure 7B1). However, after 48 h, the cytotoxic effects showed similar trend to that observed for hNLCs at day 3-early differentiated (Figure 7B2).

2.4.3. Mitochondrial Metabolism Function by ATP Evaluation

A concentration-dependent reduction of the ATP intracellular content was observed in both hNLCs exposed to 10–100 µg/mL Fe_3O_4NPs up to 48 h (Figure 7C1,C2).

Specifically, in 3-day transdifferentiated hNLCs, the ATP intracellular content was reduced from 20% to 35% compared to control following exposure from 10 to 100 µg/mL after 24 h (Figure 7C1) and 20–52% after 48 h (Figure 7C2).

Fe_3O_4NP treatments induced similar ATP intracellular depletion (ATP reduction: 15–35% at 10–100 µg/mL) after 24 (Figure 7C1) and 48 h (Figure 7C2) in 8-day transdifferentiated hNLCs.

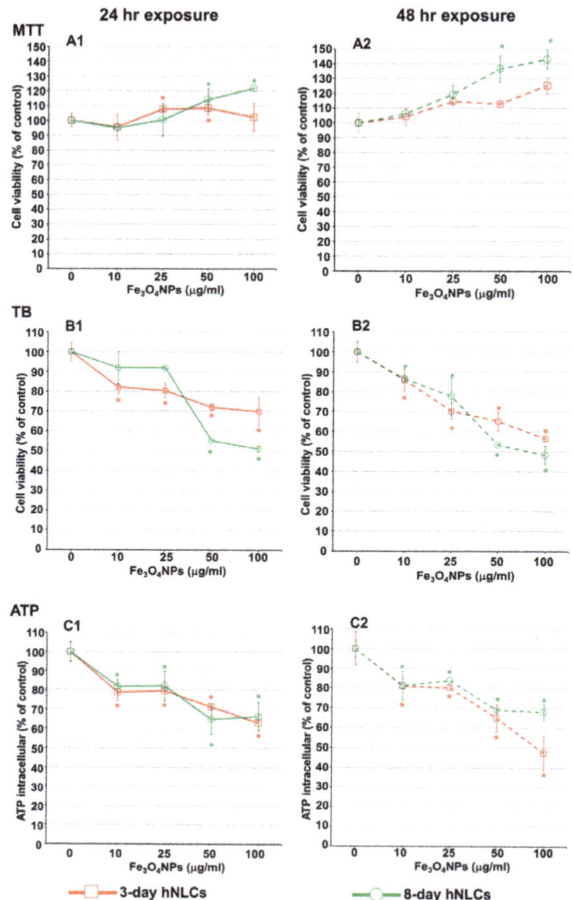

Figure 7. Cytotoxic effects of Fe_3O_4NPs on hNLCs treated, at day 3 and 8 of transdifferentiation, for 24 and 48 h with increasing Fe_3O_4NPs concentrations (10–100 µg/mL). Evaluation by three cell viability assays. (**A1,A2**) Mitochondrial activity assessed by MTT assay; (**B1,B2**) cell viability evaluation by Trypan blue (TB) test; and (**C1,C2**) evaluation of intracellular ATP content. Data are expressed as percentage of viable cells (% of each control) and represent the mean ± S.D. * $p < 0.05$, statistical analysis by two-way ANOVA followed by Dunnett's test.

2.4.4. Evaluation of Caspase-3/7 Activity and Apoptotic Features in Neuron-Like Cells

Caspase-3/7 activity and apoptotic cells were assessed after Fe_3O_4NPs treatments in both neuron-like cells (Figure 8A,B). The results obtained on hNLCs after 3 days of transdifferentiation showed that the levels of caspase-3/7 activity increased about 1.33-fold (compared to control) at 10 and 25 µg/mL (Figure 8A1).

The percentage of condensed apoptotic cells was enhanced following treatment with 10 µg/mL and 25 µg/mL of Fe_3O_4NPs: about 22% versus 9% detected in control (Figure 8A2). As observed in pictures, the control cells (untreated hNLCs) appeared oval-shaped and the nuclei stained uniformly blue fluorescent (due to the Hoechst 33258 dye). On the contrary, hNLCs (3 days) treated with 10 and 25 µg/mL exhibited typical apoptosis features such as cell shrinkage, chromatin condensation and fragmentation, formation of apoptotic cell bodies, and decrease of cell density (Figure 8A3). The effects lasted for up to 48 h.

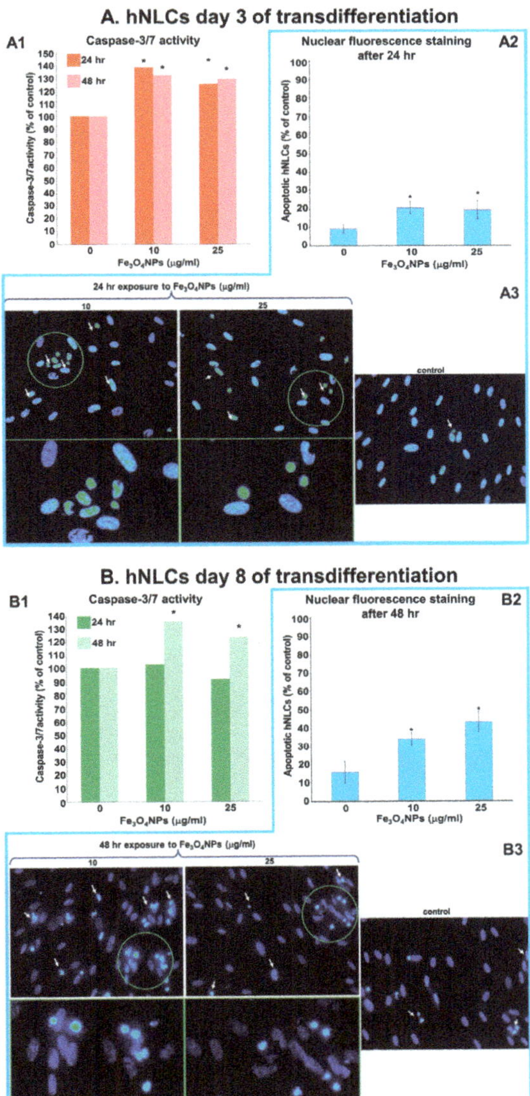

Figure 8. Caspase-3/7 activity and apoptotic cells evaluation after exposure to Fe_3O_4NPs (10 and 25 µg/mL). (**A**) Composite images that display the caspase-3/7 activity evaluated after 24 and 48 h exposure to Fe_3O_4NPs (**A1**) and apoptotic cells detected by Hoechst 33258 staining after 24 h exposure to Fe_3O_4NPs in hNLCs at day 3 of transdifferentiation (**A2,A3**). (**B**) Composite images that show the caspase-3/7 activity evaluated after 24 and 48 h exposure to Fe_3O_4NPs (**B1**) and apoptotic cells detected by Hoechst 33258 staining after 48 h to Fe_3O_4NPs in hNLCs at day 8 of transdifferentiation (**B2,B3**). The caspase-3/7 activity of the control cells was set to 100% and the data are expressed as mean ± S.D. * $p < 0.05$. Statistical analysis by two-way ANOVA followed by Dunnett's test. Apoptotic cells were expressed as % of total cell counted. Data represent the mean ± S.D. * $p < 0.05$. Statistical analysis by two-way ANOVA followed by Dunnett's test. Representative images in fluorescence microscope are taken using magnification 40×, the areas indicated by the circle show the magnification 2.5-fold. Arrows indicate nuclear morphological changes such as chromatin condensation, fragmentation, and formation of apoptotic bodies.

In hNLCs (8 days of transdifferentiation) the caspase-3/7 activity was not affected by 24 h treatment with Fe_3O_4NPs (Figure 8B1). A significant increase of caspase-3/7 activity, about 1.35-fold (respected to control) at 10 and 25 µg/mL, was instead observed after 48 h exposure, as well as well increase of apoptotic cells percentage (34.15% and 43.36%, at 10 and 25 µg/mL, respectively, compared to control: 15.76%; Figure 6B2). Typical apoptotic morphological changes (including condensation of chromatin and nuclear fragmentation) were observed after 48 h (Figure 8B3).

2.4.5. Morphological Analysis after Fe_3O_4NPs Exposure

Cell morphology in both transdifferentiated hNLCs following Fe_3O_4NPs exposure was examined by phase-contrast microscopy: no marked morphological changes/alterations have been detected on both hNLCs (early and full differentiated) after Fe_3O_4NPs treatments for each time-point considered (Figure 9).

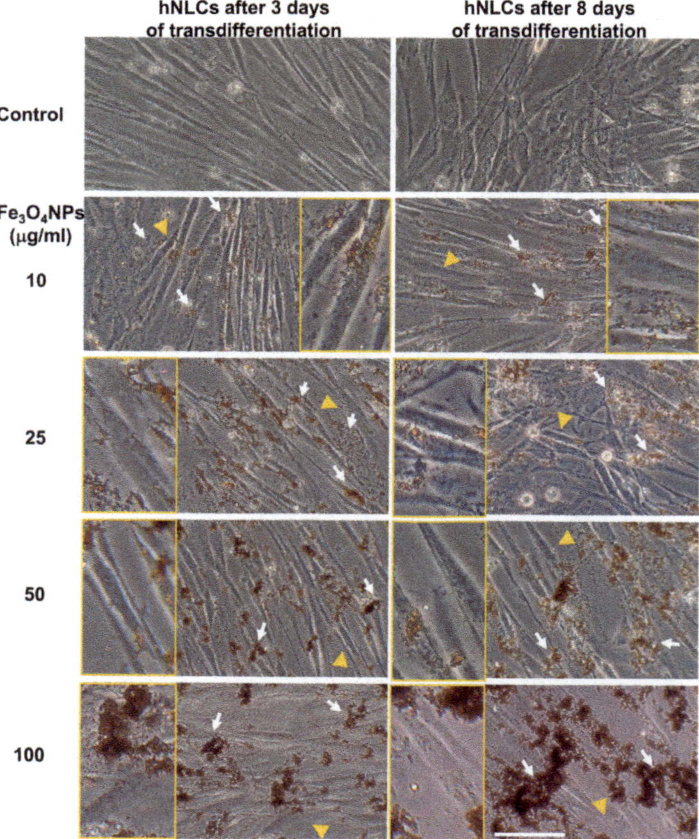

Figure 9. Representative micrographs, by phase-contrast microscopy of early and full transdifferentiated hNLCs after 24 h exposure to increasing Fe_3O_4NPs concentrations (10–100 µg/mL). No morphological alterations were observed after Fe_3O_4NPs exposure up to 48 h, however, a cell density decrease was observed from 25 µg/mL Fe_3O_4NPs at both time points considered for both hNLCs. Brownish aggregates/agglomerates of Fe_3O_4NPs in culture medium are indicated by white arrows. Inserts show the magnifications (2×) of the areas indicated by the yellow arrowheads where Fe_3O_4NPs are visible inside of the hNLCs. Scale bar: 100 µm.

However, cell density decrease was observed from 25 μg/mL Fe$_3$O$_4$NPs after 24 and 48 h exposure in both neuron-like cell types.

Fine brownish sediments of Fe$_3$O$_4$NPs were extracellularly visible in both hNLCs at the lowest concentration (10 μg/mL), which became large aggregates/agglomerates at the higher concentrations tested (25–100 μg/mL).

3. Discussion

Human MSCs derived from umbilical cord are increasingly under investigation as a promising source for a stem cell-based therapy as well as in vitro model/methods for toxicity screening of drugs/chemicals/nanoparticles, either in studying target organ toxicity or developmental toxicity [52].

The present study supported the evidence that human umbilical CL could be used as an enriched source for the isolation of pluripotent mesenchymal stem cells (hCL-MSCs). In particular, hCL-MSCs are abundantly available and can easily be isolated and expanded from healthy human subjects. These cells met the phenotypical and functional criteria for the definition of MSCs in that they (i) show the typical spindle-shape morphology forming a highly homogeneous monolayer adherent to plastic surfaces; (ii) express adhesion markers CD73, CD105, and CD90 and are devoid of hematopoietic and endothelial markers such as CD31, CD34, and CD45; express positive marker for HLA-I and negative marker for HLA-DR; and (iii) allow adipogenic and osteogenic differentiation in vitro.

Normally, around 65 million of hCL-MSCs are generated at passage 1 from 3 square centimeters of umbilical cord amniotic membrane. Cells grow rapidly, in about 3–4 days after passage adhere completely to plastic, the typical fibroblast morphology and homogeneous monolayer appear between the 10th–17th days, and the confluence (80–90%) is reached approximately 21 days after the seeded CL pieces. The cell cultures with hCL-MSCs require splitting approximately every 3 days from P1 to P9.

All these characteristics are particularly relevant since hUC is considered as a promising source of MSCs due to several advantages compared with other sources of MSCs [48]. Specifically, compared to bone marrow stem cells (the gold standard), hUC-MSCs have a painless collection procedure and faster self-renewal properties. Importantly, the number of mesenchymal stem cells that can be obtained from one UC greatly exceeds the mesenchymal stem cells that can be derived from bone marrow, cord blood, and adipose tissue [46].

By virtue of the inbuilt capacity of stem cells to differentiate, in the present study, the ability of MSCs isolated from human CL to transdifferentiate into hNLCs has been demonstrated applying an easy protocol by using commercially available neurogenic medium, which allow for efficient transdifferentiation observed as phenotypic changes.

The neuronal differentiation process was divided into two stages: early differentiated stage (at day 3) and fully differentiated stage (mature, at day 8). According to our results, cells, that have entered the transdifferentiation pathway, have changed their biological and metabolic properties. The neurogenically induced cells presented a reduced proliferation rate. These hNLCs were positive for enolase and Nissl bodies, and exhibited dendrite-like features of long spikes extending into other adjacent cells and lower cell densities especially after 8 days of transdifferentiation. An early increased and later downregulated level of early differentiation marker proteins nestin and SOX-2 further confirmed the differentiation process. While, increases on neuron marker expression, namely MAP-2 and β-tubulin III were observed during the progression of the differentiation process of the hCL-MSCs into hNCLs. These hNLCs were also positive for the pre- and postsynaptic markers SYN, PSD95, and GAP43. However, whether the obtained hNLCs have synapses capable of transmitting information needs further neuro-electrophysiological studies.

Results were clearly indicating that human umbilical CL-MSCs successfully differentiated into neuronal cells, which would be a predictive preclinical screening tool to evaluate neuronal toxicity in humans.

The ability of hMSCs to transdifferentiate into neurogenic lineage has been confirmed earlier by some authors [36–38,54,55] and this capacity has started recently to be investigated for hUC derived-MSCs [40–43,45,47]. In particular, human cord lining stem cells (hCL-MSCs) overcoming the preexisting difficulties inherent to MSCs from the bone marrow, offer not only a realistic, practical, and affordable alternative for tissue repair and regeneration [46], but also a promising cell-based model for in vitro screening and predictive studies for xenobiotic insults with particularly gained traction due to differentiation ability into several cell lineages including their potential towards converting into neural phenotypes [51,52].

After transdifferentiation and characterization of the hNLCs from hCL-MSCs, the cytotoxicity effects caused by short-term exposure to Fe_3O_4NPs have been evaluated on this cellular model. The overall effects were the following:

i. MTT data (culture medium with/without the neuron-like cells from hCL-MSCs plus Fe_3O_4NPs) indicated increments (artifacts) in cell viability after Fe_3O_4NPs exposure. These observations did not fit with the morphological analysis (by phase-contrast microscopy) and the TB and ATP data that overall showed cell density decrease at each time point considered.

ii. When applying the TB test: (a) on full differentiated neurons (day 8), the cytotoxicity effects appeared at high concentrations (50 μg/mL) after 24 h exposure producing about 50% cell death, (b) while on early differentiated neurons (day 3), a significant cell death (about 20%) appeared at the lowest concentration such as 10 μg/mL Fe_3O_4NPs after 24 h exposure, indicating a more susceptibility of the immature compared the mature neurons. After 48 h, the cell reduction became more pronounced compared to that observed after 24 h exposure with similar pattern of toxicity for cells at both stages of the differentiation process.

iii. A concentration- and time-dependent reduction of the ATP intracellular content (20–35%) was also observed in both hNLCs exposed to 10–100 μg/mL Fe_3O_4NPs up to 48 h. The early-differentiated neurons appeared more susceptible in that more marked cell death (about 50%) were detected after 48 h exposure at 100 μg/mL.

iv. Morphological analyses, parallelly, confirmed cell density decreases after 24 and 48 h exposure to Fe_3O_4NPs starting at 25 μg/mL in both neuron-like cell types without alterations of cell morphology.

v. Apoptosis induced by low concentrations of Fe_3O_4NPs (i.e., 10 and 25 μg/mL) was confirmed by the elevation of caspase-3/7 activity and nuclear staining. The effects appeared early (24 h) on 3-day-differentiated hNLCs, and later (at 48 h) on 8-day-differentiated cells, suggesting a more susceptible of the immature hNLCs than the full differentiated cells.

vi. Fe_3O_4NPs aggregate immediately after dispersion in the culture medium (i.e., mesenchymal stem cell neurogenic differentiation medium). The stability of Fe_3O_4NPs suspension (tendency to agglomerate in a specific culture medium) as well as Fe_3O_4NP optical properties represent factors that limit in vitro result interpretation depending of the applied methodology. As demonstrated in the present study, the MTT data (artifacts) suggest the not applicability of the spectrophotometric assays for hNLC culture conditions, while TB and ATP are accurate methods for determining cell viability after Fe_3O_4NPs exposure in this model.

Altogether these data indicated that Fe_3O_4NPs determined a concentration- and time-dependent reduction of human neuron-like cell viability. Cell density decrease (20–50%) were observed at the early time point (24 h) and started at ≥10 μg/mL in both types of differentiated hNLCs in association with apoptotic effects. The 3-day differentiated hNLCs were more susceptible (toxicity effects on both ATP content and membrane integrity started at 10 μg/mL already after 24 h exposure, apoptotic effects also appeared early) than the 8-day differentiated cells (effects on membrane integrity started at higher concentrations after 24 h, and apoptosis at 48 h).

This study enhanced the existing stock of knowledge on the impact of metal oxide NPs on human nervous system in particular. The potential harmful effects of Fe_3O_4NPs may be due to iron ion release, which may lead to a disruption of normal iron metabolism/homeostasis in the brain, a characteristic hallmark resembling that of several neurodegenerative disorders (e.g., Alzheimer's and Parkinson's) [20,56–58]. Indeed, it is well known that too much iron can compromise cell viability, as well as the transport and storage of iron [59]. It could be that Fe_3O_4NPs have produced iron liberation that exceeded the iron homeostasis capacity. Our preliminary data using cytochemical Perls' iron staining are indicating the uptake of Fe_3O_4NPs in hNLCs with accumulation in a time- and concentration-dependent manner (data not shown).

Our earlier investigations demonstrated Fe_3O_4NP-induced toxicity in 2D mono-cultures of human SH-SY5Y neuroblastoma cells, extensively used in vitro model for CNS toxicity evaluation, after short-term exposure to different concentrations (1–100 µg/mL): cytotoxicity occurred after 48 h only with 35–45% mortality from 10 to 100 µg/mL, whereas no effect was observed at the earlier time point (i.e., 24 h) [60]. Comparatively, the hNLCs of human primary cultures, that are applied in this study, have been shown to be more sensitive to Fe_3O_4NPs exposure compared to the SH-SY5Y cell line as toxicity effects were observable early (already after 24 h exposure in the mature cells). Furthermore, by applying this novel in vitro model, information related to NP-induced effects on the early differentiated neurons have also been achieved, revealing the more susceptibility of the immature cells compared the mature ones. In vitro neurotoxicity of Fe_3O_4NPs was already evaluated [53,61–63], even if the studies on human cell lines are limited [29], and significant differences in cytotoxicity results were observed comparing rodent versus human cells [53], suggesting the importance of the cell type and the species-specific model used to evaluate Fe_3O_4NPs toxicological profile. Notably, the findings of the present study further support the use of these human primary cells especially when considering the results obtained in cerebral cells from laboratory animals. In fact, recent studies evidenced a relative resistance of the rat astrocytes and neurons against short-term SPION exposure [25,64–67], although the experimental conditions [23,65] and the type of SPIONs used [27,67,68] have been shown to influence the toxicity in terms of reactive oxygen species formation and delayed toxic effects in these types of CNS cells.

Our study evidenced that the critical concentrations of Fe_3O_4NPs capable to induce in vitro neuronal cytotoxicity are comparable to those measured in peripheral blood and brain tissue of laboratory animals treated with SPIONs. In particular in ICR mice treated with a single intragastric administration of Fe_3O_4NPs (13 mg/mouse), the concentrations of Fe_3O_4NPs were 375 and 350 µg/mL in the peripheral blood, and 58 and 37 µg/g in brain, respectively at 3 and 10 days post exposure, indicating a relationship between the blood concentration-time changes and brain distribution of Fe_3O_4NPs [69]. Moreover, rat intranasally administered iron oxide nanoparticles have been shown to be preferentially distributed in striatum and hippocampus, causing oxidative damage in striatum [20]. The particles deposited at concentrations of 0.040 µg/g and 0.050 µg/g in striata and in hippocampi, respectively. More recently, an in vivo investigation showed that rabbit treated with Fe_3O_4NPs exhibited the mitochondrial disease and dysfunction with elevated oxidative stress in the brain [18], and another study provides the direct evidence that locally administered SPIONs in the striatum and hippocampus of mice were able to induce both apoptosis and deficits in some behavioral performance [19].

In humans, elevated iron levels seem to be associated with many types of neurodegenerative disease, such as Alzheimer's, Parkinson's, and Huntington's diseases [70–74]. In this respect, several studies have shown high levels of iron in brain autopsy section from patients with Alzheimer's disease, Huntington's disease, and Parkinson's disease [75,76], leading to the suggestion that elevated iron concentrations within the brain may contribute to the development of neurodegenerative disorders. Moreover, it has also been proposed that some of the excess iron in neurodegenerative tissue may be in the form of the magnetic Fe_3O_4 [57]. In support of this hypothesis, concentrations of Fe_3O_4 were found to be significantly higher in samples of Alzheimer's disease tissue [58]. Recently, brain magnetite nanospheres were detected in human subjects and were consistent with an external, rather than an endogenous, source [77]. Their presence proves that externally sourced iron-bearing NPs can be transported directly into the brain, where they can pose hazard to human health.

4. Materials and Methods

4.1. Chemicals and Reagents

Mesenchymal stem cell growth medium 2 (Ready-to-use; PromoCell, Heidelberg, Germany), mesenchymal stem cell neurogenic differentiation medium (Ready-to-use; PromoCell), human fibronectin solution (1 mg/mL; PromoCell), and all cell culture reagents were purchased from Carlo Erba Reagents (Carlo Erba Reagents S.r.l., Cornaredo, Italy), 75 cm^2 tissue culture flask with vented filter caps and 96- and 6-well plates (Corning), Trypan blue solution (0.4%), cresyl violet acetate, and MTT ((3-(4,5-dimethylthiazol-2-yl)-2,5-diphenyltetrazolium bromide) were purchased from VWR International PBI (Milan, Italy). CellTiter-Glo® 3D Cell Viability and Caspase-Glo® 3/7 assays were acquired from Promega (Milan, Italy). Primary antibodies (Santa Cruz Biotechnology, Dallas, TX, USA) conjugated to alexa-fluo®488 or 594 (Santa Cruz Biotechnology) for nestin, SOX-2, enolase (NSE), MAP-2, β-Tubulin III, synaptophysin (SYN), growth-associated protein 43 (GAP43), post-synaptic density 95 (PSD95), and primary antibody for glial fibrillary acidic protein (GFAP; Santa Cruz Biotechnology) were purchased from D.B.A. Italia s.r.l (Segrate (MI), Italy). Hoechst 33258 (Invitrogen, Waltham, MA, USA) was provided by Life Technologies Italia (Monza, Italy). Polyvinylpyrrolidone coated Fe$_3$O$_4$NPs were obtained from nanoComposix (San Diego, CA, USA; lot no. ECP1475).

4.2. Isolation and Primary Culture of MSCs from Human Umbilical Cord Lining Membrane (hCL-MSCs)

Human umbilical cords (UC, $n = 9$) were obtained from healthy donors, after patients' provided informed consent (Internal Ethics Committee—Prot. No. 2017000038067, 23 December 2016), who underwent full-term caesarian sections (38–40 weeks' gestation) at the Hospital Fondazione IRCCS Policlinico San Matteo in Pavia, Italy (from January 2017 to January 2019). In aseptic condition, UC samples were harvested in a collection cup containing physiological solution, store at 4 °C and immediately transported to the lab (within 2 h from collection). The UC samples were washed multiple times with ice-cold phosphate-buffered saline (PBS) to remove blood clots.

From each UC, mesenchymal stem cells (MSCs) derived from human umbilical cord lining membranes (hCL-MSCs) were obtained by the explant method and characterized [49]. Approximately 3 cm in length of UC were collected and the hCL-MSCs were isolated from the subamnion region by dissecting out the Wharton's jelly followed by removal of cord vessels. The minced pieces of the outer envelope membranes were plated in a 75 cm^2 tissue culture flask (about 20 tissue pieces/flask) and were maintained in mesenchymal stem cell growth medium 2 at 37 °C, 5% CO$_2$ in a humidified atmosphere approximately 3 weeks. In particular CL was minced into 1–2 mm pieces using sterile scissors and CL fragments were incubated in trypsin solution for 30 min at 37 °C for partial digestion, stopped by adding DMEM medium supplemented with 10% heat-inactivated fetal bovine serum (FBS), 2 mM L-glutamine, 50 IU/mL penicillin, and 50 µg/mL streptomycin. About 20 partially digested CL pieces were plated in a 75 cm^2 tissue culture flask in 10 mL of mesenchymal stem cell growth medium 2. Medium was changed every 3-4 days. Cultures were maintained at 37 °C, 5% CO$_2$ in a humidified atmosphere.

The cell growth was monitored every 3–4 days under an inverted phase-contrast microscope. When the mononuclear cells achieved 80–90% confluency, the medium was removed, the cells were rinsed with PBS and then digested with Accutase (up to 5 min at room temperature (r.t.)). Subsequently, the cells were reseeded on 75 cm^2 flasks at a density of 4000 cells/cm^2 (P0) and subcultured. Cells split every 3–4 days and cultured up to P4 to be used for the experiments. The remaining cells that were not used for the assay were frozen down at each passage.

Three cm of umbilical cord yielded around 65 million cells at passage 1, in line with literature data [46]. The cells (at passage 2) were characterized for purity and basic MSC characteristics following recommendations of the International Society for Cell Therapy (ISCT) [78,79]: surface markers (CD73, CD34, CD90, CD14, CD45, CD31, CD105, HLA-I, and HLA-DR) analysis and differentiation capacity into adipo- and osteo-lineages (using specific media) were performed as previously described [49].

Briefly, analysis of cell populations was performed with a FACS Navios flow-cytometer (Beckman Coulter) using fluorescein isothiocyanate- or phycoerythrin-conjugated monoclonal antibodies specific for CD73, CD34, CD90, CD14, CD45, CD31, CD105, HLA-I, and HLA-DR. The differentiation capacity of hCL-MSCs was assessed by incubating cells for 21 days with differentiation medium (α-minimal essential medium, 10% fetal bovine serum, 10^{-7} M dexamethasone, 50 mg/mL L-ascorbic acid, and 5 mM β-glycerol phosphate). For adipogenic differentiation 100 mg/mL insulin, 50 mM isobutyl methylxanthine, and 0.5 mM indomethacin were also added. The assessment of adipogenic differentiation was based on the morphological appearance of fat droplets after staining with Oil Red O. To detect osteogenic differentiation, cells were stained for alkaline phosphatase activity using Fast Blue and for calcium deposition with Alizarin Red.

4.3. Transdifferentiation of hCL-MSCs into Neuron-Like Cells (hNLCs)

The hCL-MSCs, obtained from $n = 4$ donors, at the 3rd–4th passage according to the logarithmic growth phase, were used for the transdifferentiation into hNLCs and neuronal characterization. The hCL-MSCs were transdifferentiated for 3 days according to the protocol supplied by the manufacturer and up to 8 days to evaluate different stages of maturity (i.e., at day 3-early differentiated and day 8-full differentiated (Figure 1). Specifically, the hCL-MSCs were cultured in standard conditions on 75 cm^2 tissue culture flask and when about 80% of the cell confluences was reached, the cells were detached by Accutase and reseeded on multiwell plates (at 4000 or 1500 cells per cm^2) coated with 10 µg/mL human fibronectin, in order to facilitate the cell attachment to the wells, in mesenchymal stem cell growth medium 2.

After 72 h, the cell became subconfluent (about 80%) then the whole culture solution (mesenchymal stem cell growth medium 2) was discarded and replaced with a ready-to-use mesenchymal stem cell neurogenic differentiation medium. Thereafter, the hCL-MSCs were induced in standard growth condition (37 °C/5% CO$_2$) and the medium changes were made every 48 h. The cells were cultured for least 3 days up to a maximum of 8 days and they were monitored for the morphological changes using an inverted phase contrast microscope (Zeiss Axiovert 25 microscope equipped with a 32× objective). Parallelly, hCL-MSCs (without passage) were further cultured (for 3 days up to 8 days) in the same condition using growth medium 2 and applied as a control of undifferentiated cells (i.e., hCL-MSCs).

4.4. Characterization of the hNLCs

4.4.1. Morphological Changes and Calculation of Positive Rate of hNLCs

On day 3 and 8 of transdifferentiation, the hNLCs were observed under the inverted microscope (Zeiss Axiovert 25 microscope equipped with a 32× contrast phase objective), in order to analyze the morphological changes induced by transdifferentiation. Afterwards, the rate of hNLCs was calculated: three non-overlapping fields were randomly selected and the total cell number and the number of hNLCs were counted.

4.4.2. Immunochemistry Analysis

Nissl Body Staining

The transdifferentiation of hCL-MSCs into the hNLCs, on day 3 and 8, were confirmed by the specific staining of neuronal Nissl bodies, characteristic granular structures composed of RNA-rich rough endoplasmic reticulum (rER) unique to the somata of neurons, as follows.

After discarding the culture medium from the hNLCs, they were washed (twice) with phosphate-buffered saline (PBS) prewarmed, and fixed with paraformaldehyde (PF 4%) for 30 min at room temperature (r.t.). Subsequently the hNLCs were washed twice with PBS and stained with Nissl staining solution (0.5% cresyl violet) for 30 min at r.t. After, the staining solution was aspirated and the monolayer was washed three times with PBS, finally mounted with Fluoroshield. The presence of

Nissl body into the somata of the hNLCs (3 and 8 days post transdifferentiation) was examined under a CX41 Olympus microscope. Digital images were captured using oil immersion objective (100×) lens.

Neuron and Synaptic Markers by Immunofluorescence Analysis

On day 3 and 8 of transdifferentiation, the hNLCs were rinsed with PBS gently and fixed in 4% PF (30 min at r.t.), then permeabilized with 0.1% Triton-100 for 5 min. Subsequently, the hNLCs were washed three times with PBS (containing Ca^{2+} and Mg^{2+}) and the samples were incubated with blocking solution (2% dry milk in D-PBS with Ca^{2+} and Mg^{2+}) for 30 min at r.t., and then removed without washing followed by incubation with primary antibodies conjugated to alexa-fluo®488 or 594 against: nestin (1:100), SOX-2 (2 µg/mL), NSE (1:100), MAP-2 (1:500); β-tubulin III (1:200), SYN (1:100), GAP43 (1:100), and PSD95 (1:100), diluted in 2% dry milk PBS incubated on orbital shaker plate in dark for 60 min at r.t. Next, the monolayers were washed (three times with PBS containing Ca^{2+} and Mg^{2+}; 5 min for each washing) and the nuclei were detected using Hoechst 33258 (5 µM for 10 min at r.t.) and finally mounted with Fluoroshield. Only for GFAP, the primary antibody (1:100), was incubated over night at +4 °C, then the cells were washed (three times with PBS containing Ca^{2+} and Mg^{2+}; 5 min for each washing), stained with secondary antibody (Alexa 488-labeled; dilution 1:100), washed again with PBS, counterstained for the nuclei and finally mounted with Fluoroshield. For each marker, three separate experiments were performed, each experiments was carried out in three replicates.

Fluorescence images were acquired using a CX41 Olympus fluorescence microscope, excitation light being provided by EPI LED Cassette and combined with digital camera.

Digital images of the eight randomly selected microscopic fields (for each neuronal and synaptic marker) were captured using oil immersion objective (100×) lens, and measurement conditions were the following: 470 nm excitation (T% = 40), 505 nm dichroic beamsplitter, and 510 nm long pass filter.

4.5. Fe_3O_4NPs Stock Suspension for the Treatment of hNLCs at Different Stages of Maturation

The physico-chemical properties of Fe_3O_4NPs stock suspension in a 2 mM citrate solution were provided by the Company (Figure 10).

Briefly, morpho-dimensional analysis using transmission electron microscopy showed that Fe_3O_4NPs, dark brown in color, had a roughly spherical, almost non-agglomerated particles, an average diameter of 20.3 ± 5 nm (by transmission electron microscopy) and a hydrodynamic diameter of 42 nm (by dynamic light scattering measurements). The surface of Fe_3O_4NPs is functionalized with polyvinylpyrrolidone (PVP), a polymer that offers steric stability and exhibit superparamagnetic properties at ambient temperatures.

Physico-chemical Fe_3O_4NPs properties in mesenchymal stem cell neurogenic differentiation medium was also performed: the size of the Fe_3O_4NPs and the zeta potential at 10 and 25 µg/mL after 30 min, 24 and 48 h were analyzed by dynamic light scattering using the Malvern Zetasizer Nano ZS90 (N.A.M. S.r.l., NANO-Analysis and Materials, Gazzada Schianno, Varese, Italy).

Fe_3O_4NP suspensions were prepared by diluting the stock suspension (20.3 mg/mL) in mesenchymal stem cell neurogenic differentiation medium to obtain increasing Fe_3O_4NPs concentrations to be used for neuron-like cells treatments. Fresh solutions of Fe_3O_4NPs were prepared and vortexed immediately before each treatment.

A

Physico-chemical properties of Fe$_3$O$_4$NPs	
Diameter (TEM)	20.3± 5 nm
Coefficient of Variation	24.6%
Surface Area (TEM)	50.2 m²/g
Mass Concentration	20.3 mg/ml
Hydrodynamic Diameter	42 nm
Zeta Potential	- 51 mV
pH Solution	7.4
Particle Surface	Polyvinylpyrrolidone
Solvent	Aqueous 2 mM Citrate

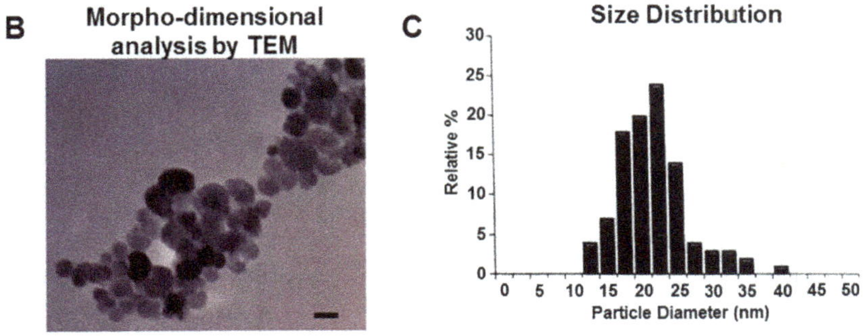

B Morpho-dimensional analysis by TEM

C Size Distribution

Figure 10. Physico-chemical properties of the Fe$_3$O$_4$NPs stock suspension in aqueous 2 mM citrate (**A**). Morpho-dimensional analysis of Fe$_3$O$_4$NPs stock suspension by TEM indicates roughly spherical almost non-agglomerated particles. Scale bar: 20 nm (**B**). Size distribution by dynamic light scattering measurements shows an average diameter of 20.3 ± 5 nm (**C**). Data provided by the Company nanoComposix.

4.6. Cell Viability Assays

For the Fe$_3$O$_4$NPs cytotoxicity studies, initially the hCL-MSCs (at P3/4) were seeded in six-well plates at the cell density of 4000 cells/cm^2 (for the Trypan blue (TB) exclusions test, double staining for apoptosis, morphology) or in 96-well plates at the cell density of 1500 cells/cm^2 (for the MTT assay, ATP intracellular content, and caspase-3/7 activity), and then following the protocol described in the paragraph "Transdifferentiation of hCL-MSCs into hNLCs", to obtain the 3-day early and 8-day full differentiated cells.

4.6.1. MTT Assay (Tetrazolium-Based Colorimetric Test Systems Utilizing MTT ((3-(4, 5-Dimethylthiazolyl-2)-2,5-Diphenyltetrazolium Bromide)

The tetrazolium derivatives are reduced in cells mainly by dehydrogenase enzymes, producing intracellular formazan products of specific color that can be measured photometrically after solubilization.

The activity of the dehydrogenase enzymes requires NAD$^+$ or NADPH as co-factors, and can therefore be seen as metabolic indicators of cell proliferation. As the result depends on mitochondrial integrity and activity, the sensitivity and general outcome of these assays is dependent on the cellular metabolic activity.

The hCL-MSCs at P3/4 were seeded in 96-well plates (1500 cells per cm^2) in 100 μL mesenchymal stem cell growth medium 2 per well and incubated (37 °C/5% CO$_2$) for 3 days then induced for neuron-like cells transdifferentiation.

Then hCL-MSCs, transdifferentiated into hNLCs, for 3 days up to 8 days, were exposed at both time points to Fe$_3$O$_4$NPs at concentrations ranging from 10 to 100 μg/mL for 24 and 48 h. After incubation time with Fe$_3$O$_4$NPs, the cell culture medium was carefully aspirated and hNLCs (early and full differentiated) were washed with PBS (200 μL per well) in order to remove unbound Fe$_3$O$_4$NPs and to avoid interference with the spectrophotometric analysis. Next, fresh medium (mesenchymal stem cell neurogenic differentiation) plus 10 μL MTT (5 mg/mL) were added to each well and incubated for 3 h. The resulting formazan crystals were solubilized by dimethyl sulfoxide 100 μL per well and quantified by measuring absorbance at 550 nm (measurement) and 655 nm (reference) using a microplate reader (Bio-Rad, Segrate, Milan, Italy). Data were expressed as a percentage of control.

4.6.2. Trypan Blue (TB) Exclusions Test

A loss of integrity of the plasma membrane is seen as one of the hallmarks of necrosis: the uncontrolled and enhanced trans-membrane flow of cytosolic elements or indicator dyes are used as endpoints. One common dye used in toxicity tests is the trypan blue.

On day 3 and 8 of transdifferentiation, the hNLCs were treated with Fe$_3$O$_4$NPs (ranging from 10 to 100 μg/mL) for 24 and 48 h, untreated control was incubated with mesenchymal stem cell neurogenic differentiation medium only. After each time point, the hNLCs were harvested: Fe$_3$O$_4$NPs suspensions were removed and cells washed with prewarmed PBS (1 mL/well), detached from the bottom of well by Accutase (500 μL; 5 min. at r.t.), pipetted gently, collected and counted manually using Bürker chamber. TB solution (0.4%) was used in a ratio of 1:10 (viable cells do not take up TB).

4.6.3. ATP Evaluation by CellTiter-Glo® 3D Assay

Another viability-related readout of cellular metabolism is represented by cytosolic ATP content, which can be detected and quantified by luciferin/luciferase assays that generate a luminescence signal proportional to the amount of cytosolic ATP present, which in turn is directly proportional to the number of viable cells present in culture.

On day 3 and 8 of the transdifferentiation, the hNLCs were treated with Fe$_3$O$_4$NPs (ranging from 10 to 100 μg/mL, for 24 and 48 h) and the ATP content was evaluated using CellTiter-Glo® 3D Cell Viability assay provided as a ready-to-use solution without no additional preparation in according to the protocol supplied by the manufacturer (Promega). Briefly, the CellTiter-Glo® 3D reagent was equilibrated as well as the cells at r.t., for 30 min. Next, the medium was carefully removed and the cells were washed with PBS (200 μL/well) and fresh mesenchymal stem cell neurogenic differentiation medium was added (100 μL/well). Then, 100 μL CellTiter-Glo® 3D reagent were added at each well and the contents were vigorously mixed for 5 min on an orbital shaker plate in order to induce cell lysis. The plate was incubated at r.t. for an additional 25 min (in the dark) to stabilize the luminescent signal. The ATP was quantified by measuring the luminescence signal using a Fluoroskan microplate fluorometer (Thermo Scientific, Milan, Italy) combined with PC software.

The blank reaction (culture medium plus CellTiter-Glo® 3D reagent) was also used in order to measure background luminescence associated with the specific cell culture medium and CellTiter-Glo® 3D reagent. Then, the blank value was subtracted from experimental values.

4.7. Assessment of Apoptosis

4.7.1. Caspase-3/7 Activity

Caspase-Glo® 3/7 assay is a luminescence-based test system designed for quantification of caspase-3 and -7 activities. The luminescence signal produced is proportional to the amount of caspase activity present. Caspase-Glo® 3/7 assay is provided as ready-to-use solution without any additional preparation.

On day 3 and 8 of transdifferentiation, the hNLCs were treated with Fe_3O_4NPs (ranging from 10 to 100 μg/mL) for 24 and 48 h, then the caspase-3/7 activity was evaluated in according to the protocol supplied by the manufacturer (Promega). Briefly, Caspase-Glo® 3/7 reagent was reconstituted and together with the cells were equilibrated at r.t., for 30 min. Next, the medium was carefully removed and the cells were washed with PBS (200 μL/well) and fresh mesenchymal stem cell neurogenic differentiation medium was added (100 μL/well) as well as the same volume (100 μL/well) of Caspase-Glo® 3/7 reagent. The contents were gently mixed using an orbital shaker plate for 2 h and 30 min at r.t. in the dark. The caspase-3/7 activity was quantified by measuring the luminescence signal using a Fluoroskan microplate fluorometer (Thermo Scientific, Milan, Italy) combined with PC software.

The blank reaction (culture medium plus Caspase-Glo® 3/7 reagent) was also used in order to measure background luminescence associated with the specific cell culture medium and Caspase-Glo® 3/7 reagent. Then, the blank value was subtracted from experimental values.

4.7.2. Nuclear Fluorescence Staining of hNLCs

The hNLCs cell cultures were fixed in PF 4% (20 min at r.t.) and after two washing (PBS) were supravitally stained with the fluorescent nuclear dye, Hoechst 33342 (5 μM for 10 min at r.t.). Afterwards the hNLCs were washed (PBS before then H_2O), let dry and scored under fluorescence microscope (CX41 Olympus fluorescence microscope equipped with a 40× objective). The microscopic fields were photographed and stored on PC.

4.8. Morphology Analysis by Light Phase-Contrast Microscopy

hNLCs, on day 3 and 8 of transdifferentiation, were observed under inverted phase-contrast microscopy after Fe_3O_4NPs (10–100 μg/mL) exposure for 24 and 48 h in order to evaluate the healthy status of the cells and their growth. Live-cell microscopy was performed using a Zeiss Axiovert 25 microscope equipped with a 32× contrast phase objective. Images were taken using a digital camera (Canon Powershot G8). Digital photographs were taken and stored on the PC.

4.9. Statistical Analysis

Data of the cytotoxicity effects (MTT, ATP, TB, caspase-3/7 activity, and apoptotic cells) were expressed as the mean ± S.D. of three separate experiments, each carried out in four replicates.

Statistical analysis was performed by two-way ANOVA followed by Dunnett's test. Only *p* values less than 0.05 were considered to be significant.

The intra-assay variation (CV) was <5–8% for the immunophenotyping analysis of hCL-MSCs derived from the same donors. The inter-assay precision ranged between 8% and 10% for the immunophenotyping analysis of the hCL-MSCs derived from different donors.

5. Conclusions

The study demonstrated that human umbilical CL-MSCs easily differentiated into neuronal-like cells. These in vitro findings add value to the relevance of using new in vitro human cell-based models in toxicology and, specifically, for the identification of the cytotoxicity of Fe_3O_4NPs.

The hCL-MSCs are versatile stem cells easily to be obtained and expanded from healthy human subjects. They have the advantages of strong proliferative ability (self-renewal), long-term proliferation, stable amplification in vitro, wide source without any ethical restriction, and can be easily differentiated into specific cells such as hNLCs. These properties make hCL-MSCs as potential gold standard tool for establishing in vitro models of neurotoxicity and development NT.

With regard to Fe$_3$C$_4$NPs, the present findings: (i) demonstrate the human neuronal-like cells susceptibility to Fe$_3$O$_4$NPs exposure, and (ii) support the use of these human primary cultures of neurons as new in vitro species-specific cell model for the evaluation of the NP safety.

The proposed in vitro cell-based model (hNLCs) and the multiple-assays approach, is readily available, either for use as stand-alone method or as a part of integrated strategies, and efficiently valuable to be applied for neurotoxicity testing of nanoparticles.

In addition, this approach is in accordance with the new toxicology paradigm, which emphasizes predictive toxicology (i) more based on in vitro models other than computational systems, bioengineering, automated micro-systems, (ii) boosting basic research more focused to obtain relevant and rapid tools for man, and (iii) understanding the disease cellular mechanisms.

Supplementary Materials: The following are available online at http://www.mdpi.com/1422-0067/21/1/271/s1, Figure S1: Optical properties of Fe$_3$O$_4$NPs.

Author Contributions: T.C.: conceived and designed the study and supervised the research, involved in the results interpretation and data analyses, edited the manuscript, performed the final version to be submitted. U.D.S.: planned and performed all the in vitro experiments, acquired, elaborated the data, performed image analyses and drew the figures, and drafted the manuscript. A.S., F.C., L.G.: contributed to study conception and design, and manuscript revision. A.S.: helped with statistical analysis. All authors discussed the results and commented on the manuscript. All authors have read and agreed to the published version of the manuscript.

Funding: This work was supported by Grant from the Italian Ministries of Health, Research and Education.

Conflicts of Interest: The authors report no declarations of interest.

References

1. Dulińska-Litewka, J.; Łazarczyk, A.; Hałubiec, P.; Szafrański, O.; Karnas, K.; Karewicz, A. Superparamagnetic iron oxide nanoparticles—Current and prospective medical applications. *Materials* **2019**, *12*, 617. [CrossRef]

2. Jahangirian, H.; Kalantari, K.; Izadiyan, Z.; Rafiee-Moghaddam, R.; Shameli, K.; Webster, T.J. A review of small molecules and drug delivery applications using gold and iron nanoparticles. *Int. J. Nanomed.* **2019**, *14*, 1633–1657. [CrossRef]

3. Gutierrez, A.M.; Dziubla, T.D.; Hilt, J.Z. Recent advances on iron oxide magnetic nanoparticles as sorbents of organic pollutants in water and wastewater treatment. *Rev. Environ. Health* **2017**, *32*, 111–117. [CrossRef]

4. Liu, X.; Zhong, Z.; Tang, Y.; Liang, B. Review on the synthesis and applications of Fe$_3$O$_4$ nanomaterials. *J. Nanomater.* **2013**, *7*. [CrossRef]

5. Kornberg, T.G.; Stueckle, T.A.; Antonini, J.A.; Rojanasakul, Y.; Castranova, V.; Yang, Y.; Wang, L. Potential toxicity and underlying mechanisms associated with pulmonary exposure to iron oxide nanoparticles: Conflicting literature and unclear risk. *Nanomaterials* **2017**, *7*, 307. [CrossRef]

6. Shi, D.; Mi, G.; Bhattacharya, S.; Nayar, S.; Webster, T.J. Optimizing superparamagnetic iron oxide nanoparticles as drug carriers using an in vitro blood-brain barrier model. *Int. J. Nanomed.* **2016**, *11*, 5371–5379. [CrossRef] [PubMed]

7. Willmann, W.; Dringen, R. How to study the uptake and toxicity of nanoparticles in cultured brain cells: The dos and don't forgets. *Neurochem. Res.* **2019**, *44*, 1330–1345. [CrossRef] [PubMed]

8. Hu, Y.; Gao, J. Potential neurotoxicity of nanoparticles. *Int. J. Pharm.* **2010**, *394*, 115–121. [CrossRef] [PubMed]

9. Cupaioli, F.A.; Zucca, F.A.; Boraschi, D.; Zecca, L. Engineered nanoparticles. How brain friendly is this new guest? *Prog. Neurobiol.* **2014**, *119–120*, 20–38. [CrossRef] [PubMed]

10. Win-Shwe, T.; Fujimaki, H. Nanoparticles and neurotoxicity. *Int. J. Mol. Sci.* **2011**, *12*, 6267–6280. [CrossRef]

11. Feng, X.L.; Chen, A.; Zhang, Y.; Wang, J.; Shao, L.; Wei, L. Central nervous system toxicity of metallic nanoparticles. *Int. J. Nanomed.* **2015**, *10*, 4321–4340.

12. Bencsik, A.; Lestaevel, P.; Canu, I.G. Nano- and neurotoxicology: An emerging discipline. *Prog. Neurobiol.* **2018**, *160*, 45–63. [CrossRef]

13. Ge, D.; Du, Q.; Ran, B.; Liu, X.; Wang, X.; Ma, X.; Cheng, F.; Sun, B. The neurotoxicity induced by engineered nanomaterials. *Int. J. Nanomed.* **2019**, *14*, 4167–4186. [CrossRef]

14. Wang, Y.; Xiong, L.; Tang, M. Toxicity of inhaled particulate matter on the central nervous system: Neuroinflammation, neuropsychological effects and neurodegenerative disease. *J. Appl. Toxicol.* **2017**, *37*, 644–667. [CrossRef]

15. Song, B.; Zhang, Y.; Liu, J.; Feng, X.; Zhou, T.; Shao, L. Is neurotoxicity of metallic nanoparticles the cascades of oxidative stress? *Nanoscale Res. Lett.* **2016**, *11*, 291. [CrossRef]

16. Valdiglesias, V.; Fernandez-Bertolez, N.; Kilic, G.; Costa, C.; Costa, S.; Fraga, S.; Bessa, M.J.; Pasaro, E.; Teixeira, J.P.; Laffon, B. Are iron oxide nanoparticles safe? Current knowledge and future perspectives. *J. Trace Elem. Med. Biol.* **2016**, *38*, 53–63. [CrossRef]

17. Laffon, B.; Fernández-Bertólez, N.; Costa, C.; Brandão, F.; Teixeira, J.P.; Pásaro, E.; Valdiglesias, V. Cellular and Molecular Toxicity of Iron Oxide Nanoparticles. In *Advances in Experimental Medicine and Biology 1048*; Saquib, Q., Faisal, M., Al-Khedhairy, A.A., Alatar, A.A., Eds.; Springer: Berlin/Heidelberg, Germany, 2019; pp. 199–213.

18. Chahinez, T.; Rachid, R.; Salim, G.; Lamia, B.; Ghozala, Z.; Nadjiba, T.; Aya, S.; Sara, H.; Hajer, C.; Samira, B.; et al. Toxicity of Fe_3O_4 nanoparticles on oxidative stress status, stromal enzymes and mitochondrial respiration and swelling of Oryctolagus cuniculus brain cortex. *Toxicol. Environ. Health Sci.* **2016**, *8*, 349–355. [CrossRef]

19. Liu, Y.; Li, J.; Xu, K.; Gu, J.; Huang, L.; Zhang, L.; Liu, N.; Kong, J.; Xing, M.; Zhang, L.; et al. Characterization of superparamagnetic iron oxide nanoparticle-induced apoptosis in PC12 cells and mouse hippocampus and striatum. *Toxicol. Lett.* **2018**, *292*, 151–161. [CrossRef]

20. Wu, J.; Ding, T.; Sun, J. Neurotoxic potential of iron oxide nanoparticles in the rat brain striatum and hippocampus. *Neurotoxicology* **2013**, *34*, 243–253. [CrossRef]

21. Kim, Y.; Kong, S.D.; Chen, L.H.; Pisanic, T.; Jin, S.; Shubayev, V.I. In vivo nanoneurotoxicity screening using oxidative stress and neuroinflammation paradigms. *Nanomedicine* **2013**, *9*, 1057–1066. [CrossRef]

22. Hwang, S.R.; Kim, K. Nano-enabled delivery systems across the blood-brain barrier. *Arch. Pharm. Res.* **2014**, *37*, 24–30. [CrossRef]

23. Petters, C.; Irrsack, E.; Koch, M.; Dringen, R. Uptake and metabolism of iron oxide nanoparticles in brain cells. *Neurochem. Res.* **2014**, *39*, 1648–1660. [CrossRef]

24. Yarjanli, Z.; Ghaedi, K.; Esmaeili, A.; Soheila Rahgozar, S.; Zarrabi, A. Iron oxide nanoparticles may damage to the neural tissue through iron accumulation, oxidative stress, and protein aggregation. *BMC Neurosci.* **2017**, *18*, 51. [CrossRef] [PubMed]

25. Petters, C.; Thiel, K.; Dringen, R. Lysosomal iron liberation is responsible for the vulnerability of brain microglial cells to iron oxide nanoparticles: Comparison with neurons and astrocytes. *Nanotoxicology* **2016**, *10*, 332–342. [CrossRef] [PubMed]

26. Feng, Q.; Liu, Y.; Huang, J.; Chen, K.; Huang, J.; Xiao, K. Uptake, distribution, clearance, and toxicity of iron oxide nanoparticles with different sizes and coatings. *Sci. Rep.* **2018**, *8*, 2082. [CrossRef]

27. Sun, Z.; Yathindranath, V.; Worden, M.; Thliveris, J.A.; Chu, S.; Parkinson, F.E.; Hegmann, T.; Miller, D.W. Characterization of cellular uptake and toxicity of aminosilane-coated iron oxide nanoparticles with different charges in central nervous system relevant cell culture models. *Int. J. Nanomed.* **2013**, *8*, 961–970. [CrossRef]

28. Chen, J.; Zhu, J.M.; Cho, H.H.; Cui, K.M.; Li, F.H.; Zhou, X.B.; Rogers, J.T.; Wong, S.T.C.; Huang, X.D. Differential cytotoxicity of metal oxide nanoparticles. *J. Exp. Nanosci.* **2008**, *3*, 321–328. [CrossRef]

29. Valdiglesias, V.; Kiliç, G.; Costa, C.; Fernández-Bertólez, N.; Pásaro, E.; Teixeira, J.P.; Laffon, B. Effects of iron oxide nanoparticles: Cytotoxicity, genotoxicity, developmental toxicity, and neurotoxicity. *Environ. Mol. Mutagen.* **2015**, *56*, 125–148. [CrossRef]

30. Bal-Price, A.; Pistollato, F.; Sachana, M.; Bopp, S.K.; Munn, S.; Worth, A. Strategies to improve the regulatory assessment of developmental neurotoxicity (DNT) using in vitro methods. *Toxicol. Appl. Pharmacol.* **2018**, *354*, 7–18. [CrossRef]

31. National Research Council. *Toxicity Testing in the 21st Century: A Vision and a Strategy*; National Academy Press: Washington, DC, USA, 2007.

32. Knudsen, T.B.; Keller, D.A.; Sander, M.; Carney, E.W.; Doerrer, N.G.; Eaton, D.L.; Fitzpatrick, S.C.; Hastings, K.L.; Mendrick, D.L.; Tice, R.R.; et al. FutureTox II: In vitro data and in silico models for predictive toxicology. *Toxicol. Sci.* **2015**, *143*, 256–267. [CrossRef]

33. Fong, C.Y.; Chak, L.L.; Biswas, A.; Tan, J.H.; Gauthaman, K.; Chan, W.K.; Bongso, A. Human Wharton's jelly stem cells have unique transcriptome profiles compared to human embryonic stem cells and other mesenchymal stem cells. *Stem Cell Rev. Rep.* **2011**, *7*, 1–16. [CrossRef] [PubMed]

34. Hsieh, J.Y.; Fu, Y.S.; Chang, S.J.; Tsuang, Y.H.; Wang, H.W. Functional module analysis reveals differential osteogenic and stemness potentials in human mesenchymal stem cells from bone marrow and Wharton's jelly of umbilical cord. *Stem Cells Dev.* **2010**, *19*, 1895–1910. [CrossRef] [PubMed]

35. Troyer, D.L.; Weiss, M.L. Wharton's jelly-derived cells are a primitive stromal cell population. *Stem Cells* **2008**, *26*, 591–599. [CrossRef] [PubMed]

36. Jang, S.; Cho, H.H.; Cho, Y.B.; Park, J.S.; Jeong, H.S. Functional neural differentiation of human adipose tissue-derived stem cells using bFGF and forskolin. *BMC Cell Biol.* **2010**, *11*, 25. [CrossRef]

37. Taran, R.; Mamidi, M.K.; Singh, G.; Dutta, S.; Parhar, I.S.; John, J.P.; Bhonde, R.; Pal, R.; Das, A.K. In vitro and in vivo neurogenic potential of mesenchymal stem cells isolated from different sources. *J. Biosci.* **2014**, *39*, 157–169. [CrossRef]

38. Zack-Williams, S.D.; Butler, P.E.; Kalaskar, D.M. Current progress in use of adipose derived stem cells in peripheral nerve regeneration. *World J. Stem Cells* **2015**, *7*, 51–64. [CrossRef]

39. Buzanska, L.; Sypecka, J.; Nerini-Molteni, S.; Compagnoni, A.; Hogberg, H.T.; del Torchio, R.; Domanska-Janik, K.; Zimmer, J.; Coecke, S. A human stem cell-based model for identifying adverse effects of organic and inorganic chemicals on the developing nervous system. *Stem Cells* **2009**, *27*, 2591–2601. [CrossRef]

40. Shi, Y.; Nan, C.; Yan, Z.; Liu, L.; Zhou, J.; Zhao, Z.; Li, D. Synaptic plasticity of human umbilical cord mesenchymal stem cell differentiating into neuron-like cells in vitro induced by edaravone. *Stem Cells Int.* **2018**. [CrossRef]

41. Czarnecka, J.; Porowińska, D.; Bajek, A.; Hołysz, M.; Roszek, K. Neurogenic differentiation of mesenchymal stem cells induces alterations in extracellular nucleotides metabolism. *J. Cell Biochem.* **2017**, *118*, 478–486. [CrossRef]

42. Kil, K.; Choi, M.Y.; Park, K.H. In vitro differentiation of human wharton's jelly-derived mesenchymal stem cells into auditory hair cells and neurons. *J. Int. Adv. Otol.* **2016**, *12*, 37–42. [CrossRef]

43. Shahbazi, A.; Safa, M.; Alikarami, F.; Kargozar, S.; Asadi, M.H.; Joghataei, M.T.; Soleimani, M. Rapid induction of neural differentiation in human umbilical cord matrix mesenchymal stem cells by camp-elevating agents. *Int. J. Mol. Cell Med.* **2016**, *5*, 167–177.

44. Singh, A.K.; Kashyap, M.P. An overview on human umbilical cord blood stem cell-based alternative in vitro models for developmental neurotoxicity. *Mol. Neurobiol.* **2016**, *53*, 3216–3226. [CrossRef]

45. Cortés-Medina, L.V.; Pasantes-Morales, H.; Aguilera-Castrejon, A.; Picones, A.; Lara-Figueroa, C.O.; Luis, E.; Montesinos, J.J.; Cortés-Morales, V.A.; De la Rosa Ruiz, M.P.; Hernández-Estévez, E.; et al. Neuronal transdifferentiation potential of human mesenchymal stem cells from neonatal and adult sources by a small molecule cocktail. *Stem Cells Int.* **2019**. [CrossRef]

46. Lim, I.J.; Phan, T.T. Epithelial and mesenchymal stem cells from the umbilical cord lining membrane. *Cell Transplant.* **2014**, *23*, 497–503. [CrossRef] [PubMed]

47. Deuse, T.; Stubbendorff, M.; Tang-Quan, K.; Phillips, N.; Kay, M.A.; Eiermann, T.; Phan, T.T.; Volk, H.D.; Reichenspurner, H.; Robbins, R.C.; et al. Immunogenicity and immunomodulatory properties of umbilical cord lining mesenchymal stem cells. *Cell Transplant.* **2011**, *20*, 655–667. [CrossRef] [PubMed]

48. Ding, D.C.; Chang, Y.H.; Shyu, W.C.; Lin, S.Z. Human umbilical cord mesenchymal stem cells: A new era for stem cell therapy. *Cell Transplant.* **2015**, *24*, 339–347. [CrossRef] [PubMed]

49. Coccini, T.; De Simone, U.; Roccio, M.; Croce, S.; Lenta, E.; Zecca, M.; Spinillo, A.; Avanzini, M.A. In vitro toxicity screening of magnetite nanoparticles by applying mesenchymal stem cells derived from human umbilical cord lining. *J. Appl. Toxicol.* **2019**, *39*, 1320–1336. [CrossRef]

50. Handral, H.K.; Tong, H.J.; Islam, I.; Sriram, G.; Rosa, V.; Cao, T. Pluripotent stem cells: An in vitro model for nanotoxicity assessments. *J. Appl. Toxicol.* **2016**, *36*, 1250–1258. [CrossRef]

51. Stueckle, T.A.; Roberts, J.R. Perspective on current alternatives in nanotoxicology research. *Appl. In Vitro Toxicol.* **2019**, *5*, 111–113. [CrossRef]

52. Suma, R.N.; Mohanan, P.V. Stem cells, a new generation model for predictive nanotoxicological assessment. *Curr. Drug Metab.* **2015**, *16*, 932–939. [CrossRef]

53. Ma, W.; Gehret, P.M.; Hoff, R.E.; Kelly, L.P.; Suh, W.H. The Investigation into the Toxic Potential of Iron Oxide Nanoparticles Utilizing Rat Pheochromocytoma and Human Neural Stem Cells. *Nanomaterials* **2019**, *9*, 453. [CrossRef] [PubMed]

54. Woodbury, D.; Schwarz, E.J.; Prockop, D.J.; Black, I.B. Adult rat and human bone marrow stromal cells differentiate into neurons. *J. Neurosci. Res.* **2000**, *61*, 364–370. [CrossRef]

55. Martens, W.; Sanen, K.; Georgiou, M.; Struys, T.; Bronckaers, A.; Ameloot, M.; Phillips, J.; Lambrichts, I. Human dental pulp stem cells can differentiate into Schwann cells and promote and guide neurite outgrowth in an aligned tissue-engineered collagen construct in vitro. *FASEB J.* **2014**, *28*, 1634–1643. [CrossRef] [PubMed]

56. Cengelli, F.; Maysinger, D.; Schudi-Monnet, F.T.; Montet, X.; Corot, C.; Petri-Fink, A.; Hofmann, H.; Juillerat-Jeanneret, L. Interaction of functionalized superparamagnetic iron oxide nanoparticles with brain structures. *J. Pharmacol. Exp. Ther.* **2006**, *318*, 108–116. [CrossRef] [PubMed]

57. Dobson, J. Nanoscale biogenic iron oxides and neurodegenerative disease. *FEBS Lett.* **2001**, *496*, 1–5. [CrossRef]

58. Hautot, D.; Pankhurst, Q.A.; Khan, N.; Dobson, J. Preliminary evaluation of nanoscale biogenic magnetite in Alzheimer's disease brain tissue. *Proc. Biol. Sci.* **2003**, *270*, S62–S64. [CrossRef]

59. Hohnholt, M.C.; Dringen, R. Uptake and metabolism of iron and iron oxide nanoparticles in brain astrocytes. *Biochem. Soc. Trans.* **2013**, *41*, 1588–1592. [CrossRef]

60. Coccini, T.; Caloni, F.; Ramírez Cando, L.J.; De Simone, U. Cytotoxicity and proliferative capacity impairment induced on human brain cell cultures after short- and long-term exposure to magnetite nanoparticles. *J. Appl. Toxicol.* **2017**, *7*, 361–373. [CrossRef]

61. Au, C.; Mutkus, L.; Dobson, A.; Riffle, J.; Lalli, J.; Aschner, M. Effects of nanoparticles on the adhesion and cell viability on astrocytes. *Biol. Trace Elem. Res.* **2007**, *120*, 248–256. [CrossRef]

62. Wu, J.; Sun, J. Investigation on mechanism of growth arrest induced by iron oxide nanoparticles in PC12 cells. *J. Nanosci. Nanotechnol.* **2011**, *11*, 11079–11083. [CrossRef]

63. Wu, H.Y.; Chung, M.C.; Wang, C.C.; Huang, C.H.; Liang, H.J.; Jan, T.R. Iron oxide nanoparticles suppress the production of IL-1beta via the secretory lysosomal pathway in murine microglial cells. *Part Fibre Toxicol.* **2013**, *10*, 46. [CrossRef] [PubMed]

64. Geppert, M.; Hohnholt, M.C.; Thiel, K.; Nürnberger, S.; Grunwald, I.; Rezwan, K.; Dringen, R. Uptake of dimercaptosuccinate-coated magnetic iron oxide nanoparticles by cultured brain astrocytes. *Nanotechnology* **2011**, *22*, 145101. [CrossRef] [PubMed]

65. Geppert, M.; Hohnholt, M.C.; Nürnberger, S.; Dringen, R. Ferritin upregulation and transient ROS production in cultured brain astrocytes after loading with iron oxide nanoparticles. *Acta Biomater.* **2012**, *8*, 3832–3839. [CrossRef]

66. Geppert, M.; Petters, C.; Thiel, K.; Dringen, R. The presence of serum alters the properties of iron oxide nanoparticles and lowers their accumulation by cultured brain astrocytes. *J. Nanopart. Res.* **2013**, *15*, 1349–1364. [CrossRef]

67. Petters, C.; Dringen, R. Accumulation of iron oxide nanoparticles by cultured primary neurons. *Neurochem. Int.* **2015**, *81*, 1–9. [CrossRef]

68. Rivet, C.J.; Yuan, Y.; Borca-Tasciuc, D.A.; Gilbert, R.J. Altering iron oxide nanoparticle surface properties induce cortical neuron cytotoxicity. *Chem. Res. Toxicol.* **2012**, *25*, 153–161. [CrossRef]

69. Wang, J.; Chen, Y.; Chen, B.; Ding, J.; Xia, G.; Gao, C.; Cheng, J.; Jin, N.; Zhou, Y.; Li, X.; et al. Pharmacokinetic parameters and tissue distribution of magnetic Fe_3O_4 nanoparticles in mice. *Int. J. Nanomed.* **2010**, *5*, 861–866.

70. Beard, J.L.; Connor, J.R.; Jones, B.C. Iron in the brain. *Nutr. Rev.* **1993**, *51*, 157–170. [CrossRef]

71. Goodman, L. Alzheimer's disease—A clinicopathologic analysis of 23 cases with a theory on pathogenesis. *J. Nerv. Ment. Dis.* **1953**, *118*, 97–130. [CrossRef]

72. Smith, M.A.; Harris, P.L.R.; Sayres, L.M.; Perry, G. Iron accumulation in Alzheimer disease is a source of redox-generated free radicals. *Proc. Natl. Acad. Sci. USA* **1997**, *94*, 9866–9868. [CrossRef]

73. Thompson, K.J.; Shoham, S.; Connor, J.R. Iron and neurodegenerative disorders. *Brain Res. Bull.* **2001**, *55*, 155–164. [CrossRef]

74. Zecca, L.; Youdim, M.B.; Riederer, P.; Connor, J.R.; Crichton, R.R. Iron, brain ageing and neurodegenerative disorders. *Nat. Rev. Neurosci.* **2004**, *5*, 863–873. [CrossRef] [PubMed]

75. Collins, J.F.; Prohaska, J.R.; Knutson, M.D. Metabolic crossroads of iron and copper. *Nutr. Rev.* **2010**, *68*, 133–147. [CrossRef] [PubMed]

76. Zheng, W.; Monnot, A.D. Regulation of brain iron and copper homeostasis by brain barrier systems: Implication in neurodegenerative diseases. *Pharmacol. Ther.* **2012**, *133*, 177–188. [CrossRef]

77. Maher, B.A.; Ahmed, I.A.; Karloukovski, V.; MacLaren, D.A.; Foulds, P.G.; Allsop, D.; Mann, D.M.; Torres-Jardón, R.; Calderon-Garciduenas, L. Magnetite pollution nanoparticles in the human brain. *Proc. Natl. Acad. Sci. USA* **2016**, *113*, 10797–10801. [CrossRef]

78. Bosch, J.; Houben, A.P.; Radke, T.F.; Stapelkamp, D.; Bünemann, E.; Balan, P.; Buchheiser, A.; Liedtke, S.; Kögler, G. Distinct differentiation potential of "MSC" derived from cord blood and umbilical cord: Are cord-derived cells true mesenchymal stromal cells? *Stem Cells Dev.* **2012**, *21*, 1977–1988. [CrossRef]

79. Dominici, M.; Le Blanc, K.; Mueller, I.; Slaper-Cortenbach, I.; Marini, F.; Krause, D.; Deans, R.; Keating, A.; Prockop, D.J.; Horwitz, E. Minimal criteria for defining multipotent mesenchymal stromal cells. The International Society for Cellular Therapy position statement. *Cytotherapy* **2006**, *8*, 315–317. [CrossRef]

Review

A Review on the Environmental Fate Models for Predicting the Distribution of Engineered Nanomaterials in Surface Waters

Edward Suhendra, Chih-Hua Chang *, Wen-Che Hou and Yi-Chin Hsieh

Department of Environmental Engineering, National Cheng Kung University, Tainan City 701, Taiwan; edsuhendra@gmail.com (E.S.); whou@mail.ncku.edu.tw (W.-C.H.); likelife3169@gmail.com (Y.-C.H.)
* Correspondence: chihhua@mail.ncku.edu.tw; Tel.: +886(6)275-7575 (ext. 65826)

Received: 28 March 2020; Accepted: 16 June 2020; Published: 26 June 2020

Abstract: Exposure assessment is a key component in the risk assessment of engineered nanomaterials (ENMs). While direct and quantitative measurements of ENMs in complex environmental matrices remain challenging, environmental fate models (EFMs) can be used alternatively for estimating ENMs' distributions in the environment. This review describes and assesses the development and capability of EFMs, focusing on surface waters. Our review finds that current engineered nanomaterial (ENM) exposure models can be largely classified into three types: material flow analysis models (MFAMs), multimedia compartment models (MCMs), and spatial river/watershed models (SRWMs). MFAMs, which is already used to derive predicted environmental concentrations (PECs), can be used to estimate the releases of ENMs as inputs to EFMs. Both MCMs and SRWMs belong to EFMs. MCMs are spatially and/or temporally averaged models, which describe ENM fate processes as intermedia transfer of well-mixed environmental compartments. SRWMs are spatiotemporally resolved models, which consider the variability in watershed and/or stream hydrology, morphology, and sediment transport of river networks. As the foundation of EFMs, we also review the existing and emerging ENM fate processes and their inclusion in recent EFMs. We find that while ENM fate processes, such as heteroaggregation and dissolution, are commonly included in current EFMs, few models consider photoreaction and sulfidation, evaluation of the relative importance of fate processes, and the fate of weathered/transformed ENMs. We conclude the review by identifying the opportunities and challenges in using EFMs for ENMs.

Keywords: engineered nanomaterials; environmental fate models; surface waters; ENM fate processes

---

1. Introduction

Engineered nanomaterials (ENMs) are intentionally produced or manufactured materials with at least one dimension in the size range of 1–100 nm [1–3]. At this size, materials behave differently to their bulk forms, which is of interest for novel applications [1,2]. Nanotechnology that is the application of scientific knowledge to manipulate and fabricate ENMs and associated products is intensively applied in fields, such as consumer products [4–7], sporting goods [8], water treatment [9–11], photocatalysis [12–14], electrocatalysis [15], medicine [16], and renewable energy [17]. ENMs are widely used in consumer products, such as clothes, sox, surface treatment, cosmetics, and packaging [4–7], potentially contributing to significant ENMs' release to the environment. For instance, silver nanoparticles (AgNPs) are used in clothes, sox, and packaging due to its antimicrobial properties [18,19]. Titanium dioxide nanoparticles (TiO$_2$ NPs) and zinc oxide nanoparticles (ZnO NPs) have applications in surface treatment, e.g., paint and coating, due to its antifouling properties, as well as cosmetics, e.g., sunscreen as ultraviolet (UV)-absorber [5,6,20]. The use of TiO$_2$ NPs and ZnO NPs in photocatalysis in the degradation of contaminants, such as •OH, O$_2$•-, H$_2$O$_2$ [12,13], and renewable energy, e.g., H$_2$

production [14], is also common. Recent research exists about the applications of nanotechnology in water remediation. For example, due to its ultrahigh surface areas, nanoscale nickel metal organic frameworks (MOFs) decorated over graphene oxide (GO) and carbon nanotubes (CNTs) have been proposed as new adsorbents in water treatment to remove methylene blue (MB) [21], and MOFs-derived magnetic Ni and Cu nanoparticles have been studied in the catalysis of a sodium borohydride ($NaBH_4$)-mediated reduction of environmental pollutants, such as 4-nitrophenol, methyl orange, and MB [22]. The upward trend of ENMs' global production, as well as the widespread applications of nanotechnology could significantly increase the release of ENMs to the environment [23,24].

ENMs are considered as a new class of pollutant due to the unique properties and extraordinary behaviors, which leads to unknown environmental risks [25,26]. While the paradigm used in the risk assessments of regular substances is generally considered to be applicable for ENMs, it is necessary to update the current risk assessment approaches in the evaluation of physicochemical properties, environmental fate and exposure, and (eco)toxicological effects applied for traditional chemicals in regulations in order to be compatible with ENMs' unique characteristics [27–29]. In the past decade, there has been strong progress in the development of methods to evaluate ENMs' (eco)toxicological effects and bioaccumulation potentials [30,31]. In terms of environmental exposure assessment, this can be done by empirically measuring ENMs in environmental matrices and/or by predicting ENMs' fate and release in environmental compartments. Despite the considerable progresses made in the detection methods of ENMs, the current analytical technologies could not readily characterize low concentrations of ENMs and distinguish natural nanomaterials from engineered ones in complex environmental matrices [31,32]. Regarding this situation, environmental exposure models could be used as an alternative method to quantitatively predict the environmental distribution of ENMs [28,31,33,34].

Traditionally, exposure models have been used as a tool in the evaluation of environmental concentrations of chemical substances [28,35,36]. Given the recognized properties of ENMs that are distinct from molecular chemicals, it is necessary to review the fate processes of existing exposure models for their relevancy for ENMs and update them accordingly [31,37,38]. For example, the equilibrium partitioning of organic chemicals is not generally considered to be relevant for ENMs [38,39]. The studies on the environmental fate processes of ENMs (e.g., heteroaggregation, dissolution, wastewater treatment removal, and transformation, etc.) have considerably increased in the past one to two decades [31,37]. There is an important opportunity to consider and include these advances in exposure models to allow for improved prediction of ENMs' environmental distribution.

Prior reviews on the environmental fate and exposure modeling of ENMs exist. Dale et al. [33] reviewed the nano-specific processes that can be included in engineered nanomaterial (ENM) fate models of aquatic systems. The review of Baalousha et al. [40] discussed the modeling of ENM fate and biological uptake in aquatic and terrestrial environments and also emphasized the considerations of nano-specific processes. Nowack [28] reviewed environmental exposure models for ENMs in a regulatory context with a large focus on the material flow analysis models (MFAMs). He concluded that current MFAMs and environmental fate models (EFMs) are both generally compatible with the framework of exposure assessment accepted in regulatory and that both models require proper validation. William et al. [34] reviewed the nano-specific processes as well as the releases, forms, and particle sizes of ENMs considered in recent aquatic fate models. Since then, significant progresses in ENM fate modeling in the aquatic environment have been made, especially in surface waters, that provide good spatial and temporal resolution with the consideration of new nano-specific processes, such as sunlight-driven photoreactions; however, these recent advances have not generally been included in existing reviews [41,42].

The potential of ENMs released to groundwater exists from groundwater remediation using nano zero-valent iron (nZVI) [43]. While there are few nZVI groundwater modeling studies, these current studies are limited to nZVI rather than other ENMs [44–46]. As such, this review does not focus on groundwater and this could be a direction for future reviews when more related work emerges.

The purpose of this paper is to comprehensively review the recent advances in the environmental exposure assessment of ENMs using computational approaches. While ENM exposure models in other environmental media, such as soil and air, are also covered, we focus on the latest advances of surface water models.

We also comprehensively review recent studies on the environmental fate processes of ENMs to identify emerging processes that can be included in the fate models towards improved prediction of ENMs' distribution in environmental matrices. Current exposure models of ENMs may be largely classified into two major types, which are MFAMs and EFMs [28,33,40]. MFAMs quantify the release of ENMs in life-cycle stages from production to waste disposal/recycling into the environment [47–51], while EFMs consider the mechanistic fate and transport processes (i.e., nano-specific fate processes) in the prediction of ENMs' concentrations in or across environmental compartments (e.g., water, sediment, soil, or air) [52,53]. EFMs could be further classified to multimedia compartmental models (MCMs) [36,54–57] and spatial river/watershed models (SRWMs) [35,41,42,58,59]. We discuss and compare the functions and utilities of recent ENM exposure models. We conclude the review by identifying the opportunities and challenges of using EFMs for predicting the environmental exposure of ENMs.

2. Exposure Models for the Prediction of Engineered Nanomaterials' (ENMs') Concentrations in Surface Waters

Obtaining the temporal–spatial distribution information of ENMs in a water environment is the basis of ENMs' exposure and risk assessment. For example, there is no potential risk of a chemical substance on the ecosystem if the predicted levels of ENMs in the environment (e.g., predicted environmental concentrations, PECs) are lower than the predicted no effect concentrations (PNECs) [36,47,60]. Before the direct and quantitative measurement methods of ENMs in natural systems are developed, current information of PECs for risk assessment is mainly estimated by exposure models. The exposure models that have been intensively applied in conventional chemical risk assessments are starting to be used for ENMs with modifications [28,33,34,40].

In the last decade, several types of exposure models have been applied to estimate environmental releases and concentrations of ENMs. These models can be classified to MFAMs that predict releases from production, fate in technical systems, and ultimate releases to the environment, and EFMs that predict ENMs' concentrations by including ENM fate processes as well as the distribution within environmental compartments [28]. The recent development of ENM exposure models are described as follows.

2.1. Material Flow Analysis Models (MFAMs)

Material flow analysis (MFA) is a mass balance-based system approach, which tracks and quantifies inventories and flows of materials throughout the entire life cycle in a well-defined geographic boundary or time span that is covered by the system. Many nanomaterial models (e.g., MFAMs) rely on MFA to predict releases from products, fate (i.e., removal) in technical systems, and final releases to the environment [28]. The outputs provided by MFAMs can be linked to EFMs as system input data. For example, the more realistic ENMs' release information resulting from MFAMs, e.g., the released forms and particle size distributions of ENMs, can be used by EFMs as a basis for choosing relevant environmental fate processes and hence allowing EFMs to better quantify the transport and distribution between and within environmental compartments. Generally, MFAMs are only ENM release models, although some extended MFAMs are also simplified EFMs, which estimate environmental concentrations by assuming standard sizes of environmental compartments and complete mixing [28]. The current MFAMs of ENMs are shown in Table 1 with the following review.

Mueller and Nowack [47] used an MFA approach for quantifying the releases of AgNPs, TiO_2 NPs, and CNTs from products, and predict their expected concentrations in soil, air, and water compartments of Switzerland. The resulting environmental concentrations from MFAMs are mainly controlled by how the system was defined, e.g., boundary and time span, as well as how the ENM specific input parameters were determined, such as the estimated production volume, the allocation of the production volume to product categories, the particle discharge from products, and the transfer coefficients between environmental compartments. As a significant amount of input information was estimated, not available, or have uncertainty, the system was usually assumed in a steady state, ENMs were released completely from products in one year, and environmental compartments were considered to be homogeneous and well mixed [47].

Probabilistic material flow analysis models (P-MFAMs) are then used with the employment of probability distributions for model input and/or output to address the inconsistency and variability of model input parameters, as well as to model PECs for ENMs, such as TiO_2 NPs, ZnO NPs, AgNPs, CNTs, and fullerenes [48,49,61,62].

Dynamic probabilistic MFAMs (DP-MFAMs) address the time-dynamic behavior of the system for material flows over several consecutive periods, considering changes in the inflow to the system and intermediate delays in local stocks. Unlike the static MFA, the dynamic MFA considers a reasonable time frame of a period rather than a single year and time-dependent ENMs' release from products over the life-cycle of the products rather than the simple assumption that ENMs are released completely from all applications in the single production year [50,51,63].

Table 1. Review of existing environmental exposure models for engineered nanomaterials (ENMs).

Model Classification	Model Name	Model Features	Compartments Considered	Fate Processes	References
Material flow analysis models	MFAMs	Steady state, less information required, simplified structure	Air, water, soil	-	Mueller and Nowack (2008) [47]
	P-MFAMs	Accounting for the uncertainty of model input parameters using probabilistic distribution	Air, water, soil, sediment	-	Gotschalk et al. (2009) [61], Gotschalk et al. (2010) [48], Gotschalk et al. (2011) [62], Sun et al. (2015) [49], Liu et al. (2015) [56]
	DP-MFAMs	Accounting for time-dependent changes in the system behavior	Air, water, soil, sediment	-	Bornhöft et al. (2016) [50], Sun et al. (2016) [63], Wang and Nowack (2018) [51]
Multimedia compartmental models	MendNano	Intermedia transport processes included partitioning ratios	Air, water, soil, sediment, biota	Homoaggregation, heteroaggregation, dissolution	Liu and Cohen (2014) [54]
	SimpleBox4 Nano (SB4N)	Steady state environmental ENM fate processes are modeled mechanistically using first-order rate constants	Air, water, soil, sediment	Heteroaggregation, dissolution	Meesters et al. (2014) [36]
	RedNano	A model system which combines a P-MFAMs based release model (LearNano) and a multimedia fate model (MendNano)	Air, water, soil, sediment, biota	Homoaggregation, heteroaggregation, dissolution	Liu et al. (2015) [56]
	SimpleBox4 Nano (SB4N)	Steady state environmental ENMs' fate, probabilistic distribution	Air, water, soil, sediment	Heteroaggregation, dissolution	Meesters et al. (2016) [55]
	nanoFate	Dynamic environmental ENMs' fate	Air, water, soil, sediment	Heteroaggregation, dissolution	Garner et al. (2017) [57]
Spatial river/watershed models	Rhine river box model	Steady state box model	Water, sediment	Heteroaggregation	Praetorius et al. (2012) [35]
	Rhone river box model	Cluster analysis, steady state box model	Water, sediment	Heteroaggregation	Sani-Kast et al. (2015) [64]
	Diagenesis model	1-D sediment diagenesis model	Freshwater sediment	Dissolution, sulfidation	Dale et al. (2013) [65]
	GWAVA	Gridded probability distribution	Water	Dissolution	Dumont et al. (2015) [66]
	Nano DUFLOW	1-D unsteady flow in open-channel systems	Water, sediment	Homoaggregation, heteroaggregation, dissolution	Quik et al. (2015) [58], Klein et al. (2016) [67]
	SOBEK river-DELWAQ	A model system integrates open channel hydraulics and water quality models	Water	Homoaggregation, heteroaggregation, dissolution	Markus et al. (2016) [68]
	WASP7–HSPF	Dynamic, mass-balance, spatially resolved differential fate and transport modeling framework	Water, sediment	Dissolution, sulfidation	Dale et al. (2015) [59]
	WASP8	A detailed surface water quality model with ENM fate and transport processes	Water, sediment	Dissolution, sulfidation, heteroaggregation, photoreaction	Bouchard et al. (2017) [41], Han et al. (2019) [42]
	SWMM-EFDC	Suitable for urban stormwater and sewage systems; coupling both surface hydrology and hydrodynamic models	Water, sediment	Heteroaggregation, dissolution	Saharia et al. (2019) [69]

2.2. Environmental Fate Models (EFMs)

EFMs are exposure models, which apply a quantitative and mechanistic description of fate processes to predict the behavior as well as the concentration of the contaminant in the environment. The basic concept governing of EFMs is the integration of pollutant transport, transfer, and degradation processes into mass balance equations for the pollutant. EFMs consist of mass balance equations for a chemical in a system of coupled environmental compartments in which the coefficients are generally based on extensive empirical data [52,53].

EFMs of ENMs can be classified into MCMs and SRWMs. MCMs comprise the mechanistic description of fate and transport processes of ENMs as intermedia transfer incorporating well-mixed environmental compartments (e.g., air, water, soil, and sediment) while SRWMs are dynamic and spatially resolved models, which incorporate variation in stream hydrology and morphology as well as sediment transport in river networks, and have the capability to consider the description of water quality linked to the watershed hydrology [28]. The current EFMs of ENMs (MCMs and SRWMs) are shown in Table 1 with the review as follows.

2.2.1. Multimedia Compartmental Models (MCMs)

MCMs treat various environmental media (e.g., surface water, groundwater, and atmosphere) as an integrated system, synthesizing information about chemical partitioning, transformation, and intermedia transport. The models have been used to predict the distribution of the contaminant based on mass balance equations at a regional and global scale. The environment is described as a set of well-mixed compartments, with each acting as a particular medium (e.g., air, water, and soil). The model consists of the compartmental mass balance ordinary differential equations (ODEs), which can be solved either at steady state or dynamically with the time steps selected [52–54]. The existing multimedia compartmental models are summarized in Table 1 with the review as follows.

MendNano was developed by Liu and Cohen [54] to predict environmental exposure concentrations of ENMs dynamically. The model consists of the compartmental mass balance ODEs, which can be solved with the time steps selected, with the consideration of ENM fate processes such as homoaggregation, heteroaggregation, and dissolution as intermedia transport between bordering compartments. For example, attachment factors are used in homoaggregation and heteroaggregation, which are defined as fractions of ENMs attached to themselves or ambient particles [54]. Later, RedNano is developed as an improvement of MendNano by coupling the model with LearNano as P-MFAMs in order to derive more realistic ENMs' release rates [56].

Then, SimpleBox4Nano (SB4N), which is the modification of a classical multimedia mass balance model in European Regulation on Registration, Evaluation, Authorization, and Restriction of Chemicals (REACH) called the SimpleBox model, was introduced by Meesters et al. [36]. It is solved at steady state with the consideration of ENM fate processes such as heteroaggregation and dissolution. Unlike MendNano, SB4N used first order rate constants for ENM fate processes, with the exception of dissolution, which was considered as a loss mechanism. Heteroaggregation is modeled between ENMs with natural colloids (< 450 nm) and larger suspended particles (> 450 nm). The ENMs' input is only free dispersive ENMs. However, the output has three ENMs' states as a result of heteroaggregation, which are: (1) Freely dispersed ENMs; (2) ENMs attached with natural colloids; and (3) ENMs heteroaggregated to larger suspended particles [36]. Subsequently, Meesters et al. [55] analyzed the sensitivity of SB4N to uncertainties in emission estimations; physicochemical properties of cerium oxide nanoparticles (CeO_2 NPs), ZnO NPs, and TiO_2 NPs; and natural variability of environmental systems by incorporating Monte Carlo simulations as a probabilistic approach.

Recently, Garner et al. [57] proposed a multimedia dynamic model called nanoFate, which incorporates a broader range of ENM processes (e.g., emissions from their manufacturing, use, and disposal) and climate variability (e.g., the use of observed daily hydrometeorological data) with the consideration of heteroaggregation and dissolution using first rate order rate constants. NanoFate can also quantify the concentration of dissolved ions as a result of the dissolution process. Furthermore, nanoFate considers not only the point source but also the non-point source, such as ENMs' loading from the use of biosolids in agriculture. Runoff was calculated using the runoff equation as well as soil loss resulting from erosion, and a variety of land uses were considered.

2.2.2. Spatial River/Watershed Models (SRWMs)

Spatial river models are dynamic and spatially resolved models that consider variability in stream hydrology and morphology as well as sediment transport in river networks. Water quality functions of these models are provided by linking hydrologic and water quality models. Hydrologic and water quality models can be either integrated as one model or separated into hydrologic models and water quality models. Watershed models underline the description of watershed hydrology and water quality, including runoff, erosion, and wash off of sediments and pollutants. When coupled to hydrologic and water quality models, watershed models offer more realistic input parameters for hydrology as well as sediment and pollutant loading [59,70]. The current SRWMs are shown in Table 1 with the review as follows.

Praetorius et al. [35] modified the river box model for ENMs in Rhine river based on established multimedia box models for organic chemicals with the inclusion of nano-specific process descriptions. The model framework is similar to MCMs, which can be solved at steady state or as a function of time, but with the consideration of spatial resolution (e.g., Rhine river is divided into 520 "boxes" representing river sections). Heteroaggregation and dissolution are considered as ENM fate processes, with the assumption of attachment efficiency of heteroaggregation in several scenarios [35]. Sani-Kast et al. [64] improved this river box model by incorporating spatial variability in environmental conditions for the Rhone river with cluster analysis.

Subsequently, Dale et al. [65] applied a simple 1-D diagenetic model for predicting AgNPs' distribution and silver speciation of resulting Ag ions in freshwater sediments. The model is a modification of a mass balance model by Di Toro et al. [71] that quantifies the speciation of cadmium in sediments considering the sediment as a function of the oxygen depletion in the time of mineralization or organic carbon diagenesis. The model considered the sulfidation and oxidative dissolution process of AgNPs as a function of dissolved oxygen, sulfide, and temperature. The model was then calibrated to data collected from AgNPs-dosed large-scale freshwater wetland mesocosms, and estimated ion release as a result of AgNPs' sulfidation [65].

DUFLOW simulates 1-D unsteady flow in open-channel systems with the calculation of water levels and flow rates based on the St. Venant equations of continuity and momentum using initial and boundary conditions. NanoDUFLOW was developed by Quik et al. [58] as a modification of the DUFLOW hydrology model, which allows the inclusion of ENM fate process descriptions such as homoaggregation, heteroaggregation, and dissolution to a spatially explicit hydrological model. The model was toward validation, which compared measured and modeled concentrations of < 450 nm cerium (Ce), aluminum (Al), titanium (Ti), and zirconium (Zr)-based particles for the Dommel river, Netherlands [67].

The Global Water Availability Assessment Model (GWAVA) is a catchment gridded hydrology model, which was applied to simulate the spatial distributions of PECs by assigning distinguished ENM loading inputs into each grid [66]. ENM fate processes in GWAVA such as heteroaggregation and dissolution are computed by first-order sink processes.

With the SOBEK-River hydrology model for the simulation of water flows and morphological changes in open channel systems, and the DELWAQ for water quality modelling in 1-D, 2-D, or 3-D surface water systems, Markus et al. [68] developed the SOBEK-River-DELWAQ model system for estimating the temporally and spatially varying ZnO NPs concentration as different ENMs' states (free, homoaggregates, and heteroaggregates) in Rhine river. The model was then validated by the comparison between the simulation of the total concentration of zinc over time to the measured total concentration of zinc for the Rhine river.

Dale et al. [59] used a dynamic water quality model Water Quality Analysis Simulation Program (WASP) to simulate ZnO NPs and AgNPs and their transformation byproducts in James River Basin, Virginia. WASP version 7 (WASP7) can be coupled with a watershed model to simulate ENMs' daily stream loads from effluent discharges. WASP is a dynamic generalized modeling framework based on the finite-volume concept to quantify the fate and transport of water quality variables in surface waters, including toxic substances in 1-D, 2-D, or 3-D. WASP7 considers transformation processes, such as dissolution and sulfidation; however, it excludes ENM fate processes, such as heteroaggregation. Hydrology Simulation Program Fortran (HSPF) is an integrated basin-scale model that combines watershed processes with in-stream fate and transport in one-dimensional stream channels. WASP7 was coupled with Chesapeake Bay Watershed Model (WSM) as an adaption of HSPF to simulate AgNPs and ZnO NPs as well as the total concentration of Ag^+ and Zn^{2+} as their speciation in James River Basin, Virginia.

WASP7 was updated to WASP version 8 (WASP8) with the inclusion of ENM fate processes, such as heteroaggregation and photoreaction. In WASP8, ENM fate processes can be parametrized using experiment-based kinetics. Heteroaggregation is described as the attachment efficiency (α_{het}). Bouchard et al. [41] used WASP8 to simulate multiwalled carbon nanotubes (MWCNTs) fate and transport in Brier Creek. In this study, α_{het} was derived from lab-scale experiments by using natural surface water with a range of ionic strengths. The phototransformation rate constant is calculated using the average light intensity in the surface water, which depends on light attenuation. Light attenuation is calculated for each wavelength band of UV and visible light. Han et al. [42] applied WASP8 to model predicted environmental concentrations of GO and its major reaction phototransformation product reduced GO (rGO) in Brier Creek, GA.

In a recent work, Saharia et al. [69] integrated an urban hydrologic-hydraulic model (Storm Water Management Model, SWMM), and a 3-D hydrodynamic and water quality model (Environmental Fluid Dynamics Code, EFDC) to estimate TiO_2 NPs concentrations in combined sewer overflows and receiving rivers. SWMM is a dynamic rainfall-runoff simulation model applied to urban areas. EFDC is a three-dimensional modeling system having hydrodynamic, water quality-eutrophication, sediment transport, and toxic contaminant transport components. The modeling results show that heteroaggregation and sedimentation are the key fate processes of TiO_2 NPs, while dissolution is almost negligible.

3. Engineered Nanomaterial (ENM) Fate Processes in Surface Waters

The understanding of ENMs' environmental behaviors is fundamentally important in establishing fate models for ENMs in the environment. In this section, we review the state of the art in the studies of ENM fate processes in the aquatic environment. Possible fate processes of ENMs are usually first investigated in laboratory experiments using simulated environmental conditions to evaluate their potential to occur in natural systems. The current literature suggests that ENMs, depending on specific types, can undergo physical transformation processes, including aggregation (homoaggregation and heteroaggregation), as well as chemical transformation processes, including dissolution, sulphidation, and photoreactions [31,37,72,73]. Figure 1 shows the environmental fate processes of ENMs in surface waters. Here, AgNPs are used as an example to illustrate the ENM fate processes because AgNPs have a wide variety of fate processes that are also common to other ENMs, such as aggregation (homoaggregation and heteroaggregation), dissolution, and sulphidation, and/or specific to themselves, such as photoreactions. These ENM fate processes are discussed as follows.

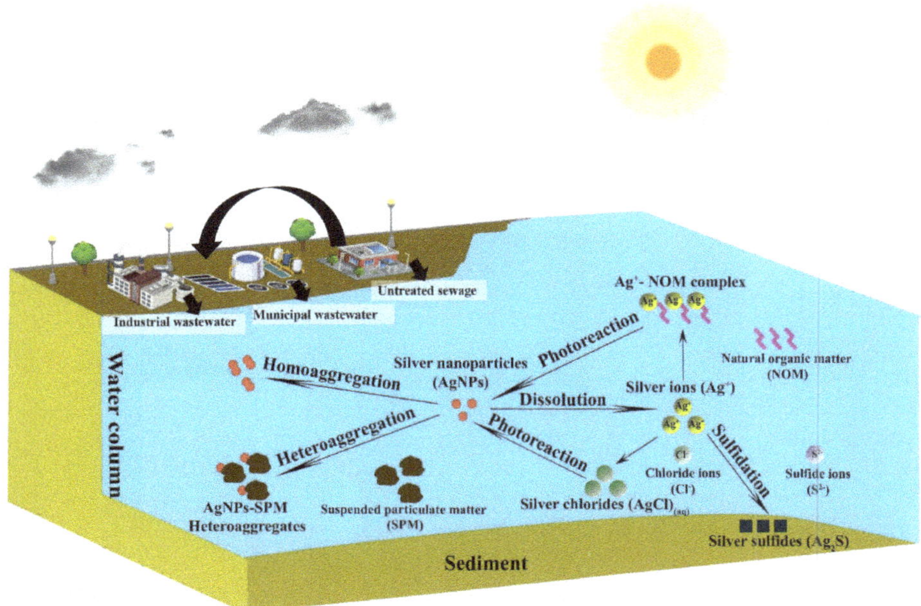

Figure 1. Engineered nanomaterial (ENM) fate processes in surface waters.

3.1. Aggregation

Aggregation is a generally recognized ENM fate process that is controlled by particle–particle interactions where particles collide and stick to each other to form larger clusters [74,75]. The classical Derjaguin–Landau–Verwey–Overbeek (DLVO) theory describes the tendency of colloids to aggregate, a process that is controlled by the combination of attractive and repulsive forces between charged colloid surfaces in a liquid medium [74]. Currently, this DLVO theory is being improved for diverse complexity, such as the electric double layer interaction energies, polyelectrolyte coating, effects of uncharged polymeric coating, and elastic steric stabilization on the van der Waals force [75]. The application of this theory to ENMs overall works well in helping to understand charge destabilization. The attachment efficiency α ($0 \leq \alpha \leq 1$) that determines if the collision between particles leads to the attachment of particles (i.e., forming aggregates) is usually used to model the aggregation kinetics of ENMs and is used in ENM fate models [35,36,41,75]. Figure 1 shows that two forms of aggregation may occur: homoaggregation among the same ENMs, or heteroaggregation among ENMs and other particles [37,76–78]. Due to the complexity of heteroaggregation, the appropriate method to derive the attachment efficiency of heteroaggregation (α_{het}) is still challenging [38,79,80]. Nevertheless, experimental parameterization could be a promising method in determining α_{het} [41,42,80].

Due to the larger concentrations of environmental particles (e.g., suspended particulate matters (SPMs)) compared to ENMs, heteroaggregation with natural particles has been suggested to play a key role in ENM aggregation in the aquatic environment compared to homoaggregation [31,37]. Greater cluster sizes due to aggregation cause increasing settling velocity, thereby more quickly removing ENMs as heteroaggregates from environmental compartments (air, water) into soil or sediments [77,78]. Heteroaggregation also tends to affect the bioavailability and toxicity of ENMs [81,82], for example, heteroaggregation of GO and aluminum oxide (Al_2O_3) particles could diminish the bioavailability and toxicity of GO to freshwater algae [81]. As a result, heteroaggregation of ENMs are predominantly considered in recent EFMs compared to homoaggregation as indicated by Table 1.

3.2. Dissolution

Dissolution of ENMs can be considered as chemical transformation processes of ENMs from particulate forms into dissolved ions as shown in Figure 1. Dissolution is particularly relevant for metallic and metal oxide ENMs, such as AgNPs, ZnO NPs, and copper oxide nanoparticles (CuO NPs), which have been shown to dissolve to Ag^+, Zn^{2+}, and Cu^{2+}, respectively, in aqueous solutions [83–87]. In contrast, metal oxide nanoparticles, such as TiO_2 NPs, CeO_2 NPs, and silicon dioxide nanoparticles (SiO_2 NPs), are fairly insoluble in environmental waters [31]. Water chemistry factors also affect the dissolution of ENMs. Generally, decreased pH increases the dissolution of metallic and metal oxide ENMs [83–87]. Additionally, agglomeration of ENMs tends to diminish the dissolution rate by reducing the diffusion of species involved in dissolution reactions [82,87]. On the other hand, the presence of natural organic matters (NOMs), which are ubiquitous in surface waters, could increase dissolution by enhancing metal solubility through complexation with dissolved metal ions from ENMs [87]. For carbon-based ENMs, dissolution and aqueous solubility are generally not applicable, although fullerene C_{60}'s hypothetical aqueous solubility has been reported at a fairly low concentration of 7.96 ng/L [88,89]. Dissolution could be important in the fate and toxicity for ENMs, such as AgNPs. Generally, dissolved ions exert a greater impact to living creatures compared to their particulate forms. Therefore, for some metallic and metal oxide ENMs, dissolution correlates with increased toxicity [31,37]. The important role of dissolution in ENMs' fate and toxicity has been reflected in its increased inclusion in current exposure models. For example, the dissolution of AgNPs, ZnO NPs, and CuO NPs has been considered in recent fate models as shown in Table 1.

3.3. Sulfidation

Sulfidation is a predominant transformation process for some metal or metal oxide ENMs, such as ZnO NPs and AgNPs, especially in an environment with high sulfide concentrations, such as in wastewater treatment plants (WWTPs) or other anaerobic environment [90–93] as described in Figure 1. The reaction mechanism of sulfidation requires both sulfide and dissolved oxygen and may be either a fast direct reaction or a slower indirect reaction, which could produce metal sulfide ENMs [94]. The low solubility of metal sulfide ENMs, such as silver sulfide nanoparticles (Ag_2S-NPs), mostly reduce their toxicity in the short term [95,96]; however, a recent study by He et al. [96] has shown transformation of Ag_2S-NPs potentially resulting in Ag^+ and in situ formation of Ag^0 and Ag^0/Ag_2S-NPs hetero-nanostructures, which may enhance the toxicity in the long term. Sulfidation may impact metallic ENMs' toxicity by altering their physicochemical properties. For instance, ZnO NPs' sulfidation does not form a protective shell of zinc sulfide (ZnS) on the ZnO core or impact its dissolution rate; however, AgNPs' sulfidation forms a passivating shell of silver sulfide (Ag_2S), which makes partially sulfidized AgNPs exhibit a far lower solubility than untransformed AgNPs [59]. Generally, the high insolubility of sulfide form, e.g., the formed Ag_2S shell, could reduce AgNPs' toxicity compared to the pristine one. However, the sulfidation process is not commonly included in current EFMs. Two studies conducted by Dale et al., using a simple diagenesis model to quantify AgNPs' sulfidation in freshwater sediments [65], and using WASP7 for modeling sulfidation of AgNPs and ZnO NPs in the rivers [59].

3.4. Photoreaction

Photoreaction can be considered as an emerging fate process in the modeling of ENMs' fate. In a 2019 study, Han et al. [42] first considered sunlight-driven phototransformation in modeling the fate and transport of GO in surface water. While the incorporation of ENM photoreaction in fate modeling is relatively new, there have been many existing process studies on ENM photoreaction. We and other groups have reported on the phototransformation of fullerenes C_{60} and C_{70} [97–100], CNTs [101–103], GO [104,105], AgNPs [72,73,106], and metal dichalcogenides [107]. Photoreaction of

ENMs is important, because the process is driven by environmentally relevant sunlight; therefore, it plays a role in the transformation fate of ENMs in surface waters as shown in Figure 1.

Photoreactions under sunlight conditions can result in the oxidation (i.e., photo-oxidation) and/or reduction (i.e., photo-reduction) of ENMs and/or their products and cause structural alterations or degradation of ENMs. For example, Hou et al. ([97,108]) found that photo-oxidation of C_{60} aggregates (nC_{60}) results in polyhydroxylated C_{60} photoproducts' formation driven by nC_{60}'s strong light absorption within the solar spectrum. Another example is GO, which undergoes rapid phototransformation under sunlight conditions, resulting in more persistent rGO-like photoproducts. It has been shown that rGO photoproducts exhibit transport and toxic properties unique from pristine GO [105,109,110]. Therefore, it is also important to be able to model the fate of transformed ENMs, due to their fate and biological properties distinct from pristine materials.

There has been a plethora of studies regarding ENM-enabled photocatalysis in the degradation/transformation of pollutants (e.g., water/wastewater treatment) [12,13,111] or in the production of valuable chemicals (e.g., H_2) [14]. ENMs, such as TiO_2 NPs and ZnO NPs, are probably the most widely studied photocatalysts [12,13]. Photocatalysis depends on the photocatalyst's capability to absorb light, create electron–hole pairs, and generate a range of reactive oxygen species (ROS), such as •OH, $O_2^{•-}$, H_2O_2, etc., able to undergo further redox reactions that ultimately degrade/transform pollutants [12,13,111]. Regardless, these results may not be directly applicable to evaluate the photochemical fate of ENMs in the aqueous environment. For example, photocatalytic studies mainly concern the degradation of pollutants or the product formation (e.g., H_2) [12–14,111], with less emphasis on the physicochemical transformation and associated kinetics of the nano-catalysts under environmentally relevant sunlight conditions, which is the information needed in evaluating the photochemical fates of ENMs in the environment.

4. Path Forward

Based on the review of recent advances in environmental exposure models for ENMs and studies of emerging ENM fate processes, in this section, we discuss and identify the opportunities and challenges in using exposure modeling for risk evaluation of ENMs. The opportunities are related to the consideration of emerging ENM fate processes in current and future EFMs, the evaluation of the relative importance of ENM fate and transport processes within EFMs, modeling the fate and transport of weathered/transformed ENMs, and realistic parametric input of ENM fate processes. The challenge is related to the model validation. We also discuss the model applicability in regulatory context.

This review indicates that studies on the emerging ENM fate processes, such as photoreaction and sulfidation, that are relevant for ENMs, such as AgNPs, carbon-based ENMs, and ZnO [72,73,91–94,96, 97,100,102–105], have grown significantly in the past 5–10 years, but these processes are not commonly included in recent EFMs [42,59,65]. For most EFMs, either MCMs or SRWMs, the most common ENM fate processes considered have been heteroaggregation and dissolution [36,41,54,57,59,68,69]. In the future, as the studies on emerging ENM fate processes continue to increase, there is a valuable opportunity to consider and evaluate them in EFMs. This would also require the translations of ENM fate process studies that are usually conducted in the laboratory under simulated environmental conditions into mathematical expressions and computer codes.

The addition of emerging ENM fate processes into EFMs needs to be accompanied by the evaluation of their relative importance to the more commonly included processes, such as heteroaggregation, dissolution, and others. The specific processes considered are also dependent on the environmental compartments (e.g., surface water columns, anaerobic sediments, etc.) to be modeled. Recent studies have begun to consider this aspect. For example, Meesters et al. [55] analyzed SB4N's sensitivity using probability distributions to environmental concentrations, distributions, and speciations of TiO_2 NPs, CeO_2 NPs, and ZnO NPs in Europe, with the results showing that heteroaggregation is the dominant process for TiO_2 NPs and CeO_2 NPs as well as dissolution for ZnO NPs. A similar approach was also used to determine the most crucial physicochemical properties governing the environmental fate and transport of ENMs in this study. Transformation rate constants and attachment efficiency are found to be the most sensitive physicochemical properties, which can lead to the dominant ENM fate processes governing the environmental fate and transport of ENMs [112]. Future research could have similar considerations that are expected to result in a better understanding of the relative importance of ENM fate and transport processes, particularly when new processes are being evaluated. The relative importance of the ENM fate processes may not always be easy to be measured quantitatively in simulated experiments or in the field, as they often occur simultaneously.

It has been widely shown that some ENMs can transform physically or chemically into products or species that have distinct environmental and ecotoxicological properties from their parent materials [37]. This would include ENMs that are weathered and released from products, such as polymeric nanocomposites, from which released ENMs are often embedded in polymer matrices [113–116]. Some have argued that transformed ENMs are most environmentally relevant [31,37]. Therefore, modeling the fate and transport of transformed ENMs is equally, if not more, important. Recent research has begun to model the fate and transport of transformed ENMs, rather than just considering transformation as a sink process. For instance, the fate of dissolved products of ZnO NPs and CuO NPs (i.e, Zn^{2+} and Cu^{2+}) has been modeled in San Francisco Bay [57]. Other studies modeled the fate and transport of ions (Ag^+ and Zn^{2+}), Ag_2S, and rGO as the dissolved, sulphidation, and phototransformation products of AgNPs, ZnO NPs, and GO in river systems [42,59]. Further research could continue to model these existing transformed ENMs and also consider other less common ones, such as ENMs released from industrial or consumer products that are embedded in product matrices (e.g., carbon nanotubes embedded in polymers).

Using realistic parametric inputs/estimations to model ENM fate and transport processes creates opportunities to improve the accuracy of EFMs, particularly for site-specific models, such as SRWMs. ENM fate process parameters used in EFMs can be assumed, obtained from the literature, or from experiments. Extensive process studies have shown that environmental factors, such as the ionic strength, particulate, and NOMs, can affect the heteroaggregation and dissolution of ENMs [31,79,87]. These factors often vary depending on the specific sites [117]. Modeling work in the early development of EFMs used assumed parameters, such as attachment efficiencies, given that the information about ENM fate parameters was limited [35], so these parameters are assumed as necessary. Some model studies also extracted ENM fate parameters from the literature [36,54,58,59]. Recent EFMs started to use experimentally derived fate parameters as inputs to models. For instance, studies based on WASP8 derived the attachment efficiency (α_{het}) for heteroaggregation processes using laboratory experiments where water and sediment samples collected from the specific sites (i.e., site to be modeled) were mixed with CNTs and GO [41,42]. Experimental parameterization could result in more accurate ENM fate parameters, reflective of specific sites [80]. It is noted that while the experimental derivation of fate parameters can be used for local scale models, it may not be applicable to larger regional or global-scale models where the fate behaviors are usually averaged.

All models require appropriate validation, but this remains a challenge for many current ENM exposure models. The validation of PECs derived from exposure models remains limited due to the lack of suitable analytical methods for ENMs' measurement in complex environmental matrices [28,40]. Nevertheless, there are recent efforts in the validation of other model outcomes (e.g., dissolved ions) or input parameters in several spatial river/watershed models. For example, Quik et al. [58,67] tried to validate the NanoDUFLOW model by comparing the concentrations of ENMs predicted by the model and the measured concentrations of particles with sizes < 450 nm. The measurement concentrations of < 450 nm size particles are deliberate as an important step in the ENMs' validation of EFMs [58,67]. Another study validated the SOBEK river-DELWAQ by comparing the simulated total zinc concentration over time to the measurement of the total zinc concentration [68]. Dale et al. [65] validated AgNPs' distribution and silver speciation in freshwater sediments predicted by a simple one-dimensional diagenetic model, as well as calibrated the model using data collected from the mesocosm scale. There are few studies about the validation of EFMs for ENMs in mesocosm-scale studies, because accurate descriptions of fate processes at the mesocosm-scale are often challenging owing to the predominance of highly heterogeneous conditions [118,119]. Despite the validation of other model outcomes or input parameters, EFMs should further be validated at the mesocosm-scale by comparing PECs to the measurement of ENMs.

This review covers three types of existing exposure models for ENMs that have varied complexity and spatial and temporal resolutions. In regulatory applicability, the multimedia box models (i.e., the SimpleBox) have been indicated as the first-tier model in the European Chemical Agency's (ECHA's) guidance document in exposure assessment of chemicals [60]. Given the similarity, the SB4N and other related box models are likely directly applicable in the regulatory context for ENMs. The MFAMs has been suggested to be compatible with the material flow concept adopted by ECHA and would likely be applicable [28]. The site-specific models, such as WASP8, describe in greater detail the local environment, such as hydrology, and river cross-section with temporal consideration and is traditionally used to model nutrients and other pollutants in water quality management [120]. This model seems to align with the high-tier models indicated by the ECHA document that can be used when the refinement of chemical risk assessment is warranted [60].

5. Summary

The recent development of EFMs and the understanding of emerging ENM fate processes were comprehensively reviewed. We found that ENM fate processes, such as heteroaggregation and dissolution, are the most common ones that are considered in recent EFMs. With the strong advances in emerging ENM fate processes, such as photoreaction and sulfidation, future EFMs should begin to incorporate and evaluate the relative importance of these emerging processes. Sensitivity analysis can be a useful tool to evaluate the relative importance of ENM fate processes in EFMs that comprise multiple concurrent fate and hydrological processes. The consideration of weathered/transformed ENMs, including ENMs released from products that are embedded in product matrices, also represent an important aspect in ENMs' modeling, as these transformed ENMs often have fate and toxic behaviors distinct from parent ENMs. Recent advances in surface water fate and transport models for ENMs have significantly enhanced our ability in the aspects mentioned above. Opportunities exist in using experimentally derived fate parameters and by coupling MFAMs (i.e., ENM environmental release) and EFMs to improve the accuracy of PECs' prediction. Nevertheless, further development of EFMs for ENMs especially regarding model validation is urgently needed.

Funding: Financial support provided by the Ministry of Science and Technology (MOST) of Taiwan under grant number MOST 109-2923-E-006-003-MY4 is acknowledged. The paper is a part of the European Union Horizon 2020 project NanoInformaTIX (Grant agreement ID: 814426).

Conflicts of Interest: The authors declare no conflict of interest.

References

1. European Commission. RECOMMENDATIONS: COMMISSION RECOMMENDATION of 18 October 2011 on the Definition of Nanomaterial (2011/696/EU). *Off. J. Eur. Union* **2011**, 38–40.
2. Dolez, P.I. Nanomaterials Definitions, Classifications, and Applications. *Nanoengineering* **2015**, 3–40. [CrossRef]
3. Auffan, M.; Rose, J.; Bottero, J.-Y.; Lowry, G.V.; Jolivet, J.-P.; Wiesner, M.R. Towards a definition of inorganic nanoparticles from an environmental, health and safety perspective. *Nat. Nanotechnol.* **2009**, *4*, 634–641. [CrossRef] [PubMed]
4. Huang, Y.; Mei, L.; Chen, X.; Wang, Q. Recent Developments in Food Packaging Based on Nanomaterials. *Nanomaterials* **2018**, *8*, 830. [CrossRef] [PubMed]
5. Fytianos, G.; Rahdar, A.; Kyzas, G.Z. Nanomaterials in Cosmetics: Recent Updates. *Nanomaterials* **2020**, *10*, 979. [CrossRef] [PubMed]

6.   AZoNano.com. Nanotechnology and Consumer Products—Opportunities for Nanotechnology in Consumer Products. Available online: https://www.azonano.com/article.aspx?ArticleID=2364 (accessed on 13 June 2020).

7.   Izzyfortiz. Nano-textiles: The Fabric of the Future. Sustainable Nano. Available online: http://sustainable-nano.com/2018/11/28/nano-textiles/ (accessed on 13 June 2020).

8.   Says, S.W.; AZoNano.com. Sport and Nanotechnology: Are the Big Sports Looking to Go Small? Available online: https://www.azonano.com/article.aspx?ArticleID=4859 (accessed on 13 June 2020).

9.   Alvarez, P.J.J.; Chan, C.K.; Elimelech, M.; Halas, N.J.; Villagrán, D. Emerging opportunities for nanotechnology to enhance water security. *Nat. Nanotechnol.* **2018**, *13*, 634–641. [CrossRef]

10. Westerhoff, P.; Alvarez, P.; Li, Q.; Gardea-Torresdey, J.; Zimmerman, J. Overcoming implementation barriers for nanotechnology in drinking water treatment. *Environ. Sci. Nano* **2016**, *3*, 1241–1253. [CrossRef]

11. Mauter, M.S.; Zucker, I.; Perreault, F.; Werber, J.R.; Kim, J.-H.; Elimelech, M. The role of nanotechnology in tackling global water challenges. *Nat. Sustain.* **2018**, *1*, 166–175. [CrossRef]

12. Luo, C.; Ren, X.; Dai, Z.; Zhang, Y.; Qi, X.; Pan, C. Present Perspectives of Advanced Characterization Techniques in TiO_2-Based Photocatalysts. *ACS Appl. Mater. Interfaces* **2017**, *9*, 23265–23286. [CrossRef]

13. Azzouz, I.; Habba, Y.G.; Capochichi-Gnambodoe, M.; Marty, F.; Vial, J.; Leprince-Wang, Y.; Bourouina, T. Zinc oxide nano-enabled microfluidic reactor for water purification and its applicability to volatile organic compounds. *Microsyst. Nanoeng.* **2018**, *4*, 17093. [CrossRef]

14. Singh, R.; Dutta, S. A review on H_2 production through photocatalytic reactions using TiO_2/TiO_2-assisted catalysts. *Fuel* **2018**, *220*, 607–620. [CrossRef]

15. Farooq, U.; Phul, R.; Alshehri, S.M.; Ahmed, J.; Ahmad, T. Electrocatalytic and Enhanced Photocatalytic Applications of Sodium Niobate Nanoparticles Developed by Citrate Precursor Route. *Sci. Rep.* **2019**, *9*, 4488. [CrossRef] [PubMed]

16. Guo, J.; Zeng, H.; Chen, Y. Emerging Nano Drug Delivery Systems Targeting Cancer-Associated Fibroblasts for Improved Antitumor Effect and Tumor Drug Penetration. *Mol. Pharm.* **2020**, *17*, 1028–1048. [CrossRef] [PubMed]

17. Ashraf, M.; Khan, I.; Usman, M.; Khan, A.; Shah, S.S.; Khan, A.Z.; Saeed, K.; Yaseen, M.; Ehsan, M.F.; Tahir, M.N.; et al. Hematite and Magnetite Nanostructures for Green and Sustainable Energy Harnessing and Environmental Pollution Control: A Review. *Chem. Res. Toxicol.* **2020**, *33*. [CrossRef] [PubMed]

18. Schneider, G. Antimicrobial silver nanoparticles—Regulatory situation in the European Union. *Mater. Today Proc.* **2017**, *4*, S200–S207. [CrossRef]

19. Potter, P.M.; Navratilova, J.; Rogers, K.R.; Al-Abed, S.R. Transformation of silver nanoparticle consumer products during simulated usage and disposal. *Environ. Sci. Nano* **2019**, *6*, 592–598. [CrossRef]

20. AZoNano.com. Metal Oxide Nanoparticles—Are they Safe? Available online: https://www.azonano.com/article.aspx?ArticleID=5444 (accessed on 13 June 2020).

21. Ahsan, M.d.A.; Jabbari, V.; Imam, M.A.; Castro, E.; Kim, H.; Curry, M.L.; Valles-Rosales, D.J.; Noveron, J.C. Nanoscale nickel metal organic framework decorated over graphene oxide and carbon nanotubes for water remediation. *Sci.Total Environ.* **2020**, *698*, 134214. [CrossRef]

22. Ahsan, M.A.; Jabbari, V.; El-Gendy, A.A.; Curry, M.L.; Noveron, J.C. Ultrafast catalytic reduction of environmental pollutants in water via MOF-derived magnetic Ni and Cu nanoparticles encapsulated in porous carbon. *Appl. Surf. Sci.* **2019**, *497*, 143608. [CrossRef]

23. Camboni, M.; Hanlon, J.; Pérez García, R.; Floyd, P.; European Chemicals Agency. A State of Play Study of the Market for so Called "Next Generation" Nanomaterials. 2019. Available online: https://op.europa.eu/publication/manifestation_identifier/PUB_ED0219746ENN (accessed on 11 March 2020).

24. Calipinar, H.; Ulas, D. Development of Nanotechnology in the World and Nanotechnology Standards in Turkey. *Procedia Comput. Sci.* **2019**, *158*, 1011–1018. [CrossRef]

25. Peng, Z.; Liu, X.; Zhang, W.; Zeng, Z.; Liu, Z.; Zhang, C.; Liu, Y.; Shao, B.; Liang, Q.; Tang, W.; et al. Advances in the application, toxicity and degradation of carbon nanomaterials in environment: A review. *Environ. Int.* **2020**, *134*, 105298. [CrossRef]

26. Romeo, D.; Salieri, B.; Hischier, R.; Nowack, B.; Wick, P. An integrated pathway based on in vitro data for the human hazard assessment of nanomaterials. *Environ. Int.* **2020**, *137*, 105505. [CrossRef] [PubMed]

27. Sayre, P.G.; Steinhäuser, K.G.; van Teunenbroek, T. Methods and data for regulatory risk assessment of nanomaterials: Questions for an expert consultation. *NanoImpact* **2017**, *8*, 20–27. [CrossRef]

28. Nowack, B. Evaluation of environmental exposure models for engineered nanomaterials in a regulatory context. *NanoImpact* **2017**, *8*, 38–47. [CrossRef]

29. Gao, X.; Avellan, A.; Laughton, S.; Vaidya, R.; Rodrigues, S.M.; Casman, E.A.; Lowry, G.V. CuO Nanoparticle Dissolution and Toxicity to Wheat (*Triticum aestivum*) in Rhizosphere Soil. *Environ. Sci. Technol.* **2018**, *52*, 2888–2897. [CrossRef] [PubMed]

30. Hou, W.-C.; Westerhoff, P.; Posner, J.D. Biological accumulation of engineered nanomaterials: A review of current knowledge. *Environ. Sci. Process. Impacts.* **2013**, *15*, 103–122. [CrossRef] [PubMed]

31. Lead, J.R.; Batley, G.E.; Alvarez, P.J.J.; Croteau, M.-N.; Handy, R.D.; McLaughlin, M.J.; Judy, J.D.; Schirmer, K. Nanomaterials in the environment: Behavior, fate, bioavailability, and effects-An updated review: Nanomaterials in the environment. *Environ. Toxicol. Chem.* **2018**, *37*, 2029–2063. [CrossRef]

32. Johnston, L.J.; Gonzalez-Rojano, N.; Wilkinson, K.J.; Xing, B. Key challenges for evaluation of the safety of engineered nanomaterials. *NanoImpact* **2020**, *18*, 100219. [CrossRef]

33. Dale, A.L.; Casman, E.A.; Lowry, G.V.; Lead, J.R.; Viparelli, E.; Baalousha, M. Modeling Nanomaterial Environmental Fate in Aquatic Systems. *Environ. Sci. Technol.* **2015**, *49*, 2587–2593. [CrossRef]

34. Williams, R.J.; Harrison, S.; Keller, V.; Kuenen, J.; Lofts, S.; Praetorius, A.; Svendsen, C.; Vermeulen, L.C.; van Wijnen, J. Models for assessing engineered nanomaterial fate and behaviour in the aquatic environment. *Curr. Opin. Environ. Sustain.* **2019**, *36*, 105–115. [CrossRef]

35. Praetorius, A.; Scheringer, M.; Hungerbühler, K. Development of Environmental Fate Models for Engineered Nanoparticles—A Case Study of TiO$_2$ Nanoparticles in the Rhine River. *Environ. Sci. Technol.* **2012**, *46*, 6705–6713. [CrossRef]

36. Meesters, J.A.J.; Koelmans, A.A.; Quik, J.T.K.; Hendriks, A.J.; van de Meent, D. Multimedia Modeling of Engineered Nanoparticles with SimpleBox4nano: Model Definition and Evaluation. *Environ. Sci. Technol.* **2014**, *48*, 5726–5736. [CrossRef] [PubMed]

37. Lowry, G.V.; Gregory, K.B.; Apte, S.C.; Lead, J.R. Transformations of Nanomaterials in the Environment. *Environ. Sci. Technol.* **2012**, *46*, 6893–6899. [CrossRef] [PubMed]

38. Praetorius, A.; Tufenkji, N.; Goss, K.-U.; Scheringer, M.; von der Kammer, F.; Elimelech, M. The road to nowhere: Equilibrium partition coefficients for nanoparticles. *Environ. Sci. Nano* **2014**, *1*, 317–323. [CrossRef]

39. Cornelis, G. Fate descriptors for engineered nanoparticles: The good, the bad, and the ugly. *Environ. Sci. Nano* **2015**, *2*, 19–26. [CrossRef]

40. Baalousha, M.; Cornelis, G.; Kuhlbusch, T.A.J.; Lynch, I.; Nickel, C.; Peijnenburg, W.; van den Brink, N.W. Modeling nanomaterial fate and uptake in the environment: Current knowledge and future trends. *Environ. Sci. Nano* **2016**, *3*, 323–345. [CrossRef]

41. Bouchard, D.; Knightes, C.; Chang, X.; Avant, B. Simulating Multiwalled Carbon Nanotube Transport in Surface Water Systems Using the Water Quality Analysis Simulation Program (WASP). *Environ. Sci. Technol.* **2017**, *51*, 11174–11184. [CrossRef] [PubMed]

42. Han, Y.; Knightes, C.D.; Bouchard, D.; Zepp, R.; Avant, B.; Hsieh, H.-S.; Chang, X.; Acrey, B.; Henderson, W.M.; Spear, J. Simulating graphene oxide nanomaterial phototransformation and transport in surface water. *Environ. Sci. Nano* **2019**, *6*, 180–194. [CrossRef]

43. Zou, Y.; Wang, X.; Khan, A.; Wang, P.; Liu, Y.; Alsaedi, A.; Hayat, T.; Wang, X. Environmental Remediation and Application of Nanoscale Zero-Valent Iron and Its Composites for the Removal of Heavy Metal Ions: A Review. *Environ. Sci. Technol.* **2016**, *50*, 7290–7304. [CrossRef]

44. Krol, M.; Khan, U.T.; Asad, M.A. Factors affecting nano scale zero valent iron (nZVI) travel distance in heterogeneous groundwater aquifers: A statistical modeling approach. In Proceedings of the AGU Fall Meeting, Washington, DC, USA, 10–14 December 2018.

45. Asad, M.A.; Krol, M.; Briggs, S. Nano zero valent iron (nZVI) remediation: A COMSOL modelling approach. In Proceedings of the Canadian Geotechnical Society (CGS) Conference (GeoEdmonton 2018), Edmonton, AB, Canada, 23–26 September 2018.

46. Yu, Z.; Hu, L.; Lo, I.M.C. Transport of the arsenic (As)-loaded nano zero-valent iron in groundwater-saturated sand columns: Roles of surface modification and As loading. *Chemosphere* **2019**, *216*, 428–436. [CrossRef]

47. Mueller, N.C.; Nowack, B. Exposure Modeling of Engineered Nanoparticles in the Environment. *Environ. Sci. Technol.* **2008**, *42*, 4447–4453. [CrossRef]

48. Gottschalk, F.; Scholz, R.W.; Nowack, B. Probabilistic material flow modeling for assessing the environmental exposure to compounds: Methodology and an application to engineered nano-TiO$_2$ particles. *Environ. Model. Softw.* **2010**, *25*, 320–332. [CrossRef]

49. Sun, T.Y.; Conroy, G.; Donner, E.; Hungerbühler, K.; Lombi, E.; Nowack, B. Probabilistic modelling of engineered nanomaterial emissions to the environment: A spatio-temporal approach. *Environ. Sci. Nano* **2015**, *2*, 340–351. [CrossRef]

50. Bornhöft, N.A.; Sun, T.Y.; Hilty, L.M.; Nowack, B. A dynamic probabilistic material flow modeling method. *Environ. Model. Softw.* **2016**, *76*, 69–80. [CrossRef]

51. Wang, Y.; Nowack, B. Dynamic probabilistic material flow analysis of nano-SiO$_2$, nano iron oxides, nano-CeO$_2$, nano-Al$_2$O$_3$, and quantum dots in seven European regions. *Environ. Pollut.* **2018**, *235*, 589–601. [CrossRef] [PubMed]

52. Scheringer, M.; Praetorius, A.; Goldberg, E.S. Environmental Fate and Exposure Modeling of Nanomaterials. *Front. Nanosci.* **2014**, *7*, 89–125. [CrossRef]

53. Di Guardo, A.; Gouin, T.; MacLeod, M.; Scheringer, M. Environmental fate and exposure models: Advances and challenges in 21st century chemical risk assessment. *Environ. Sci. Process. Impacts* **2018**, *20*, 58–71. [CrossRef]

54. Liu, H.H.; Cohen, Y. Multimedia Environmental Distribution of Engineered Nanomaterials. *Environ. Sci. Technol.* **2014**, *48*, 3281–3292. [CrossRef]

55. Meesters, J.A.J.; Quik, J.T.K.; Koelmans, A.A.; Hendriks, A.J.; van de Meent, D. Multimedia environmental fate and speciation of engineered nanoparticles: A probabilistic modeling approach. *Environ. Sci. Nano* **2016**, *3*, 715–727. [CrossRef]

56. Liu, H.H.; Bilal, M.; Lazareva, A.; Keller, A.; Cohen, Y. Simulation tool for assessing the release and environmental distribution of nanomaterials. *Beilstein J. Nanotechnol.* **2015**, *6*, 938–951. [CrossRef]

57. Garner, K.L.; Suh, S.; Keller, A.A. Assessing the Risk of Engineered Nanomaterials in the Environment: Development and Application of the nanoFate Model. *Environ. Sci. Technol.* **2017**, *51*, 5541–5551. [CrossRef]

58. Quik, J.T.K.; de Klein, J.J.M.; Koelmans, A.A. Spatially explicit fate modelling of nanomaterials in natural waters. *Water Res.* **2015**, *80*, 200–208. [CrossRef]

59. Dale, A.L.; Lowry, G.V.; Casman, E.A. Stream Dynamics and Chemical Transformations Control the Environmental Fate of Silver and Zinc Oxide Nanoparticles in a Watershed-Scale Model. *Environ. Sci. Technol.* **2015**, *49*, 7285–7293. [CrossRef] [PubMed]

60. ECHA. *Guidance on Information Requirements and Chemical Safety Assessment Chapter R.16: Environmental Exposure Assessment, Version 3.0—February 2016*; ISBN 978-92-9247-775-2. European Chemicals Agency (ECHA): Helsinki, Finland, 2016.

61. Gottschalk, F.; Sonderer, T.; Scholz, R.W.; Nowack, B. Modeled Environmental Concentrations of Engineered Nanomaterials (TiO$_2$, ZnO, Ag, CNT, Fullerenes) for Different Regions. *Environ. Sci. Technol.* **2009**, *43*, 9216–9222. [CrossRef] [PubMed]

62. Gottschalk, F.; Ort, C.; Scholz, R.W.; Nowack, B. Engineered nanomaterials in rivers—Exposure scenarios for Switzerland at high spatial and temporal resolution. *Environ. Pollut.* **2011**, *159*, 3439–3445. [CrossRef] [PubMed]

63. Sun, T.Y.; Bornhöft, N.A.; Hungerbühler, K.; Nowack, B. Dynamic Probabilistic Modeling of Environmental Emissions of Engineered Nanomaterials. *Environ. Sci. Technol.* **2016**, *50*, 4701–4711. [CrossRef]

64. Sani-Kast, N.; Scheringer, M.; Slomberg, D.; Labille, J.; Praetorius, A.; Ollivier, P.; Hungerbühler, K. Addressing the complexity of water chemistry in environmental fate modeling for engineered nanoparticles. *Sci. Total. Environ.* **2015**, *535*, 150–159. [CrossRef]

65. Dale, A.L.; Lowry, G.V.; Casman, E.A. Modeling Nanosilver Transformations in Freshwater Sediments. *Environ. Sci. Technol.* **2013**, *47*, 12920–12928. [CrossRef]

66. Dumont, E.; Johnson, A.C.; Keller, V.D.J.; Williams, R.J. Nano silver and nano zinc-oxide in surface waters —Exposure estimation for Europe at high spatial and temporal resolution. *Environ. Pollut.* **2015**, *196*, 341–349. [CrossRef]

67. Klein, J.J.M.; de Quik, J.T.K.; Bäuerlein, P.S.; Koelmans, A.A. Towards validation of the NanoDUFLOW nanoparticle fate model for the river Dommel, The Netherlands. *Environ. Sci. Nano* **2016**, *3*, 434–441. [CrossRef]

68. Markus, A.A.; Parsons, J.R.; Roex, E.W.M.; de Voogt, P.; Laane, R.W.P.M. Modelling the transport of engineered metallic nanoparticles in the river Rhine. *Water Res.* **2016**, *91*, 214–224. [CrossRef]

69. Saharia, A.M.; Zhu, Z.; Aich, N.; Baalousha, M.; Atkinson, J.F. Modeling the transport of titanium dioxide nanomaterials from combined sewer overflows in an urban river. *Sci. Total Environ.* **2019**, *696*, 133904. [CrossRef]

70. Shoemaker, L.; Dai, T.; Koenig, J. *TMDL Model Evaluation and Research Needs*; EPA/600/R-05/149; National Risk Management Research Laboratory, Office Of Research And Development, U.S. Environmental Protection Agency: Cincinnati, OH, USA, 2005.

71. Di Toro, D.M.; Mahony, J.D.; Hansen, D.J.; Berry, W.J. A model of the oxidation of iron and cadmium sulfide in sediments. *Environ. Toxicol. Chem.* **1996**, *15*, 2168–2186. [CrossRef]

72. Hou, W.-C.; Stuart, B.; Howes, R.; Zepp, R.G. Sunlight-Driven Reduction of Silver Ions by Natural Organic Matter: Formation and Transformation of Silver Nanoparticles. *Environ. Sci. Technol.* **2013**, *47*, 7713–7721. [CrossRef]

73. Singh, A.; Hou, W.-C.; Lin, T.-F.; Zepp, R.G. Roles of Silver–Chloride Complexations in Sunlight-Driven Formation of Silver Nanoparticles. *Environ. Sci. Technol.* **2019**, *53*, 11162–11169. [CrossRef] [PubMed]

74. Elimelech, M.; Gregory, J.; Jia, X.; Williams, R.A. *Particle Deposition and Aggregation: Measurement, Modelling and Simulation*; Butterworth-Heinemann: Oxford, UK, 1998.

75. Petosa, A.R.; Jaisi, D.P.; Quevedo, I.R.; Elimelech, M.; Tufenkji, N. Aggregation and Deposition of Engineered Nanomaterials in Aquatic Environments: Role of Physicochemical Interactions. *Environ. Sci. Technol.* **2010**, *44*, 6532–6549. [CrossRef]

76. Quik, J.T.K.; Velzeboer, I.; Wouterse, M.; Koelmans, A.A.; van de Meent, D. Heteroaggregation and sedimentation rates for nanomaterials in natural waters. *Water Res.* **2014**, *48*, 269–279. [CrossRef]

77. Velzeboer, I.; Quik, J.T.K.; van de Meent, D.; Koelmans, A.A. Rapid settling of nanoparticles due to heteroaggregation with suspended sediment: Nanoparticle settling due to heteroaggregation with sediment. *Environ. Toxicol. Chem.* **2014**, *33*, 1766–1773. [CrossRef]

78. Bouchard, D.; Chang, X.; Chowdhury, I. Heteroaggregation of multiwalled carbon nanotubes with sediments. *Environ. Nanotechnol. Monit. Manag.* **2015**, *4*, 42–50. [CrossRef]

79. Clavier, A.; Praetorius, A.; Stoll, S. Determination of nanoparticle heteroaggregation attachment efficiencies and rates in presence of natural organic matter monomers. Monte Carlo modelling. *Sci. Total Environ.* **2019**, *650*, 530–540. [CrossRef]

80. Praetorius, A.; Badetti, E.; Brunelli, A.; Clavier, A.; Gallego-Urrea, J.A.; Gondikas, A.; Hassellöv, M.; Hofmann, T.; Mackevica, A.; Marcomini, A.; et al. Strategies for determining heteroaggregation attachment efficiencies of engineered nanoparticles in aquatic environments. *Environ. Sci. Nano* **2020**, *7*, 351–367. [CrossRef]

81. Zhao, J.; Dai, Y.; Wang, Z.; Ren, W.; Wei, Y.; Cao, X.; Xing, B. Toxicity of GO to Freshwater Algae in the Presence of Al_2O_3 Particles with Different Morphologies: Importance of Heteroaggregation. *Environ. Sci. Technol.* **2018**, *52*, 13448–13456. [CrossRef] [PubMed]

82. Wang, R.; Dang, F.; Liu, C.; Wang, D.-J.; Cui, P.-X.; Yan, H.-J.; Zhou, D.-M. Heteroaggregation and dissolution of silver nanoparticles by iron oxide colloids under environmentally relevant conditions. *Environ. Sci. Nano* **2019**, *6*, 195–206. [CrossRef]

83. Franklin, N.M.; Rogers, N.J.; Apte, S.C.; Batley, G.E.; Gadd, G.E.; Casey, P.S. Comparative Toxicity of Nanoparticulate ZnO, Bulk ZnO, and $ZnCl_2$ to a Freshwater Microalga (*Pseudokirchneriella subcapitata*): The Importance of Particle Solubility. *Environ. Sci. Technol.* **2007**, *41*, 8484–8490. [CrossRef]

84. Liu, J.; Hurt, R.H. Ion Release Kinetics and Particle Persistence in Aqueous Nano-Silver Colloids. *Environ. Sci. Technol.* **2010**, *44*, 2169–2175. [CrossRef] [PubMed]

85. Bian, S.-W.; Mudunkotuwa, I.A.; Rupasinghe, T.; Grassian, V.H. Aggregation and Dissolution of 4 nm ZnO Nanoparticles in Aqueous Environments: Influence of pH, Ionic Strength, Size, and Adsorption of Humic Acid. *Langmuir* **2011**, *27*, 6059–6068. [CrossRef] [PubMed]

86. Peretyazhko, T.S.; Zhang, Q.; Colvin, V.L. Size-Controlled Dissolution of Silver Nanoparticles at Neutral and Acidic pH Conditions: Kinetics and Size Changes. *Environ. Sci. Technol.* **2014**, *48*, 11954–11961. [CrossRef] [PubMed]

87. Hedberg, J.; Blomberg, E.; Odnevall Wallinder, I. In the Search for Nanospecific Effects of Dissolution of Metallic Nanoparticles at Freshwater-Like Conditions: A Critical Review. *Environ. Sci. Technol.* **2019**, *53*, 4030–4044. [CrossRef]

88. Petersen, E.J.; Huang, Q.; Weber, W.J. Relevance of octanol-water distribution measurements to the potential ecological uptake of multi-walled carbon nanotubes. *Environ. Toxicol. Chem.* **2010**, *29*, 1106–1112. [CrossRef]

89. Jafvert, C.T.; Kulkarni, P.P. Buckminsterfullerene's (C_{60}) Octanol–Water Partition Coefficient (K_{ow}) and Aqueous Solubility. *Environ. Sci. Technol.* **2008**, *42*, 5945–5950. [CrossRef]

90. Kaegi, R.; Voegelin, A.; Sinnet, B.; Zuleeg, S.; Hagendorfer, H.; Burkhardt, M.; Siegrist, H. Behavior of Metallic Silver Nanoparticles in a Pilot Wastewater Treatment Plant. *Environ. Sci. Technol.* **2011**, *45*, 3902–3908. [CrossRef]

91. Kaegi, R.; Voegelin, A.; Ort, C.; Sinnet, B.; Thalmann, B.; Krismer, J.; Hagendorfer, H.; Elumelu, M.; Mueller, E. Fate and transformation of silver nanoparticles in urban wastewater systems. *Water Res.* **2013**, *47*, 3866–3877. [CrossRef] [PubMed]

92. Ma, R.; Levard, C.; Judy, J.D.; Unrine, J.M.; Durenkamp, M.; Martin, B.; Jefferson, B.; Lowry, G.V. Fate of Zinc Oxide and Silver Nanoparticles in a Pilot Wastewater Treatment Plant and in Processed Biosolids. *Environ. Sci. Technol.* **2014**, *48*, 104–112. [CrossRef]

93. Liu, J.; Zhang, F.; Allen, A.J.; Johnston-Peck, A.C.; Pettibone, J.M. Comparing sulfidation kinetics of silver nanoparticles in simulated media using direct and indirect measurement methods. *Nanoscale* **2018**, *10*, 22270–22279. [CrossRef]

94. Liu, J.; Pennell, K.G.; Hurt, R.H. Kinetics and Mechanisms of Nanosilver Oxysulfidation. *Environ. Sci. Technol.* **2011**, *45*, 7345–7353. [CrossRef] [PubMed]

95. Zhang, J.; Guo, W.; Li, Q.; Wang, Z.; Liu, S. The effects and the potential mechanism of environmental transformation of metal nanoparticles on their toxicity in organisms. *Environ. Sci. Nano* **2018**, *5*, 2482–2499. [CrossRef]

96. He, D.; Garg, S.; Wang, Z.; Li, L.; Rong, H.; Ma, X.; Li, G.; An, T.; Waite, T.D. Silver sulfide nanoparticles in aqueous environments: Formation, transformation and toxicity. *Environ. Sci. Nano* **2019**, *6*, 1674–1687. [CrossRef]

97. Hou, W.-C.; Jafvert, C.T. Photochemical Transformation of Aqueous C_{60} Clusters in Sunlight. *Environ. Sci. Technol.* **2009**, *43*, 362–367. [CrossRef]

98. Kong, L.; Tedrow, O.; Chan, Y.F.; Zepp, R.G. Light-Initiated Transformations of Fullerenol in Aqueous Media. *Environ. Sci. Technol.* **2009**, *43*, 9155–9160. [CrossRef]

99. Isaacson, C.W.; Bouchard, D. Asymmetric flow field flow fractionation of aqueous C_{60} nanoparticles with size determination by dynamic light scattering and quantification by liquid chromatography atmospheric pressure photo-ionization mass spectrometry. *J. Chromatogr. A* **2010**, *1217*, 1506–1512. [CrossRef]

100. Hou, W.-C.; Huang, S.-H. Photochemical reactivity of aqueous fullerene clusters: C_{60} versus C_{70}. *J. Hazard. Mater.* **2017**, *322*, 310–317. [CrossRef]

101. Qu, X.; Alvarez, P.J.J.; Li, Q. Photochemical Transformation of Carboxylated Multiwalled Carbon Nanotubes: Role of Reactive Oxygen Species. *Environ. Sci. Technol.* **2013**, *47*, 14080–14088. [CrossRef] [PubMed]

102. Hou, W.-C.; BeigzadehMilani, S.; Jafvert, C.T.; Zepp, R.G. Photoreactivity of Unfunctionalized Single-Wall Carbon Nanotubes Involving Hydroxyl Radical: Chiral Dependency and Surface Coating Effect. *Environ. Sci. Technol.* **2014**, *48*, 3875–3882. [CrossRef] [PubMed]

103. Chen, C.-Y.; Zepp, R.G. Probing Photosensitization by Functionalized Carbon Nanotubes. *Environ. Sci. Technol.* **2015**, *49*, 13835–13843. [CrossRef] [PubMed]

104. Hou, W.-C.; Chowdhury, I.; Goodwin, D.G.; Henderson, W.M.; Fairbrother, D.H.; Bouchard, D.; Zepp, R.G. Photochemical Transformation of Graphene Oxide in Sunlight. *Environ. Sci. Technol.* **2015**, *49*, 3435–3443. [CrossRef]

105. Hou, W.-C.; Henderson, W.M.; Chowdhury, I.; Goodwin, D.G., Jr.; Chang, X.; Martin, S.; Fairbrother, D.H.; Bouchard, D.; Zepp, R.G. The contribution of indirect photolysis to the degradation of graphene oxide in sunlight. *Carbon* **2016**, *110*, 426–437. [CrossRef]

106. Yin, Y.; Liu, J.; Jiang, G. Sunlight-Induced Reduction of Ionic Ag and Au to Metallic Nanoparticles by Dissolved Organic Matter. *ACS Nano* **2012**, *6*, 7910–7919. [CrossRef]

107. Lee, T.-W.; Chen, C.-C.; Chen, C. Chemical Stability and Transformation of Molybdenum Disulfide Nanosheets in Environmental Media. *Environ. Sci. Technol.* **2019**, *53*, 6282–6291. [CrossRef]

108. Hou, W.-C.; Kong, L.; Wepasnick, K.A.; Zepp, R.G.; Fairbrother, D.H.; Jafvert, C.T. Photochemistry of Aqueous C_{60} Clusters: Wavelength Dependency and Product Characterization. *Environ. Sci. Technol.* **2010**, *44*, 8121–8127. [CrossRef]

109. Hou, W.-C.; Lee, P.-L.; Chou, Y.-C.; Wang, Y.-S. Antibacterial property of graphene oxide: The role of phototransformation. *Environ. Sci. Nano* **2017**, *4*, 647–657. [CrossRef]

110. Chowdhury, I.; Hou, W.-C.; Goodwin, D.; Henderson, M.; Zepp, R.G.; Bouchard, D. Sunlight affects aggregation and deposition of graphene oxide in the aquatic environment. *Water Res.* **2015**, *78*, 37–46. [CrossRef]

111. Fanourakis, S.K.; Peña-Bahamonde, J.; Bandara, P.C.; Rodrigues, D.F. Nano-based adsorbent and photocatalyst use for pharmaceutical contaminant removal during indirect potable water reuse. *NPJ Clean Water* **2020**, *3*, 1. [CrossRef]

112. Meesters, J.A.J.; Peijnenburg, W.J.G.M.; Hendriks, A.J.; Van de Meent, D.; Quik, J.T.K. A model sensitivity analysis to determine the most important physicochemical properties driving environmental fate and exposure of engineered nanoparticles. *Environ. Sci. Nano* **2019**, *6*, 2049–2060. [CrossRef]

113. Nowack, B.; David, R.M.; Fissan, H.; Morris, H.; Shatkin, J.A.; Stintz, M.; Zepp, R.; Brouwer, D. Potential release scenarios for carbon nanotubes used in composites. *Environ. Int.* **2013**, *59*, 1–11. [CrossRef]

114. Harper, S.; Wohlleben, W.; Doa, M.; Nowack, B.; Clancy, S.; Canady, R.; Maynard, A. Measuring Nanomaterial Release from Carbon Nanotube Composites: Review of the State of the Science. *J. Phys. Conf. Ser.* **2015**, *617*, 012026. [CrossRef]

115. Wohlleben, W.; Neubauer, N. Quantitative rates of release from weathered nanocomposites are determined across 5 orders of magnitude by the matrix, modulated by the embedded nanomaterial. *NanoImpact* **2016**, *1*, 39–45. [CrossRef]

116. Wohlleben, W.; Kingston, C.; Carter, J.; Sahle-Demessie, E.; Vázquez-Campose, S.; Acrey, B.; Chen, C.-Y.; Walton, E.; Egenolf, H.; Müller, P.; et al. NanoRelease: Pilot interlaboratory comparison of a weathering protocol applied to resilient and labile polymers with and without embedded carbon nanotubes. *Carbon* **2017**, *113*, 346–360. [CrossRef]

117. Praetorius, A.; Arvidsson, R.; Molander, S.; Scheringer, M. Facing complexity through informed simplifications: A research agenda for aquatic exposure assessment of nanoparticles. *Env. Sci. Process. Impacts* **2013**, *15*, 161–168. [CrossRef]

118. Geitner, N.K.; Bossa, N.; Wiesner, M.R. Formulation and Validation of a Functional Assay-Driven Model of Nanoparticle Aquatic Transport. *Environ. Sci. Technol.* **2019**, *53*, 3104–3109. [CrossRef]

119. Stegemeier, J.P.; Avellan, A.; Lowry, G.V. Effect of Initial Speciation of Copper- and Silver-Based Nanoparticles on Their Long-Term Fate and Phytoavailability in Freshwater Wetland Mesocosms. *Environ. Sci. Technol.* **2017**, *51*, 12114–12122. [CrossRef]

120. Ambrose, B.; Avant, B.; Han, Y.; Knightes, C.; Wool, T. *Water Quality Assessment Simulation Program (WASP8): Upgrades to the Advanced Toxicant Module for Simulating Dissolved Chemicals, Nanomaterials, and Solids*; EPA/600/R-17/326; Office of Research and Development, National Exposure Research Laboratory, U.S. Environmental Protection Agency: Washington, DC, USA, 2017.

MDPI
St. Alban-Anlage 66
4052 Basel
Switzerland
Tel. +41 61 683 77 34
Fax +41 61 302 89 18
www.mdpi.com

International Journal of Molecular Sciences Editorial Office
E-mail: ijms@mdpi.com
www.mdpi.com/journal/ijms